Active Volcanoes of the World

About 500 active volcanoes presently exist on the Earth's surface, of which around 50 erupt each year. Volcanoes played a crucial role in the evolution of the planet and early life, and are constantly reshaping the morphology of Planet Earth. Many active volcanoes are located in dense settlement areas, with over 500 million people living in close proximity of still active or dormant volcanoes.

On one side, volcanoes provide valuable soil and rock basis for agriculture, but often the "mountains of fire" cause disastrous societal and economical disasters caused by ash clouds, lahars, lava flows, and pyroclastic flows. Eruptions are still difficult to predict, although volcanologists around the world are constantly working on new ways to understand the character and behavior of volcanoes.

Active Volcanoes of the World is an official book series of the International Association of Volcanology and Chemistry of the Earth's Interior (IAVCEI). The series aims to be a scientific library of monographs that provide authoritative and detailed reviews of state-of-the art research on individual volcanoes or a volcanic area that has been active in the last 10.000 years, e.g. the Teide Volcano or the Chiapas Region. The books in the series cover the geology, eruptive history, petrology and geochemistry, volcano monitoring, risk assessment and mitigation, volcano and society, and specific aspects related to the nature of each described volcano.

The Active Volcanoes of the World series contains single and multi-authored books as well as edited volumes. The Series Editors, Dr. Corrado Cimarelli and Dr. Sebastian Müller are currently accepting proposals and a proposal document can be obtained from the Publisher, Dr. Annett Buettner (annett.buettner@springer.com).

Shigeo Aramaki
Masaki Takahashi
Minoru Takeo • Maya Yasui
Yosuke Aoki
Editors

The Asama Volcano

The Most Active and Hazardous Andesitic Volcano near the Tokyo Metropolitan Area

Springer

Editors
Shigeo Aramaki
University of Tokyo
Tokyo, Japan

Minoru Takeo
Earthquake Research Institute
University of Tokyo
Tokyo, Japan

Yosuke Aoki
Earthquake Research Institute
University of Tokyo
Tokyo, Japan

Masaki Takahashi
Nihon University
Tokyo, Japan

Maya Yasui
Nihon University
Tokyo, Japan

ISSN 2195-3589 ISSN 2195-7029 (electronic)
Active Volcanoes of the World
ISBN 978-981-95-5798-1 ISBN 978-981-95-5799-8 (eBook)
https://doi.org/10.1007/978-981-95-5799-8

This Springer imprint is published by the registered company Springer Nature Singapore Pte Ltd.
The registered company address is: 152 Beach Road, #21-01/04 Gateway East, Singapore 189721, Singapore

Preface

Asama volcano is one of the most active composite volcanoes in Japan. It is situated in a zone crowded with the Quaternary volcanoes running along the Northeast Japan volcanic front and extending to the Izu-Mariana volcanic arcs: both are associated with subduction of the Pacific plate. The basement rocks of Asama volcano are mainly composed of the Miocene to Pliocene volcanic rocks and sediments and Quaternary volcanic rocks older than Asama volcano. Asama volcano rests on a flat depressed area around the altitude of 1000 m, spreading its skirt toward north and south with a negative gravity anomaly of more than 20 mgal. This depression was probably formed within the Neogene volcanogenic formations constituting the eastern margin of Fossa Magna, which is a great tectonic zone separating northeastern Honshu Island from the southwestern part.

The overall geological characteristics of Asama volcano were revealed and compiled by pioneering research in Aramaki (1963). On the west, there is a row of ancient late Quaternary volcanoes, which are collectively called the Eboshi volcano group. Volcanism appears to have progressed eastward, with Asama volcano group as its eastern and youngest member of the row (Aramaki, 1963). Asama volcano itself is a complex of four volcanic edifices partly overlapping one another: Kurofu, Sennin, Hotokeiwa, and Maekake; each one is a different eruptive type. Kurofu is the oldest large conical stratovolcano and is composed of an alternation of pyroclastic beds and lava flows of pyroxene andesites. After Kurofu ceased to grow, the edifice was dissected by erosion and Sennin volcano began to be constructed by explosive eruptions. The eastern part of the cone of Kurofu and Sennin volcanoes was destroyed concurrent with the large explosive eruption of Sennin. Hotokeiwa is the third oldest medium-sized volcanic edifice with a rhyolitic to andesitic composition, whose vent situated about 2 km east of the vent of Kurofu and Sennin. Maekake is the youngest volcano and is composed of a medium-sized andesitic pyroclastic cone with pyroclastic flows and lava flows (Aramaki, 1963; Takahashi et al., 2007).

Maekake began to form just after the cessation of the activity at Hotokeiwa about 10,000 years ago. Its lava is mainly andesite, which is typical of arc volcanoes. Hence, its eruptions are sometimes explosive. The eruptive history of Maekake consists of three repetitions of active-dormant phases. The third active phase has been continuing about 2000 years and includes the historical Plinian eruptions of the second century, AD1108, AD1128, and AD1783. A Volcanic Explosivity Index (VEI) of 5 and 4 occurred in AD1108 and

AD1783, respectively. Recent activity in the first half of the twenty century was high and the eruptions of VEIs were up to 2. In the past few decades, the eruptions have been less explosive. However, moderate-sized eruptions occurred in 1973, 1982, 1983, and 2004, and minor eruptions were recorded in 2008, 2009, and 2015.

Asama volcano is located about 150 km northwest of the Tokyo metropolitan area with a population of more than 20 million. Therefore, the Tokyo metropolitan area is potentially at risk of volcanic hazards from Asama volcano during future larger eruptions. With this background, Asama volcano has attracted public interest, and seismic observations in Asama volcano have been collected for over 100 years. Eruptions and seismic swarms in 1909 motivated seismic observations of Asama volcano in 1910 (Omori, 1910). Omori established the Asama-yama seismological observatory in 1911 about 2.3 km west-southwest from the summit crater (Omori, 1914), and the characteristic features of the explosive eruptions of Asama volcano have been closely monitored since 1911. This was one of the earliest seismic observations made of an active volcano in the world. Since then, the volcano has been monitored and scientifically investigated, and a traditional classification of volcanic earthquakes was archived based on the seismic observation at Asama volcano by Minakami (1960). This classification has been made use in other volcanoes for a certain length of time before monitoring networks of volcanic earthquake were modernized.

After the turn of the twenty-first century, a modern monitoring network was established in and around Asama volcano to gain more insights into the internal structure of the volcano and the mechanisms of magma transport beneath the volcano. As of 2017, it was comprised of 30 seismometers including 19 broad-band sensors and 22 continuous global navigation satellite system (GNSS) sites. In addition to the seismic and geodetic sensors, other kinds of sensors such as microphones, video and thermal cameras, and a detector of chemical components in volcanic gas have been installed in the small caves located on the west and east rims of the summit crater. The seismic structure of the upper crust beneath Asama volcano was revealed using these modern observation network data and a regional seismic network data around Asama volcano. The magma pathway beneath Asama volcano was synthesized by combining the seismic structure with a hypocenter distribution of volcanic earthquakes and ground deformations before and after eruptions. Magma rising to the upper crust was stored in the magma chamber at a depth of 5–10 km. The magma chamber was offset by approximately 8 km to the west from the summit and was represented by a low *S* wave velocity region. Then the magma moved to a depth of about 1 km below sea level as a dyke. The shallow seismic and resistivity structures suggested that the magma pathway is subject to structural controls, generating a winding path to the surface. Temporal evolutions of multidiscipline data at Asama volcano revealed that the supplied amount of magma from the magma chamber and the internal condition in the shallow regions of the conduit controlled the recent eruptions.

The neighboring areas of Asama volcano is one of the most popular tourist resort in Japan. Many tourists visit these areas and tens of thousands of people climb Asama volcano every year. Therefore, it has been a pressing task at Asama volcano to progress technologies of eruption prediction and volcanic disaster mitigation. Before the end of World War II, at least 45 casualties of eruption had been recorded; these people died within a radius of 4 km from the summit crater. A pioneering task of eruption prediction had been exhibited by Minakami (1960) based on the frequency of volcanic earthquakes. After the modern monitoring network has been established, we obtain the knowledge about the magma pathway beneath Asama volcano and the multidisciplinary observation data. These progressions make it possible to enforce the volcanic alert level by the Japan Meteorological Agency; there has been no casualty in the last 50 years. The local communities surrounding Asama volcano formed a joint consortium to prepare an overall disaster mitigation management program to deal with large-scale eruption in the future, such as those that occurred in 1108 and 1783.

References

Aramaki S (1963) Geology of Asama Volcano. J Fac Sci, University of Tokyo, Section II. 14:229–443

Minakami T (1960) Fundamental research for predicting volcanic eruptions (Part 1). Bulletin of Earthquake Research Institute University of Tokyo. 38:497–544

Omori F (1910) The eruptions of Asama-yama. Report of the Imperial Earthquake Investigation Committee. 67:1–26 (in Japanese)

Omori F (1914) The eruptions and earthquakes of Asama-Yama II: list of the volcanic earthquakes instrumentally registered at the Asama-Yama observatory in 1911 and 1912. Bulletin of the Imperial Earthquake Investigation Committee. 6: 149–226

Takahashi M, Yasui M, Ichikawa Y, Kamioka Y, Asaka N, Sakagami M, Tanaka E (2007) Whole-rock major element chemistry for eruptive products of Asama-Maekake Volcano, central Japan: summary for data of 288 samples. Proceeding Institute of Nature Science Nihon University. **42**:55–70 (in Japanese with English abstract)

Tokyo, Japan Shigeo Aramaki
Tokyo, Japan Masaki Takahashi
Tokyo, Japan Minoru Takeo
Tokyo, Japan Maya Yasui
Tokyo, Japan Yosuke Aoki

Introducing Chief Editor Shigeo Aramaki

Dr. Shigeo Aramaki, the Chief Editor, is a well-known researcher in the field of geology, especially for his pioneering work on the study of Asama volcano. His paper titled "Geology of Asama Volcano," published in 1963, is highly regarded in the Japanese volcanological society. He is also known for his early contributions to the study of pyroclastic flows in Japan, and his vigorous works on the various fields of volcanic geology and petrology. Dr. Aramaki has made significant contributions to both Japanese and international volcanology, with a particular emphasis on volcanic geology and disaster prevention. He helped raise awareness of the importance of volcanic disaster prevention and played an important role in creating volcanic hazard maps for Mt. Asama and Mt. Fuji in Japan. He has held several prestigious positions, including President of the Volcanological Society of Japan and the International Association of Volcanology and the Chemistry of the Earth's Interior (IAVCEI). He is now an Honorary Member of IAVCEI and has received several awards, including the Geological Society of America 1998 awards and the Krafft Medal for IAVCEI: Volcanology and Hazard Mitigation in 2013.

Dr. Shigeo Aramaki, the Chief Editor, conducted field work in Iceland in 1999.

(Photo by M. Yasui)

Masaki Takahashi
Institute of Natural Sciences,
College of Humanities and Sciences
Nihon University
Tokyo, Japan

Contents

Contributors

Koki Aizawa Faculty of Science, Institute of Seismology and Volcanology, Kyushu University, Fukuoka, Japan

Yosuke Aoki Earthquake Research Institute, The University of Tokyo, Bunkyo, Tokyo, Japan

Andreas Auer Department of Geoscience, Shimane University, Matsue, Shimane, Japan

Hiroshi Shinohara Geological Survey of Japan, National Institute of Advanced Industrial Science and Technology (AIST), Tsukuba, Ibaraki, Japan

Sei Iijima Maebashi Local Meteorological Office, Maebashi, Gunma, Japan

Hikaru Iwamori Earthquake Research Institute, The University of Tokyo, Tokyo, Japan

Tatsuo Kanamaru Department of Earth and Environmental Sciences, College of Humanities and Sciences, Nihon University, Tokyo, Japan

Koji Kato Volcanic Observation Division, Japan Meteorological Agency, Minato, Tokyo, Japan

Takahito Kazama Graduate School of Science, Kyoto University, Kyoto, Japan

Takao Koyama Earthquake Research Institute, The University of Tokyo, Tokyo, Japan

Isoji Miyagi Geological Survey of Japan, National Institute of Advanced Industrial Science and Technology (AIST), Tsukuba, Ibaraki, Japan

Masayuki Murase Department of Earth and Environmental Sciences, College of Humanities and Sciences, Nihon University, Tokyo, Japan

Masashi Nagai Volcano Disaster Resilience Research Division, National Research Institute for Earth Science and Disaster Resilience, Tsukuba, Ibaraki, Japan

Yutaka Nagaoka Meteorological College, Japan Meteorological Agency, Chiba, Japan

Junichi Nakajima Department of Earth and Planetary Sciences, School of Science, Institute of Science Tokyo, Tokyo, Japan

Hitomi Nakamura Geological Survey of Japan, National Institute of Advanced Industrial Science and Technology (AIST), Tsukuba, Ibaraki, Japan

Kiwamu Nishida Earthquake Research Institute, The University of Tokyo, Tokyo, Japan

Takao Ohminato Earthquake Research Institute, The University of Tokyo, Tokyo, Japan

Teruki Oikawa Geological Survey of Japan, AIST, Tsukuba, Ibaraki, Japan

Janire Prudencio Department of Theoretical Physics and Cosmos, Universidad de Granada, Granada, Spain

Andalusian Institute of Geophysics and Earthquake Disaster Prevention, Universidad de Granada, Granada, Spain

Masaki Takahashi Institute of Natural Sciences, College of Humanities and Sciences, Nihon University, Tokyo, Japan

Minoru Takeo Earthquake Research Institute, The University of Tokyo, Tokyo, Japan

Japan Meteorological Agency, Tokyo, Japan

Hiroyuki K. M. Tanaka Earthquake Research Institute, The University of Tokyo, Tokyo, Japan

Takashi Tsutsumi Center for Obsidian and Lithic Studies, Meiji University, Tokyo, Japan

Yoshiya Usui Earthquake Research Institute, The University of Tokyo, Tokyo, Japan

Xiaowen Wang Department of Surveying and Geo-Informatics, Faculty of Geosciences and Environmental Engineering, Southwest Jiaotong University, Chengdu, China

Yoshiaki Yamaguchi Department of Geology, Shinshu University, Matsumoto, Nagano, Japan

Hitoshi Yamasato CTI Engineering Co., Ltd., Chuo, Tokyo, Japan

Maya Yasui Department of Earth and Environmental Sciences, College of Humanities and Sciences, Nihon University, Tokyo, Japan

Tectonic Setting

Deep Seismic Structure, Plate Geometry, and Magmatism Around Asama Volcano

Junichi Nakajima

1 Introduction

The Japanese Islands are located at a plate-convergent margin, where two oceanic plates, the Pacific (PAC) and Philippine Sea (PHS) plates, are subducting beneath the continental plates (Fig. 1a). The PAC plate is subducting beneath northern Japan at a rate of 8–9 cm/year, while the PHS plate is subducting northwestward beneath an area from Kanto to southwestern Japan at a rate of 3–5 cm/year. The convergences of the two oceanic plates cause extremely high seismic activity, which include shallow crustal earthquakes, thrust-type (interplate) earthquakes along the upper surface of the subducting plates, and intermediate and deep-focus earthquakes (intraslab earthquakes) in the subducting plates.

Figure 1 summarizes the geometry of the subducting PAC and PHS slabs that are delineated from seismological observations. The upper surface of the PAC slab is estimated from the locations of low-angle thrust-type and intraslab (Watati-Benioff) earthquakes (e.g., Igarashi et al. 2001; Nakajima et al. 2009a), analyses of mode-converted waves at the upper surface of the plate (e.g., Matsuzawa et al. 1986; Zhao et al. 1997), and the distribution of a low-velocity layer interpreted as the subducted crust (e.g., Shiina et al. 2013). In contrast, low-angle thrust-type earthquakes are rarely observed for the PHS slab in southwestern Japan, and the geometry of the PHS slab is primarily determined from the distribution of intraslab earthquakes (e.g., Nakajima and Hasegawa 2007), the location of a low-velocity layer interpreted as the subducted crust (e.g., Hirose et al. 2008), the analyses of mode-converted waves at the upper surface of the plate (e.g., Hori et al. 1985), and the locations of a high-velocity anomaly interpreted as the subducted mantle (Huang et al. 2013; Miyazaki et al. 2023). The subduction of the PHS slab can be traced to ~140 km depth in central Japan and >400 km depth in southwestern Japan.

The most notable tectonic feature is a dual plate subduction beneath Kanto, where the PHS plate is subducting in a tiny space between the overlying continental and underlying PAC plates (Uchida et al. 2009; Nakajima et al. 2009a). The dual plate subduction results in a unique tectonic setting, such that the lower surface of the PHS slab is in contact with the upper surface of the underlying PAC slab, forming an area of slab-slab contact beneath Kanto. The extent of the slab-slab contact area has been inferred from seismological observations (a yellow-shaded area in Fig. 1). The mechanical and frictional interactions between the PHS and PAC slabs affect seismotectonics beneath the Tokyo Metropolitan area (e.g., Nakajima and Hasegawa

J. Nakajima (✉)
Department of Earth and Planetary Sciences, School of Science, Institute of Science Tokyo, Tokyo, Japan
e-mail: nakajima@eps.sci.isct.ac.jp

S. Aramaki et al. (eds.), *The Asama Volcano*, Active Volcanoes of the World,
https://doi.org/10.1007/978-981-95-5799-8_1

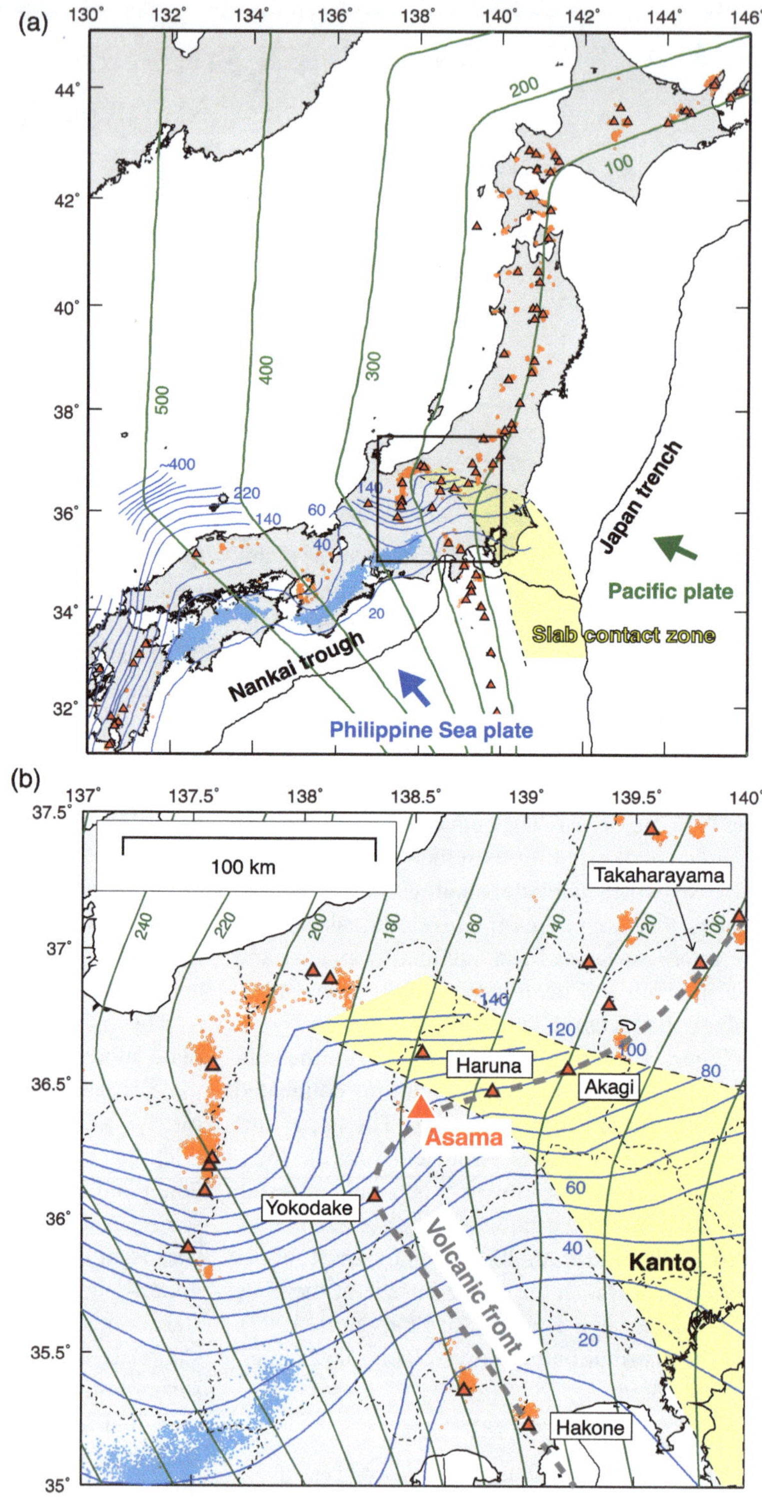

Fig. 1 (**a**) Geometries of the PAC (green) and PHS (blue) slabs (Hirose et al. 2008; Nakajima et al. 2009a; Huang et al. 2013; Asamori and Zhao 2015). A yellow-shaded area denotes a contact zone of the two slabs (Uchida et al. 2009; Nakajima et al. 2009a). Active volcanoes are shown by red triangles. Volcanic or semi-volcanic low-frequency earthquakes (LFEs) are shown by orange dots, while tectonic LFEs are shown by light blue dots. A black rectangle denotes an area shown in (**b**). (**b**) An enlarged map around Asama volcano. The volcanic front is shown by a gray dashed line. Names of volcanoes discussed in the text are indicated

2010), producing the most active seismicity beneath the Japanese Islands.

Subduction of oceanic plates generates convection in the overlying mantle wedge (Mckenzie 1969). Mantle materials on the topside of a subducting plate are entrained with the plate subduction by viscous coupling, and materials in the middle of the mantle wedge flow inward to fill the potential void left by the downward advection. Although the flow patterns depend largely on the viscosity structure of the mantle and the depth range of viscous coupling between the slab and mantle materials, the inflow (upwelling flow or return flow) generated by the slab subduction plays an essential role in transferring heat flux from the deep mantle wedge. This process produces a locally high-temperature area in the core of the mantle wedge where arc magmatism potentially occurs.

The most fundamental feature of arc volcanism is the formation of the volcanic front that is defined as a trench-ward boundary of volcanoes, which is generally formed sub-parallel to the trench axis. The depth to a slab (depth of deep seismic zone) beneath the volcanic front varies in subduction zones, ranging from 70 to 160 km but is almost constant in individual arcs, with the maximum frequency of 90–110 km (e.g., England et al. 2004; Syracuse and Abers 2006; England and Katz 2010).

The volcanic front is well developed in the northeastern Japan arc as a result of the subduction of the PAC slab, and the depth to the PAC slab beneath the volcanic front is ~100 km in the northeastern Japan arc and ~130 km in the Izu-Bonin arc (Fig. 1a). However, the volcanic front in central Japan at the west of Kanto evidently deflects westward, and the depth to the PAC slab beneath the volcanic front gradually deepens toward the south, reaching a maximum depth of 170 km beneath Yokodake (Fig. 1b). The westward deflection of the volcanic front is interpreted to be caused by the dual plate subduction. It is likely that the presence of the PHS slab immediately above the PAC slab inhibits the thermal recovery of the PAC slab, thereby shifting the dehydration reactions to greater depth and resulting in the westward deflection of the volcanic front (Iwamori 2000). Asama volcano is formed in such a tectonically and volcanologically unique area.

This chapter first provides an overview of the large-scale tectonic framework and unique features of the plate subduction in central Japan. We then carry out travel-time tomography to determine three-dimensional (3D) P- and S-wave velocity structures and discuss the upper mantle structure beneath active volcanoes. We highlight fluid migration paths from the subducting PAC slab to the continental crust and propose a plausible model of melt generation in the uppermost mantle with the focus on the dual slab subduction. This chapter does not detail seismic velocity structures in the continental crust because of the limited spatial resolution of tomographic images. Readers can refer to seismological studies that focused on fine crustal structure beneath Asama volcano (e.g., Aoki et al. 2009, 2013; Savage et al. 2010; Nagaoka et al. 2012; Prudencio et al. 2017, and references therein).

2 Plate Subduction and Deep Seismic Zone

The depth to the PAC slab beneath Asama volcano is 150–160 km, and the westward deflection of the volcanic front appears to occur to the south of Takaharayama, which is located ~100 km northeast of Asama volcano (Fig. 1b). The northern limit of the PHS plate corresponds almost to the area where the westward deflection of the volcanic front starts to occur. Interestingly, the strike of the volcanic front from Akagi to Asama volcano is sub-parallel to the iso-depth contours of the PHS slab, and these volcanoes are located where the PHS slab reaches a depth of ~100 km.

Figure 2 shows across-arc vertical cross sections of earthquakes along seven profiles in the inset map. We plotted earthquakes ($M \geq 1$) that are reported in the unified catalog of the Japan Meteorological Agency (JMA) in the period of 2003–2020. We set four profiles (A–D) sub-parallel to the subduction direction of the PAC plate and three (E–G) profiles sub-parallel to that

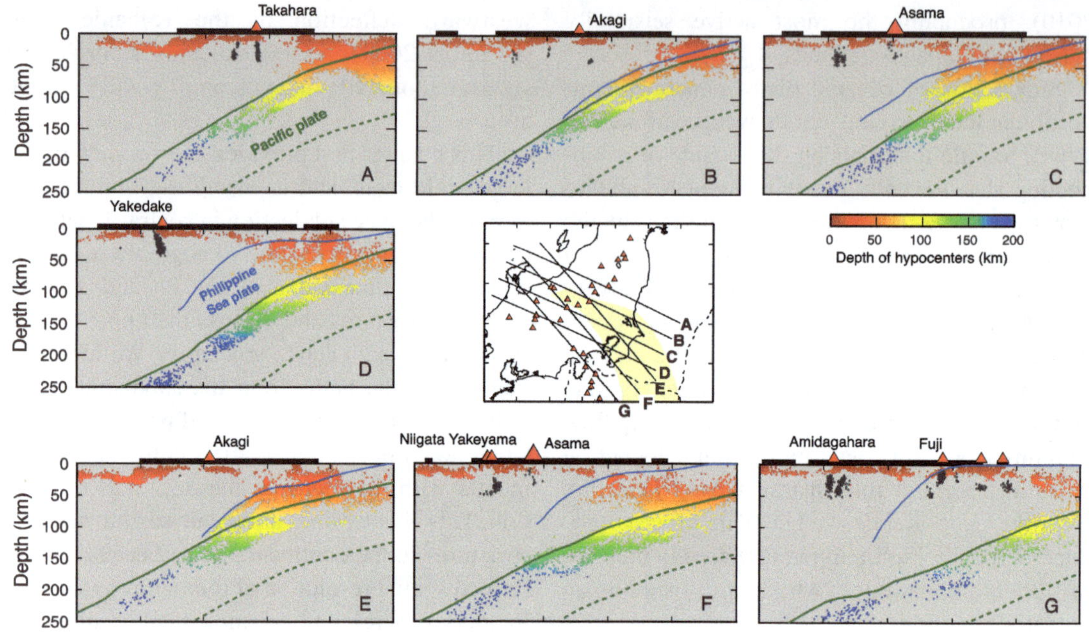

Fig. 2 Across-arc vertical cross sections of earthquakes along seven profiles (A–G) in the inset map. Hypocenters (colored dots), LFEs (black dots), and active volcanoes (red triangles) within ±10 km from each profile are shown. Colors of the hypocenters represent the focal depths. The upper surfaces of the PHS and PAC slabs are shown by blue and green lines, respectively. The inferred bottom of the PAC slab from the assumed slab thickness of 100 km is shown by a green dashed line. Black bars on the top denote the land area

of the PHS plate. Horizontal separations between the profiles are 50 km for profiles A–D and 60 km for profiles E–G. We here summarize the characteristics of the Wadati-Benioff zone (deep seismic zone) along each profile. It is noted that the bottom of the PHS cannot be well constrained except the slab contact area where the bottom of the PHS slab is in contact with the upper surface of the PAC slab (yellow shading area in the inset map of Fig. 2).

Profile A: This profile illustrates an area where only the PAC slab is subducting. Intraslab seismicity dips at ~30°, and the double seismic zone, which consists of the upper and lower seismic planes separated by ~30 km, is clearly observed at 60–150 km depth. A volcano (Takaharayama) is located in the middle of the arc, where the deep seismic zone reaches ~100 km. These features show typical distributions of earthquakes and volcanoes beneath the northeastern Japan arc (e.g., Hasegawa and Nakajima 2017; Igarashi et al. 2001; Nakajima et al. 2001).

Profile B: The double seismic zone of the PAC slab is visible, but another seismicity appears above the PAC slab at 50–100 km depth. This seismicity is related to the subduction of the PHS slab immediately above the PAC slab. The depth to the PAC plate beneath the volcanic front becomes slightly deeper (~120 km) than in profile A.

Profile C: The subduction of the PHS slab becomes more distinct, and inclined seismicity is distributed above the PAC slab to ~100 km depth. Although the PHS slab is aseismic beyond ~100 km depth, the subduction of the PHS slab can be evidently traced as a continuous high-velocity anomaly to a depth of at least ~140 km. The depths to the PAC and PHS slabs beneath Asama volcano are 150–160 and ~100 km, respectively.

Profile D: The dual plate subduction is evidenced from thickened seismicity at 20–80 km depth above the PAC slab. While the PHS slab is almost flat at the eastern part of the profile, the dip of the PHS slab becomes as steep as

40–50° at the middle of the profile. A volcano, Yokodake, is formed where the PHS slab is steeply subducting.

Profile E: The distribution of earthquakes is similar to that along profile C, and the subduction of the PHS slab beneath Kanto is well visualized from active seismicity above the PAC slab. The PHS slab is in contact with the PAC slab over a wide area along this profile.

Profile F: The PHS slab is seismic to ~70 km depth but becomes aseismic at greater depths. The dip of the PHS slab beneath Asama volcano is moderately steep, and the mantle wedge above the PHS slab is well developed.

Profile G: The PHS slab is completely aseismic along this profile, and it has been a long debate whether the PHS slab is subducted to the north of the Izu Peninsula. Although we have no definite evidence, comprehensive interpretations of seismicity in the PHS slab, results of regional tomographic studies, and the inferred deformation history of the PHS slab suggest a continuous subduction of the PHS slab between Kanto and Tokai without any disruption or splitting at the north of the Izu Peninsula (e.g., Nakajima et al. 2009a; Nakajima 2018).

3 Travel-Time Tomography

3.1 Data and Methods

We first selected earthquakes ($M \geq 1$) from the JMA catalog that occurred at 0–300 km depth between January 2003 and December 2021 and kept earthquakes whose focal depth is larger than the epicentral distance to the nearest station to ensure the reliable depth constraints of hypocenters. This procedure yielded 11,675 earthquakes (Fig. 3a). We then collected P- and S-wave arrival-time data for these earthquakes from the JMA catalog. The total number of phase data was 305,322 for P waves and 207,939 for S waves, which were recorded at 397 seismograph stations located in the study area (squares in Fig. 3b).

We applied the ray tracing technique and tomographic method of Zhao et al. (1992) to estimate 3D seismic velocity structure. We set up horizontal grid nodes with a spacing of 0.15° (~15 km) and vertical ones at depths of 0, 10, 20, 30, 40, 60, 80, 100, 120, 150, 180, 210, 240, 270, and 300 km (Fig. 3b). The initial 1D P-wave velocity model was constructed from the JMA 2001 model, which is used in the JMA routine work (Ueno et al. 2002) (Fig. 3c). A constant Vp/Vs value of 1.73 was used to calculate 1D S-wave model. We introduced a priori information for three discontinuities: the Conrad and Moho discontinuities in the continental crust (Katsumata 2010) and the upper surface of the PAC slab (Nakajima et al. 2009a). The initial velocity in the PAC slab was assigned to be 5% faster than that of the mantle wedge at the same depth. The root mean squares of the arrival-time residuals were reduced from 0.27 to 0.16 s for P wave and from 0.47 to 0.26 s for S wave after four iterations. After we obtained the final 3D P- and S-wave velocity model, we calculated the average P- and S-wave velocities at each depth. The deviations (in %) from the average values are shown in subsequent figures of tomographic images.

3.2 3D Seismic Velocity Structure

Figure 4 shows the obtained Vp and Vs structures at four representative depths (10, 20, 40, and 80 km). We observe two areas with strong low-velocity anomalies at 10 km (the upper crust) in the southeastern and northwestern parts of the study area, both of which probably present the thick sedimentary layers. The upper crust beneath active volcanoes partly shows weak low-velocity anomalies. At 20 km depth (the lower crust), we observe strong low-velocity anomalies beneath volcanoes. Marked S-wave low-velocity anomalies are distributed along the volcanic front and in the back-arc side at 40 km depth, and these anomalies are probably a deep magmatic source of volcanoes. Conversely, a marked E–W trending P- and S-wave low-velocity anomaly is observed beneath Kanto where no volcanism occurs. This low-velocity anomaly has been reported in previous tomographic studies and

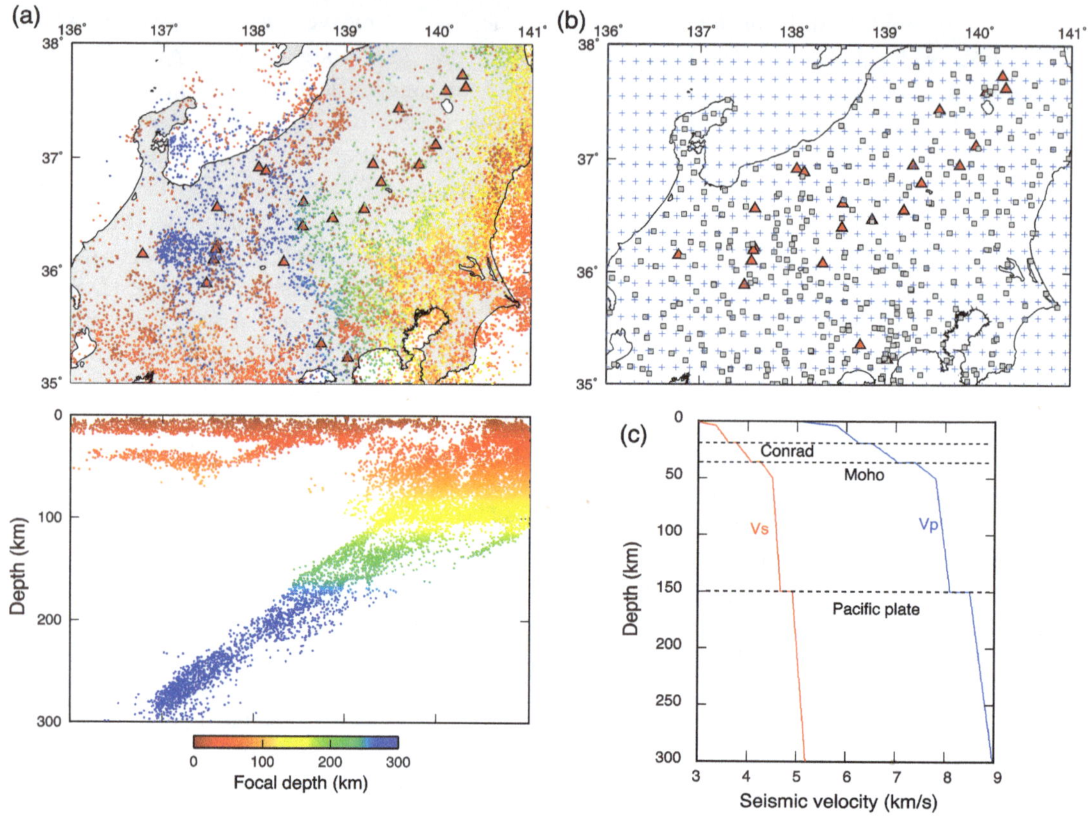

Fig. 3 (**a**) Distribution of earthquakes (N = 11,675) used in the tomographic inversion. Colors represent the focal depth. (**b**) The configuration of horizontal grid nodes (blue crosses) and the locations of seismograph stations (gray squares). Vertical grid notes are set at 0, 10, 20, 30, 40, 60, 80, 100, 120, 150, 180, 210, 240, 270, and 300 km depth. (**c**) The initial seismic velocity model for the tomographic inversion for the case with the depths of the Conrad and Moho discontinues, and the upper surface of the Pacific slab of 17, 35, and 150 km, respectively

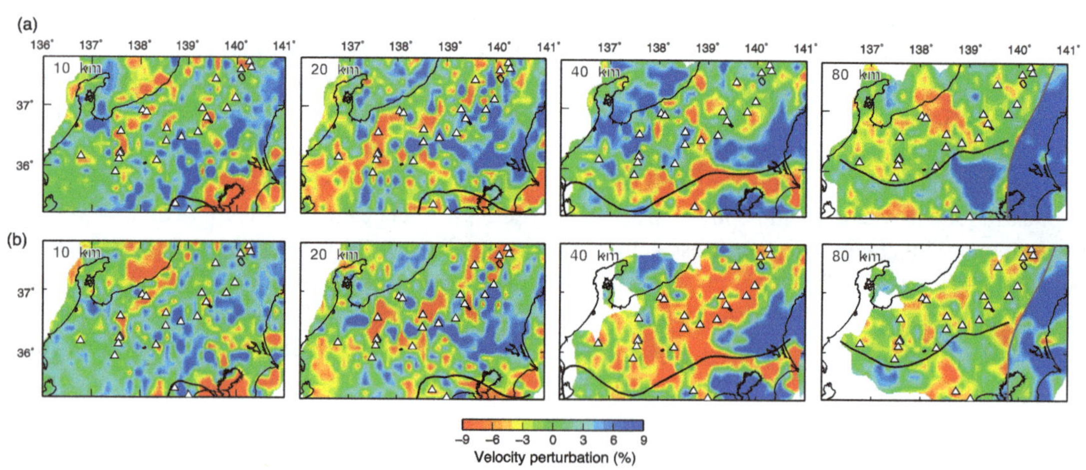

Fig. 4 Map showing distributions of (**a**) P- and (**b**) S-wave velocity perturbations (in %) at depths of 10, 20, 40, and 80 km. The velocity perturbations represent the deviation of the average velocity at each depth. White triangles denote active volcanoes, and black and gray lines represent the iso-depth contours of the PHS and PAC slab, respectively, at that depth

interpreted as the hydrated crust of the PHS slab (Matsubara et al. 2005; Nakajima and Hasegawa 2010). Low-velocity anomalies observed beneath volcanoes are relatively weak at 80 km depth compared to those at 40 km depth. This indicates that the low-velocity anomalies beneath volcanoes are strongest immediately below the continental Moho.

The observed velocity anomalies are well represented along across-arc vertical cross sections. Figure 5 shows that the PAC slab, which is imaged as moderate- to high-velocity anomalies, is subducting from the east and possesses the inclined Wadati-Benioff seismicity. The PHS slab subducting immediately above the PAC slab is imaged as a high-velocity anomaly to ~150 km depth in profiles B and C and to a depth of ~100 km in profile D. The PHS slab becomes aseismic at >100 km depth (profiles B, C, and D). Moderate low-Vs areas are imaged immediately above the PAC slab at 150–200 km, which appears to derive from the upper surface of the PAC slab (Fig. 5b). We observe large low-Vs areas above the PHS slab at 40–100 km depth beneath volcanoes.

Figure 6 shows vertical cross sections of Vp and Vs structures along the volcanic front over a length of 450 km, covering from Hakone in the Izu Peninsula to Azuma in the central part of Tohoku. The depth to the PAC slab beneath the volcanic front is almost constant (~100 km) to the north of Takahara, but it deepens toward the southwest and reaches a maximum depth of ~170 km beneath Yokodake. The uppermost mantle below the continental Moho shows a marked S-wave low velocity along the entire profile. However, the depth extent of the low-velocity zones is different between areas to the northeast and south of Yokodake, and the upper mantle to the south of Yokodake possesses the distinct low-velocity zones over a wide depth range of 20–120 km.

3.3 Resolution Tests

We carried out several kinds of synthetic tests to assess the reliability of the obtained images and to evaluate estimation errors of the velocity structure. First, we conducted two checkerboard resolution tests (CRTs) with different grid spacing, in which positive and negative velocity perturbations of 10% were assigned, respectively, to alternate grid nodes (CRT1) or to every two grid

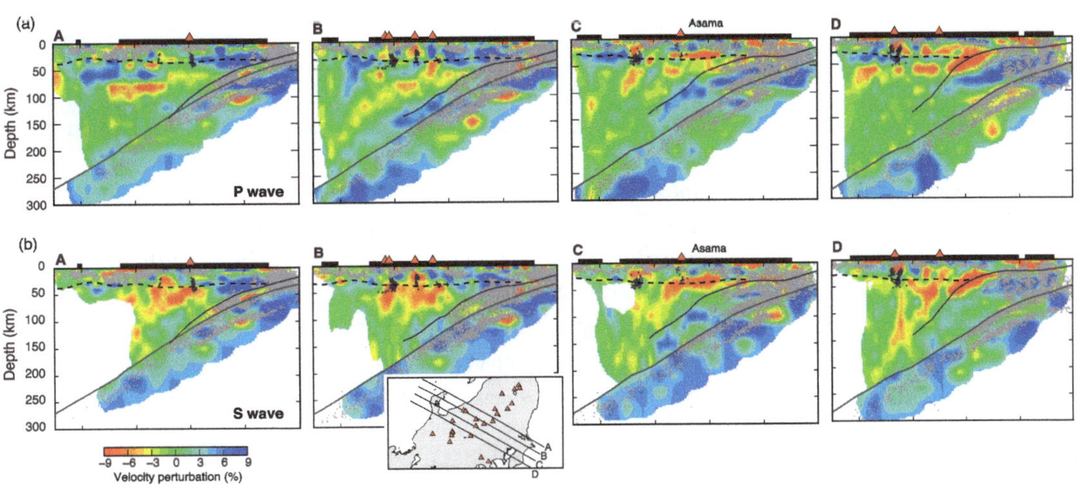

Fig. 5 Across-arc vertical cross sections of (**a**) P- and (**b**) S-wave velocity perturbations along four profiles shown in the inset map. Gray and black dots denote earthquakes and LFEs within ±10 km from each profile. Black and gray lines represent the upper surface of the PHS and PAC slabs. The continental Moho (Katsumata 2010) are shown by black dashed lines

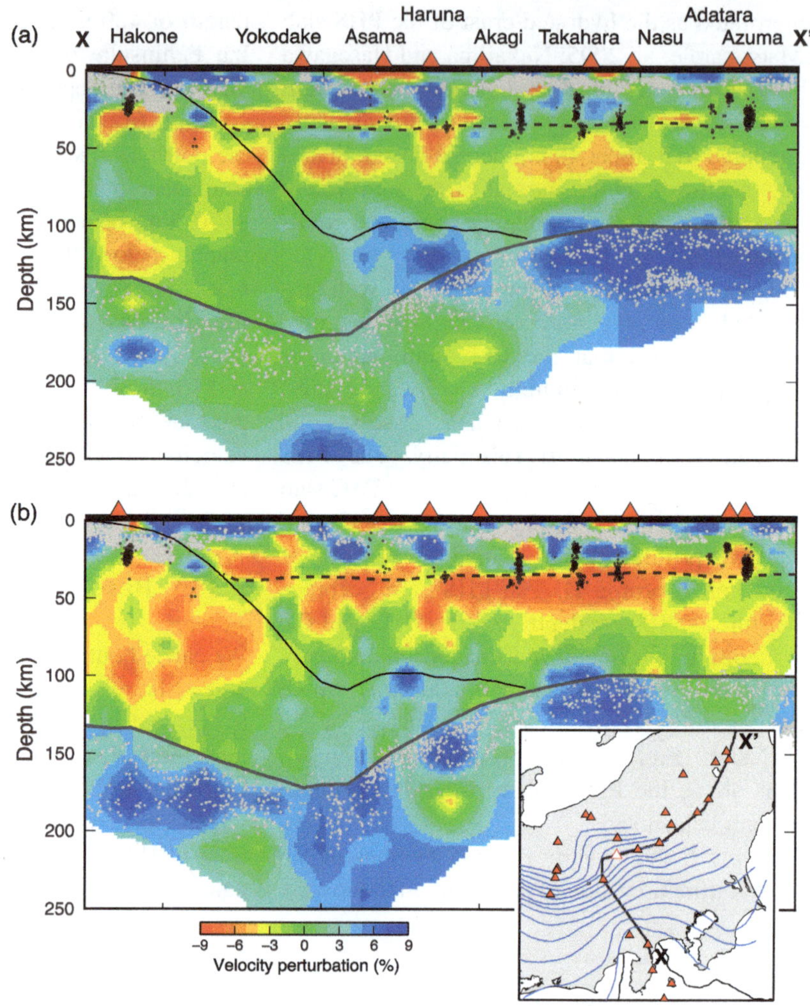

Fig. 6 Vertical cross sections of (**a**) P- and (**b**) S-wave velocity perturbations along X–X′ profile along the volcanic front. The names of volcanoes are shown in (**a**). Blue lines in the inset map are iso-depth contours of the PHS slab at an interval of 10 km. Other symbols are the same as in Fig. 5

nodes (CRT2). Travel times were then calculated for the two models to produce synthetic arrival-time data. Random noises with a standard deviation of 0.1 s for the P wave and 0.2 s for the S wave, which corresponds to the phase-picking errors, were added to the synthetic data, and the synthetic arrival-time data thus generated were inverted for the same earthquake–station pairs and model parameterizations as in the real inversion. We confirmed from the results of the CRT1 and CRT2 that the checkerboard patterns are well reproduced in the fore-arc side, while the checkerboard patterns in the back-arc side are smeared at >20 km depth in the CRT1 but are considerably recovered in the CRT2. Therefore, we consider that seismic velocity anomalies with a size of ~30 km can be resolved by the seismic tomography conducted in this chapter.

We carried out another synthetic test, restoring resolution test (RRT) (Zhao et al. 1992), in which the 3D tomographic results obtained by inverting the real data were assumed as a synthetic model and synthetic data were generated for that 3D velocity model. When the restoring image obtained by inverting the synthetic data is similar to the real image, we consider that our data set can resolve the complicated but realistic velocity anomalies that are not introduced in CRTs. The results of RRT showed that the overall shape of the observed low-velocity anomaly is well recovered for both P and S waves down to ~200 km depth. Since the average velocity differences between real data inversion and RRT are 2–3% at depths of 0–50 km and 1–2% at greater depths, we infer that these values are comparable to the estimation errors in Vp and Vs structures obtained in this chapter.

It is generally considered that Vp/Vs values are useful to quantify the origin of the observed low-velocity anomaly because Vp/Vs changes little for the thermal heterogeneities but is variable for the presence of fluids or melts. However, Vp/Vs values are very sensitive to the estimation errors of Vp and Vs, which easily overestimate or underestimate Vp/Vs values. For example, for a rock with Vp/Vs of 1.75 (Vp of 7.0 km/s and Vs of 4.0 km/s), an estimation error of 2% for Vp and Vs yields Vp/Vs of 1.68 (Vp of −2% and Vs of +2%) or 1.82 (Vp of +2% and Vs of −2%). Although these estimates are an extreme example, Vp/Vs values are inherently unstable and are difficult to estimate precisely without a sophisticated strategy for reliable estimates of Vp/Vs values. Therefore, we here do not highlight Vp/Vs images to discuss the deep source of magmatism.

4 Magmatism in Central Japan

Low-velocity anomalies are often interpreted as being caused by either high-temperature anomaly, or the presence of fluids and melts, or their combinations. It is known that Vp and Vs are reduced, respectively, by ~1% for an increase of temperatures by 100 °C (Duffy and Anderson 1989; Nishimoto et al. 2008). This suggests that a typical low-velocity anomaly of −5% beneath active volcanoes corresponds to a high temperature anomaly of ~500 °C when only the thermal effect is considered. In contrast, Vp and Vs are significantly reduced by the presence of a small volume of fluids or melts, even though the degree of the velocity reduction depends on the kinds of inclusions (fluids or melt) and the shape of pores (Takei 2002). Nakajima et al. (2005) argued that a combined effect of high temperatures by 100–200 °C and melt inclusions of 1–5 vol% is enough to explain the low-velocity anomalies (~−6%) observed in the mantle wedge (40–90 km depth) beneath northeastern Japan. Therefore, we here interpret that low-velocity anomalies in the uppermost mantle beneath volcanic areas are caused by a combination of higher temperatures and some amount of partial melts.

The release of slab-derived fluids to the mantle wedge is an essential control on the generation of arc magmatism, and the depth of fluid release from the PAC slab to the mantle wedge has been discussed from seismological studies (see Hasegawa and Nakajima 2017, for details). It is known that earthquakes in the upper plane of the double seismic zone are concentrated at 70–90 km depth in Tohoku and 100–150 km in Kanto, and such concentration of seismicity (the upper-plane seismic belt) is interpreted as being caused by fluid-related embrittlement as a result of dehydration reactions of hydrous minerals in the subducting crust (Kita et al. 2006; Hasegawa et al. 2007; Nakajima et al. 2009b). Therefore, it is inferred that dehydrated fluids are potentially released to the mantle wedge in a depth range of 70–150 km. However, it is further suggested that a hydrous, probably serpentinized layer can be formed immediately above the PAC slab as a result of metamorphism of the peridotite by the slab-derived fluids. The hydrous layer can act as a secondary storage of the fluids, whereby the slab-derived fluids are absorbed in the hydrous layer and carried to greater depths along the slab surface. The fluids are eventually released to the mantle wedge at 120–150 km depth in Tohoku (Iwamori 1998; Kawakatsu and Watada 2007; Tsuji et al. 2008) and at >150 km in central Japan, where the PHS slab overlies on the PAC slab and the thermal recovery of the PAC slab is delayed (Iwamori 2000).

Figure 5b shows that the moderate low-Vs areas elongated in the vertical direction are distributed immediately above the PAC slab at 150–200 km depth. Since the temperatures immediately above the PAC slab are unlikely to be high enough for melting of the peridotite (Yoshioka et al. 2015), the low-velocity anomalies probably represent migration paths of the slab-derived fluids (e.g., Iwamori 2000). In addition, marked, large low-Vs areas above the PHS slab at 40–100 km depth are probably interpreted as areas of partial melting due to the supply of the slab-derived fluids to the high-temperature mantle wedge. Nakamura et al. (2008) revealed that the fluids originating from both the PAC and PHS slabs contribute to magmatism in central Japan (including Asama volcano). This suggests that the PHS slab-derived fluids are added to the areas of partial melting above the PHS slab even though this chapter cannot reliably resolve fluid migration paths that originate from the PHS slab. Nakamura et al. (2008) further argued that the PHS slab does not inhibit upward fluid flow from the underlying PAC slab, and their inference is supported by our observations that the PAC slab-derived fluids can migrate upward in a detour route through the leading edge of the PHS slab (profiles B, C, and D in Fig. 5b).

We here emphasize again that Asama volcano is located above the area of two overlappingly subducting oceanic plates (Figs. 5 and 6), which is a tectonic situation different from northern Japan where only the PAC plate is subducting (see profile A in Fig. 2). Figure 7 shows schematic cartoons of arc magmatism along two representative profiles, based on the 3D Vp and Vs structure estimated in this chapter. In the across-arc profile C (the same profile as profile C in Fig. 5), the PHS slab is evidently subducted beneath Asama volcano, and partial melting occurs in the mantle wedge above the PHS slab (Fig. 7a), which is a different situation from the northeastern Japan arc where magmatism occurs in the mantle wedge of the PAC slab (Nakajima et al. 2001). These observations confirm that the westward deflection of the volcanic front in the western part of Kanto is caused by the subduction of the PHS slab above the underlying PAC slab.

Along profile X–X′ in Fig. 7b, we infer that melts are widely distributed at depths of 20–120 km between Hakone and Yokodake but are limited to 30–60 km beneath the volcanic

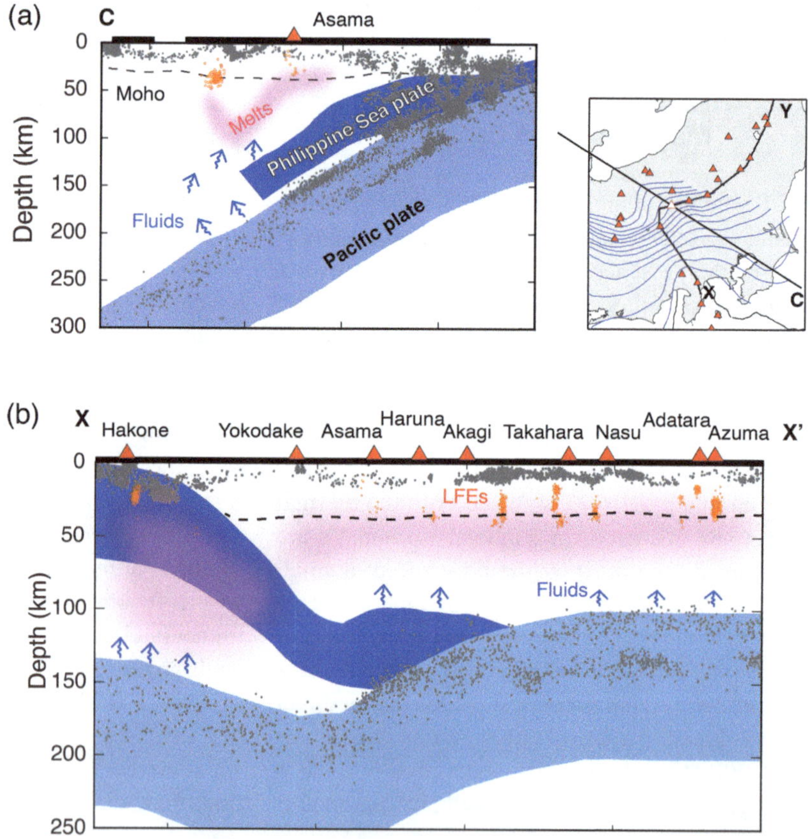

Fig. 7 Schematic illustrations of magmatism along (**a**) profile C in Fig. 5 and (**b**) profile X–X′ in Fig. 6. Note that the bottom of the PHS slab in (**b**) has not been constrained by observations

front to the north of Yokodake. Kashiwagi and Nakajima (2019) discussed this difference in magmatic system to be caused by the subduction of the aseismic PHS slab between Hakone and Yokodake. Seismic tomography studies have revealed that the PHS slab subducting toward the north of the Izu Peninsula shows a distinct low-velocity anomaly (Fig. 6) (e.g., Matsubara et al. 2005; Nakajima et al. 2009a). It is noted that we interpret in Fig. 7b that partial melting occurs even in the PHS slab beneath Hakone because the low-velocity zone is continuously distributed from depths of 100–120 km to depths of 20–30 km through the PHS slab. However, it is highly likely that the incoming PHS plate has long been heated and hydrated by magmatism along the Izu-Bonin volcanic arc. Therefore, the low-velocity zone in the PHS slab may partly represent the metamorphosed PHS slab. Further investigations are required to discuss the origin of the large low-velocity anomaly at depths of 20–120 km beneath the north of the Izu Peninsula.

Low-frequency earthquakes (LFEs) are recognized as earthquakes that radiate seismic waves dominantly in the 2–8 Hz frequency band, which is lower than the characteristic frequency of ordinary earthquakes with similar magnitudes. LFEs that occur around the continental Moho are often called either deep LFEs or volcanic (semi-volcanic) LFEs (Aso et al. 2013) and have been interpreted as being facilitated by the accumulation of fluids or magma near the Moho (Aki and Koyanagi 1981; Hasegawa and Yamamoto 1994; Oikawa et al. 2021). We found that there is a large variability in LFE activity along the volcanic front (Figs. 1 and 6). For example, we observe the occurrence of many LFEs beneath Hakone, Fuji, Takahara, Nasu, Adatara, and Azuma, while few LFEs occur beneath Yokodake, Asama, and Akagi. Even though it is not conclusive what causes the regional variation in LFE activity, LFE activity may reflect the physical, chemical, or frictional process during fluid accumulation around the Moho. A systematic analysis of deep LFEs will provide additional constraints on the magmatic process around the Moho and in the lower crust.

It is known that the inflow (upwelling flow) in the back-arc mantle wedge is generated perpendicular to the trench axis beneath northeastern Japan (Nakajima and Hasegawa 2004). However, the flow patterns become highly complex to the south of ~37°N (Nakajima et al. 2006), and such a complex flow is probably generated by the dual subduction of the PAC and PHS slabs with different convergent directions. A comprehensive and integrated study of geophysical, geochemical, and petrological observations could update the pseudo-3D cartoon of magmatism proposed in this chapter. The real 3D model of arc magmatism will deepen our understanding of where fluids are released from the two overlapping slabs, how the fluids migrate in the mantle wedge, where melts are generated, and how melts are migrated to the continental crust and are eventually supplied to the surface.

5 Summary

This chapter provided a brief review of the large-scale tectonic framework of dual plate subduction around Asama volcano and discussed arc magmatism in central Japan, based on 3D seismic velocity structures derived from the local tomographic study. The major conclusions are as follows:

1. The westward deflection of the volcanic front occurs due to the subduction of the PHS slab immediately above the PAC slab. The depth to the PAC slab beneath Asama volcano is 150–160 km, which is deeper by 50–60 km than that beneath volcanoes in the northeastern Japan arc where only the PAC slab is subducted.
2. The strike of the volcanic front from Akagi to Asama is sub-parallel to the iso-depth contours of the PHS slab, and the depth to the PHS slab is ~100 km.
3. The PAC and PHS slabs are overlappingly subducted beneath Asama volcano. The fluids derived from the PAC slab appear to migrate upward in a detour route through the leading edge of the PHS slab.
4. Fluids derived from both the PHS and PAC slabs are supplied to the uppermost mantle beneath Asama volcano, and partial melting occurs at depths of 40–100 km.

This chapter revealed that the subduction of the PAC and PHS slabs influences magmatism in central Japan. However, our tomographic images are a present-day snapshot and do not provide any information on temporal evolution of slab subduction and arc magmatism. Petrological and geochemical signatures, as well as numerical simulations of plate subaudition, will resolve time-dependent magmatic processes of this tectonically very complex area that is located at the arc-arc junction with the dual slab subduction.

Acknowledgments We would like to thank D. Zhao for providing us with the tomography codes. Arrival-time data of earthquakes reported in the JMA-unified catalogue are used in this chapter. Discussion with A. Hasegawa was very fruitful. This work was supported by JSPS KAKENHI Grant Numbers JP20K20939 and JP21H01176 and by the Ministry of Education, Culture, Sports, Science and Technology (MEXT) of Japan, under its The Second and Third Earthquake and Volcano Hazards Observation and Research Programs (Earthquake and Volcano Hazard Reduction Research).

References

Aki K, Koyanagi R (1981) Deep volcanic tremor and magma ascent mechanism under Kilauea, Hawaii. J Geophys Res 86:7095–7109. https://doi.org/10.1029/JB086iB08p07095

Aoki Y, Takeo M, Aoyama H et al (2009) P-wave velocity structure beneath Asama Volcano, Japan, inferred from active source seismic experiment. J Volcanol Geotherm Res 187:272–277. https://doi.org/10.1016/j.jvolgeores.2009.09.004

Aoki Y, Takeo M, Nagaoka Y, Nishida K (2013) Magma pathway and its structural controls of Asama Volcano, Japan. Geol Soc Spec Pub 380:67–84. https://doi.org/10.1144/SP380.6

Asamori K, Zhao D (2015) Teleseismic shear wave tomography of the Japan subduction zone. Geophys J Int 203:1752–1772. https://doi.org/10.1093/gji/ggv334

Aso N, Ohta K, Ide S (2013) Tectonic, volcanic, and semi-volcanic deep low-frequency earthquakes in western Japan. Tectonophysics 600:27–40. https://doi.org/10.1016/j.tecto.2012.12.015

Duffy TS, Anderson DL (1989) Seismic velocities in mantle minerals and the mineralogy of the upper mantle. J Geophys Res 94:1895. https://doi.org/10.1029/JB094iB02p01895

England PC, Katz RF (2010) Melting above the anhydrous solidus controls the location of volcanic arcs. Nature 467:700–703. https://doi.org/10.1038/nature09417

England P, Engdahl R, Thatcher W (2004) Systematic variation in the depths of slabs beneath arc volcanoes. Geophys J Int 156:377–408. https://doi.org/10.1111/j.1365-246X.2003.02132.x

Hasegawa A, Nakajima J (2017) Seismic imaging of slab metamorphism and genesis of intermediate-depth intraslab earthquakes. Prog Earth Planet Sci 4:12. https://doi.org/10.1186/s40645-017-0126-9

Hasegawa A, Yamamoto A (1994) Deep, low-frequency microearthquakes in or around seismic low-velocity zones beneath active volcanoes in northeastern Japan. Tectonophysics 233:233–252. https://doi.org/10.1016/0040-1951(94)90243-7

Hasegawa A, Nakajima J, Kita S et al (2007) Anomalous deepening of a belt of intraslab earthquakes in the Pacific slab crust under Kanto, central Japan: possible anomalous thermal shielding, dehydration reactions, and seismicity caused by shallower cold slab material. Geophys Res Lett 34:161–177. https://doi.org/10.1029/2007GL029616

Hirose F, Nakajima J, Hasegawa A (2008) Three-dimensional seismic velocity structure and configuration of the Philippine Sea slab in southwestern Japan estimated by double-difference tomography. J Geophys Res 113:B09315. https://doi.org/10.1029/2007JB005274

Hori S, Inoue H, Fukao Y, Ukawa M (1985) Seismic detection of the untransformed "basaltic" oceanic crust subducting into the mantle. Geophys J Int 83:169–197. https://doi.org/10.1111/j.1365-246X.1985.tb05162.x

Huang Z, Zhao D, Hasegawa A et al (2013) Aseismic deep subduction of the Philippine Sea plate and slab window. J Asian Earth Sci 75:82–94. https://doi.org/10.1016/j.jseaes.2013.07.002

Igarashi T, Matsuzawa T, Umino N, Hasegawa A (2001) Spatial distribution of focal mechanisms for interplate and intraplate earthquakes associated with the subducting Pacific plate beneath the northeastern Japan arc: a triple-planed deep seismic zone. J Geophys Res Solid Earth 106:2177–2191. https://doi.org/10.1029/2000JB900386

Iwamori H (1998) Transportation of H2O and melting in subduction zones. Earth Planet Sci Lett 160:65–80. https://doi.org/10.1016/S0012-821X(98)00080-6

Iwamori H (2000) Deep subduction of H2O and deflection of volcanic chain towards backarc near triple junction due to lower temperature. Earth Planet Sci Lett 181:41–46. https://doi.org/10.1016/S0012-821X(00)00180-1

Kashiwagi H, Nakajima J (2019) Three-dimensional seismic attenuation structure of central Japan and deep sources of arc magmatism. Geophys Res Lett 46:13746–13755. https://doi.org/10.1029/2019GL084793

Katsumata A (2010) Depth of the Moho discontinuity beneath the Japanese islands estimated by traveltime analysis. J Geophys Res 115:B04303. https://doi.org/10.1029/2008JB005864

Kawakatsu H, Watada S (2007) Seismic evidence for deep-water transportation in the mantle.

Science 316:1468–1471. https://doi.org/10.1126/science.1140855
Kita S, Okada T, Nakajima J et al (2006) Existence of a seismic belt in the upper plane of the double seismic zone extending in the along-arc direction at depths of 70-100 km beneath NE Japan. Geophys Res Lett 33:161–177. https://doi.org/10.1029/2006GL028239
Matsubara M, Hayashi H, Obara K, Kasahara K (2005) Low-velocity oceanic crust at the top of the Philippine Sea and Pacific plates beneath the Kanto region, central Japan, imaged by seismic tomography. J Geophys Res Solid Earth 110:1–15. https://doi.org/10.1029/2005JB003673
Matsuzawa T, Umino N, Hasegawa A, Takagi A (1986) Upper mantle velocity structure estimated from Ps-converted wave beneath the north-eastern Japan arc. Geophys J R Astron Soc 86:767–787. https://doi.org/10.1111/j.1365-246X.1986.tb00659.x
Mckenzie DP (1969) Speculations on the consequences and causes of plate motions. Geophys J R Astron Soc 18:1–32
Miyazaki K, Nakajima J, Suenaga N, Yoshioka S (2023) Deep subduction of the Philippine Sea slab and formation of slab window beneath central Japan. Earth Planets Space 75:93. https://doi.org/10.1186/s40623-023-01846-z
Nagaoka Y, Nishida K, Aoki Y et al (2012) Seismic imaging of magma chamber beneath an active volcano. Earth Planet Sci Lett 333–334:1–8. https://doi.org/10.1016/j.epsl.2012.03.034
Nakajima J, Matsuzawa T, Hasegawa A, Zhao, D (2001) Three-dimensional structure of V p V s and V p / V s beneath northeastern Japan: Implications for arc magmatism and fluids. Journal of Geophysical Research: Solid Earth 106(B10) 21843–21857. https://doi.org/10.1029/2000JB000008
Nakajima J (2018) Isolated intermediate-depth seismicity north of the Izu peninsula, Japan: implications for subduction of the Philippine Sea Plate. Earth Planets Space 70:11. https://doi.org/10.1186/s40623-018-0779-7
Nakajima J, Hasegawa A (2004) Shear-wave polarization anisotropy and subduction-induced flow in the mantle wedge of northeastern Japan. Earth Planet Sci Lett 225:365–377. https://doi.org/10.1016/j.epsl.2004.06.011
Nakajima J, Hasegawa A (2007) Subduction of the Philippine Sea plate beneath southwestern Japan: slab geometry and its relationship to arc magmatism. J Geophys Res 112:B08306. https://doi.org/10.1029/2006JB004770
Nakajima J, Hasegawa A (2010) Cause of M ~ 7 intraslab earthquakes beneath the Tokyo metropolitan area, Japan: possible evidence for a vertical tear at the easternmost portion of the Philippine Sea slab. J Geophys Res Solid Earth 115:1–16. https://doi.org/10.1029/2009JB006863
Nakajima J, Takei Y, Hasegawa A (2005) Quantitative analysis of the inclined low-velocity zone in the mantle wedge of northeastern Japan: a systematic change of melt-filled pore shapes with depth and its implications for melt migration. Earth Planet Sci Lett 234:59–70. https://doi.org/10.1016/j.epsl.2005.02.033
Nakajima J, Shimizu J, Hori S, Hasegawa A (2006) Shear-wave splitting beneath the southwestern Kurile arc and northeastern Japan arc: a new insight into mantle return flow. Geophys Res Lett 33:161–177. https://doi.org/10.1029/2005GL025053
Nakajima J, Hirose F, Hasegawa A (2009a) Seismotectonics beneath the Tokyo metropolitan area, Japan: effect of slab-slab contact and overlap on seismicity. J Geophys Res 114:B08309. https://doi.org/10.1029/2008JB006101
Nakajima J, Tsuji Y, Hasegawa A (2009b) Seismic evidence for thermally-controlled dehydration reaction in subducting oceanic crust. Geophysical Research Letters 36(3). https://doi.org/10.1029/2008GL036865
Nakamura H, Iwamori H, Kimura J-I (2008) Geochemical evidence for enhanced fluid flux due to overlapping subducting plates. Nat Geosci 1:630. https://doi.org/10.1038/ngeo290
Nishimoto S, Ishikawa M, Arima M et al (2008) Simultaneous high P-T measurements of ultrasonic compressional and shear wave velocities in Ichino-megata mafic xenoliths: their bearings on seismic velocity perturbations in lower crust of northeast Japan arc. J Geophys Res Solid Earth 113:161–177. https://doi.org/10.1029/2008JB005587
Oikawa G, Aso N, Nakajima J (2021) Depth dependent focal mechanisms of volcanic deep low-frequency earthquakes in Northeast Japan. J Geophys Res Solid Earth 126(12):e2021JB022666. https://doi.org/10.1029/2021JB022666
Prudencio J, Aoki Y, Takeo M et al (2017) Separation of scattering and intrinsic attenuation at Asama Volcano (Japan): evidence of high volcanic structural contrasts. J Volcanol Geotherm Res 333–334:96–103. https://doi.org/10.1016/j.jvolgeores.2017.01.014
Savage MK, Ohminato T, Aoki Y et al (2010) Stress magnitude and its temporal variation at Mt. Asama Volcano, Japan, from seismic anisotropy and GPS. Earth Planet Sci Lett 290:403–414. https://doi.org/10.1016/j.epsl.2009.12.037
Shiina T, Nakajima J, Matsuzawa T (2013) Seismic evidence for high pore pressures in the oceanic crust: implications for fluid-related embrittlement. Geophys Res Lett 40:2006–2010. https://doi.org/10.1002/grl.50468
Syracuse EM, Abers GA (2006) Global compilation of variations in slab depth beneath arc volcanoes and implications. Geochem Geophys Geosyst 7:161–177. https://doi.org/10.1029/2005GC001045
Takei Y (2002) Effect of pore geometry on VP/VS: from equilibrium geometry to crack. J Geophys Res 107:2043. https://doi.org/10.1029/2001JB000522
Tsuji Y, Nakajima J, Hasegawa A (2008) Tomographic evidence for hydrated oceanic crust of the Pacific slab beneath northeastern Japan: implications for water transportation in subduction zones. Geophys Res Lett 35:161–177. https://doi.org/10.1029/2008GL034461

Uchida N, Nakajima J, Hasegawa A, Matsuzawa T (2009) What controls interplate coupling?: evidence for abrupt change in coupling across a border between two overlying plates in the NE Japan subduction zone. Earth Planet Sci Lett 283:111–121. https://doi.org/10.1016/j.epsl.2009.04.003

Ueno H, Hatakeyama S, Aketagawa T et al (2002) Improvement of hypocenter determination procedures in the Japan meteorological agency. Q J Seism 65:123–134

Yoshioka S, Takagi R, Matsumoto T (2015) Relationship between temperatures and fault slips on the upper surface of the subducting Philippine Sea plate beneath the Kanto district, central Japan. Geophys J Int 201:878–890. https://doi.org/10.1093/gji/ggv032

Zhao D, Hasegawa A, Horiuchi S (1992) Tomographic imaging of P and S wave velocity structure beneath northeastern Japan. J Geophys Res 97:19909–19928. https://doi.org/10.1029/92jb00603

Zhao D, Matsuzawa T, Hasegawa A (1997) Morphology of the subducting slab boundary in the northeastern Japan arc. Phys Earth Planet Inter 102:89–104. https://doi.org/10.1016/S0031-9201(96)03258-X

The Deflection of Volcanic Front Around Asama Volcano and Deep Dehydration Beneath the Noto Peninsula: Constraints from Regional Magma Geochemistry

Hikaru Iwamori and Hitomi Nakamura

1 Introduction

The Japanese archipelago is located in one of the most geologically active zones on Earth, sandwiched between large continental and oceanic plates (the Eurasian and Pacific plates). Active geological phenomena, including earthquakes and volcanic eruptions in the Japanese archipelago, are attributed to the interaction of four plates, including the North American (or Okhotsk) plate and the Philippine Sea plate, in addition to the Eurasian and Pacific plates (Seno et al. 1996; Bird 2003; Argus et al. 2011).

Corresponding to these plate configurations, the Japanese archipelago can be divided into several arc segments (Fig. 1): Kuril, Northeast Japan (NE Japan), Central Japan, Southwest Japan (SW Japan), Ryukyu, and Izu-Bonin arcs. The Kuril, NE Japan, and Izu-Bonin arcs correspond to subduction of the Pacific plate (PAP), and the SW Japan and Ryukyu arcs correspond to subduction of the Philippine Sea plate (PHP). Figure 1 shows the depth and shape of the subducting plate (hereafter referred to as "slab"). The Central Japan and SW Japan arcs correspond to the same PHP subduction, but beneath the Central Japan arc, the two seismic slabs, PAP and PHP, overlap, where the contour lines of PHP slab surface are dense (i.e., steep slab subduction).

The magma compositions of the arc segments shown in Fig. 1 have their own characteristics and can be broadly distinguished (Nakamura et al. 2019): not only unique activities such as the alkaline rocks in the SW Japan arc (Iwamori 1991), but also other island arc-type magmas exhibit geographical provenance. Based on the isotopic composition of lava, magmas are broadly classified geochemically into the arc segments, and their characteristics are correlated with differences in subducting materials and subduction parameters (Ishikawa and Nakamura 1994; Taylor and Nesbitt 1998; Shinjo et al. 2000; Hochstaedter et al. 2001; Ishizuka et al. 2003; Hauff et al. 2003; Moriguti et al. 2004; Straub et al. 2004; Kimura and Yoshida 2006; Hanyu et al. 2006; Nakamura et al. 2008; Nakamura and Iwamori 2009, 2013).

Supplementary Information The online version contains supplementary material available at https://doi.org/10.1007/978-981-95-5799-8_2.

H. Iwamori (✉)
Earthquake Research Institute, The University of Tokyo, Tokyo, Japan
e-mail: hiwamori@eri.u-tokyo.ac.jp

H. Nakamura
Geological Survey of Japan, National Institute of Advanced Industrial Science and Technology (AIST), Tsukuba, Ibaraki, Japan
e-mail: hitomi-nakamura@aist.go.jp

S. Aramaki et al. (eds.), *The Asama Volcano*, Active Volcanoes of the World,
https://doi.org/10.1007/978-981-95-5799-8_2

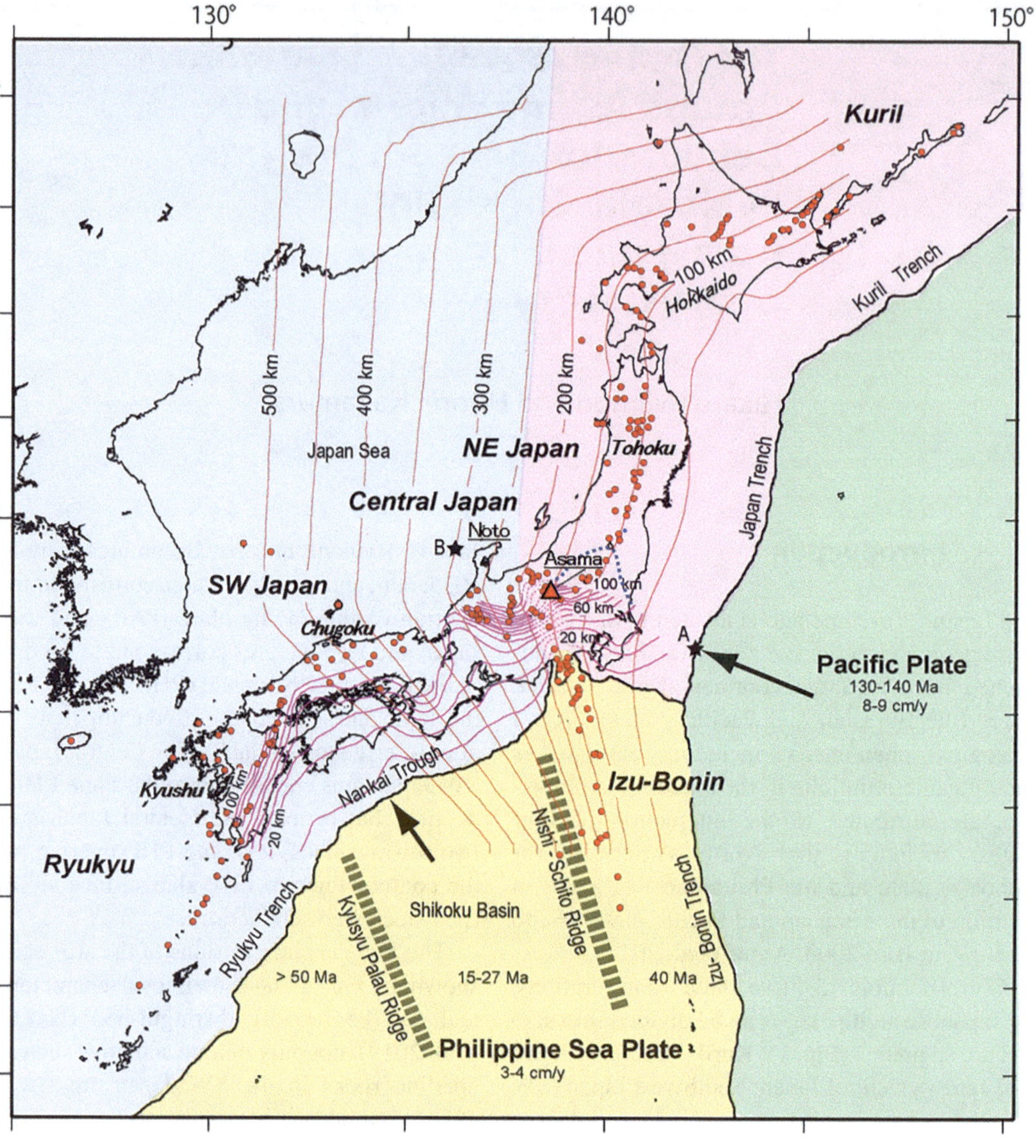

Fig. 1 Tectonic setting of the Japan arcs: Kuril, Northeast (NE) Japan, Central Japan, Southwest (SW) Japan, Ryukyu, and Izu-Bonin arcs. The depth contours of the Pacific slab are from Kita et al. (2008) and Nakajima and Hasegawa (2006), while those of the Philippine Sea slab are from Hirose et al. (2008) after Nakajima and Hasegawa (2007). A part of the Philippine Sea slab is observed as an aseismic slab in the collision zone of the Izu–Bonin and Central Japan arcs (thin dotted lines). The two arrows represent the plate motions of the Pacific plate and the Philippine Sea plate. The two stars "A" and "B" indicate the location of the across-arc section through Asama volcano and the Noto Peninsula, corresponding to Fig. 11. The thick dotted line in the southern Tohoku area represents the extent of an aseismic Philippine Sea slab north of the seismically defined Philippine Sea slab (Nakamura et al. 2018)

Among the arc segments, the Central Japan arc has distinctive features such as presence of adakites in the Ryohaku Mountains (Nakamura and Iwamori 2013) and a remarkable bend of the volcanic front (Figs. 1 and 2). Asama volcano is located almost at the inflection point of this bend, along the across-arc section from point A in the Japan trench to point B near the Noto Peninsula

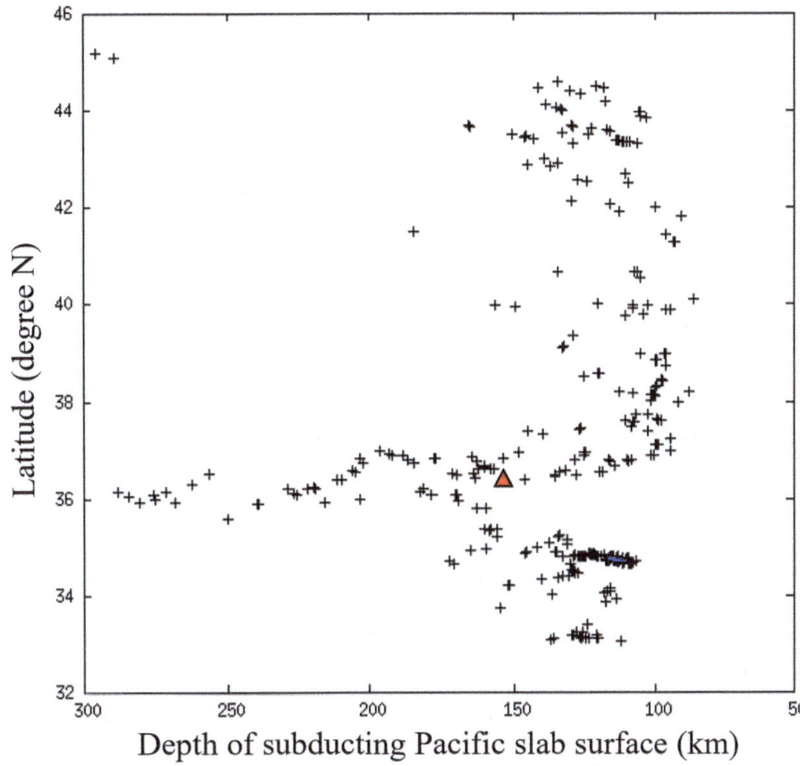

Fig. 2 Depth of subducting Pacific slab surface beneath the location of the lava samples used in this chapter. The depth of the Pacific slab surface straight beneath each lava sample (horizontal axis in the figure) is obtained based on the seismically determined slab surface geometry (Fig. 1) and its bicubic interpolation. The depth of the slab surface is then plotted against the latitude of the sample location (vertical axis). The red triangle represents Asama volcano

(Fig. 1). It has been proposed that these features are related to the low-temperature structure associated with overlapping subduction (Fig. 1), the dehydration of the slab, and the melting conditions of the mantle wedge (Iwamori 2000; Nakamura et al. 2008; Nakamura and Iwamori 2013).

Using an expanded isotopic dataset and trace element compositions of lavas, this chapter revisits the segment-specific characteristics of the entire Japan arcs and the position of the Central Japan arc within that context. In particular, from the basaltic lava compositions, we derive geochemical characteristics of the source mantle involved in magma generation, fluid distribution from subducted slabs, and constraints on melting degree and depth, and discuss conditions and magma production mechanisms beneath the Japan arcs, in particular the Central Japan arc.

2 Whitened Data-Based k-Means Cluster Analysis

Geochemical data of basaltic lavas with SiO_2 less than 56 wt% from the Japan arcs were used in this chapter, based on Nakamura et al. (2019) and GEOROC (http://georoc.mpch-mainz.gwdg.de/georoc/). A total of 747 samples with five isotopic ratios ($^{87}Sr/^{86}Sr$, $^{143}Nd/^{144}Nd$, $^{206}Pb/^{204}Pb$, $^{207}Pb/^{204}Pb$, $^{208}Pb/^{204}Pb$) are used, of which 382 samples are from Nakamura et al. (2019) and an additional 365 samples from GEOROC. The geographical distribution and the isotopic ratios are shown in Fig. 3. Most of these data sets also include trace element concentrations, such as large-ion lithophile and high-field strength elements (LILE and HFSE, respectively), which will also be utilized to discuss the magma genesis in this chapter.

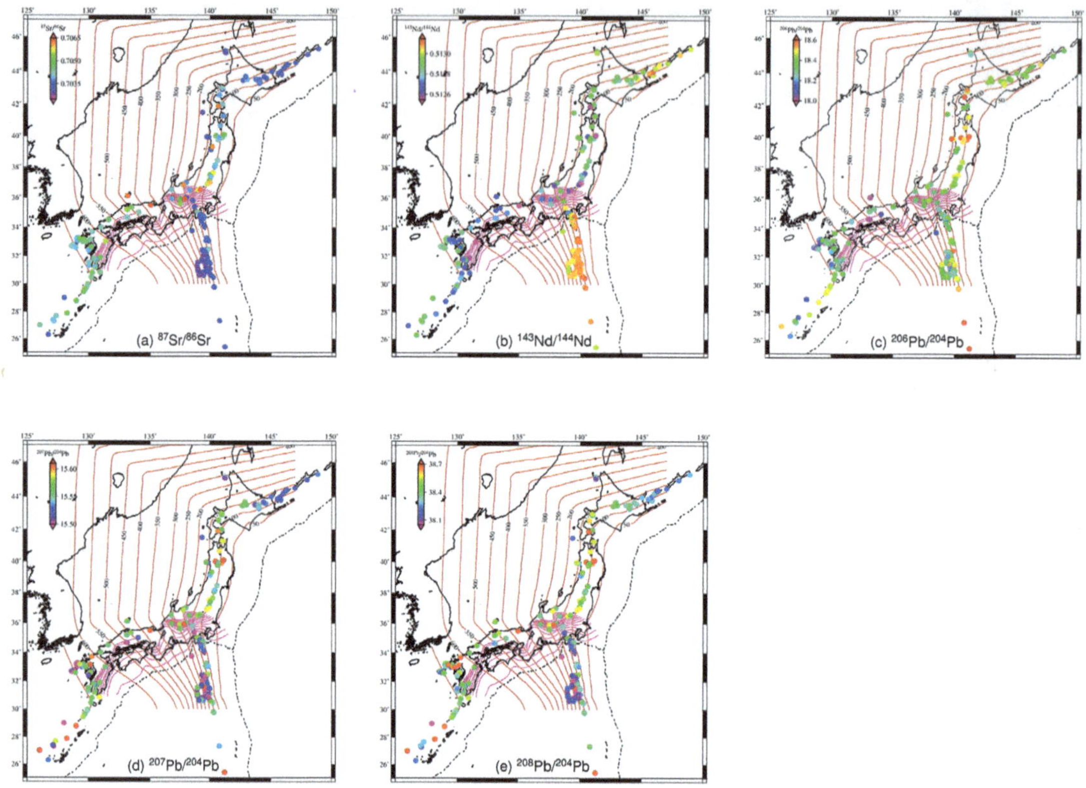

Fig. 3 The five isotopic ratios of 747 samples used for the statistical analysis and their geographical distributions. (**a**) $^{87}Sr/^{86}Sr$, (**b**) $^{143}Nd/^{144}Nd$, (**c**) $^{206}Pb/^{204}Pb$, (**d**) $^{207}Pb/^{204}Pb$, and (**e**) $^{208}Pb/^{204}Pb$. The data sources are Nakamura et al. (2019) and GEOROC (http://georoc.mpch-mainz.gwdg.de/georoc/). The trenches and the depth contours of the subducted slabs are illustrated as in Fig. 1

To investigate the independent sources and processes that produced these lavas, we used an unsupervised machine learning approach with multivariate statistical analysis for this high-dimensional data set. The statistical method used is "whitened data-based k-means cluster analysis" (hereafter referred to as KCA) and is suitable for extracting independent features hidden in geochemical data (Iwamori et al. 2017, 2023). Of the standard multivariate analyses, principal component analysis (PCA) is common to find the principal components (PCs) that account for the sample variance most effectively. The extracted PCs are uncorrelated with each other, but are not necessarily independent. When the data constitute a (joint) non-Gaussian distribution, PCA always fails to extract the independent features in the data (Hyvärinen et al. 2001). Instead, KCA is suitable for data showing non-Gaussian distribution. As will be shown in the next chapter, the concentration data used in this chapter show non-Gaussian distributions, and KCA is used in the following analysis.

First, the isotopic data set consisting of 747 samples with 5 isotopic ratios was preprocessed by performing PCA for whitening of the data (i.e., normalizing the data along the eigenvectors by the square root of the eigenvalues to make the new transformed variables uncorrelated with equalized variance of unity). Then k-means cluster analysis was applied to the data set for identifying the clusters after Iwamori et al. (2017). The distribution of clusters in geochemical compositional space and the geographical space provides insights into the origin of clusters and magma genetic conditions, which will be combined later with the trace element information.

3 Compositional Clusters Based on Isotopic Ratios

As a result of data preprocessing by PCA, 89.2% and 94.0% of the sample variance are accounted for by two and three principal components (61.4%, 27.8%, and 4.8% by the first three components), respectively. In this case, four to eight clusters are in principle suitable for describing the observed variations, since two and three components span four and eight quadrants, respectively (Iwamori et al. 2017). In this chapter, five clusters are found to discriminate the geographical provenance as shown later and will be used in the following arguments. It is noted that the five clusters are broadly and consistently subdivided into smaller clusters when the number of clusters is increased (e.g., to seven clusters as in Fig. S1), or merged to a larger cluster when decreased (e.g., to three clusters, Fig. S1), although the details of cluster division are variable depending on the cluster number due to the intrinsic behavior of KCA.

The five clusters were labeled from C1 to C5 in order of increasing average $^{87}Sr/^{86}Sr$ ratio (i.e., the lowest for C1 [black solid circle] and the highest for C5 [light blue cross mark]). Figure 4 shows the average isotopic ratios of C1 to C5 and the spread of the data. Although the average $^{143}Nd/^{144}Nd$ ratio generally decreases from C1 to C5, the average $^{143}Nd/^{144}Nd$ value of C2 is higher than that of C1. $^{206}Pb/^{204}Pb$ has no simple correlation with $^{87}Sr/^{86}Sr$, and C3 shows the minimum average value. The three Pb isotopic ratios, especially $^{206}Pb/^{204}Pb$ and $^{207}Pb/^{204}Pb$, show good correlation globally (e.g., Iwamori and Nakamura 2015), but this is not necessarily the case in Fig. 4. For example, the average values of $^{206}Pb/^{204}Pb$ are almost the same (or slightly decreased in this order) in the three clusters C2, C4, and C5, but the average values of $^{207}Pb/^{204}Pb$ are higher in C4 and C5 than in C2. In addition, the average values of $^{206}Pb/^{204}Pb$ for C1 to C3 decrease in this order, but the average values of $^{208}Pb/^{204}Pb$ do not follow this order, and C3 is higher than C2 and almost the same as C1. These variations indicate that each cluster has a characteristic isotopic composition distinct from the others.

Figure 5 shows the distribution of the five clusters in the binary diagrams of isotopic ratio. In the $^{87}Sr/^{86}Sr$-$^{143}Nd/^{144}Nd$ diagram (Fig. 5a), there is an overall negative correlation, as with the global trend. However, C2 (red plus sign) is slightly above the trend and $^{143}Nd/^{144}Nd$ is slightly higher. Since C5 is located in the lower rightmost position, the influence of crustal material can be considered, but there is no significant correlation between SiO_2 and $^{87}Sr/^{86}Sr$, either in the data as a whole or within individual clusters (Fig. 5b). Although the average SiO_2 of each cluster is slightly lower for C3, which has the third highest $^{87}Sr/^{86}Sr$, the average values for C1, C2, C4, and C5 are almost the same.

In the $^{87}Sr/^{86}Sr$-$^{143}Nd/^{144}Nd$ diagram (Fig. 5a), a broad trend appears as a whole, but the binary diagrams with Pb isotopic ratios (Fig. 5c, d) show roughly two trends with some overlap. For example, in the relationship between $^{207}Pb/^{204}Pb$ and $^{143}Nd/^{144}Nd$ (Fig. 5c), two broad trends formed by C1–C2–C4 and C3–C5 are identified. In the former, $^{207}Pb/^{204}Pb$ increases and $^{143}Nd/^{144}Nd$ decreases in the order C2–C1–C4, and in the latter, C3–C5. The slopes of the two trends are different, and they tend to overlap more as $^{207}Pb/^{204}Pb$ increases. This trend is more pronounced in the $^{207}Pb/^{204}Pb$-$^{208}Pb/^{204}Pb$ diagram (Fig. 5g), where the two trends seem to converge at $^{207}Pb/^{204}Pb$ around 15.6. The binary diagram including $^{206}Pb/^{204}Pb$ shows yet another relationship (e.g., Fig. 5e), where C1–C2 and C3–C4–C5 seem to form broad but two distinct trends, unlike the apparent combination of C1–C2–C4 and C3–C5 in Fig. 5c.

As described above, the relationship among the five isotopic ratios is complex; up to a combination of three isotopic ratios and their relationship can be visualized, and the structure can be grasped by actually rotating a three-dimensional plot. However, it is not possible to visualize the relationships in four or more dimensions or to accurately grasp the relationships in combinations of plots of three or fewer dimensions. In such a case, a mathematical method based on

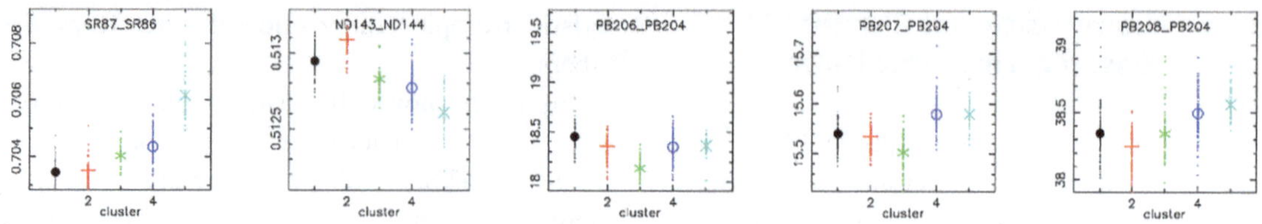

Fig. 4 The average isotopic ratios of the five clusters for the five isotopic ratios ($^{87}Sr/^{86}Sr$, $^{143}Nd/^{144}Nd$, $^{206}Pb/^{204}Pb$, $^{207}Pb/^{204}Pb$, $^{208}Pb/^{204}Pb$, from left to right) are represented by large symbols, and the spread of the data is expressed by the small dots

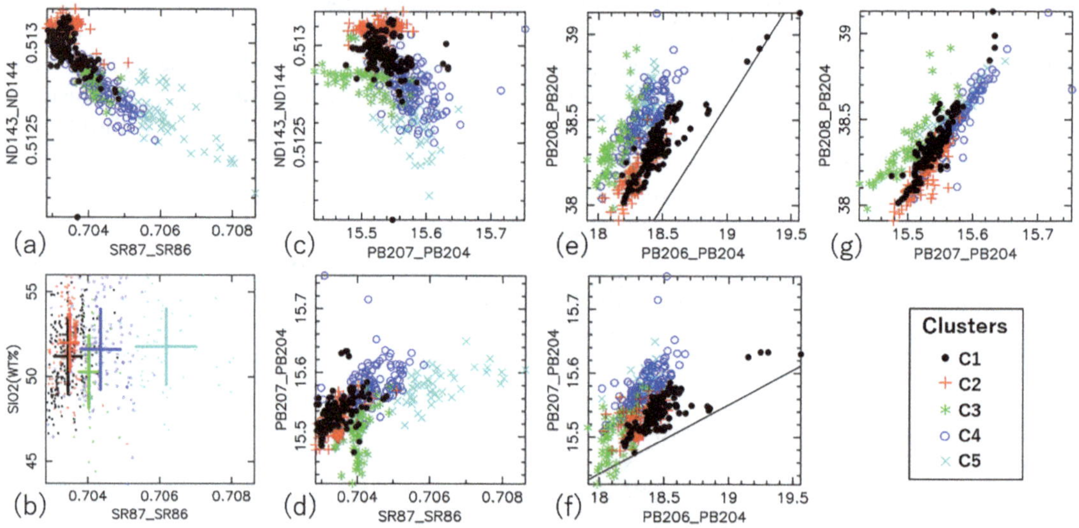

Fig. 5 Five clusters in the selected binary diagrams of isotopic ratios: (**a**) $^{87}Sr/^{86}Sr$ vs. $^{143}Nd/^{144}Nd$, (**b**) $^{87}Sr/^{86}Sr$ vs. SiO_2, (**c**) $^{207}Pb/^{204}Pb$ vs. $^{143}Nd/^{144}Nd$, (**d**) $^{87}Sr/^{86}Sr$ vs. $^{207}Pb/^{204}Pb$, (**e**) $^{206}Pb/^{204}Pb$ vs. $^{208}Pb/^{204}Pb$, (**f**) $^{206}Pb/^{204}Pb$ vs. $^{207}Pb/^{204}Pb$, (**g**) $^{207}Pb/^{204}Pb$ vs. $^{208}Pb/^{204}Pb$. In (**b**), each cluster is represented by the cross symbol for average and standard deviation (1 sigma) along with the individual data points. The straight lines in (**e, f**) represent the Northern Hemisphere Reference Line (NHRL) from Hart (1984)

multivariate statistical analysis, such as the one used here, is essential and important.

The five clusters identified in the compositional space also show some systematic distribution in the geographical space (Fig. 6). C1 and C2 occur mostly along the Kuril, NE Japan, and Izu-Bonin arcs. C1 characterizes the NE Japan arc as well as the back-arc area of Izu arc, while both C1 and C2 occur along the Kuril arc. On the other hand, most of C3, C4, and C5 basalts occur along the Central Japan (Fig. 6), including the bend of volcanic front around Asama volcano (Figs. 1 and 2), and SW Japan arcs, while C1 and C4 characterize the Ryukyu arc (Fig. 6). C3, which exhibits the lowest $^{206}Pb/^{204}Pb$ and $^{207}Pb/^{204}Pb$ average ratios (Fig. 4) with relatively low $^{143}Nd/^{144}Nd$ and high $^{208}Pb/^{204}Pb$ in the compositional trends (Fig. 5), appears mostly along a geographically linear trend including basalts from Fuji and Ueno, as well as alkaline basalts from the Chugoku and northern Kyushu districts of the SW Japan arc (Fig. 6).

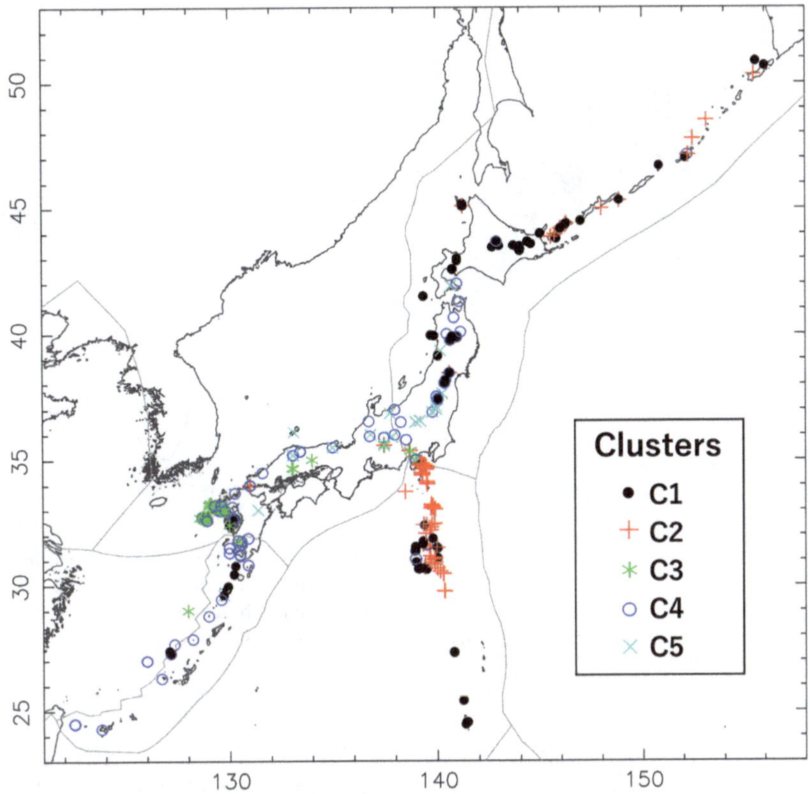

Fig. 6 Geographical distribution of the five compositional clusters from C1 to C5. The plate boundaries are after Bird (2003)

4 Magma Source and Slab Fluid

The isotopic ratios and the clusters of basaltic magmas described in Sect. 3 mainly reflect the nature of the magma source material and are independent of the degree of melting and the mineral assemblage of the residual solid phase, provided that isotopic equilibrium is established. Taking advantage of this relationship, numerous studies have been conducted based on isotopic ratios of basaltic magmas, and the mantle composition beneath the Japan arcs and the fluid supply from subducting slabs have been discussed (e.g., Ishikawa and Nakamura 1994; Taylor and Nesbitt 1998; Shinjo et al. 2000; Hochstaedter et al. 2001; Ishizuka et al. 2003; Hauff et al. 2003; Moriguti et al. 2004; Straub et al. 2004; Kimura and Yoshida 2006; Hanyu et al. 2006; Nakamura et al. 2008; Nakamura and Iwamori 2009, 2013).

Ishikawa and Nakamura (1994) revealed the source materials of magma and the contribution of fluid supplied by the Pacific slab beneath the Izu arc based on B and Pb isotopic ratios, and discussed the across-arc variations of magma genesis. Nakamura and Iwamori (2009) and Nakamura et al. (2019) quantified the amount of fluid added to the mantle from the subducting Pacific and Philippine Sea slabs based on Sr-Nd-Pb isotopic ratios. At the same time, they found that the composition of the mantle wedge prior to fluid addition varied on the scale of the entire Japan arcs. In these studies, the $^{207}Pb/^{204}Pb$-$^{143}Nd/^{144}Nd$ system was found to be particularly useful for quantifying the mantle composition and the amount of fluid added to the mantle.

Figure 7 shows the isotopic ratios of basaltic lavas for the five clusters identified in this chapter, as well as the compositional range of the mantle wedge beneath the Japan arcs ("mantle")

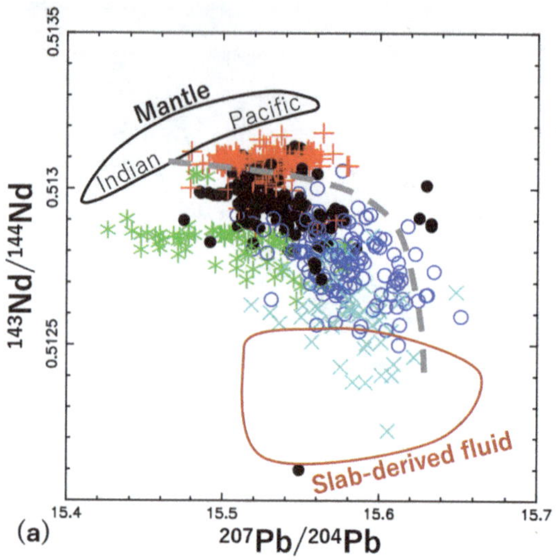

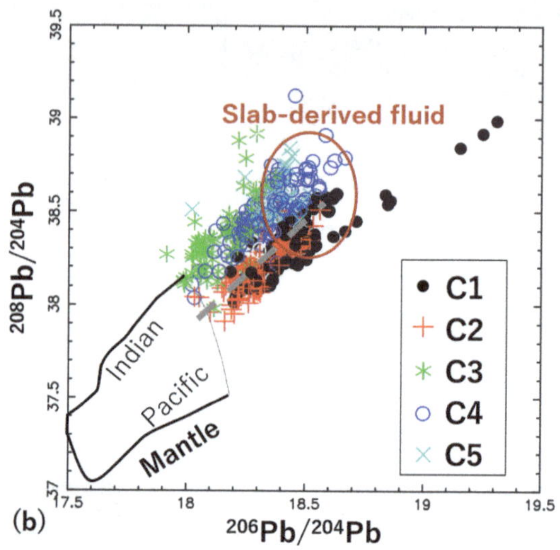

Fig. 7 (**a**) ^{207}Pb/^{204}Pb-^{143}Nd/^{144}Nd, and (**b**) ^{206}Pb/^{204}Pb-^{208}Pb/^{204}Pb diagrams of the five clusters (C1 to C5) with the end-members of the mantle and the slab-derived fluid beneath the Japan arcs, based on Hickey-Vargas (1998), Nakamura et al. (2008), and Nakamura and Iwamori (2009). Thick grayish broken lines are examples of mixing line between the mantle and the slab-derived fluid, based on Nakamura et al. (2008) and Nakamura and Iwamori (2009)

and fluids supplied from the subducting Pacific and Philippine Sea slabs ("Slab-derived fluid") as revealed by previous studies (Hickey-Vargas 1998; Nakamura et al. 2008; Nakamura and Iwamori 2009). The two broad trends recognized in several figures in Fig. 5 (including ^{207}Pb/^{204}Pb-^{143}Nd/^{144}Nd and ^{206}Pb/^{204}Pb-^{208}Pb/^{204}Pb shown in Fig. 7) can be interpreted as trends where the Pacific-type and Indian Ocean-type mantle mix with the slab-derived fluid (e.g., thick grayish broken lines in Fig. 7). Nakamura et al. (2019) quantified the amount of fluid added to the mantle for each arc segment based on the same ^{207}Pb/^{204}Pb vs. ^{143}Nd/^{144}Nd systematics and mass balance as in Fig. 7a, and provided the geofluid map in Fig. 8. The dataset in this chapter also includes data from Nakamura et al. (2019), and the evaluation in Fig. 8 applies, although the number of data has increased.

There is a systematic relationship between the compositional clusters and the fluid addition; the lower right in Fig. 7a and the upper right in Fig. 7b approximately have greater fluid addition. Conversely, clusters C2 and C3 near the upper left in Fig. 7a and lower left in Fig. 7b reflect the intrinsic mantle composition prior to fluid addition, and there is a distinct difference between C2 and C3 (Figs. 5 and 7). C3 shows a composition closer to the Indian Ocean-type mantle, while C2 shows a composition closer to the Pacific-type mantle. More precisely, in both Fig. 7a, b, the composition of C2 is close to the intermediate between the Pacific-type and Indian Ocean-type, which is different from the typical Pacific-type. This is consistent with the fact that in Fig. 5e, f, most of the data for the Japan arcs, including C2, fall above the NHRL. In addition, within the context of the global hemispheric structure of the mantle geochemistry, the Japan arcs are located within the Eastern Hemisphere domain (a domain enriched in hydrophilic components developed beneath the supercontinent, involving the Indian Ocean-type mantle; Iwamori and Nakamura 2015), which is harmonious with the above arguments for Figs. 5 and 7.

With increasing the amount of fluid addition, the isotopic composition deviates from C2 or C3 roughly toward the lower right in Fig. 7a and upper right in Fig. 7b. In Fig. 7a, C3–C5 appears to form such a trend with increasing fluid amount starting from the Indian Ocean-type source mantle, and together with Fig. 7b, C4 appears to form

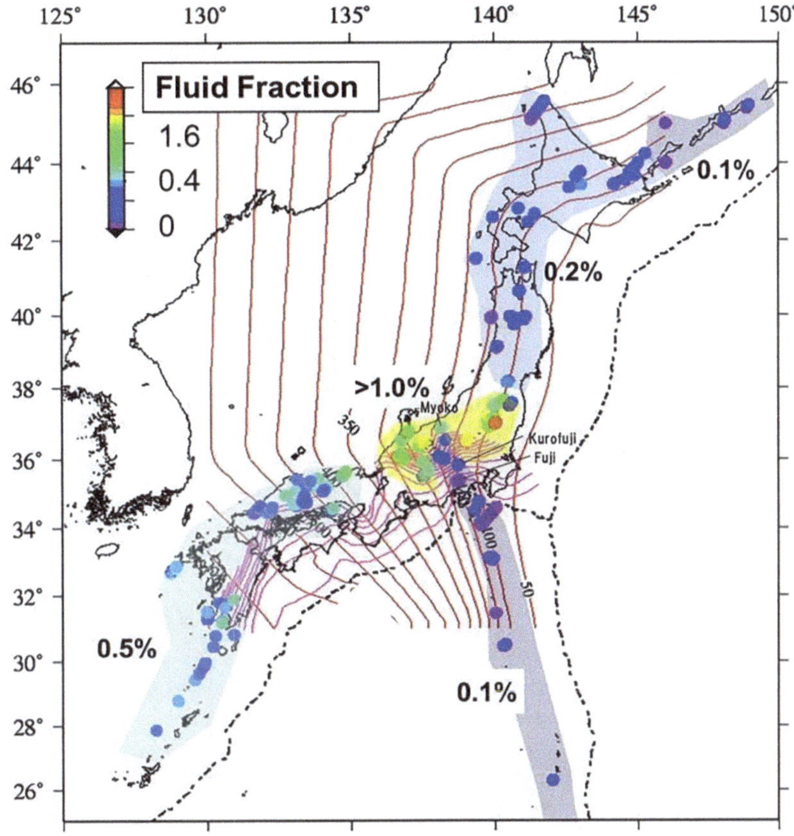

Fig. 8 Geographical distribution of the amount of slab-derived fluid that is added to the mantle wedge, after Nakamura et al. (2019)

a broader group with C3–C5 rather than with C2–C1. This is coherent with the geographic distribution of the compositional clusters (Fig. 6): C3, C4, and C5 are commonly distributed in the western part of the Japan arcs (Central Japan, Southwest Japan, and Ryukyu arcs), whereas C2 and C1 are mainly distributed in the eastern part (Izu-Bonin, Northeast Japan, and Kuril arcs).

In Fig. 7, the compositional range labeled "Slab-derived fluid" includes the fluids from both Pacific and Philippine Sea slabs, partly because the individual compositional clusters C1 to C5 cover relatively large geographical areas and are difficult to distinguish the fluids from the two slabs. However, regression analyses based on the individual arc segments can distinguish between the two slab-derived fluids (Nakamura et al. 2008, 2018). Based on such analyses, Nakamura et al. (2018) found that the Philippine Sea slab-derived fluid (PHP fluid) extends to the southern part of the NE Japan arc (north beyond the northern limit of the seismologically inferred Philippine Sea slab (Fig. 1)) and that the amount of PHP fluid progressively decreases toward the north. Based on these findings, Nakamura et al. (2018) inferred that an aseismic slab may extend into the southern Tohoku region (thick dotted line in Fig. 1). A transition in magma composition from the Central Japan arc to the southern part of NE Japan arc is seen in the distribution of the clusters, especially in C5 (Fig. 6). The relationship between the origin of compositional clusters and geographical distribution will be discussed in Sect. 6.

5 Constraints on P-T Condition of Melt Generation

In addition to isotope-based knowledge for the source mantle and the slab-derived fluids, trace element concentrations and ratios are sensitive to

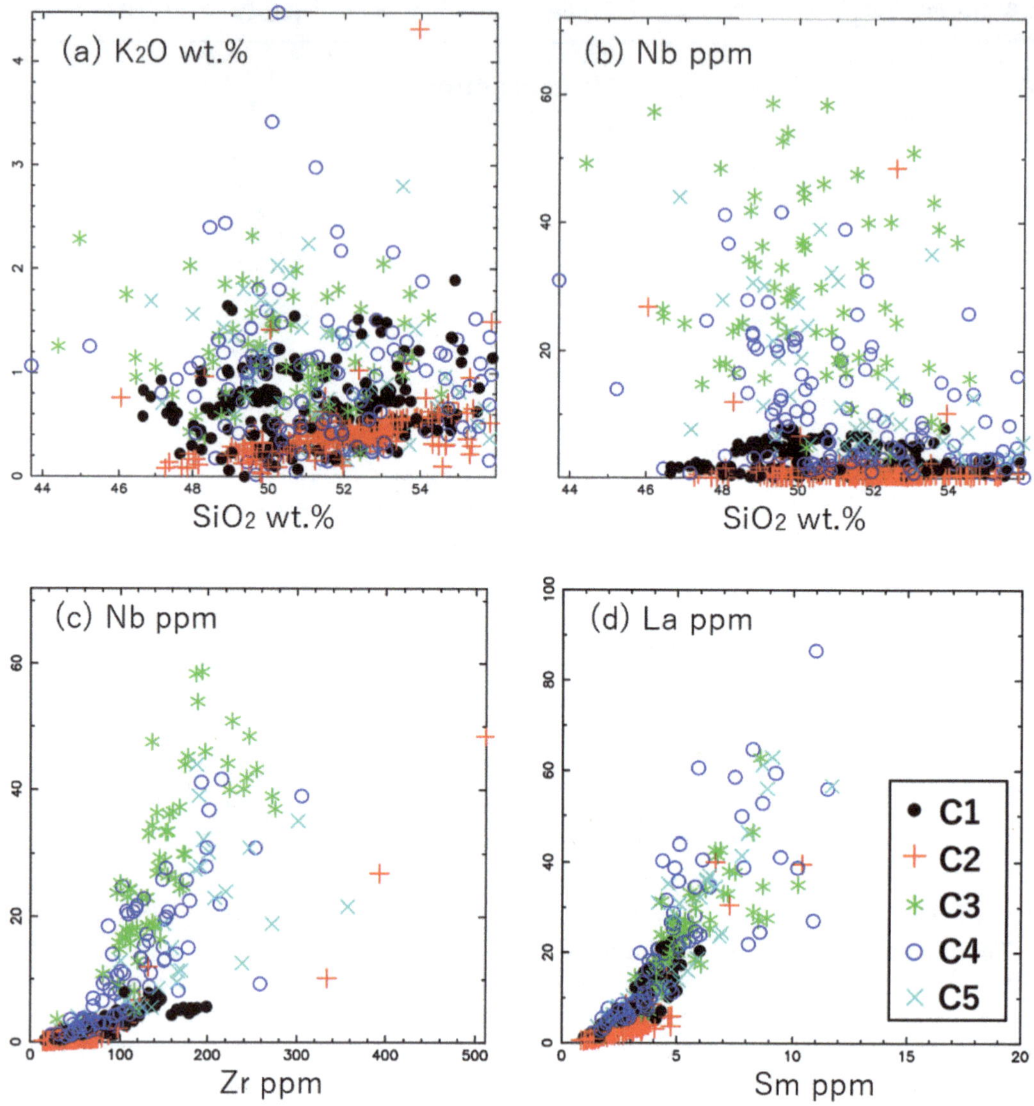

Fig. 9 Incompatible element concentrations of the lavas used in this chapter. (**a**) SiO_2-K_2O, (**b**) SiO_2-Nb, (**c**) Zr-Nb, and (**d**) Sm-La

the degree of melting and the residual mineral assemblage during melting (e.g., proportion of spinel-garnet in peridotite involved in melting) and can provide information on the temperature and pressure of melting (e.g., McKenzie and O'Nions 1991). There are differences in incompatible element concentrations and their ranges among the clusters C1 to C5 (Fig. 9), with C2 in particular having the lowest concentration of many incompatible elements and C3 having the highest.

Among these trace elements, high field strength elements (HFSE) and rare-earth elements (REE) are also affected by the fluid, but to a lesser extent than large-ion lithophile elements (LILE) for the magma generation beneath the Japan arcs (Nakamura and Iwamori 2013). Possibly due to this characteristic, there is somewhat less HFSE and REE variations from cluster to cluster (Fig. 9b) compared to LILE (Fig. 9a). In particular, the fractionation between light-REE and heavy-REE (Fig. 9d) varies systematically and significantly depending on the proportion of spinel-garnet in the residual peridotite, and can be a good indicator of melting temperature and pressure conditions.

Figure 10 shows the composition of melt originated from Depleted MORB Mantle (DMM) as a function of melting degree and proportion of spinel-garnet in DMM. La/Sm and Sm/Yb are not strongly dependent on the mantle type, for example, 0.803 and 0.655 for average DMM (A-DMM), 0.813 and 0.715 for enriched DMM (E-DMM), and 0.638 and 0.603 for depleted DMM (D-DMM) (Workman and Hart 2005), which are within or slightly larger than the size of the DMM symbol in Fig. 10.

The gray line shows the DMM-derived melt composition without fluid addition, the near-vertical line extending in the direction of increasing La/Sm while keeping Sm/Yb nearly constant on the horizontal axis shows the melting of spinel peridotite ("sp + no fluid"), and the line extending in the direction of increasing Sm/Yb from DMM along the horizontal axis corresponds to melting of garnet peridotite ("gt + no fluid").

The compositions of the Japan arcs fall roughly between these two curves, but are closer to the melting curve of the spinel peridotite. C2, the most incompatible element-depleted cluster, is plotted in the lowest La/Sm range and can be explained by ~20% melting of DMM with no or

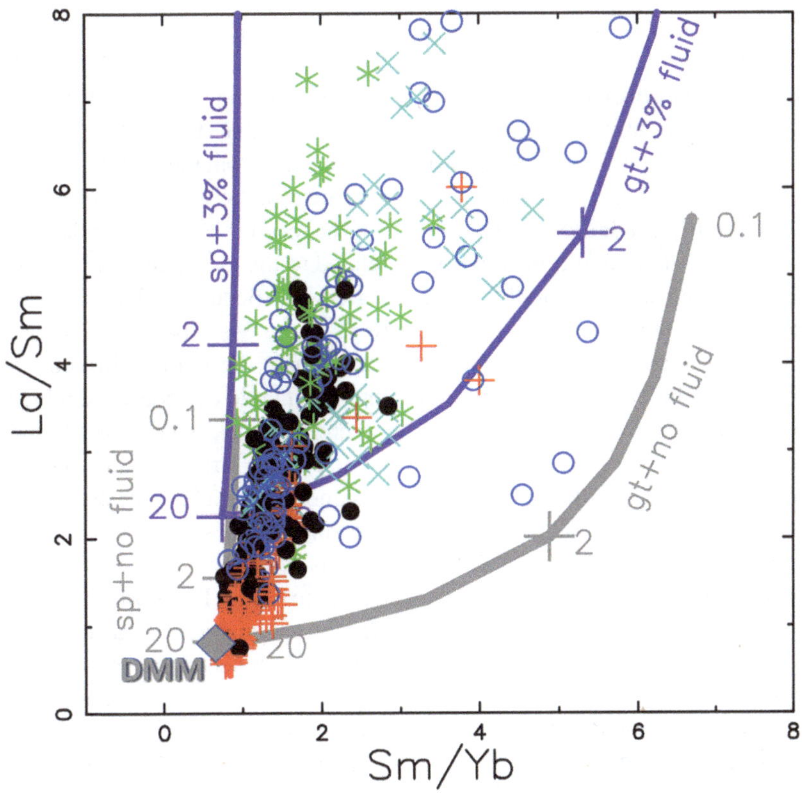

Fig. 10 Calculated partial melting curves assuming non-modal batch melting of garnet (gt) and spinel (sp) peridotite in Sm/Yb vs. La/Sm diagram. The mineral proportions of garnet peridotite: 0.598 olivine (ol), 0.211 orthopyroxene (opx), 0.076 clinopyroxene (cpx), 0.115 garnet (gt); melting proportions: 0.05 ol, 0.2 opx, 0.3 cpx, 0.45 gt; mineral proportions of spinel lherzolite: 0.578 ol, 0.27 opx, 0.119 cpx, 0.033 sp.; melting proportions: 0.1 ol, 0.27 opx, 0.5 cpx, 0.13 sp. (Thirlwall et al. 1994; Shaw et al. 2003; Peters et al. 2008). The partition coefficients are from McKenzie and O'Nions (1991) and Nakamura and Iwamori (2013). The labels "20," "2," "0.1" indicate the degree of melting in percent. "sp + no fluid" and "sp + 3% fluid" indicate melting curves of spinel peridotite with 0% fluid added to the peridotite and 3% fluid added, respectively. "gt + no fluid" and "gt + 3% fluid" indicate those of garnet peridotite

few fluid addition. On the other hand, C3, C4, and C5, which are rich in incompatible elements, have many samples with La/Sm values between 3 and 8, and part of C1 is difficult to explain by melting of DMM with no fluid: theoretically, at ~0% incipient melting, La/Sm in the melt reaches a maximum but below 4. The purple curves in Fig. 10 correspond to the melting of DMM with 3% added fluid, and in this case, the melt composition with La/Sm exceeding 4 is reproduced at melting degree of a few percent. Conversely, C2 is difficult to reproduce from such fluid-rich source peridotite that corresponds to the purple curves, and requires a relatively dry source.

C2 is also low with Sm/Yb <~1.5, except for a very few outliers, and the contribution of garnet peridotite melting is small. C2 is characterized by relatively low fluid volume, high melting degree, and greater contribution from spinel peridotite (i.e., relatively shallow melting). C3, C4, and C5, on the other hand, have relatively high Sm/Yb, with some samples exceeding 3. This feature of C3, C4, and C5 suggests a larger contribution of melting at greater depths, in addition to the relatively large amount of fluid addition and low melting degree described above. For the same temperature conditions, the degree of melting increases with the amount of fluid addition (Iwamori 1998). Thus, the contrast between C2 and C3/C4/C5 suggests that there are differences in mantle melting temperatures, as well as melting pressure and fluid amount in the source. How these conditions vary along the Japan arcs and how the Central Japan arc is positioned within them are discussed in the following.

6 Mantle Structure and Processes Beneath the Central Japan Arc

As described in the previous sections, the information on isotopic ratios and trace element concentrations provides constraints on (1) the type of DMM as source mantle (Pacific-type or Indian Ocean-type mantle), (2) the amount of fluid added to DMM, (3) the degree of melting, and (4) the proportion of spinel-garnet in source peridotite. The isotopic ratios provide estimates on (1) and (2), and the trace elements provide estimates on (2) through (4). Both the isotopic ratios and the trace element concentrations reflect (2) independently, and can be used to mutually check the individual results regarding the amount of added fluid.

Concerning (1), the Indian Ocean-type mantle-derived magmas are represented by the cluster group of C3–C4–C5 (Fig. 7), most of which are distributed in the Southwest Japan arc (the Kyushu and Chugoku regions and the Central Japan arc, Figs. 1 and 6). On the other hand, C2 and C1 represent magmas derived from the Pacific-type mantle (Fig. 7), and their distribution is mainly restricted to the Izu-Bonin and Kuril arcs (Fig. 6). With fewer data compared to this study, Nakamura and Iwamori (2009) pointed out the existence of both the Indian Ocean-type and Pacific-type mantle beneath the Japan arcs, and speculated that the Indian Ocean-type mantle may extend from the mantle beneath the Eurasian continent in the west.

The origin of this mantle is thought to be related to the global and long-term circulation of material associated with former subducting plates (e.g., Hart 1984; Iwamori and Albarède 2008), particularly the hydrophilic component brought into the mantle by focused subduction toward the supercontinents, which has created the east-west hemispheric structure of mantle composition (Iwamori and Nakamura 2012, 2015). The more data and analysis with denser spatial distribution in this chapter support the point made by Nakamura and Iwamori (2009) that an Indian Ocean-type mantle (or Eastern Hemisphere mantle) exists beneath the Eurasian plate that once constituted the supercontinent Pangea, extending toward the Southwest and Central Japan arcs.

In the Izu arc, C2 and C1 derived from the Pacific-type mantle are dominant (Fig. 6), and the transition occurs rather discontinuously from the Central Japan arc where C3–C4–C5 are dominant. This would indicate that the mantle beneath the Central Japan and the Izu arcs is physically separated, although C2 occurs along the southernmost margin of Central Japan, including Fuji and Ueno volcanoes, suggesting some northward leakage of the Pacific-type component through the Izu collision zone, where the Philippine Sea slab may not work as a physical barrier between the two distinct mantle domains.

On the other hand, the transition from the Central Japan to the Northeast Japan arc (the Tohoku and Hokkaido regions) is relatively gradual. C5, the main cluster in Central Japan, is also scattered from Tohoku to southern Hokkaido, and the overall transition occurs with the emergence of C1 in the Northeast Japan arc and C2 in the Kuril Arc. To the north of the Kuril arc, the northern edge of the Pacific plate is present near the Kamchatka, and the Pacific-type mantle may flow in from the western hemisphere toward the Kuril arc and can be in contact with the Indian Ocean-type mantle beneath the Northeast Japan arc. There is a caveat for "Pacific-type mantle" beneath the Japan arcs; as was argued earlier, it involves a significant Indian Ocean-type component and is different from the typical Pacific-type (e.g., Hickey-Vargas 1998). Most of the data used in this chapter plot on the upper side of NHRL (Fig. 5e, f) and globally belong to the "Indian Ocean-type" or "Eastern Hemisphere" mantle domain. Within this domain, we have found that there is a remarkable heterogeneity beneath the Japan arcs, such as C2 and C3.

In relation to (2) above (the amount of fluid addition), C2 is distributed mainly in the Izu and Kuril arcs, where the estimated amount of fluid addition is small based on the isotopic systematics (Figs. 7 and 8). At the same time, the estimates based on the trace elements (Fig. 10), which are independent from Fig. 8, also suggest that C2 corresponds to a small amount of fluid addition. On the other hand, C3, C4, and C5 are distributed mainly in the Central Japan and Southwest Japan arcs (Fig. 6). Evaluation based on the isotopic ratios (Fig. 8) indicates that the amount of fluid added to the mantle is relatively large in these two arcs, especially in the Central Japan arc (>1%). The trace element analysis in Fig. 10 also suggests that C3, C4, and C5 have a greater amount of fluid addition (a few percent), indicating that the independent constraints are consistent and the results in Figs. 8 and 10 show the robust feature.

Based on the estimates for (2), we are now able to discuss the relationship between the amount of added fluid and the degree of melting ((3) above), in which a seemingly enigmatic relationship is observed between (2) and (3). In principle, for the same mantle potential temperature, the higher the amount of fluid added to the mantle, the higher the degree of melting (Iwamori 1998). However, the characteristics of each cluster suggest a rather inverse relationship, e.g., C2 with the least amount of added fluid has the highest melting degree of ~20%, while C4 and C5 with the highest fluid amount exhibit melting of a few percent.

The highest amount of added fluid for the Central Japan arc (Fig. 8) is thought to be due to fluid supply from the two overlapping slabs (Nakamura et al. 2008). Nakamura and Iwamori (2013) found that adakite is widely distributed in the Central Japan arc. They revealed that with the large amount of slab-derived fluid that increases Sr concentration in the magma source, a small amount of melting of garnet peridotite occurs at approximately 2.5–3.0 GPa and 1000–1100 °C, based on an integrated analysis of major elements, isotopic ratios, and trace element concentrations. This is consistent with Fig. 10 in that the proportion of garnet peridotite melting (related to item (4) listed at the beginning of this Sect. 6) is large for C4 and C5 that are found commonly in the Central Japan arc (Fig. 6). In this case, the solidus temperature is lowered by the large amount of fluid and magma is generated slightly above the solidus when the mantle potential temperature is low.

Such a fluid-rich cold mantle wedge has been proposed by a numerical simulation that includes plate subduction, slab dehydration, fluid migration, and chemical reactions (including melting) between migrating fluids and the convecting mantle (Fig. 11). Since the subducting Philippine Sea slab overlies the Pacific slab and impedes convection in the large mantle wedge formed by the Pacific slab, the potential temperature in the mantle wedge is low and the subducting Pacific slab remains cold, resulting in dehydration of the Pacific slab at greater depths (compared to the case without the

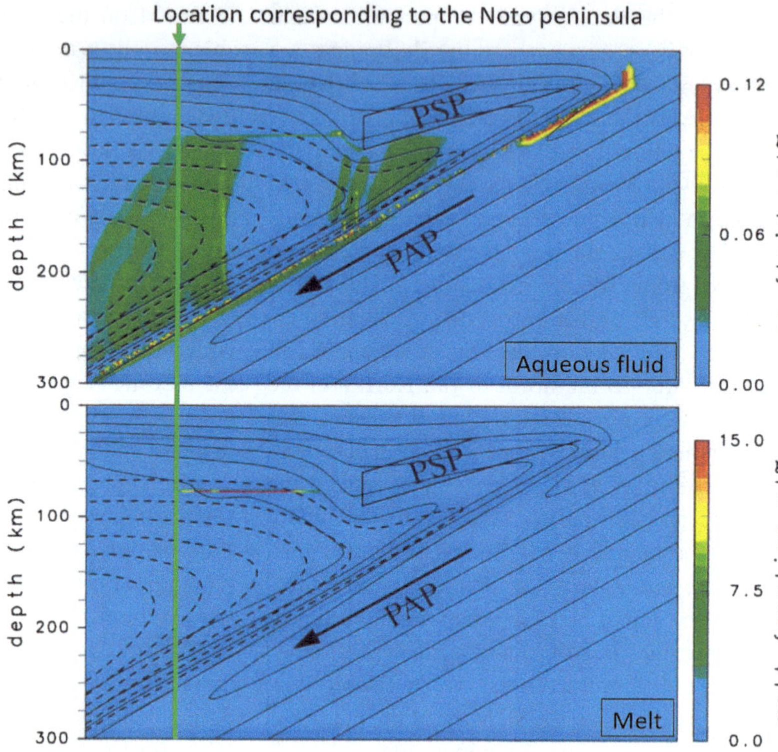

Fig. 11 Model predictions for (**a**) aqueous fluid distribution (volume%) and (**b**) melt (wt%) beneath the Central Japan arc along the across-arc section A-B in Fig. 1, based on the numerical simulation of plate subduction, slab dehydration, fluid migration, and chemical reactions (including melting) between migrating fluids and the convecting mantle after Iwamori (2000). PSP and PAP indicate the Philippine Sea and the Pacific slabs, respectively

Philippine Sea slab). This is thought to be the mechanism to bend the volcanic front toward the back-arc, and to have extended the entire volcanic activity to the back-arc area (Figs. 1 and 2) where the slab surface depth reaches greater than 250 km (Iwamori 2000).

The mantle potential temperature is low beneath the Central Japan arc, and as shown in Fig. 11, the extent of the melting zone is determined by a combination of large melting point depression and temperature structure due to abundant slab-derived fluids. At certain depths (e.g., ~80 km depth, corresponding to the melting zone under Ryohaku Volcanoes in Central Japan), temperatures increase slightly toward the back-arc, but the melting point depression is insufficient because of less slab-derived fluid and no melting occurs. The Noto Peninsula, located in the back arc of the Central Japan arc (Fig. 1), may correspond to such an "ascent of deep fluid without melting" region (along the green vertical line in Fig. 11), and may be part of a larger picture of geological phenomena such as the overlap of two subducting slabs, the cold mantle wedge and slab, and the remarkable bend of volcanic fronts and volcanic zone toward the back-arc side as seen in Fig. 2.

The deep dehydration associated with the overlapping subduction of the two slabs may also be the cause of crustal deformation. In the Noto Peninsula, intense seismic activity began in December 2020, and became more active around July 2021, with a M5.4 earthquake occurring on June 19, 2022, and a M6.5 earthquake on May 5, 2023. The seismic activity is still active as of December 2023 (Japan Meteorological Agency: https://www.data.jma.go.jp/kanazawa/shosai/notojishinportal.html). In addition, crustal deformation with uplift of up to 7 cm has been observed with these seismic activities, which can be explained by a volume increase of about 3×10^7 m^3 in total at a depth of ~16 km with an aseismic reverse-type slip and the opening of a southeast-dipping fault zone at a depth of 14–16 km (Nishimura et al. 2023). Upwelling fluid from depths greater than 16 km has been suggested as a cause of such expansion and deformation (Nishimura et al. 2022, 2023; Nakajima 2022). Nishimura et al. (2023) argued that the fluid spread through an existing shallow-dipping permeable fault zone and then diffused into the fault zone under a regional compressional stress field, triggering the aseismic slip and intense earthquake swarms since 2020.

The 3×10^7 m^3 volume increase can be attributed to a ~1000-year accumulation of fluid flux of ~0.3 m^3/year per unit area, calculated from the numerical model in Fig. 11, assuming a 10 km × 10 km area comparable to the seismically active area after 2020 was fluid-fluxed. If the fluid flux has continued over a geological time scale with the regional compressional stress field, significant deformation and uplift are expected. In the Noto Peninsula, Hiramatsu et al. (2008) estimated that ~40 m uplift occurred over the last 120,000 years by the recurrence of coseismic vertical deformation and the background tectonic uplift. Based on the geological structures, the compressional regional stress field around Noto occurred since ~7 million years ago after the extensional stress period, including the formation of Japan Sea and the Green Tuff volcanism that constitutes the basement geology of the present-day Noto Peninsula (e.g., Okamura 2007).

The timing of the onset of deep fluid flux to this area is unknown, but the present plate configuration with overlapping subduction of the Pacific and the Philippine Sea slabs was almost established in 4 Ma (Seno and Maruyama 1984), and deep dehydration is thought to have begun in the back-arc region and beneath the Noto at that time. A regional compressive stress field with the deep fluid flux since 4 Ma could result in a ~1330 m uplift of the Noto area, assuming a 40 m uplift per 120,000 years (Hiramatsu et al. 2008), which accounts for the highest peak on the Noto Peninsula (637 m, Hodatsu-san) when the erosion of ~700 m is considered. The long-term uplift under the compressional stress field with the deep fluid flux may be the mechanism of formation of the Noto Peninsula, as a singular point associated with the overlapping subduction of the two plates and the deep slab dehydration. As long as this tectonic configuration is maintained, the uplift and the associated seismic activity in the Noto Peninsula are expected to continue.

7 Summary

Based on the five isotopic ratios ($^{87}Sr/^{86}Sr$, $^{143}Nd/^{144}Nd$, $^{206}Pb/^{204}Pb$, $^{207}Pb/^{204}Pb$, $^{208}Pb/^{204}Pb$) of 747 basaltic lavas over the Japan arcs (including Kuril, NE Japan, Central Japan, SW Japan, Ryukyu, Izu-Bonin arcs), five geochemical clusters have been identified using the whitened data-based k-means cluster analysis, a type of unsupervised machine learning approach. The five clusters show geographical provenance, as well as correlations with the trace element concentrations and ratios. By combining the isotopic and trace element data, the mass balance and melting model provided constraints on the following points:

(1) The type of source mantle (Pacific-type or Indian Ocean-type mantle).
(2) The amount of fluid added to the mantle.
(3) The degree of melting.
(4) The proportion of spinel-garnet in the source mantle.

The isotopic compositions provided constraints on (1) and (2), suggesting that there are two distinct mantle sources of Indian Ocean-type and Pacific-type beneath the Japan arcs, which are variably fluxed by slab-derived fluids. It has been also found that the Pacific-type mantle beneath the Japan arc is different from the typical Pacific-type mantle, but involves more Indian Ocean-type or "Eastern Hemispheric Mantle" component that was possibly developed under the supercontinent with more hydrophilic components. Regarding (2) (the amount of fluid), both the isotopic ratios and trace element concentrations provided independent and consistent estimates: the Kuril arc and the Izu-Bonin arc are characterized by a small amount of added fluid (~0.1%), while the Central Japan, SW Japan, and Ryukyu arcs are characterized by larger amounts (>0.5%). Nakamura et al. (2019) attributed these differences to different subduction materials/parameters of the two slabs (the Pacific and the Philippine Sea slabs) and different physicochemical conditions of the mantle wedge beneath the individual arc segments. In addition to these characteristics of the arc segments, the systematic differences are found in (3) and (4) above: compared to the Kuril arc and the Izu-Bonin arc, a smaller degree of melting and more involvement of melt from garnet peridotite are found for the three clusters that are distributed mostly in the western half of the Japan arcs (Central Japan, SW Japan, and Ryukyu arcs), where more slab-derived fluids are added to the mantle wedge. In particular, the Central Japan arc is characterized by a large amount of added fluid (>1%) and a few percent of melting of spinel-garnet peridotite, suggesting a low potential temperature of the mantle wedge. Such a fluid-rich low-temperature condition was proposed also by the numerical simulation, and indicates a significant deepening of dehydration of the subducting Pacific slab, compared to the other arc segments. This deepening of slab dehydration may explain the remarkable bend of the volcanic front around Asama volcano and the volcanic zone toward the back-arc area below which the Pacific slab surface is deeper than 250 km. Such deep dehydration can supply a significant amount of fluid to the non-volcanic back-arc area such as the Noto Peninsula, possibly causing the earthquake swarm and the crustal uplift in the peninsula since December 2020. The long-term (~400 m.y.) uplift under the compressional stress field with the deep fluid flux may be the mechanism of formation of the Noto Peninsula, as a singular point associated with the overlapping subduction of the two plates and the deep slab dehydration.

References

Argus DF, Gordon RG, DeMets C (2011) Geologically current motion of 56 plates relative to the no-net-rotation reference frame. Geochem Geophys Geosyst. https://doi.org/10.1029/2011GC003751

Bird P (2003) An updated digital model of plate boundaries. Geochem Geophys Geosyst 4(3):1027. https://doi.org/10.1029/2001GC000252

Hanyu T, Tatsumi Y, Nakai S, Chang Q, Miyazaki T, Sato K, Tani K, Shibata T, Yoshida T (2006) Contribution of slab melting and slab dehydration to magmatism in the Northeast Japan arc for the last 25 Myr: constraints from geochemistry. Geochem Geophys Geosyst 7(8):1–29. https://doi.org/10.1029/2005GC001220

Hart SR (1984) A large-scale isotope anomaly in the Southern Hemisphere mantle. Nature 309:753–757

Hauff F, Hoernle K, Schmidt A (2003) Sr–Nd–Pb composition of Mesozoic Pacific oceanic crust (Site 1149 and 801, ODP Leg 185): implications for alteration of ocean crust and the input into the Izu–Bonin–Mariana subduction system. Geochem Geophys Geosyst 4(8):8913. https://doi.org/10.1029/2002GC000421

Hickey-Vargas R (1998) Origin of the Indian ocean-type isotopic signature in basalts from Philippine Sea plate spreading centers: an assessment of local versus large-scale processes. J Geophys Res 103:20963–20979

Hiramatsu Y, Moriya K, Kamiya T, Kato M, Nishimura T (2008) Fault model of the 2007 Noto Hanto earthquake estimated from coseismic deformation obtained by the distribution of littoral organisms and GPS: implication for neotectonics in the northwestern Noto Peninsula. Earth Planets Space 60:903–913

Hirose F, Nakajima J, Hasegawa A (2008) Three-dimensional seismic velocity structure and configuration of the Philippine Sea slab in southwestern Japan estimated by double-difference tomography. J Geophys Res 113:B09315. https://doi.org/10.1029/2007JB005274

Hochstaedter AG, Gill JB, Peters R, Broughton P, Holden P, Taylor B (2001) Across-arc geochemical trends in the Izu–Bonin arc: contributions from the subducting slab. Geochem Geophys Geosyst 2. https://doi.org/10.1029/2000GC000105

Hyvärinen A, Karhunen J, Oja E (2001) Independent component analysis. John Wiley & Sons, New York

Ishikawa T, Nakamura E (1994) Origin of the slab component in arc lavas from across-arc variation of B and Pb isotopes. Nature 370:205–208

Ishizuka O, Taylor RN, Milton JA, Nesbitt RW (2003) Fluid–mantle interaction in an intra-oceanic arc: constraints from high-precision Pb isotopes. Earth Planet Sci Lett 211:221–236

Iwamori H (1991) Zonal structure of Cenozoic basalts related to mantle upwelling in southwest Japan. J Geophys Res Solid Earth 96:6157–6170. https://doi.org/10.1029/90JB02399

Iwamori H (1998) Transportation of H_2O and melting in subduction zones. Earth Planet Sci Lett 160:65–80

Iwamori H (2000) Deep subduction of H_2O and deflection of volcanic chain towards back-arc near triple junction due to lower temperature. Earth Planet Sci Lett 181:41–46

Iwamori H, Albarède F (2008) Decoupled isotopic record of ridge and subduction zone processes in oceanic basalts by independent component analysis. Geochem Geophys Geosys 9. https://doi.org/10.1029/2007GC001753

Iwamori H, Nakamura H (2012) East–west geochemical hemispheres anchored to asthenosphere in the Earth's mantle. Geochem J 46:e39–e46. https://doi.org/10.2343/geochemj.2.0224

Iwamori H, Nakamura H (2015) Isotopic heterogeneity of oceanic, arc and continental basalts and its implications for mantle dynamics. Gondwana Res 27:1131–1152. https://doi.org/10.1016/j.gr.2014.09.003

Iwamori H, Yoshida K, Nakamura H, Kuwatani T, Hamada M, Haraguchi S, Ueki K (2017) Classification of geochemical data based on multivariate statistical analyses: complementary roles of cluster, principal component, and independent component analyses. Geochem Geophys Geosyst 18:994–1012. https://doi.org/10.1002/2016GC006663

Iwamori H, Nakamura H, Morikawa N, Takahashi M, Inamura A, Haraguchi S, Nishizawa T, Sakata S (2023) Groundwaters and deep-seated fluid circulation around Aso Volcano, Southwest Japan, revealed by multivariate statistical analysis of the geochemical data. J Volcanol Geotherm Res 433:107739. https://doi.org/10.1016/j.jvolgeores.2022.107739

Kimura J-I, Yoshida T (2006) Contributions of slab fluid, mantle wedge and crust to the origin of quaternary lavas in the NE Japan arc. J Petrol 47:2185–2232

Kita S, Okada T, Nakajima J, Matsuzawa T, Hasegawa A, Katsumata K (2008) Anomalous deepening of upper-plane seismic belt in the Pacific slab beneath the Hokkaido corner. In: 7th general assembly of Asian seismological commission and seismological Society of Japan, fall meeting, abstract, X2-052

McKenzie D, O'Nions RK (1991) Partial melt distributions from inversion of rare-earth element concentrations. J Petrol 32(5):1021–1091

Moriguti T, Shibata T, Nakamura E (2004) Lithium, boron and lead isotope and trace element systematics of quaternary basaltic volcanic rocks in northeastern Japan: mineralogical controls on slab-derived fluid composition. Chem Geol 212:81–100

Nakajima J (2022) Crustal structure beneath earthquake swarm in the Noto peninsula, Japan. Earth Planets Space 74:160. https://doi.org/10.1186/s40623-022-01719-x

Nakajima J, Hasegawa A (2006) Anomalous low-velocity zone and linear alignment of seismicity along it in the subducted Pacific slab beneath Kanto, Japan: reactivation of subducted fracture zone? Geophys Res Lett 33:L16309. https://doi.org/10.1029/2006GL026773

Nakajima J, Hasegawa A (2007) Subduction of the Philippine Sea plate beneath southwestern Japan: slab geometry and its relationship to arc magmatism. J Geophys Res 112:B08306. https://doi.org/10.1029/2006JB004770

Nakamura H, Iwamori H (2009) Contribution of slab-fluid in arc magmas beneath the Japan arcs. Gondwana Res 16:431–445. https://doi.org/10.1016/j.gr.2009.05.004

Nakamura H, Iwamori H (2013) Generation of adakites in a cold subduction zone due to double subducting plates. Contrib Mineral Petrol 165:1107–1134

Nakamura H, Iwamori H, Kimura J-I (2008) Geochemical evidence for enhanced fluid flux due to overlapping subducting plates. Nat Geosci 1:380–384. https://doi.org/10.1038/ngeo200

Nakamura H, Iwamori H, Ishizuka O, Nishizawa T (2018) Distribution of slab-derived fluids around the edge of the Philippine Sea plate from central to northeast Japan. Tectonophysics 723:297–308. https://doi.org/10.1016/j.tecto.2017.12.004

Nakamura H, Iwamori H, Nakagawa M, Shibata T, Kimura J, Miyazaki T, Chang Q, Vaglarov BS, Takahashi T, Hirahara Y (2019) Geochemical mapping of slab-derived fluid and source mantle along Japan arcs. Gondwana Res 70:36–49. https://doi.org/10.1016/j.gr.2019.01.007

Nishimura T, Hiramatsu Y, Ohta Y (2022) Earthquake swarms and crustal deformation in the Noto peninsula since around December 2020. Earthquake J 74:1–9. (in Japanese)

Nishimura T, Hiramatsu Y, Ohta Y (2023) Episodic transient deformation revealed by the analysis of multiple GNSS networks in the Noto Peninsula, central Japan. Sci Rep 13:8381. https://doi.org/10.1038/s41598-023-35459-z

Okamura Y (2007) Tectonic evolution and active tectonics of the Noto Peninsula (Hanto) and its surrounding area. In: Annual report on active fault and paleoearthquake researches, vol 7, pp 197–207

Peters TJ, Menzies M, Thirlwall M, Kyle P (2008) Zuni-Bandera volcanism, Rio Grande, USA -melt formation in garnet- and spinel-facies mantle straddling the asthenosphere-lithosphere boundary. Lithos 102:295–315

Seno T, Maruyama S (1984) Paleogeographic reconstruction and origin of the Philippine Sea. Tectonophysics 102:53–84

Seno T, Sakurai T, Stein S (1996) Can the Okhotsk plate be discriminated from the North American plate? J Geophys Res 101:11305–11315. https://doi.org/10.1029/96JB00532

Shaw JE, Baker JA, Menzies MA, Thirlwall MF, Ibrahim KM (2003) Petrogenesis of the largest intraplate volcanic field on the Arabian Plate (Jordan): a mixed lithosphere-asthenosphere source activated by lithospheric extension. J Petrol 44:1657–1679

Shinjo R, Woodhead JD, Hergt JM (2000) Geochemical variation within the northern Ryukyu Arc: magma source compositions and geodynamic implications. Contrib Mineral Petrol 140:263–282

Straub SM, Layne GD, Schmidt A, Langmuir CH (2004) Volcanic glasses at the Izu arc volcanic front: new perspectives on fluid and sediment melt recycling in subduction zones. Geochem Geophys Geosys 5. https://doi.org/10.1029/2002GC000408

Taylor RN, Nesbitt RW (1998) Isotopic characteristics of subduction fluids in an intra-oceanic setting, Izu–Bonin Arc, Japan. Earth Planet Sci Lett 164:79–98

Thirlwall MF, Upton BGJ, Jenkins C (1994) Interaction between continental lithosphere and the Iceland plume-Sr-Nd-Pb isotope geochemistry of Tertiary Basalts, NE Greenland. J Petrol 35:839–879

Workman RK, Hart SR (2005) Major and trace element composition of the depleted MORB mantle (DMM). Earth Planet Sci Lett 231:53–72

Geology

Geologic Background: Basement Rocks of Asama Volcano

Teruki Oikawa, Masaki Takahashi, and Maya Yasui

1 Introduction

Asama volcano is located in the eastern part of the Eboshi-Asama Volcano Group, a composite volcano that formed after 0.45 Ma (see chapter "Geology of Eboshi-Asama Volcano Group"). Asama volcano is an active volcano that forms part of the Quaternary volcanic front in central Japan, along with the Yatsugatake Volcanic Group (Kawachi 1998) to the south. Furthermore, Asama volcano is located at the junction of three island arcs: the Southwest Japan Arc, the Northeast Japan Arc, and the Izu-Bonin-Marian Arc. The junction of these three island arcs is called the Fossa Magna (Nauman 1893; Kosaka et al. 1992; Yamashita 1995; Tateishi et al. 1997), and has Neogene to Quaternary thick deposits of marine sediments. However, the area around Asama volcano within the Fossa Magna has been uplifted since the Late Miocene, and no marine sediments have been confirmed since the Middle Miocene. In addition, the area where this volcano is located is the farthest from the sea in the Japanese Islands.

Since the Late Miocene, volcanic activity has repeatedly occurred around Asama volcano, forming volcanoes that have grown over hundreds of thousands of years, alternating between dormant and active periods (Kaneko et al. 1989; Oikawa 2003; Tanase et al. 2007; Nishiki et al. 2011). Therefore, there is a high possibility that a hidden ancient volcanic body forms part of the basement rock of Asama volcano. Asama volcano and its surrounding area are covered thickly and extensively with its lavas and volcaniclastic sediments, so exposure of the basement rocks is rare. This chapter outlines the stratigraphy of the area surrounding the volcano from the middle Miocene, which is postulated to also form the basement rocks of Asama volcano.

T. Oikawa (✉)
Geological Survey of Japan, AIST, Tsukuba, Ibaraki, Japan
e-mail: teruki-oikawa@aist.go.jp

M. Takahashi
Institute of Natural Sciences, College of Humanities and Sciences, Nihon University, Tokyo, Japan
e-mail: takahashi.masaki@nihon-u.ac.jp

M. Yasui
Department of Earth and Environmental Sciences, College of Humanities and Sciences, Nihon University, Tokyo, Japan
e-mail: yasui.maya@nihon-u.ac.jp

2 Basement Rocks of Asama Volcano

In this chapter, we explain the geological units from the Middle Miocene to approximately 1 Ma, which are older than the activity of Asama volcano, in chronological order from oldest to newest. A simplified geological map of in and around Asama volcano is shown in Fig. 1. A stratigraphic relationship diagram is shown in Fig. 2.

S. Aramaki et al. (eds.), *The Asama Volcano*, Active Volcanoes of the World,
https://doi.org/10.1007/978-981-95-5799-8_3

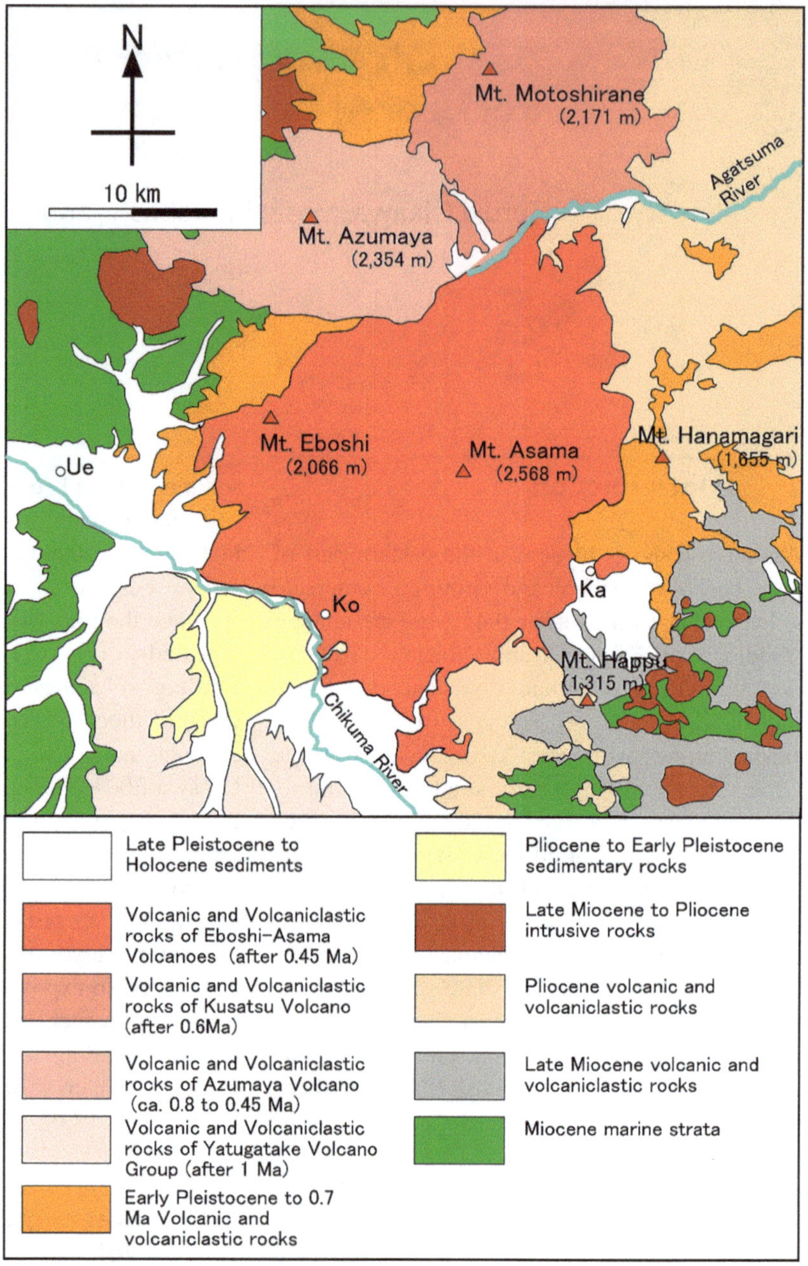

Fig. 1 A simplified geological map around Asama volcano, based on Nakano et al. (1998), Takahashi (2004), and Yasui et al. (2019). For the classification of the Eboshi-Asama Volcano Group, see chapter "Geology of Eboshi-Asama Volcano Group". Ko: Komoro, Ka: Karuizawa, Ue: Ueda

2.1 Miocene Marine Strata

The Early to Middle Miocene geological units distributed around Asama volcano are as follows: Early Miocene marine strata are distributed in the Uchiyama Formation and Komagome Formation (Kosaka et al. 1991) in the southeastern region. Middle Miocene marine strata are distributed in the Fukushima and Itahana Formations (Akima Collaborative Research Group 1996) in the eastern region, the Idozawa Formation (Nomura and Akima Collaborative Research Group 1981) and the Yaekubo Formation (Kosaka et al. 1991) in the southeastern region, and the Uchimura Formation and Bessho Formations (Nakano et al. 1998) in the western region. These layers mainly consist of sandstone and mudstone, although there are some areas with volcanic rocks in between. These formations are locally distributed

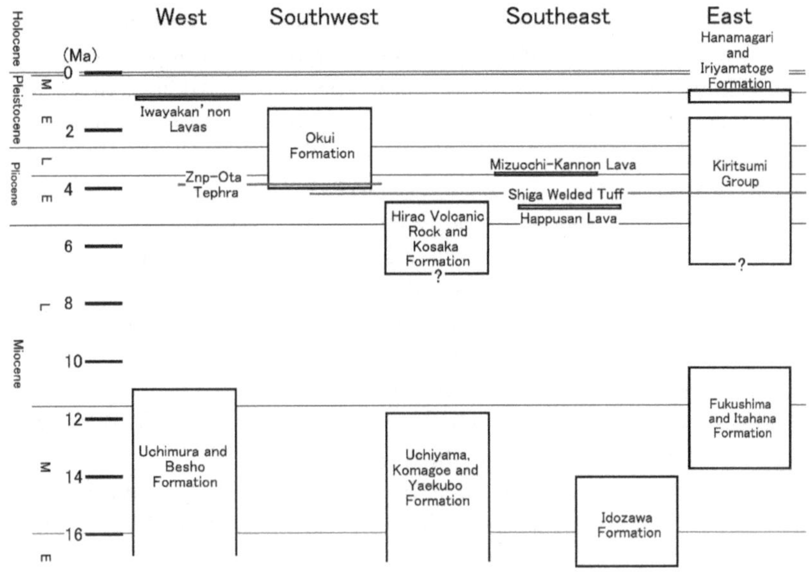

Fig. 2 Miocene to Middle Pleistocene stratigraphy around Asama volcano, based on Akima Collaborative Research Group (1996), Nakano et al. (1998), Nakamura et al. (2014), Nomura and Akima Collaborative Research Group (1981), Nomura and Ebihara (1998), Kosaka et al. (1991), Oishi and Takahashi (1990), Oikawa et al. (2005), Sato and Matsukawa (2014), and Takahashi (2004), Yano et al. (2022). E: Early, M: Middle, L: Late

after the Late Miocene volcanic and volcaniclastic rocks.

2.2 Late Miocene to Pliocene Volcanic and Volcaniclastic Rocks

After the Late Miocene, marine strata were not distributed around Asama volcano. The late Miocene and Pliocene geological units in this area consist of volcanic and volcaniclastic rocks. They are named the Kiritumi Formation (Akima Collaborative Research Group 1996) in the eastern region, the Kosaka Formation (Kosaka et al. 1991), and the Hiro Volcanics Member (Kosaka et al. 1991) or the Hiraofuji Volcanic Rock (Nakamura et al. 2014) in the southeastern region. There are no remaining volcanic landforms created by these volcanic and volcaniclastic rocks.

2.3 Pliocene Welded Tuff

In the southeastern region around Asama volcano, an andesitic welded tuff, ca. 4.4 Ma, is widely distributed. They are the Komoro Welded Tuff, the Shiga Welded Tuff, and the Yomoharayama Welded Tuff (Nakamura et al. 2014) and were formed through a series of volcanic eruptions. These welded tuffs, especially the Shiga welded tuffs, cover the Miocene to Pliocene volcanic rocks and volcaniclastic rock layers, forming a pyroclastic flow plateau and a wide, flat plateau-like ridge in the mountains in the southeast region. The maximum layer thickness exceeds 150 m. These welded tuffs are opx-cpx andesite to dacite, but only the Yomoharayama Welded Tuff contains hornblende phenocrysts. Shiga and Komoro Welded Tuff are exposed in a narrow area around the Ryugaeshi Fall in the eastern side of Asama volcano and around Higashizawa, Komoro City, in the southwestern part of Asama volcano. It has been suggested that they are widely distributed under the eastern part of Asama volcano.

During the periods before and after the formation of these welded tuffs, volcanic activities formed andesite and dacite lava, such as the Happusan Lava and the Mizuochi-Kannon Lava (Nakamura et al. 2014) in the southeast region.

2.4 Pliocene to Early Pleistocene Sedimentary Rocks

The Okui Formation of the Komoro Group (Iijima et al. 1956), which was deposited approximately 4–1 Ma, is distributed along the Chikuma

River on the southwest region around Asama volcano. Okui Formation is the lowest formation of the Komoro Group. The Okui Formation is mainly composed of river deposits consisting of conglomerate, sandstone, and mudstone. The Komoro Welded Tuff (Kawachi and Aramaki 1979), which corresponds to the Shiga Welded Tuff, is directly covered by the Okui Formation (Kawachi and Aramaki 1979; Terao 2001; Oikawa et al. 2005). Znp-Ota Tephra (Kurokawa and Tomita 1998), a Pliocene wide-area tephra that erupted at ~3.9 Ma, is sandwiched in the Okui Formation above the welded tuff (Oikawa et al. 2005). Furthermore, above the tephra, pyroclastic density currents (pyroclastic flow) deposits with an age of 2–3 Ma are sandwiched between the sedimentary rocks of the Okui Formation (Terao 2001; Nakamura et al. 2014; Sato and Matsukawa 2014; Miyasaka and Kano 2015). The Okui Formation is only exposed in a narrow area along the Chikuma River. Therefore, it is not known whether they are widely distributed under Asama volcano.

2.5 2–1 Ma Volcanoes

In the eastern and western regions around Asama volcano, andesitic volcanic and volcaniclastic rocks with ages of approximately 2–1 Ma are widely distributed. In the eastern region, these are volcanic and volcaniclastic rocks covering the Kiritumi Formation, named the Iriyamatouge and Hanamagari formations (Akima Collaborative Research Group 1996). These formations form the current mountain ridges. In the western region, it is named the Iwayakan'non Lava (Takahashi 2004), K-Ar age of approximately 0.97 Ma (Kaneko et al. 1989). These volcanic rocks and formations form topographically higher areas, such as mountains, but due to erosion, the volcanic topography is not clear. On the east side of Asama volcano, around Asama-Otaki Fall and Naka-Karuizawa, there are some lavas with an age of around 1 Ma that are covered by volcaniclastics of Asama volcano. Therefore, there is a high possibility that a volcanic edifice that was active around 1 Ma is buried beneath the eastern part of Asama volcano.

Additionally, the northern region around Asama volcano is distributed in the Azumaya volcano (ca. 0.8–0.45 Ma: Nishiki et al. 2014) and Kusatsu-Shirane volcano (ca. 0.6 Ma to present: Uto 1983; Uto et al. 1983), and it is possible that the lavas and volcaniclastics from these volcanoes constitute the base of the northern side of Asama volcano.

3 Summary

The geological units that form the basement rocks of Asama volcano are divided into Miocene marine strata, Miocene to Pliocene volcanic rocks, and welded tuffs of ca. 4.4 Ma, fluvial sedimentary rocks of the Pliocene, and volcanic rocks of approximately 2–1 Ma. Since the late Miocene, volcanic activity has continued intermittently around Asama volcano, and it is highly likely that these volcanic rocks mainly constitute the basement rocks. In particular, there is a high possibility that the Shiga welded tuff, about 4.4 Ma, and volcanic rocks, approximately 2–1 Ma, are widely distributed under the eastern part of Asama volcano.

References

Akima Collaborative Research Group (1996) Geology of the area of upper stream of the Usui River, Matsuida-machi, Gunma Prefecture, central Japan. Earth Sci (Chiku Kagaku) 50:161–176. https://doi.org/10.15080/agcjchikyukagaku.50.2_161

Iijima N, Iwasa K, Koda M, Taguti K (1956) Geology of the so-called "shiogawa formation". J Geol Soc Japan 62:622–635. https://doi.org/10.5575/geosoc.62.622

Kaneko T, Shimizu S, Itaya T (1989) K-Ar chronological study of the quaternary volcanic activity in Shin-etsu Highland. J Mineral Petrol Econ Geol 84:211–225

Kawachi S (1998) K-Ar age of Yatsugatake and its around volcanic rocks and Happuyama lava. J Faculty Educ Shinshu Univ 93:149–160

Kawachi S, Aramaki S (1979) Geology of the Komoro district. Quadrangle Series, scale 1:50,000. Geological Survey of Japan, p 39

Kosaka T, Takano T, Kitazume O (1991) The tertiary system in the north-western part of the Kanto

Mountainland, Central Japan (part 1)—the geology and geological significance of the Rivers Kousaka and Uchiyama area, especially on the Komagome Belt. Earth Sci (Chiku Kagaku) 45:203–216. https://doi.org/10.15080/agcjchikyukagaku.45.3_203

Kosaka T, Midori T, Hoyanagi K, Kubota M, Miyahigashi Y (1992) Late Cenozoic stratigraphy and paleogeographic changes in the Northern Fossa Magna, vol 37. Memoirs of the Geological Society of Japan, pp 71–83

Kurokawa K, Tomita Y (1998) The Znp-Ohta Ash; an early Pliocene widespread subaqueous tephra deposit in central Japan. J Geol Soc Japan 104:558–561. https://doi.org/10.5575/geosoc.104.558

Miyasaka A, Kano K(2015) Pliocene to Middle Pleistocene tectonics of the Komoro collapse basin in the southeastern part of North Fossa Magna, central Japan. Geoscience Reports of Shizuoka University, no. 42, pp 63–83

Nakamura H, Oikawa T, Geshi N, Matsumoto A (2014) Migration of a volcanic front inferred from K–Ar ages of late Miocene to Pliocene volcanic rocks in central Japan. Island Arc 23:236–250. https://doi.org/10.1111/iar.12073

Nakano S, Takeuchi K, Kato S, Sakai A, Hamasaki S, Hiroshima T, Komazawa M (1998) Geological map of Japan 1: 200,000, Nagano. Geological Survey of Japan, 1 sheet

Nauman E (1893) Neuere Beitrage zur Geologie and Geographie Japans. Petermanns Mitt. No. 108, 45S

Nishiki K, Takahashi K, Matsumoto A, Miyake Y (2011) Quaternary volcanism and tectonic history of the Suwa–Yatsugatake Volcanic Province, Central Japan. J Volcanol Geotherm Res 203:158–167. https://doi.org/10.1016/j.jvolgeores.2011.04.005

Nishiki K, Takeshita Y, Tanabe T, Matsumoto A (2014) K–Ar ages for lavas of Azumaya volcano, central Japan: reexamination of volcanic history of Azumaya volcano. J Geol Soc Japan 120:89–103. https://doi.org/10.5575/geosoc.2014.0004

Nomura S, Akima Collaborative Research Group (1981) Geology of the northwest margin of Kanto Plain, central Japan. Mem Geol Soc Jpn 20:161–167

Nomura S, Ebihara M (1998) K-Ar age and paleomagnetism of Cenozoic volcanic rocks in western Gunma Prefecture, Japan. Gunma J Lib Arts Sci 22:65–78

Oikawa T (2003) The spatial and temporal relationship between uplifting and magmatism in the Hida Mountain Range, Central Japan. Quat Res (Daiyonki-Kenkyu) 42:141–156. https://doi.org/10.4116/jaqua.42.141

Oikawa T, Furusawa A, Takahashi K (2005) Pliocene widespread tephra of the Komoro Group in central Japan; correlation between the Znp-Ohta Tephra Bed and the U-1 pyroclastic flow deposit in the Okui formation. J Geol Soc Japan 111:308–311. https://doi.org/10.5575/geosoc.111.308

Oishi M, Takahashi M (1990) Miocene formations in the Takasaki district, central Japan: with special reference to the developmental process of the Niwaya unconformity. In: Contributions from the Institute of Geology and Paleontology Tohoku University, no. 92, pp 109–117

Sato K, Matsukawa M (2014) Plio-Pleistocene terrestrial ecosystem of the Komoro Group in Nagano Prefecture, Japan. Bull Tokyo Gakugei Univ Div Nat Sci 66:95–131

Takahashi K (2004) Geology and volcanic history of Mt. Eboshidake and its adjacent area, Northeastern Nagano Prefecture, Central Japan. J Geol Soc Jpn 49:83–102. https://doi.org/10.18940/kazan.49.2_83

Tanase A, Oikawa T, Ninomiya A, Hayashi S, Umeda K (2007) Temporal-spatial variations of Plio-Pleistocene volcanic activity in the Ryohaku Mountains, central Japan: evidences from K-Ar ages. Bull Volcanol Soc Japan 52:39–61. https://doi.org/10.18940/kazan.52.1_39

Tateishi M, Takano O, Takashima T, Kurokawa K (1997) Depositional system and provenance of Cenozoic coarse sediments in Northern Fossa Magna. J Jpn Assoc Petrol Technol 62:35–44. https://doi.org/10.3720/japt.62.35

Terao M (2001) Formative history and volcanism of the Komoro Collapse Basin, Nagano Prefecture, Central Japan. Daishiki 33:21–33

Uto K (1983) K-Ar ages of Kusatsu-Shirane volcano. Second Ser Bull Volcanol Soc Jpn 28:416–417. https://doi.org/10.18940/kazanc.28.4_416_3

Uto K, Hayakawa Y, Aramaki S, Ossaka J (1983) Geological map of Kusatsu-Shirane volcano. Geological map of volcano, 3, 1 sheet. Geological Survey of Japan

Yamashita N (1995) Fossa Magna. Tokai University Press, Tokyo, p 310

Yano T, Kosaka T, Takahashi K (2022) Upper Cenozoic sedimentary basin in the northern Fossa Magna—its development history and generating mechanism. Earth Sci (Chikyu Kagaku) 76:161–178. https://doi.org/10.15080/agcjchikyukagaku.76.4_161

Yasui M, Takahashi M, Kanamaru T (2019) Rediscovery of the Higuruma debris avalanche deposit in Asama volcano. Proc Inst Nat Sci Nihon Univ 54:123–142

Geology of Eboshi-Asama Volcano Group

Masaki Takahashi, Maya Yasui, Masashi Nagai, and Tatsuo Kanamaru

1 Introduction

The Eboshi-Asama volcano group is located at the deflection point of the volcanic front of the Northeast Japan volcanic belt, where the volcanic front changes its direction from ENE-WSW to NNW-SSE (Fig. 1); the Northeast Japan volcanic belt is associated with the subduction of Pacific plate.

The presently active Asama-Maekake volcano belongs to the Eboshi-Asama volcano group, which extends east-west ca 24 km and north-south ca 34 km; the Eboshi-Asama volcano group comprises andesitic, dacitic to rhyolitic polygenetic volcanoes and monogenetic lava domes (Figs. 2 and 3), which consists of the Eboshi and Asama volcano groups. The start of the Eboshi-Asama volcano group was 0.45 Ma. The volcanic activity of the Eboshi-Asama volcano group comprises four stages. Each stage is divided by the episode of large-scale sector collapse excluding the boundary between Stage 3 and 4 (Fig 4).

Stage 1 (0.45–0.32 Ma) comprises the Eboshi and Kakuma-Nabebuta andesitic composite volcanoes. Stage 2 (0.32–0.17 Ma) consists of the Yunomaru andesitic lava dome, andesitic to dacitic Sanpogamine, and andesitic Takamine composite volcanoes. Stage 3 (0.17–0.10 Ma) is composed of the andesitic Mizunoto composite volcano, Murakami and Sajiki andesitic lava domes. Stage 4 (<0.10 Ma) comprises the andesitic Kagonoto lava dome and Asama volcano group (Figs. 3 and 4). The Asama volcano group consists of the Kurofu, Sennin, Hanareyama, Hotokeiwa, and Maekake volcanoes. The Kurofu volcano is an andesitic composite volcano. The Sennin volcano is an andesitic pyroclastic cone with lavas and large-scale fallout pumice deposits. The Hanareyama volcano is a dacitic to rhyolitic lava dome with a pyroclastic density current deposit. The Hotokeiwa volcano is an andesitic, dacitic to rhyolitic lava cone with large-scale pyroclastic density current deposits and fallout pumice deposits. The Maekake volcano is an andesitic pyroclastic cone with lavas, large-scale pyroclastic density current deposits, and fallout pumice deposits.

M. Takahashi (✉)
Institute of Natural Sciences, College of Humanities and Sciences, Nihon University, Tokyo, Japan
e-mail: takahashi.masaki@nihon-u.ac.jp

M. Yasui · T. Kanamaru
Department of Earth and Environmental Sciences, College of Humanities and Sciences, Nihon University, Tokyo, Japan
e-mail: yasui.maya@nihon-u.ac.jp; kanamaru.tatsuo@nihon-u.ac.jp

M. Nagai
Volcano Disaster Resilience Research Division, National Research Institute for Earth Science and Disaster Resilience, Tsukuba, Ibaraki, Japan
e-mail: mnagai@bosai.go.jp

S. Aramaki et al. (eds.), *The Asama Volcano*, Active Volcanoes of the World,
https://doi.org/10.1007/978-981-95-5799-8_4

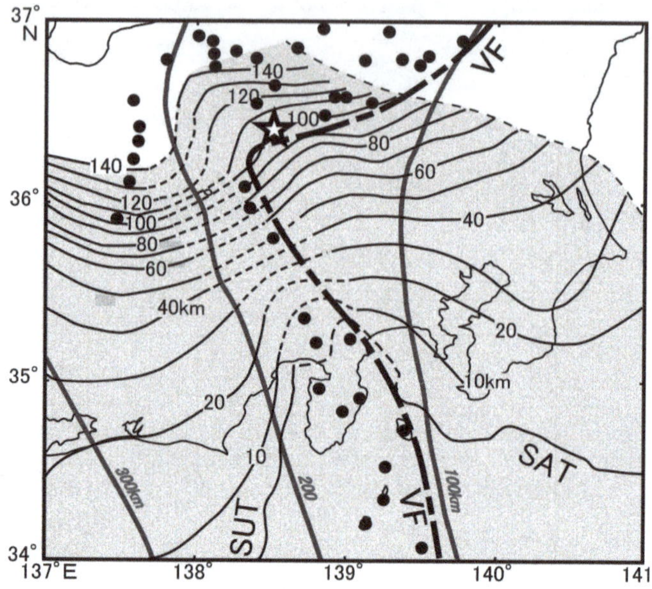

Fig. 1 Map showing the location of the Eboshi-Asama volcano group and subducting slab geometries of the Pacific and Philippine Sea plates (Hirose et al. 2008; Nakajima et al. 2009). SAT: Sagami trough; SUT: Suruga trough; black contour line with numerals: surface of the subducting Philippine Sea slab and its depth in km; bold black contour line with numerals: surface of the subducting Pacific slab and its depth in km; solid circle: Quaternary volcano; broken line: volcanic front (VF); open star: Eboshi-Asama volcano group

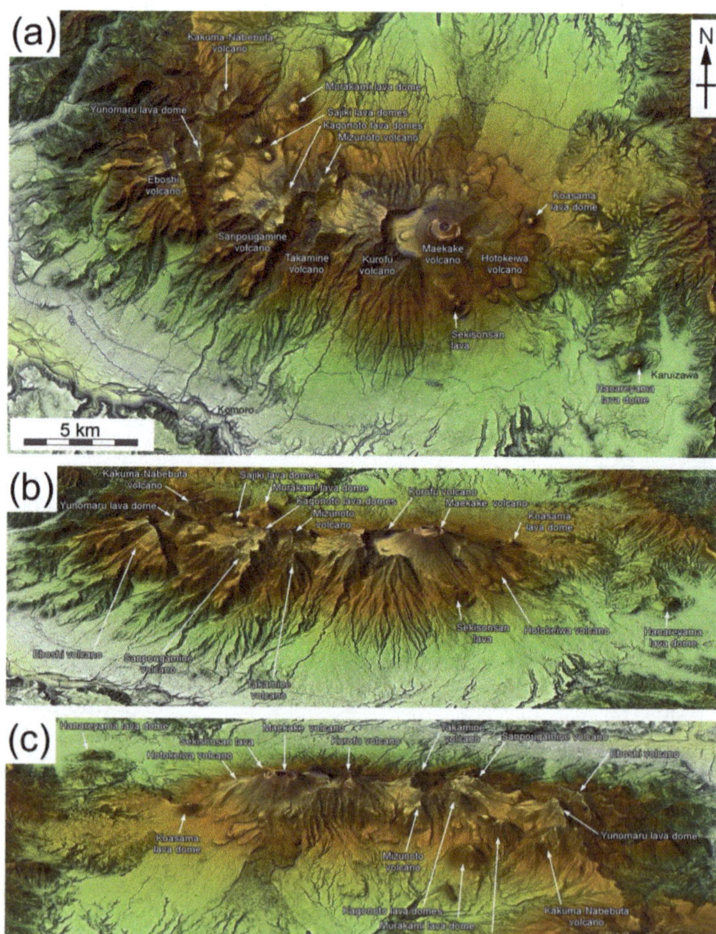

Fig. 2 Image of topography (**a**) and the view of Eboshi-Asama volcano group from the south (**b**) and north (**c**). This figure was created using Kashmir 3D (http://www.kashmir3d.com/) and DEM data provided by the Geospatial Information Authority of Japan (GSI)

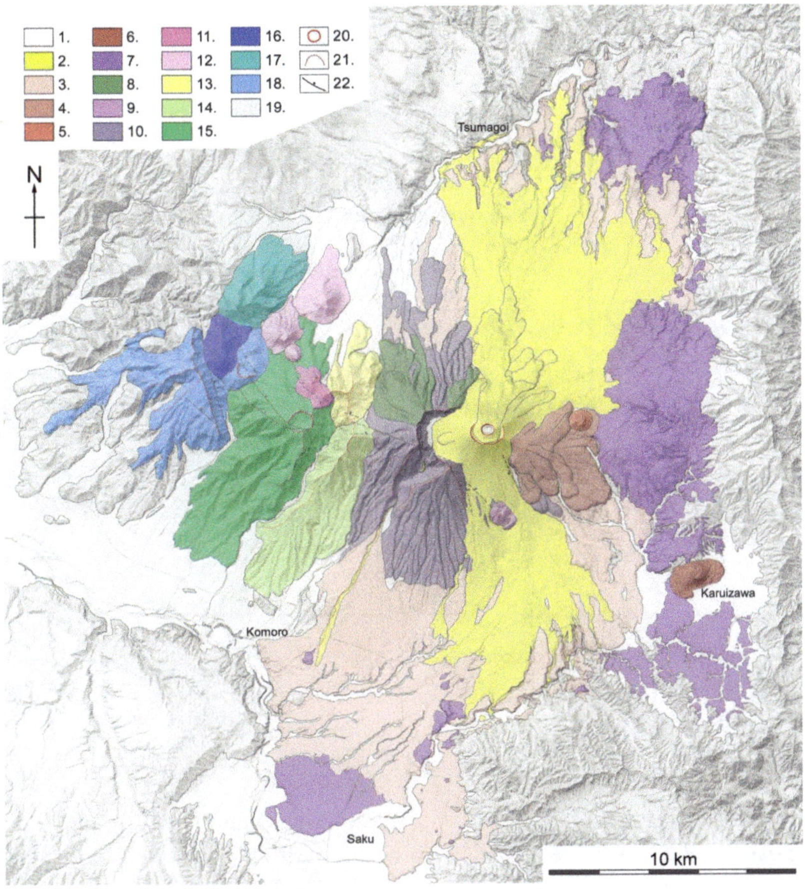

Fig. 3 Geologic map showing the distribution of eruptive products of the Eboshi-Asama volcano group. 1: alluvium. 2: Maekake volcano. 3: Komoro pyroclastic density current deposit. 4: Hotokeiwa volcano. 5: Koasama lava dome. 6: Hanareyama lava dome. 7: debris avalanche deposits formed by sector collapse of Kurofu and Sennin volcanoes. 8: Sennin volcano. 9: Sekisonsan lava. 10: Kurofu volcano. 11: Kagonoto lava dome. 12: Sajiki-Kosajiki and Murakami lava domes. 13: Mizunoto volcano. 14: Takamine volcano. 15: Sanpogamine volcano. 16: Yunomaru lava dome. 17: Kakuma-Nabebuta volcanoes. 18: Eboshi volcano. 19: Quaternary volcanic rocks and basement rocks. 20: crater. 21: collapsed caldera rim. 22: fault. This figure was created using Kashmir 3D (http://www.kashmir3d.com/) and DEM data provided by the Geospatial Information Authority of Japan (GSI)

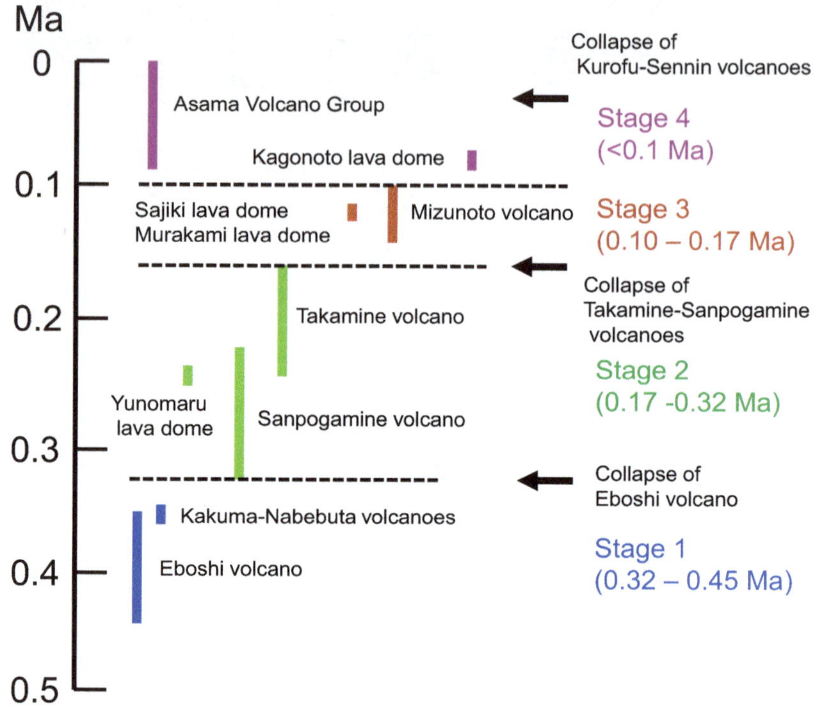

Fig. 4 Figure showing ages and stages of the volcanic activity of Eboshi-Asama volcano group

2 Volcanoes of Stage 1 (0.45–0.32 Ma)

The volcanoes of Stage 1 are distributed in the westernmost side of Eboshi-Asama volcano group and consist of the Eboshi and Kakuma-Nabebuta volcanoes (Figs. 3, 5 and 6).

2.1 Eboshi Volcano (Younger Eboshi Volcano) (Eboshi Volcano Group)

The Eboshi volcano is the largest composite volcano in the Eboshi volcano group, the age of which is 0.45–0.35 Ma by K-Ar dating (Kaneko et al. 1991; Nishiki et al. 2013). The basal area of the volcanic edifice of Younger Eboshi volcano (including the Narahara lava) is approximately 10 × 6 km, and its estimated volume is more than 9 km^3 (Takahashi 2004). The volcanic edifice is cut by a 6 km long normal fault extending in NNW to SSE; the majority of the eastern half of the volcanic edifice is lost by subsidence along the normal fault, resulting in the collapse of volcanic edifice and debris avalanche deposits (Fig. 7). However, the subsided block with a summit crater of 0.8 km in diameter remains in the eastern foot of the fault cliff, which is overlain by the Yunomaru lava dome after the collapse. The geologic details of debris avalanche deposit have not yet been revealed, but the hill of Nukazukayama located to the northeast of the Murakami lava dome may be a large toreva block of debris avalanche deposit. The collapse of Eboshi volcano occurred before the effusion of the Yunomaru lava dome (0.28 Ma). The original volume of Eboshi volcano including the missing eastern half is estimated to be more than 18 km^3.

The Eboshi volcano is underlain by older andesitic volcanoes; they are the Older Eboshi volcano (0.76 Ma), Ofuji lava dome (0.77 Ma), Gotomiki volcano, Omuro lava dome, Kokuzo lavas, Kanahara lavas (0.77 Ma), and Iwayakannon volcano (0.97 Ma) (Takahashi 2004; Takahashi and Miyake 2004). The dormant period between Eboshi volcano and underlying older volcanic rocks is 0.31 million years.

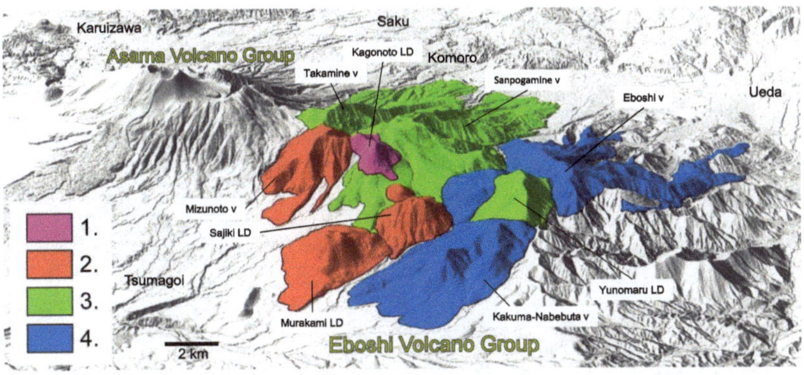

Fig. 5 Perspective topographic relief image of Eboshi volcano group from the northwest showing the distribution of volcanic edifice and their ages. 1: Stage 4 (<0.1 Ma). 2: Stage 3 (0.10–0.17 Ma). 3: Stage 2 (0.17–0.32 Ma). 4: Stage 1 (0.32–0.45 Ma). This figure was created using Kashmir 3D (http://www.kashmir3d.com/) and DEM data provided by the Geospatial Information Authority of Japan (GSI)

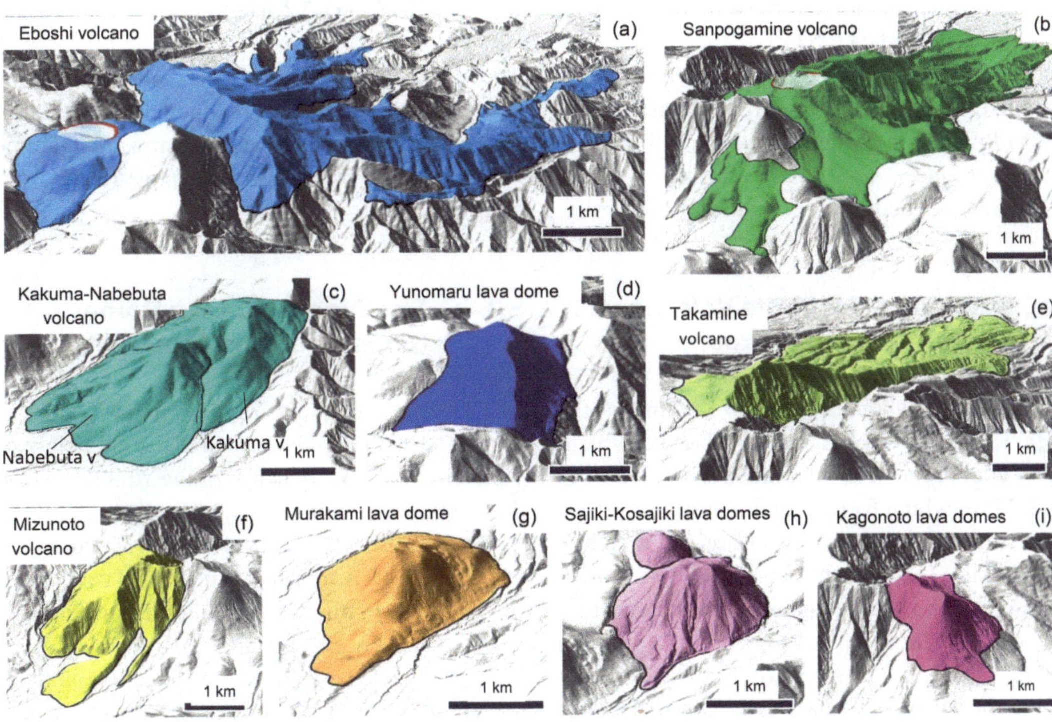

Fig. 6 Perspective topographic relief images from the northwest showing the individual volcanic edifice of Eboshi volcano group. (**a**) Eboshi volcano. (**b**) Sanpogamine volcano. (**c**) Kakuma-Nabebuta volcanoes. (**d**) Yunomaru lava dome. (**e**) Takamine volcano. (**f**) Mizunoto volcano. (**g**) Murakami lava dome. (**h**) Sajiki-Kosajiki lava domes. (**i**) Kagonoto lava dome. This figure was created using Kashmir 3D (http://www.kashmir3d.com/) and DEM data provided by the Geospatial Information Authority of Japan (GSI)

The long time interval suggests that the volcanic activity of Eboshi volcano is distinct from those of other older volcanoes and is considered to be the onset of the Eboshi-Asama volcano group which is a new stage of volcanism.

The Eboshi volcano consists of an alternation of lavas with a thickness of 5–25 m and pyroclastic rocks. Agglutinate is distributed around the peak of Mt. Eboshi (2099 m above sea level in altitude), indicating that the eruptive center of

Fig. 7 Photograph showing the Eboshi volcano and Yunomaru lava dome viewed from the summit of Kakuma volcano. The Yunomaru lava dome erupted on the eastern part of Eboshi volcano which subsided along the fault cliff

Eboshi volcano was located near the current peak of Mt. Eboshi. The net height of Eboshi volcano probably exceeded ca 0.5 km.

The volcanic rocks of Eboshi volcano are andesite (55–61 wt% SiO_2) (Fig. 10) with phenocrysts of plagioclase, clinopyroxene, orthopyroxene, ±olivine, and ±hornblende, whose representative phenocryst content is 30–40 vol%.

2.2 Kakuma-Nabebuta Volcano (Eboshi Volcano Group)

The Kakuma-Nabebuta volcano comprises the older Kakuma and younger Nabebuta volcanoes. The basal area of Nabebuta volcano is 2 × 2.3 km, which formed on the northeast flank of Kakuma volcano, the basal area of which is 2 × 3.5 km. The Kakuma and Nabebuta volcanoes are composite volcanoes mainly consisting of thick lava flows with net heights of less than ca 0.68 and 0.53 km, respectively. The eruption volumes of Kakuma and Nabebuta volcanoes are 1.6 and 0.8 km^3, respectively. The summit peak of Kakuma volcano is Mt. Kakuma (1981 above sea level in altitude) and that of the Nabebuta volcano is Mt. Nabebuta (1829 m above sea level in altitude). The age of Kakuma volcano obtained by K-Ar dating is 0.36 Ma (Kaneko et al. 1991). The volcanic rocks of both volcanoes are andesite (58–59 wt% SiO_2) (Fig. 10) with phenocrysts of plagioclase, clinopyroxene, and orthopyroxene, whose representative phenocryst content is 34 vol%.

3 Volcanoes of Stage 2 (0.32–0.17 Ma)

The volcanoes of Stage 2 are distributed to the east of volcanoes that formed in Stage 1, and consist of the Yunomaru lava dome, and Sanpogamine and Takamine composite volcanoes (Figs. 3, 5, and 6).

3.1 The Yunomaru Lava Dome (Eboshi Volcano Group)

The basal area of the Yunomaru lava dome is 2 × 1.3 km and ovoidal in shape, the net height of which is ca 0.3 km. The age of the Yunomaru lava dome ranges from 0.31 to 0.24 (average 0.28) Ma by K-Ar dating (Nishiki et al. 2013; Takahashi et al. 2013b). The north-south aligned twin peaks of Mt. Yunomaru are the summit of the lava dome, which are 2099 and 2101 m above sea level in altitude (Fig. 7). The estimated volume of the Yunomaru lava dome is 0.26 km^3. The volcanic rocks of the Yunomaru lava dome are andesite (60–61 wt% SiO_2) (Fig. 10) with phenocrysts of plagioclase, clinopyroxene, and orthopyroxene, the phenocryst content of which is 30–35 vol%.

3.2 Sanpogamine Volcano (Eboshi Volcano Group)

The Sanpogamine volcano is the next largest composite volcano in the Eboshi volcano group, whose ages by K-Ar dating are 0.32–0.22 Ma (Nishiki et al. 2013; Takahashi et al. 2013b). The basal area of Sanpogamine volcano is approximately 10 × 4 km, and its net height is less than ca 1.1 km; the estimated original volume of the volcano is approximately 15 km^3. The Sanpogamine volcano consists of the lower, middle, and upper members. The lower member comprises an alternation of thin andesitic lavas with a thickness of several meters and pyroclastic rocks. The pyroclastic density current deposits (mainly block and ash flow deposits) are abundant in the middle member. The upper member is mainly composed of thick andesitic to dacitic lava flows with a thickness of 10–70 m. A crater with a diameter of 1 km (the Ikenotaira crater) is present at the summit area of the volcano, which is surrounded by a rim of agglutinate (Fig. 8). The highest point of the rim of the crater is Mt. Sanpogamine (2041 m above sea level in altitude). The southern rim of the crater is cut by a normal fault. The volcanic rocks of Sanpogamine volcano are andesite to dacite (55–65 wt% SiO_2) (Fig. 10) with phenocrysts of plagioclase, clinopyroxene, and orthopyroxene, the phenocryst content of which is 30–42 vol%.

3.3 Takamine Volcano (Eboshi Volcano Group)

The Takamine volcano is a composite volcano comprising alternating thin andesitic lava flows with a thickness of less than several meters and pyroclastic rocks, the age of which is 0.24–0.17 Ma by K-Ar dating (Nishiki et al. 2013; Takahashi et al. 2013a, b). The basal area of the volcanic edifice is approximately 7 × 2.5 km with a net height of less than ca 1.3 km; the central vent is located near the present summit of Mt. Takamine (2106 m above sea level in altitude) where agglutinate beds are dis-

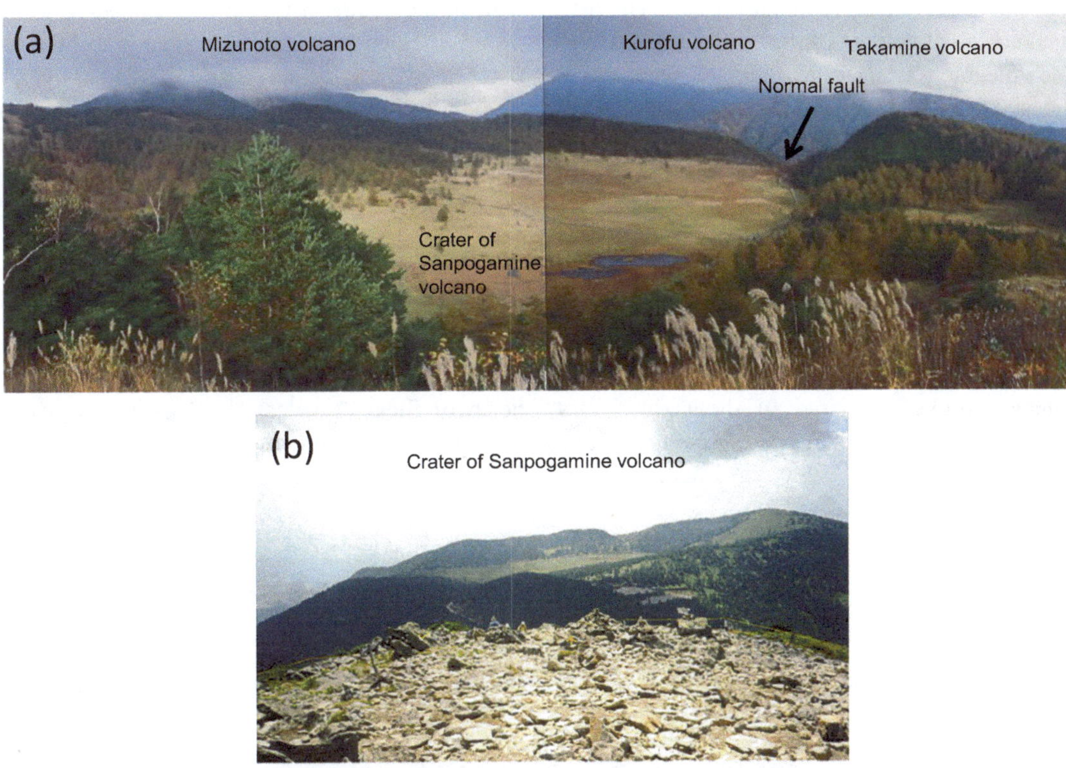

Fig. 8 Photograph showing the summit crater of Sanpogamine volcano (Ikenotaira crater). (**a**) View from the western rim of the crater. Normal fault runs cutting the southern rim of the crater. The crater is completely buried by debris. (**b**) Summit crater of Sanpogamine volcano viewed from the summit of Mizunoto volcano

tributed. The summit vent is located near the northern margin of the volcanic edifice; most lavas flow down the south flank from the summit vent to the foot of the volcano and construct an asymmetric volcanic edifice; the volcanic edifice to the southwest of the summit vent occupies the large volume of Takamine volcano. The northern end of the volcanic edifice is cut by a normal fault which continues to the Tomi fault of Kurofu volcano. The western half of the volcanic edifice and summit vent area was lost by volcano collapse and subsequent erosion; as a result, the deep valley of the source of Fukazawa river formed between the Takamine and Sanpogamine volcanoes. The geologic details of Takamine's debris avalanche deposits have not yet been revealed, but the Matsubagawa mud flow deposit in the southern foot of the Eboshi volcano group is a probable candidate for the debris avalanche of Takamine-Sanpogamine volcanoes. The northern flank of Takamine and Sanpogamine volcanoes also collapsed and formed the debris avalanche deposits of the upper Tsumagoi group. The estimated original volume of the volcano is 7.6 km^3. The volcanic rocks of the Takamine volcano are andesite (55–58 wt% SiO_2) (Fig. 10) with phenocrysts of plagioclase, clinopyroxene, orthopyroxene, and ±olivine, the phenocryst content of which is 34–48 vol%.

3.4 The Middle Tsumagoi Group

The Tsumagoi group comprises sediments and volcanic rocks deposited in the sedimentary basin surrounded by the Eboshi-Asama volcano group to the south, Azumaya volcano to the west, and Kusatsu-Shirane volcano to the north, which consists of the lowermost (0.7–0.6 Ma), lower (0.6–0.4 Ma), middle (0.4–0.2 Ma), and upper members (0.2–0.1 Ma) (Tanabe et al. 1992). The formation of the middle Tsumagoi group was concurrent with Stages 1 and 2 of the Eboshi volcano group. The Ozasa formation of the middle Tsumagoi group is composed of mud-flow, sand, and conglomerate deposits that might be sediments related to debris avalanche deposits and secondary sediments formed by the large-scale sector collapse of Eboshi volcano.

3.5 The Matsubagawa Mud Flow Deposit

The Pleistocene debris avalanche deposits (the Matsubagawa mud flow deposit) and overlying clastic sediments consisting of sand and conglomerate (the Suwayama formation) are distributed in the southern foot of the Eboshi volcano group, the age of which is estimated to be after 0.4 Ma (Miyasaka and Kano 2015). They are probably the products of the large-scale sector collapse of Eboshi or Takamine and Sanpogamine volcanoes.

4 Volcanoes of Stage 3 (0.17–0.10 Ma)

The volcanoes of Stage 3 are distributed to the north of Stage 2 volcanoes and consist of the Mizunoto composite volcano and Murakami and Sajiki lava domes (Figs. 3, 5, 6 and 9).

Fig. 9 Photograph showing the Murakami, Sajiki, and Kagonoto lava domes viewed from the summit of Kakuma volcano. A distant view is the Mizunoto, Kurofu, and Maekake volcanoes

4.1 Mizunoto Volcano (Eboshi Volcano Group)

The Mizunoto volcano is a small composite volcano, the ages of which are 0.13–0.10 Ma by K-Ar dating (Nishiki et al. 2013). The basal area of Mizunoto volcano is 3 × 1.8 km with a net height of ca 0.8 km. The estimated original volume of the volcano is 1.8 km^3 and consists of thick andesitic lava flows up to several tens of meters in thickness and pyroclastic rocks. The central vent was located in the southern end of the volcanic edifice, but it was completely lost by sector collapse and erosion; the highest peak of the rim of the collapsed crater is Mt. Mizunoto (2202 m above sea level in altitude). Most lavas flowed down the northern flank of Mizunoto volcano. The volcanic rocks of Mizunoto volcano are andesite (56–62 wt% SiO_2) (Fig. 10) with phenocrysts of plagioclase, clinopyroxene, orthopyroxene, ±olivine, and ±quartz; the representative phenocryst content of which is ca 32 vol%.

4.2 The Murakami Lava Dome (Eboshi Volcano Group)

The Murakami lava dome with small lava lobes is a monogenetic volcano, and the basal area is 2.8 × 1.8 km with a net height of ca 0.5 km (the summit peak of Mt. Murakami is 1747 m above sea level in altitude); its estimated volume is 0.83 km^3 (Fig. 9). The age of the Murakami lava dome is 0.12 Ma by K-Ar dating (Nishiki et al., 2013; Takahashi et al. 2013b). The volcanic rocks of Murakami lava dome are andesite (59–62 wt% SiO_2) (Fig. 10) with phenocrysts of plagioclase, clinopyroxene, orthopyroxene, and ±olivine, and the representative phenocryst content of which is ca 40 vol%.

4.3 The Sajiki Lava Dome (Eboshi Volcano Group)

The monogenetic Sajiki lava dome consists of the Sajiki and Kosajiki lava domes, which are twin domes aligned in a northwest to southeast direc-

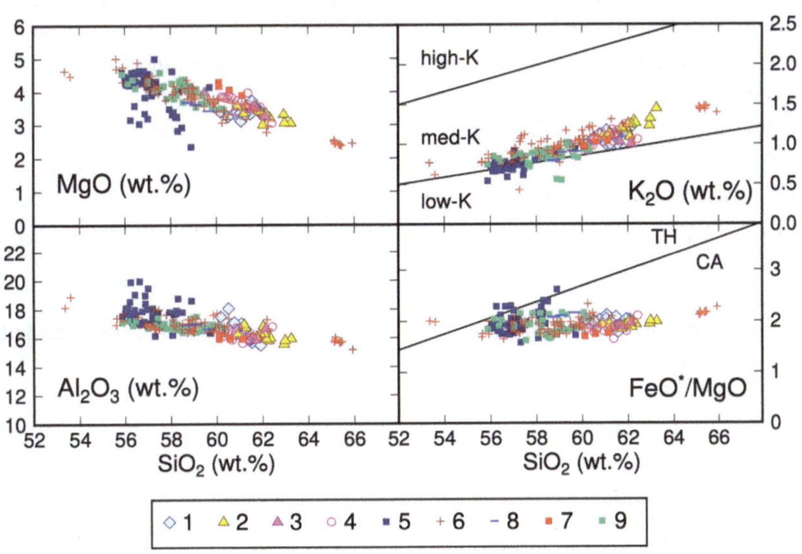

Fig. 10 SiO_2 variation diagrams for MgO, Al_2O_3, K_2O, and FeO*/MgO of Eboshi volcano group; high-K, med-K, and low-K denote the high-K_2O, medium-K_2O, and low-K_2O series of Gill (1981), and TH (tholeiitic rock-series) and CA (calc-alkalic rock-series of Miyashiro (1974). 1: Kagonoto lava dome. 2: Sajiki lava dome. 3: Murakami lava dome. 4: Mizunoto volcano. 5: Takamine volcano. 6: Sanpogamine volcano. 7: Yunomaru lava dome. 8: Kakuma-Nabebuta volcanoes. 9: Eboshi volcano

tion (Figs. 5 and 9). The K-Ar age of the Sajiki lava dome has not been obtained, but it is estimated to have been concurrent with or younger than the Murakami lava dome from the stratigraphic relation and degree of erosion. The basal area of the larger Sajiki lava dome is 1.5 × 1.3 km, and that of the smaller Kosajiki lava dome is 0.75 × 0.75 km; their net heights are ca 0.38 and 0.15 km, respectively. The summit of Sajiki lava dome (Mt. Sajiki) is 1916 m above sea level in altitude and that of the Kosajiki (Mt. Kosajiki) is 1852 m above sea level in altitude. The estimated volume of the Sajiki lava dome is 0.24 km^3 and that of the Kosajiki lava dome is 0.03 km^3. The volcanic rocks of Sajiki and Kosajiki lava domes are andesite (61–62 wt% SiO_2) (Fig. 10) with phenocrysts of plagioclase, clinopyroxene, and orthopyroxene, the phenocryst content of which is 32–34 vol%.

4.4 The Upper Tsumagoi Group

The deposition of the upper Tsumagoi group (0.2–0.1 Ma) was concurrent with the volcanic activity of Stage 3 of Eboshi volcano group, which consists of the lower Takahanezawa, upper Takahanezawa formations, and Osawa pyroclastic rocks (Tanabe et al. 1992). The lower Takahanezawa formation comprises the alternation of silt and sand deposits which is a lake deposit. The Ata-Th tephra erupted in southern Kyushu, the age of which is 0.24 Ma, is found in the lower Takahanezawa formation (Hayakawa 2010), so the age of the onset of deposition of lower Takahanezawa formation is older than 0.2 Ma. The upper Takahanezawa formation is a debris avalanche deposit formed by the sector collapse of the northern flanks of Sanpogamine and Takamine volcanoes; the sector collapse may have occurred several times and dammed the Agatsuma river forming several lakes. The Osawa pyroclastic rocks may be a member of the Mizunoto volcano.

5 Volcanoes of Stage 4 (<0.10 Ma)

Most volcanoes of Stage 4 are distributed to the east of Stages 2 and 3 volcanoes and consist of the Kagonoto lava dome and Asama volcano group (Figs. 3, 5 and 6).

5.1 The Kagonoto Lava Dome (Eboshi Volcano Group)

The Kagonoto lava dome comprises twin domes aligned in a northwest to southeast direction: the Higashikagonoto and Nishikagonoto lava domes. The Kagonoto lava dome is the youngest volcano in the Eboshi volcano group, and its age is 0.08–0.07 Ma by K-Ar dating (Nishiki et al. 2013; Takahashi et al. 2013a, b), which is nearly the same as that of the Gippa-Kengamine lava groups of Kurofu volcano. The basal area of the Kagonoto lava dome is 1.5 × 1.0 km with a net height of ca 0.43 km; the peak of Mt. Higashikagonoto is 2228 m above sea level and that of Mt. Nishikagonoto is 2212 m above sea level in altitude. The estimated volume of the Kagonoto lava dome is 0.21 km^3. The volcanic rocks of Kagonoto lava dome are andesite (60–62 wt% SiO_2) (Fig. 10) with phenocrysts of plagioclase, clinopyroxene, and orthopyroxene, the phenocryst content of which is 30–36 vol%.

5.2 Kurofu Volcano (Asama Volcano Group)

Kurofu volcano is the largest composite volcano in the Eboshi-Asama volcano group. The basal area of the present volcanic edifice is 16 × 4 km, but the estimated original basal area of the volcano which was lost by the sector collapse is estimated to be 16 × 11 km. The Kurofu volcano is estimated to have been active from 0.09 to

0.055 Ma by K-Ar dating (Nishiki et al. 2013) and tephra-stratigraphy. The eastern half including central vent areas of the original volcanic edifice were completely destroyed by sector collapse, and the U-shaped amphitheater of Yunotaira caldera (2 km in NS and 1.8 km in EW) was formed. The Kurofu volcano comprises alternating lavas and pyroclastic rocks, most of which are thin beds with a thickness of less than a few meters. The original volume of Kurofu volcano before collapse is estimated to be approximately 40 km^3 (Aramaki 1963). The height of the original cone of the composite volcano before collapse is estimated to be 2800 m above sea level in altitude by Aramaki (1963), whereas the highest peak of the caldera rim of Kurofu volcano (Mt. Kurofu) is 2404 m above sea level in altitude.

The Sekisonsan lava, which is present in the southeastern flank of Kurofu volcano, was considered to be a lateral volcano of Kurofu volcano by Aramaki (1963); the basal area is 1.3 × 0.9 km with a net height of 0.25 km and volume of 0.07km^3; the altitude of the summit of Mt. Sekisonsan is 1668 m above sea level in altitude. However, it was proposed recently that the Sekisonsan volcanic edifice is not a lava dome in situ but a toreva block of the debris avalanche deposit.

The volcanic rocks of the Kurofu volcano are andesite (54–61 wt% SiO_2) with phenocrysts of plagioclase, clinopyroxene,orthopyroxene, and ±olivine; the phenocryst content of which is 29–47 vol%.

5.3 Sennin Volcano (Asama Volcano Group)

The Sennin volcano consists of a densely welded pyroclastic cone, lava flows, and fallout pumice deposits (As-BP) (Takahashi et al. 2013a), lacking large-scale pyroclastic density current deposits. The Sennin volcano erupted at the northern flank of Kurofu volcano, and the onset of its volcanic activity was 30,000 yearBP based on tephra-stratigraphy (Nakamura et al. 1997; Tsuji et al. 2004); it commenced after a dormant period of erosion of several tens of thousands of years which had continued since the cessation of volcanic activity of Kurofu volcano. The lavas and pyroclastic rocks of Kurofu volcano are overlain by the welded pyroclastic cone of Sennin volcano with angular unconformity at the northern end of Yunotaira caldera wall. The cessation of volcanic activity of Sennin volcano was the large-scale pumice eruption associated with the large-scale sector collapse of the volcanic edifice of Kurofu and Sennin volcanoes, which occurred in ca 28,000 years calBP (Higuchi 1989); the duration of the volcanic activity of Sennin volcano was relatively short, which was only 2,000 years. The basal area of Sennin volcano is inferred to have been more than 2 × 3 km, but the main part of volcanic edifice is completely lost by the sector collapse, so the precise estimation of original volume is difficult. The inferred original volume of volcanic edifice of Sennin volcano may exceed 3 km^3. The lavas of Sennin volcano are andesite (62–63 wt% SiO_2) with phenocrysts of plagioclase, clinopyroxene, and orthopyroxene, the phenocryst content of which is 42–45 vol%.

The volcanic activity of the Asama volcano group changed from the effusive style (Strombolian to Vulcanian) of Kurofu volcano to the explosive style (sub-Plinian to Vulcanian) of Sennin volcano; the volcanic activity of the Asama volcano group has been explosive since the formation of Sennin volcano.

5.4 Hanareyama Volcano (Asama Volcano Group)

The Hanareyama volcano in the Karuizawa town is located approximately 10 km to the southeast of the summit crater of Maekake volcano. The Hanareyama volcano is a monogenetic volcano with a lava dome and a small pyroclastic density current deposit, which erupted at ca 24,000 years calBP after the sector collapse of Sennin and Kurofu volcanoes (Nakamura et al. 1997; Tsuji et al. 2004). The basal area of Hanareyama lava dome is 2.3 × 1 km with a net height of 0.26 km; the summit of Mt. Hanareyama is 1256 m above sea level in altitude. The estimated volume of Hanareyama lava dome is ca 0.2 km^3. The Kumoba pyroclastic density current deposit, the precursor of effusion of Hanareyama lava dome, is distributed in the southern foot of Hanareyama lava dome; the volume of Kumoba pyroclastic density current deposit is less than 0.01 km^3. The volcanic rocks of Hanareyama lava dome and Kumoba pyroclastic density current deposit are andesite, dacite to rhyolite (60–75 wt% SiO_2), the phenocrysts of which are plagioclase, orthopyroxene, and clinopyroxene, but rhyolite sometimes includes a minor amount of quartz, hornblende, and rare biotite; their phenocryst contents are 8–21 vol%.

5.5 Hotokeiwa Volcano (Asama Volcano Group)

The Hotokeiwa volcano is an asymmetrical flat-shaped lava cone with a maximum apparent height of less than 0.4 km; its basal area is one-third of a circle with a radius of 4 km, the center of which is located at the summit crater of the volcano. The volume of the volcano is approximately 0.78 km^3. The summit crater of Hotokeiwa volcano is estimated to be located on the eastern flank of Maekake volcano at about 2150 m above sea level in altitude, but it is now completely buried by the volcanic products of Maekake volcano. The Hotokeiwa volcano consists of lower rhyolitic (0.51 km^3), middle dacitic to rhyolitic (0.25 km^3), and upper andesitic (0.02 km^3) thick lava flows, associated with the voluminous pyroclastic density current and fallout pumice deposits. The pyroclastic density current deposits are widely distributed in the northern and southern foot of Hotokeiwa volcano, with the maximum runout of 19 km from the summit crater of Maekake volcano in the northern area and 21 km in the southern area.

The volcanic activity commenced with the eruption of a large-scale rhyolitic fallout pumice deposit (the Shiraitonotaki fallout pumice: 1.21 km^3 DRE) at ca 23,000 years calBP (Nakamura et al. 1997; Tsuji et al. 2004), followed by the effusion of the Hotokeiwa lower lava and Koasama lava dome, both of which are rhyolitic in composition. The Koasama lava dome is a small monogenetic volcano which effused at the northeastern margin of Hotokeiwa volcano, the basal diameter of which is 0.8 km with a net height of 0.2 km and volume of 0.03 km^3. The height of the summit of Koasama lava dome is 1655 m above sea level.

The large-scale eruptions of Hotokeiwa volcano produced voluminous pyroclastic density current and fallout pumice deposits, named Okubozawa 1 (21,000 years calBP), Okubozawa 2 (20,000 years calBP), Komoro 1 (16,000 years calBP), and Komoro 2 (13,000 years calBP) (Nakamura et al. 1997; Tsuji et al. 2004). Only pyroclastic density current and fallout pumice deposits were erupted in the dacitic Okubozawa 1 and 2 eruptive events; they are relatively small in eruptive volume, which are approximately 0.26 and 0.21 km^3 DRE, respectively, but lack lava flows. The dacitic Hotokeiwa middle lava effused in the Komoro 1 eruptive episode, concurrently with the eruption of voluminous pyroclastic materials. The Komoro 1 was the largest eruptive episode in the Hotokeiwa volcano, and its eruptive volume probably exceeds 2.81 km^3 DRE. The volume of eruptive products of the Komoro 2 large-scale eruption is at least 0.44 km^3 DRE.

The volcanic activity of the Hotokeiwa volcano ceased until ca 10,000 years calBP; its duration is relatively short with a time span of ca 13,000 years.

The volcanic rocks of the Hotokeiwa volcano are andesitic, dacitic to rhyolitic in composition (55–74 wt% SiO_2), the phenocrysts of which are plagioclase, orthopyroxene, and clinopyroxene, but rhyolitic eruptive products sometimes include minor amounts of hornblende; their phenocryst contents are 10–38 vol%.

5.6 Maekake Volcano (Asama Volcano Group)

The Maekake volcano is the youngest and currently active volcano in the Asama volcano group.

The summit of Maekake volcano is the highest peak of Kamayama pyroclastic cone with a height of 2568 m above sea level in altitude. The volcanic activity of Maekake volcano started somewhere between the Yunotaira caldera and Hotokeiwa volcano at approximately 9,400 years calBP (Yasui et al. 2021; Takahashi et al. 2022). The volcanic edifice of Maekake volcano consists of welded pyroclastic cones and lava flows with a basal area of 8.8 × 4.1 km and a net height of ca 0.5 km. The main lava flows are the Onioshidashi lava in 1783 AD, Kamibutai lava in 1108 AD, and Shimobutai, Kuromamegawara, Higashiizumisawa, and Maruyama lavas in the second century; most lavas flowed down the northern flank up to 6 km from the summit crater. At the foot of the volcano, fields of pyroclastic density current deposits with maximum run-outs of 19 km in the northern area and 21 km in the southern area have developed; named the 1783 AD Agatsuma pyroclastic density current deposit, 1108 AD Oiwake pyroclastic density current deposit, and second-century Kotaki pyroclastic density current deposit. The lava flows, pyroclastic density current, and fallout pumice deposits were produced mainly by large-scale sub-Plinian eruptions. The present summit area is composed of the Maekake welded pyroclastic cone with a crater of 1.5 × 1.0 km, which was formed in the large-scale eruption of 1108–1128 AD, and the relatively small Kamayama welded pyroclastic cone with a crater diameter of 0.5 × 0.44 km found in the composite crater of Maekake welded pyroclastic cone, which was formed by the large-scale sub-Plinian eruption of 1783 AD. The total erupted volume of Maekake volcano is approximately 2.6 km^3 DRE, since its onset at ca 9,400 years calBP. Presently, repeated Vulcanian eruptions often occur from the Kamayama crater of Maekake volcano. The eruption in 1783 AD was the last large-scale sub-Plinian eruption, but volcanic activity is still ongoing since then and the last intermediate-scale Vulcanian eruption occurred in 2004 AD.

The volcanic rocks of Maekake volcano are andesite (58–64 wt% SiO_2), the phenocrysts of which are plagioclase, orthopyroxene, clinopyroxene, and ±olivine; their phenocryst contents are 18–24 vol%.

6 Whole-Rock Major Element Chemistry of the Eboshi-Asama Volcano Group

The whole-rock chemistry of the Eboshi and Asama volcano groups was reported by Takahashi (2004) and Takahashi et al. (2007, 2008a, b, Takahashi et al. 2013a, b, 2018, 2024).

Dacitic rocks erupted in Stage 2 (the Sanpogamine volcano), and silicic andesitic rocks with more than 60 wt% SiO_2 increased in Stage 3 (the Mizunoto volcano, Murakami and Sajiki lava domes) (Fig. 10).

The compositional variation of oxides in the SiO_2 variation diagram for eruptive products of Eboshi volcano group is nearly the same excluding MgO and Al_2O_3; the MgO and Al_2O_3 of Takamine volcano are lower and higher than others, respectively (Fig. 11). The volcanic rocks are andesite, and most of them belong to the medium-K series and calc-alkaline rock series, but some volcanic rocks of the Takamine volcano are low-K series and tholeiitic rock series (Fig. 11).

The whole-rock SiO_2 content of volcanic rocks of the Kagonoto, Kurofu, Sennin, Hotokeiwa, and Maekake volcanoes in Stage 4 is 60–62, 54–62, 59–64, 55–74, and 57–65 wt%, respectively (Figs. 10 and 12). The SiO_2 content of volcanic rocks increased to rhyolitic in the Hanareyama and Hotokeiwa volcanoes, but those of the Maekake volcano are andesitic and similar to the Sennin volcano (Fig. 12); the style of eruption, regardless of the chemical compositions, had changed from effusive to explosive (sub-Plinian to Vulcanian) since the Sennin volcano.

The compositional variation of oxides of eruptive products in the Asama volcano group in Stage 4 is almost the same as those of Stages 1–3 in the SiO_2 variation diagram excluding MgO and Al_2O_3, though the SiO_2 content is higher in the Asama volcano group (52–76 wt%) than in the Eboshi volcano group (54–66 wt%)

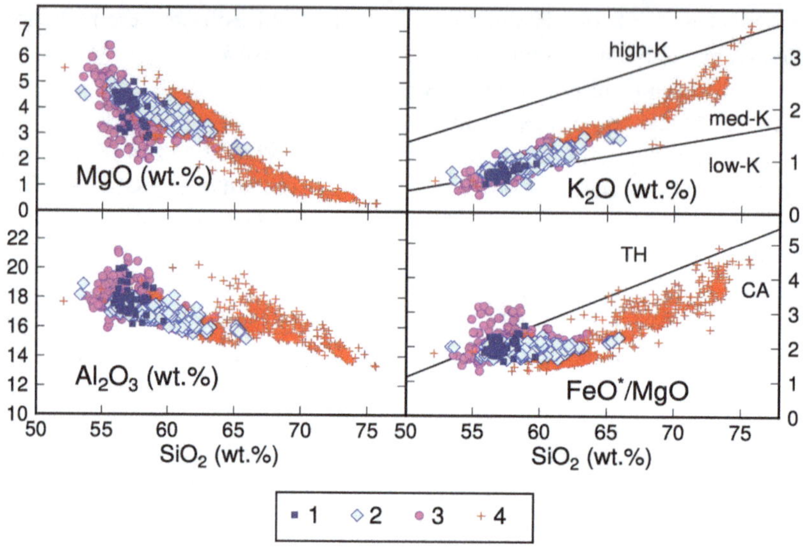

Fig. 11 SiO_2 variation diagrams for MgO, Al_2O_3, K_2O, and FeO*/MgO of Eboshi-Asama volcano group; high-K, med-K, and low-K denote the high-K_2O, medium-K_2O, and low-K_2O series of Gill (1981), and TH and CA denote the tholeiitic and calc-alkalic rock-series of Miyashiro (1974). 1: Takamine volcano. 2: Eboshi volcano group excluding Takamine volcano. 3: Gippa lava group (Kurofu volcano). 4: Asama volcano group excluding Gippa lava group

Fig. 12 SiO_2 variation diagrams for MgO, Al_2O_3, K_2O, and FeO*/MgO of the Asama volcano group; high-K, med-K, and low-K denote the high-K_2O, medium-K_2O, and low-K_2O series of Gill (1981), and TH and CA denote tholeiitic and calc-alkalic rock-series of Miyashiro (1974). 1: Kurofu volcano. 2: Sennin volcano. 3: Hanareyama-Hotokeiwa volcanoes. 4: Maekake volcano

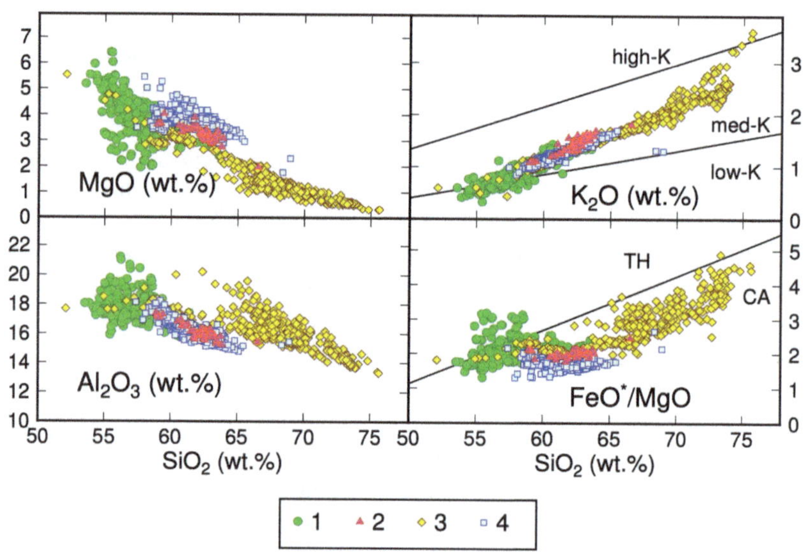

(Fig. 11). The MgO and Al_2O_3 of Gippa lava group of Kurofu volcano are lower and higher than other volcanoes in Stage 4, respectively, but they are similar to those of the Takamine volcano in Stage 2 (Fig. 11). The volcanic rocks of the Asama volcano group belong to the medium-K series and calc-alkaline rock series, but some volcanic rocks of the Kurofu volcano are low-K series and tholeiitic rock series (Fig. 12).

7 Summary

1. The Eboshi-Asama volcano group is located at the deflection point of the volcanic front of the Northeast Japan volcanic arc.
2. The Eboshi-Asama volcano group extends ca 24 km EW and ca 34 km NS, and comprises andesitic, dacitic to rhyolitic polygenetic

volcanoes with widespread pyroclastic density current deposits and monogenetic lava domes.

3. The Eboshi-Asama volcano group is divided into four stages, generally truncated by large-scale sector collapse: Stage 1 (0.45–0.32 Ma), Stage 2 (0.32–0.17 Ma), Stage 3 (0.17–0.10 Ma), and Stage 4 (<0.10 Ma). The Eboshi (0.45 Ma; 55–61 wt% SiO_2) and Kakuma-Nabebuta (0.35 Ma; 58-59 wt% SiO_2) andesitic composite volcanoes are members of Stage 1. The Yunomaru andesitic lava dome (0.28 Ma; 57–61 wt% SiO_2), Sanpogamine andesitic to dacitic composite volcano (0.32–0.22 Ma; 53–65 wt% SiO_2), and Takamine andesitic composite volcano (0.24–0.17 Ma; 55–58 wt% SiO_2) are members of Stage 2. The Mizunoto andesitic composite volcano (0.13–0.10 Ma; 56–62 wt% SiO_2), Murakami (0.12 Ma; 59–62 wt% SiO_2), and Sajiki andesitic lava domes (60–62 wt% SiO_2) are members of Stage 3. The Kagonoto andesitic lava dome (0.08–0.07 Ma; 60–62 wt% SiO_2) and Asama volcano group are members of Stage 4.
4. The Asama volcano group is made up of the Kurofu, Sennin, Hanareyama, Hotokeiwa, and Maekake volcanoes. The Kurofu volcano (0.09–0.055 Ma; 54–61 wt% SiO_2) is a large andesitic composite volcano. The Sennin (30–28 ka; 61–64 wt% SiO_2) and Maekake volcanoes (10 ka>; 58–64 wt% SiO_2) both form andesitic pyroclastic cones with lavas and large-scale pumice-fall deposits, although the Maekake volcano is associated with widely distributed pyroclastic density current deposits. The Hanareyama monogenetic volcano (24 ka; 60–75 wt% SiO_2) is an andesitic, dacitic to rhyolitic lava dome with a pyroclastic density current deposit. The Hotokeiwa volcano (23–10 ka; 52–73 wt% SiO_2) is an andesitic, dacitic to rhyolitic lava cone with large-scale pyroclastic density current deposits and pumice-fall deposits with the Koasama rhyolitic lateral lava dome.
5. The whole-rock major element compositions for eruptive products of the Eboshi volcano group are andesitic and nearly the same as those of the Asama volcano group, but those of the Asama volcano group are more silicic and abundant in dacitic and rhyolitic volcanic rocks.

Acknowledgments We are grateful to Aramaki S., Otsuka T., Hirakawa T., Mori S., and Toda Y. for their cooperation and help with the study. We also appreciate Martin A. for critical reading and revising of the manuscript.

References

Aramaki S (1963) Geology of Asama volcano. J Fac Sci Univ Tokyo Sec II XIV:229–443

Gill J (1981) Orogenic andesites and plate tectonics. Springer-Verlag, Berlin. 385p

Hayakawa Y (2010) The story behind the scenery of Asama volcano. Bull Fac Educ Gunma Univ Ser Nat Sci 58:65–81. (in Japanese with English abstract)

Higuchi K (1989) Study of volcanic activity of Asamayama. Chikuma 66:15–33. (in Japanese)

Hirose F, Nakajima J, Hasegawa A (2008) Three- dimensional seismic velocity structure and configuration of the Phillipine Sea slab in southwestern Japan estimated by double-difference tomography. J Geophys Res 113:B09315

Kaneko T, Shimizu S, Itaya T (1991) K-Ar ages of the Quaternary volcanoes in the Shin-etsu highland area, central Japan, and their formation history. Bull Earthq Res Inst Univ Tokyo 66:299–332

Miyasaka A, Kano K (2015) Pliocene to middle Pleistocene tectonics of the Komoro collapse basin in the southeastern part of North Fossa Magna, central Japan. Geosci Rep Shizuoka Univ 42:63–83. (in Japanese with English abstract)

Miyashiro A (1974) Volcanic rock series in Island arcs and active continental margin. Am J Sci 274:321–355

Nakajima J, HIrose F, Hasegawa A (2009) Seismotectonics beneath the Tokyo metropolitan area, Japan: Effect of slab-slab contact and overlap on seismicity. J Geophys Res, 114:B08309

Nakamura T, Tsuji S, Takemoto H, Ikeda A (1997) 14C age measurements with accelerator mass spectrometry of Asama tephra stratigraphic samples around Minami-Karuizawa, the latest Pleistocene, Nagano Prefecture, central Japan. J Geol Soc Japan 103:990–993. (in Japanese with English abstract)

Nishiki K, Takahashi K, Matsumoto T (2013) Spatial and temporal change of volcanic field in Asama-Eboshi volcanoes, central Japan. J Geol Soc Japan 119:474–487. (in Japanese with English abstract)

Takahashi K (2004) Geology and volcanic history of Mt. Eboshidake and its adjacent area, northeastern Nagano prefecture, central Japan. J Volcanol Soc Japan 49:83–102. (in Japanese with English abstract)

Takahashi K, Miyake Y (2004) K-Ar ages of lavas from the Mount Eboshidake, Joshin district, central Japan. J Volcanol Soc Japan 49:207–212. (in Japanese with English abstract)

Takahashi M, Yasui M, Ichikawa Y, Kamioka Y, Asaka N, Sakagami M, Tanaka E (2007) Whole-rock major element chemistry for eruptive products of Asama-Maekake volcano, central Japan. Proc Inst Nat Sci Nihon Univ 42:55–70 (in Japanese with English abstract)

Takahashi M, Nakajima T, Mukai Y, Yasui M, Kanamaru T (2008a) Major element chemistry for eruptive products of Asama-Kurofu volcano, central Japan: summary for data of 288 samples. Proc Inst Nat Sci Nihon Univ 43: 195-216 (in Japanese with English abstract)

Takahashi M, Mukai Y, Nakajima T, Yasui M, Kanamaru T (2008b) Whole-rock major element chemistry for eruptive products of Asama-Hotokeiwa volcano: summary for data of 307 samples. Proc Inst Nat Sci Nihon Univ 43:167–193 (in Japanese with English abstract)

Takahashi M, Ichikawa H, Kanamaru T, Yasui M, Maseguchi T (2013a) Proximal volcaniclastic and pyroclastic rocks of the Asama-Kurofu volcano at the Sennin-iwa located in the wall of Kurofu sector-collapsed caldera: volcanic breccia and tuff breccia of the Gippa lava group and welded pyroclastic rocks of the Sennin lava group. Proc Inst Nat Sci Nihon Univ 48:141–168 (in Japanese with English abstract)

Takahashi M, Otsuka T, Hirakawa T, Nagai M, Yasui M, Aramaki S (2013b) Whole-rock major element chemistry for eruptive products of the Eboshi volcano group, central Japan: summary of 222 analytical data. Proc Inst Nat Sci Nihon Univ 48:111–140 (in Japanese with English abstract)

Takahashi M, Kikuchi M, Orito K, Yasui M, Kanamaru T, Takemoto H (2018) Major element chemistry for clasts of volcanic rocks in the large-scale debris avalanche and volcanic mud flow deposits formed by the sector collapse of Asama-Kurofu and Asama-Sennin volcanoes: summary of 148 analytical data and the origin of clasts of volcanic rocks. Proc Inst Nat Sci Nihon Univ 53:51–75 (in Japanese with English abstract)

Takahashi M, Yasui M, Kanamaru T, Nakagawa M (2022) Eruptive volume and age of large-scale pumice eruptions of the Asama-Maekake volcano, Central Japan, as revealed by detailed trench surveys: implications for future volcanic hazards. J Disaster Res 17: 1080–1089

Takahashi M, Kanamaru T, Yasui M (2024) Whole-rock major element chemistry for eruptive products of Asama volcano group: summary of analytical data of 1,922 samples. Proc Inst Nat Sci Nihon Univ 59:47–108 (in Japanese with English abstract)

Tanabe T, Collaborative Research Group for Jo-Shin volcanic terrain (1992) Middle Pleistocene deposits around Sugadaira and Tsumagoi highland. Daishiki 25:47–56. (in Japanese with English abstract)

Tsuji S, Miyaji N, Arai F (2004) Tephrostratigraphy and chronology of Asama volcano at Minami-Karuizawa as a basis of studies on environmental and disaster history. Bull Natl Mus Jpn Hist 118:165–187. (in Japanese with English abstract)

Yasui M, Takahashi M, Kanamaru T, Nagai M (2021) High-resolution reconstruction of history of large-scale eruptions of the Asama-Maekake volcano inferred from the pyroclastic fall deposits. Bull Volcanol Soc Japan 66:293–325. (in Japanese with English abstract)

Geology of Asama-Kurofu Volcano

Masaki Takahashi, Tatsuo Kanamaru, Maya Yasui, and Masashi Nagai

1 Introduction

Kurofu volcano, the oldest member of the Asama volcano group, is the largest conical polygenetic composite volcano in the Eboshi-Asama volcano group with a basal area of ca 20 × 15 km; the original total eruption volume estimated by Aramaki (1963) is approximately 40 km^3 (Fig. 1). The Kurofu volcano was involved in the large-scale sector collapse of Sennin volcano, and the eastern half of the volcanic cone including the summit area was considerably destroyed; more than half of the original volcanic edifice was lost by the collapse. Resultantly, the internal structure of the composite volcano is well exposed on the Yunotaira caldera wall formed by the collapse (Figs. 2 and 6). The highest peak of the rim of the western wall of Yunotaira caldera is Mt. Kurofuyama (2404 m above sea level in altitude).

Aramaki (1963) studied the Kurofu volcano and grouped the lavas and pyroclastic rocks into three members stratigraphically from the lower to upper, the Gippa, Mitsuone, and Sennin groups, adding the Sekisonsan lava dome as its lateral volcano. Aramaki's study is briefly summarized as follows. The Gippa group comprises alternating thin andesitic lavas with a thickness of less than a few meters and pyroclastic rocks. The andesitic lavas are porphyritic with phenocrysts of plagioclase, clinopyroxene, orthopyroxene, and rare olivine. Large phenocrysts of clinopyroxene are a conspicuous feature of the lavas. The pyroclastic rocks consist of tuff breccia and volcanic breccia, which are more abundant than lavas in volume. The lithofacies of volcanic rocks of the Mitsuone group is similar to that of the Gippa group, but pyroclastic rocks are less abundant than in the Gippa group. The andesitic lavas are porphyritic with phenocrysts of plagioclase, clinopyroxene, orthopyroxene, and olivine. Phenocrysts of olivine are characteristic and more abundant in lavas in the Mitsuone group than in the Gippa group. The Sennin group is composed of thick lava flows with a thickness of up to 15 m and intercalated by thin layers of pyroclastic rocks. The Sennin group is distributed in the northernmost area of the caldera wall.

After Aramaki's classical study, the recent research on Kurofu volcano has revealed its geology in more detail. The results of the recent progress of geologic study are presented in this chapter.

M. Takahashi (✉)
Institute of Natural Sciences, College of Humanities and Sciences, Nihon University, Tokyo, Japan
e-mail: takahashi.masaki@nihon-u.ac.jp

T. Kanamaru · M. Yasui
Department of Earth and Environmental Sciences, College of Humanities and Sciences, Nihon University, Tokyo, Japan
e-mail: kanamaru.tatsuo@nihon-u.ac.jp; yasui.maya@nihon-u.ac.jp

M. Nagai
Volcano Disaster Resilience Research Division, National Research Institute for Earth Science and Disaster Resilience, Tsukuba, Ibaraki, Japan
e-mail: mnagai@bosai.go.jp

S. Aramaki et al. (eds.), *The Asama Volcano*, Active Volcanoes of the World,
https://doi.org/10.1007/978-981-95-5799-8_5

Fig. 1 Aerial photograph showing the Kurofu volcano viewed from the southwest (courtesy of the Asama Jomon Museum). More than half of the volcanic edifice of Kurofu volcano including the summit portion was lost by sector collapse. The Maekake and Hotokeiwa volcanoes are to the east of Kurofu volcano, and the Eboshi volcano group is to the west. The plain spreading in front of the Kurofu volcano consists of the pyroclastic density current deposits of Hotokeiwa volcano

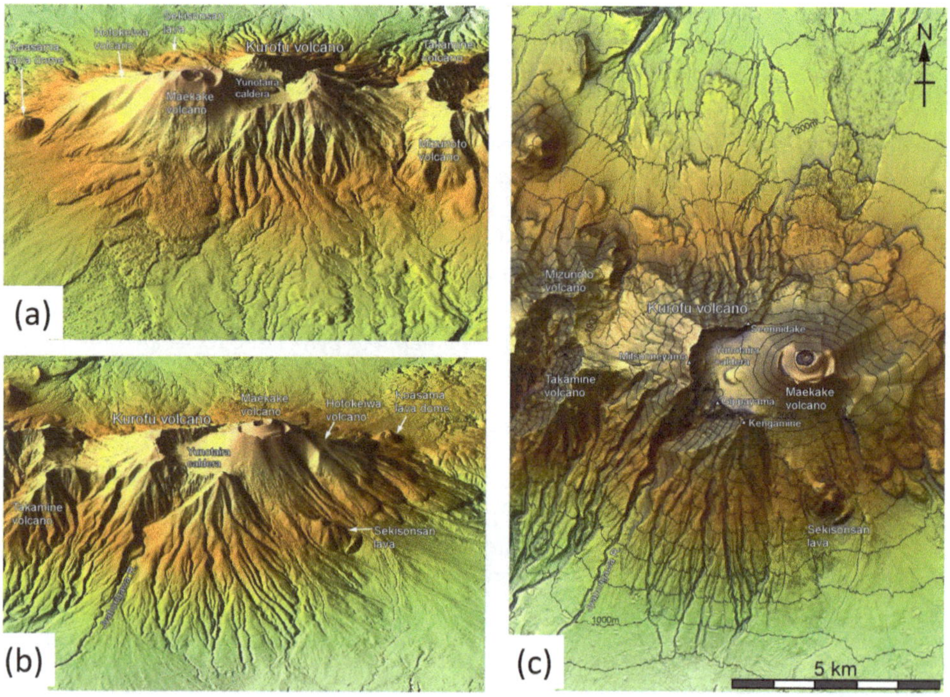

Fig. 2 Perspective view and topographic relief map of Kurofu volcano. The eastern half including the summit crater area was completely lost by the sector collapse. (**a**) Kurofu volcano viewed from the north. (**b**) Kurofu volcano viewed from the south. (**c**) Topographic relief map of the Kurofu volcano. These figures were created using Kashmir 3D (http://www.kashmir3d.com/) and DEM data provided by the Geospatial Information Authority of Japan (GSI)

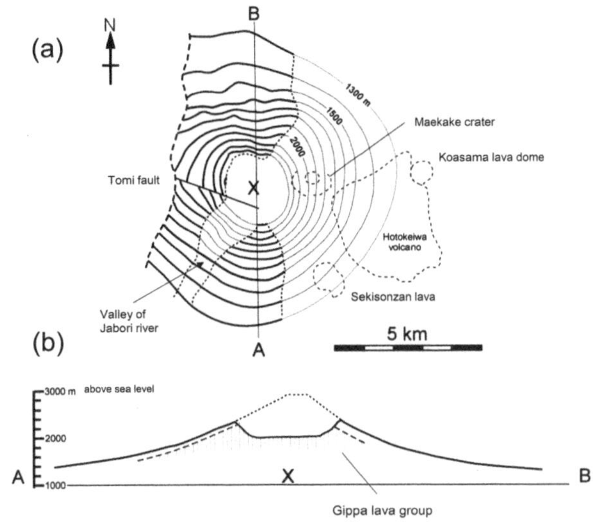

Fig. 3 Reconstructed topographic contour lines of the Kurofu volcano and its cross-section, based on Aramaki (1963). (**a**) Reconstructed topographic contour lines. (**b**) Cross section along A-B section

2 Original Shape of the Volcanic Edifice

Aramaki (1963) estimated the original shape of the volcanic edifice of Kurofu volcano. The original outline of the volcanic cone of Kurofu volcano before the destruction is restored (Fig. 3). The destroyed eastern half of the original volcanic cone is restored by extrapolating the contours to form a symmetrical shape around the eruption center. The extension toward upper restored slopes of the volcanic cone in the north-south cross-section passes the center, where the summit crater is presumed to have been present. If the diameter of the summit crater is assumed to be 500 m, the height of the cone of Kurofu volcano is approximately 2800 m above sea level.

3 History of Volcanic Activity

The age of the Gippa lava group is 0.09 Ma according to K-Ar dating (Nishiki et al. 2013). In the tephra distributed near the western foot of Haruna volcano to the east of Kurofu volcano, away from ca 20 km from the presently active Maekake volcano, no conspicuous tephra derived from the Kurofu volcano is present between the Tateyama D Pumice (DPm) (the fallout pumice of Tateyama volcano) whose age is 120,000 years BP (Machida and Arai 2003) and the Ontake-Nagawa Pumice (NgP) (the fallout pumice of Ontake volcano) the age of which is 75,000 years BP (Takemoto 1999); this provides evidence that Kurofu volcano was effusive and produced little fallout tephra tephra between 120,000 and 75,000 years BP (Takemoto 1999) (Fig. 4). However, during the period between NgP and Daisen-Kurayoshi pumice (DKP) (the fallout pumice of Daisen volcano) whose age is 55,000 years BP (Machida and Arai 2003), three thick lapilli-tuff beds, the Jizotoge, Hagiu, and Nakanojyo lapilli-tuffs, each with thickness of about a few tens of centimeters, are found, and probably derived from the Kurofu volcano by large-scale Vulcanian eruptions (Takemoto 1999); these lapilli-tuff beds with olivine crystals can be correlated with the olivine-bearing Mitsuone lava group of Kurofu volcano; it is concluded that the Mitsuone lava group is estimated to have been active between 75,000 years BP and 55,000 years BP (Fig. 4). Based on these ages, the volcanic activity of Kurofu volcano commenced ca 90,000 years BP and ceased from ca 55,000 years BP, the duration of which is approximately 35,000 years at most.

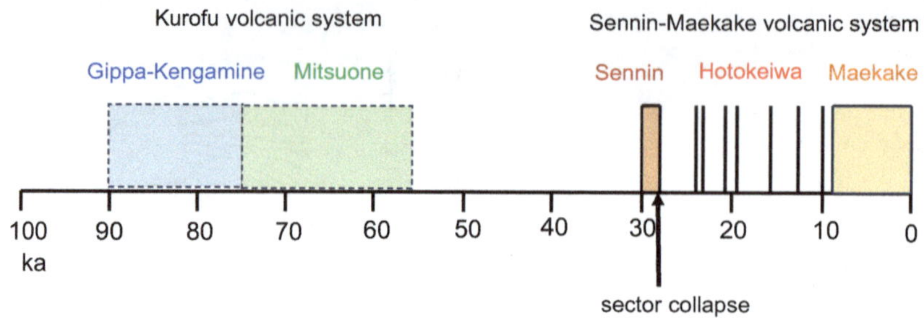

Fig. 4 Diagram showing the temporal variation of the Kurofu and Sennin-Maekake volcanic systems. The Kurofu volcanic system (90–55 ka) comprises the Gippa-Kengamine and Mitsuone lava groups, and the Sennin-Maekake volcanic system (<30 ka) consists of the Sennin, Hotokeiwa, and Maekake volcanoes. Sector collapse occurred at ca 28,000 years calBP in the last stage of volcanic activity of Sennin volcano. There was a long dormant period of approximately 25,000 years between the two volcanic systems

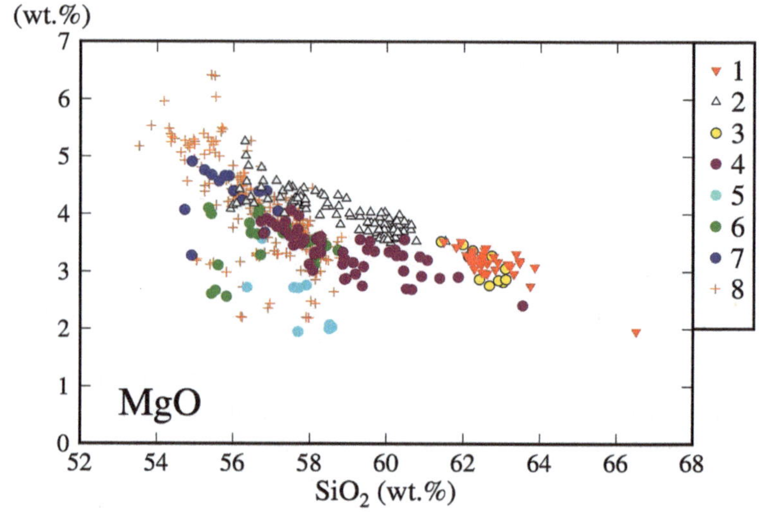

Fig. 5 MgO versus SiO_2 diagram of the lava groups of Kurofu volcano. 1: Sennin volcano. 2: Mitsuone lava group. 3: Sekisonsan lava. 4: Kengamine lava group. 5: upper Gippa lava group. 6: middle Gippa lava group. 7: lower Gippa lava group. 8: unclassified Gippa lava group

4 Lava Groups

The lavas of Kurofu volcano are classified into three groups, mainly based on whole-rock MgO contents (Fig. 5); they are, in stratigraphic order, the Gippa, Kengamine, and Mitsuone lava groups (Fig. 6). Most lavas of the Kengamine and Mitsuone lava groups belong to the Mitsuone group of Aramaki (1963). The lavas of Kurofu volcano are block and aa lavas.

4.1 The Gippa Lava Group

The Gippa lava group, the stratigraphically lowest member, occupies approximately 85 vol% of the volcanic edifice. The thickness of most lava flows is a few meters. The lavas are andesite with phenocryst contents of 30–47 vol%; phenocrysts are plagioclase, clinopyroxene, orthopyroxene, and ±olivine. The lavas characteristically contain large phenocrysts of clinopyroxene up to 0.5 cm in the major axis.

The whole-rock SiO_2 contents of lavas of the Gippa lava group range from 54 to 58 wt% (Takahashi et al. 2008, 2024). Whole-rock MgO content shows a steep decreasing trend on the MgO versus SiO_2 variation diagram (Fig. 5); as SiO_2 increases, MgO of lavas decreases remarkably from the stratigraphically lower to upper horizons (Fig. 5).

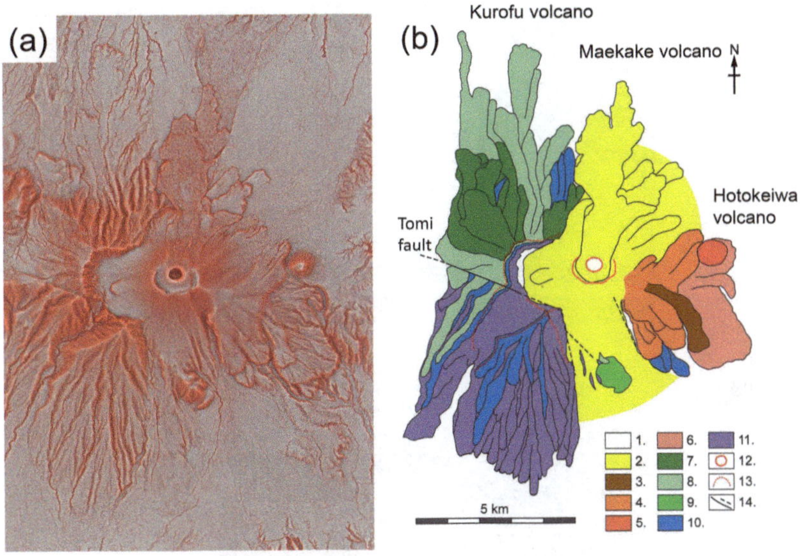

Fig. 6 Topographic relief map (**a**) and geological map (**b**) of the Kurofu volcano. 1: alluvium and volcanic rocks other than volcanic edifices of Asama volcano group. 2: eruptive products of Maekake volcano. 3: Hotokeiwa upper andesitic lava. 4: Hotokeiwa middle rhyolitic to dacitic lava. 5: Koasama rhyolitic lava dome. 6: Hotokeiwa lower rhyolitic lava. 7: lavas of Sennin volcano. 8: Mitsuone lava group of Kurofu volcano. 9: Sekisonsan lava. 10: Kengamine lava group of Kurofu volcano. 11: Gippa lava group of Kurofu volcano. 12: craters of Maekake volcano. 13: rim of collapsed caldera. 14: faults (solid line) and inferred faults (dashed line)

4.2 The Kengamine Lava Group

The volume of the Kengamine lava group, the stratigraphically middle member, is approximately 5 vol% of the volcanic edifice. The lavas are andesite with phenocryst contents of 37–43 vol%; phenocrysts are plagioclase, clinopyroxene, orthopyroxene, ±olivine.

The whole-rock SiO_2 contents of lavas of the Kengamine lava group range from 56 to 61 wt% (Takahashi et al. 2008, 2024). The MgO contents decrease as SiO_2 contents increase on the MgO versus SiO_2 variation diagram, but the decreasing trend is gentle in contrast to that of the Gippa lava group (Fig. 5).

4.3 The Mitsuone Lava Group

The Mitsuone lava group, the stratigraphically upper member, occupies approximately 10 vol% of the volcanic edifice. The lavas are andesite with phenocryst contents of 14–38 vol%; phenocrysts are plagioclase, clinopyroxene, orthopyroxene, and olivine. The olivine phenocryst content is characteristic of the lavas of Mitsuone lava group.

The whole-rock SiO_2 contents of lavas of the Mitsuone lava group range from 56 to 60 wt% (Takahashi et al. 2008; 2024). The whole-rock MgO contents of lavas of the Mitsuone lava group shows slight decrease as SiO_2 contents increase on the MgO versus SiO_2 variation diagram, just similar to the Kengamine lava group; however, the MgO content is higher than that of the Kengamine lava group. The MgO of volcanic rocks of Sennin volcano is plotted on the same variation trend as those of the Mitsuone lava group, though they are more enriched in SiO_2 than the Mitsuone lava group (Fig. 5).

4.4 The Sekisonsan Lava

The Sekisonsan lava distributes on the subsided block to the east of the eastern cliff of Mt. Kengamine (2281 m above sea level in altitude), which was the southeastern flank of Kurofu volcano (Figs. 6 and 7). Aramaki (1963) considered the Sekisonsan lava as a lateral lava dome; the

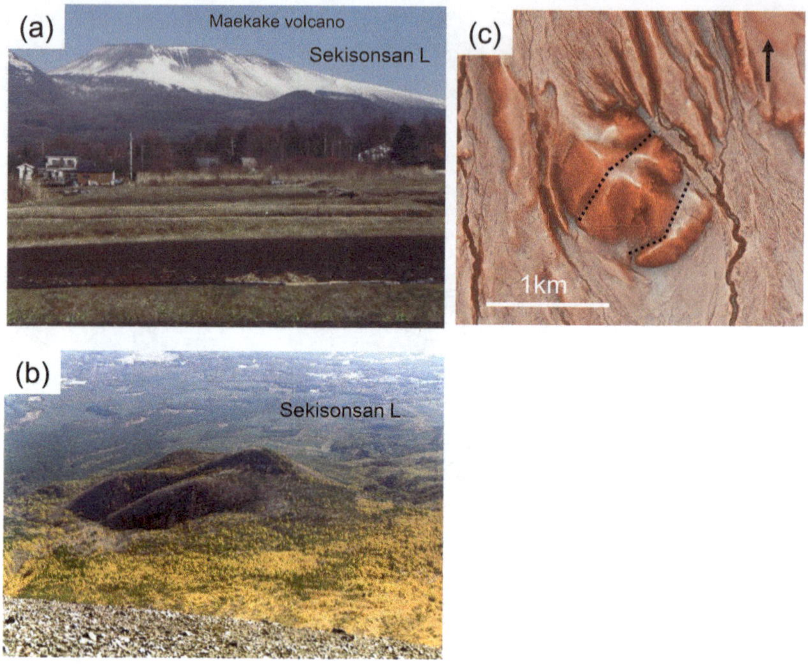

Fig. 7 Photographs and topographic relief map of the Sekisonsan lava. (**a**) Sekisonsan lava viewed from the southwestern foot of Kurofu volcano. (**b**) Sekisonsan lava viewed from the flank of Maekake volcano. (**c**) Topographic relief map of Sekisonsan lava (courtesy of T.Chiba). The Sekisonsan lava is cut by NE-SW trending faults

basal area is 1.3 × 0.9 km with an original height of 0.25 km and volume of 0.07 km^3. The Sekisonsan lava consists of a massive lava with a carapace of welded tuff breccia. The Sekisonsan lava is cut by faults trending NE to SW (Fig. 7c). The Sekisonsan lava is andesite with phenocryst contents of 36 vol%; phenocrysts are plagioclase, clinopyroxene, and orthopyroxene.

The whole-rock SiO_2 contents of Sekisonsan lava are 62–63 wt% (Takahashi et al. 2008, 2024). The whole-rock MgO contents plotted on the MgO versus SiO_2 variation diagrams align on a descending compositional trend, which is an extension of the compositional trend of the Mitsuone lava group and similar to those of the Sennin volcano (Fig. 5).

The Sekisonsan lava is underlain by lavas of the Gippa lava group and overlain by those of the Gippa and Mitsuone lava groups. There are several hypotheses regarding the origin of the Sekisonsan lava: (1) a syn-eruptive lateral lava dome of the Gippa lava group, (2) a lateral lava dome intruded into the Gippa and Mitsuone lavas, (3) a toreva block of thick lava flow erupted as a member of Sennin volcano and emplaced together with the lavas of Gippa and Mitsuone lava groups during the large-scale sector collapse of the Sennin and Kurofu volcano, (4) a toreva block of an intrusion of Sennin volcano emplaced into the volcanic edifice of Kurofu volcano, (5) a toreva block of a solidified volcanic neck that filled the conduit of Sennin volcano which was subsequently exposed at the surface and transported by a debris avalanche of the large-scale sector collapse.

5 Origin of Cone-Forming Clastic Rocks Revealed by Paleomagnetic Studies

Kurofu volcano is a polygenetic composite volcano consisting of alternating lavas and cone-forming clastic rocks. The composite volcano had been considered to be an edifice consisting of an alternation of lavas and pyroclastic rocks. However, recently most clastic rocks of composite volcanoes are thought to be not pyroclastic rocks, but autoclastic materials emplaced concurrently with lava flows, which are derived from fragmentation of lavas (Grunbensky et al. 1998; Smith et al. 1999). To reveal the origin of the cone-forming clastic rocks of Kurofu volcano, paleomagnetic measurements were made of the clastic rocks.

5.1 Cone-Forming Clastic Rocks

The cone-forming clastic rocks of Kurofu volcano are composed of clast-supported volcanic breccias and matrix-supported tuff breccias (Fig. 8). In the Sennin-iwa cliff near the northern end of the Yunotaira collapsed caldera, the alternation of lavas and cone-forming clastic rocks of the Gippa lava group can be seen in exposed outcrops (Fig. 9). Along the nearly 100 m height of the cliff, 10 andesitic lava flow sheets, the thickness of which is 2–4 m, 18 layers of volcanic breccia, and 14 layers of tuff breccia, the thickness of which is 0.5–4.5 m, are well exposed (Figs. 10 and 11). The size of clasts of volcanic breccia and tuff breccia ranges from several centimeters to 80 cm in diameter, but 5–30 cm in diameter is predominant. The volume ratios of lava, volcanic breccia, and tuff breccia are 27, 48, and 25 vol%, respectively. The volume of cone-forming clastic rocks is 73 vol% and overwhelms that of lavas in the Sennin-iwa cliff.

5.2 Paleomagnetic Study

The clasts in the volcanic breccias and tuff breccias sampled in the Sennin-iwa cliff are studied paleomagnetically. The paleomagnetic directions of clasts are projected onto equal-area stereographic diagrams (Fig. 12). The magnetizations of the clasts are classified as random or non-random types in the diagram according to Watson's test. The random type indicates that remnant magnetization was acquired prior to emplacement as the clasts which had already cooled to ambient temperature when emplaced, indicating settling and rotation of clasts after cooling or incorporation of colder clasts. The non-random type indicates acquired remnant magnetization after emplacement of clasts; the clasts had not yet cooled at the time of deposition.

Fig. 8 Photograph showing lava (L5), clast-supported volcanic breccia (V11), and matrix-supported tuff-breccia (T11) in the Gippa lava group of Kurofu volcano at the Sennin-iwa cliff

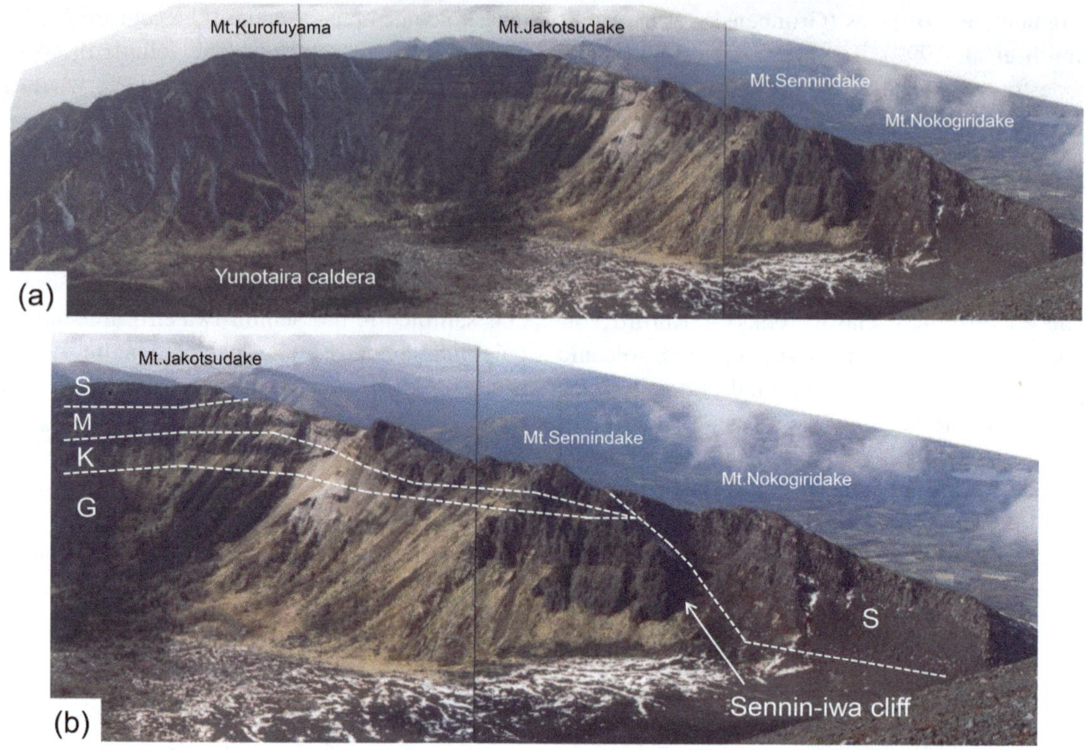

Fig. 9 Photographs of the Yunotaira collapsed caldera. (**a**) Yunotaira caldera viewed from the summit of Maekake volcano. (**b**) Close view showing the relation of eruptive products of Kurofu and Sennin volcanoes; the volcanic edifice of Kurofu volcano is overlain by that of Sennin volcano with angular unconformity. G: Gippa lava Group. K: Kengamine lava Group. M: Mitsuone lava group. S: Sennin volcano

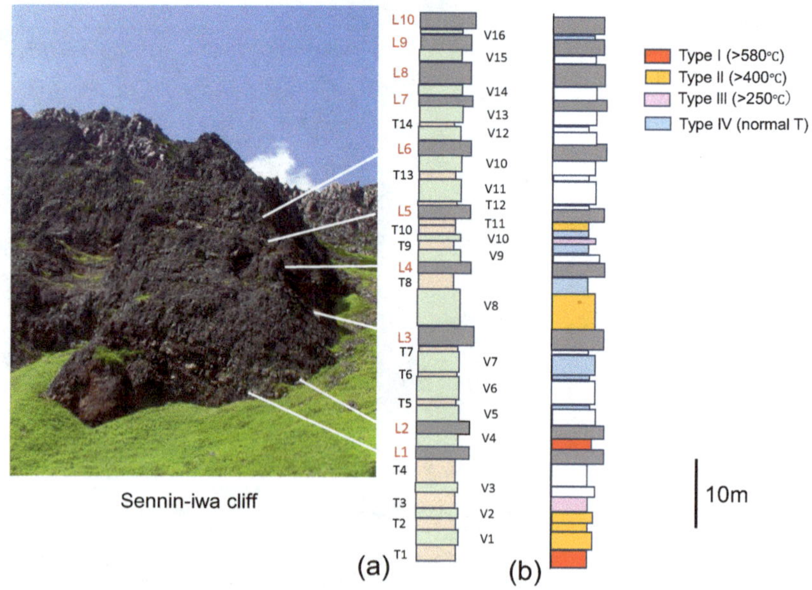

Fig. 10 Photograph of the Sennin-iwa cliff, the columnar-section of Gippa lava group (**a**) and that showing various types of clastic rocks with different temperatures (**b**). L1–L10: lava. V1–V16: volcanic breccia. T1–T14: tuff-breccia

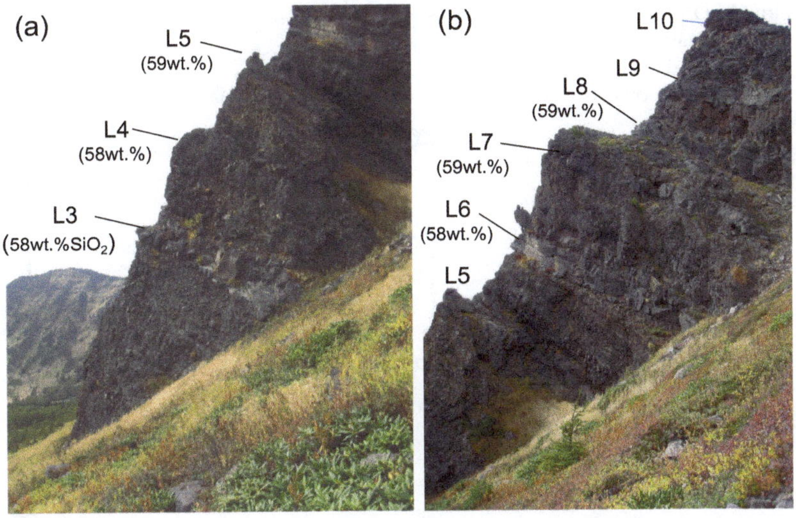

Fig. 11 Close view of the Sennin-iwa cliff showing the alternation of lavas (L) and cone-forming clastic rocks. (**a**) L3 to L5. (**b**) L5 to L10

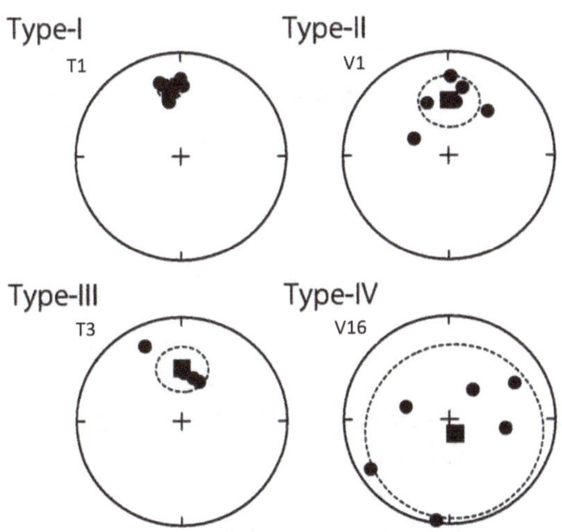

Fig. 12 Stereographic diagrams showing the paleomagnetic direction of clasts of cone-forming clastic rocks in the Sennin-iwa cliff. Type IV is a random type and others are non-random types

Progressive thermal demagnetization data of the non-random type clasts were plotted on orthogonal-demagnetization (Zijderveld) diagrams (Fig. 13); thermal demagnetization data for samples display stable single or more than two-component magnetizations. In the one-component magnetization, all data with different unblocking temperatures between room temperature and 580 °C can be fitted by a well-defined line, while in the two-component magnetization, points of high-temperature component lie on a high-temperature line with a direction different from the low-temperature component. The low-temperature line shows the acquired thermal remanent magnetization after the emplacement of a moving and rotating hot clast. The estimated emplacement temperature uses the temperature of the last point on the low-temperature line. The clastic beds can be classified based on the minimum emplacement temperature into Type-I (>580 °C), Type-II (>400 °C), Type-III (>250 °C), and Type-IV (random type magnetization, normal temperature).

5.3 Origin of Cone-Forming Clastic Rocks

Among the studied nine beds of Sennin-iwa cliff with clasts of non-random magnetization, only two beds of clastic rocks had one-component magnetization and other beds were a mixture of clasts with single- and dual-component magnetization. The percentages of numbers of each type in the cone-forming clastic rocks of Sennin-iwa cliff are as follows: Type-I:13%; Type-II: 37%; Type-III: 13%; Type-IV: 37% (Fig. 10b). The non-random magnetization type (Type-I + Type-II + Type-III) with high emplacement temperature is 63% and random magnetization type (Type-IV) with normal emplacement temperature is 37% (Fig. 10b). Type-I volcanic

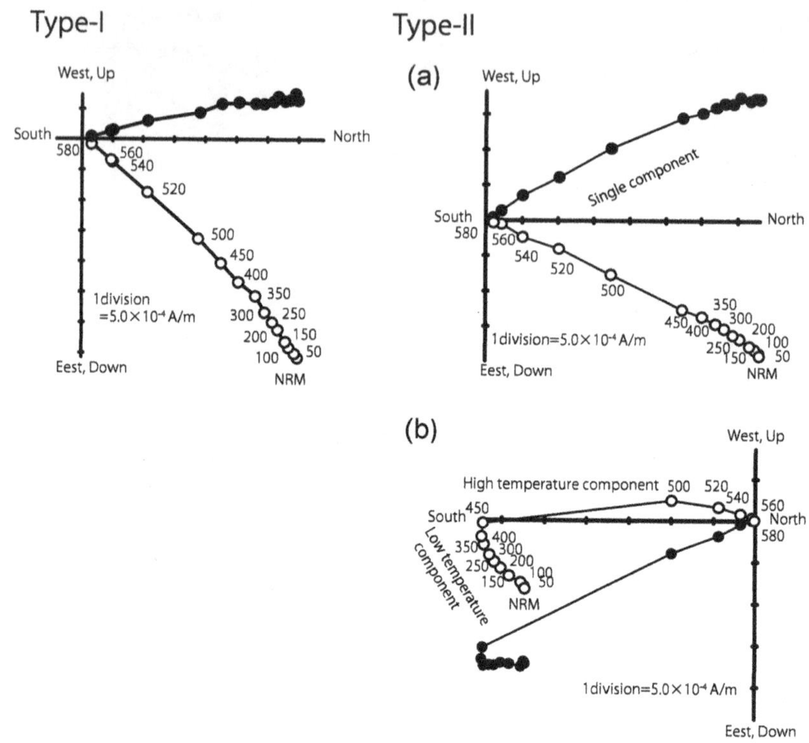

Fig. 13 Orthogonal-demagnetization (Zeidervelt) diagrams showing the thermal demagnetization of clasts of cone-forming clastic rocks in the Sennin-iwa cliff. Numerals: temperature (°C). (**a**) Diagram showing the single component clast in the Type II clastic bed. (**b**) Diagram showing the dual components clast in the Type II clastic bed

breccias were the carapace of block lavas and Type-I tuff breccias were probably high-temperature pyroclastic density current deposits. Type-IV was secondary volcaniclastic deposits consisting of cooled clasts, and 83% of them are matrix-supported tuff breccias produced by small-scale avalanches. Unstable slopes greater than approximately 30° were present in the path of block lava flows, and gravity slides of loose carapace breccia especially of its moving front were deposited far from the farthest downslope extent of the parent lava flows; most of them are not pyroclastic but autoclastic rocks (Smith et al. 1999). The crumbling autoclastic and its secondary clastic rocks tend to have accumulated thickly just like talus deposits near the place where the slope of the surface of the volcano abruptly changes from steep to gentle.

The eruption style of Kurofu volcano was probably effusive with Strombolian to Vulcanian eruptions because major cone-forming clastic rocks are not pyroclastic but autoclastic; explosive eruptions, which produce voluminous pyroclastic rocks, are essentially lacking during the construction of the volcanic edifice of Kurofu volcano.

6 Structure of the Volcanic Edifice

The internal structures of Kurofu composite volcano are well exhibited along the wall of Yunotaira collapsed caldera and the canyon along the Jabori river which breached the southern caldera wall. Particularly, the variations of lithofacies in the area near the vent to farther away from the eruption center are displayed along the valley of Jabori river (Fig. 14). Along the caldera wall and near the head of Jabori river around the southern end of the caldera floor, the cone-forming clastic rocks are overwhelming and occupies approximately 70 to 80 vol% in the alternation of lava and autoclastic and volcaniclastic rocks with minor pyroclastic rocks, which was probably where the original slope changed from steep to gentle and autoclastic and volcaniclastic rocks were accumulated to form thick piles as volcanic talus deposits

Fig. 14 Photographs showing the representative lithofacies of the internal structure of Kurofu volcano. (**a**, **b**) Cone-forming clastic rocks and lavas at the wall of Yunotaira caldera (proximal lithofacies). (**c**) Lava flows and pyroclastic or autoclastic rocks along the valley of Jabori river. (**d**) Lava flows and pyroclastic density current (PDC) or lahar deposits near the foot of Kurofu volcano (distal lithofacies)

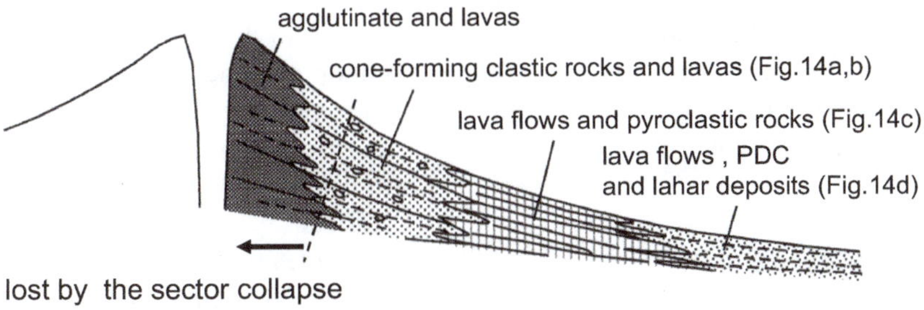

Fig. 15 Schematic cross-section of the reconstructed Kurofu composite volcano showing the internal structure of the volcanic edifice. The near vent area consists of agglutinate and lavas (at present completely lost by sector collapse), which changes to lava flows and abundant cone-forming clastic rocks. The cone-forming clastic rocks decrease and lava flows reach far from the vent area, and in the distal area near the foot of the volcanic edifice, lava flows decrease and pyroclastic density current (PDC) and lahar deposits are predominant

(Fig. 14b). The amounts of autoclastic and volcaniclastic rocks decrease away from the caldera wall along the canyon of Jabori river, and around 1.0 km from the caldera wall, lava flows with minor amounts of autoclastic, volcaniclastic, and pyroclastic rocks become dominant (Fig. 14c). Near the entrance of the valley of Jabori river about 4.5 km from the caldera wall, which is near the foot of the Kurofu volcanic cone, minor lavas are associated with pyroclastic density current and lahar deposits (Fig. 14d). A sketch of the estimated internal structure of the Kurofu composite volcanic cone is shown in Fig. 15. The near vent central portion of the cone, which is inferred to have been composed of densely welded agglutinate, is completely lacking owing to the large-scale sector collapse; blocks of the densely welded agglutinate comprising the near vent area are only found as large blocks in the debris avalanche deposit formed by the large-scale sector collapse.

7 Faults and Dikes

A few large-scale normal faults cut the volcanic edifice of Kurofu volcano (Fig. 6). The most conspicuous fault is the Tomi fault trending in the direction of WNW to ESE, which cuts the southern end of the western wall of Yunotaira caldera. The Tomi fault extends to the east and constitutes the foot of the southern wall of the caldera to the north of Mt. Kengamine. It is evident that the Tomi normal fault was active after the eruption of the Komoro 1 pyroclastic density current deposit of Hotokeiwa volcano because it is cut by the Tomi fault, but probably it had existed before the formation of Hotokeiwa volcano. The northern area of Tomi fault subsided and the net downthrow was more than 30 m (Aramaki 1963). Another fault is estimated along the foot of the eastern cliff of Mt. Kengamine which trends in the direction of NNW to SSE. Aramaki (1963) estimated another NNW to SSE trending normal fault cutting the western side of the Hotokeiwa middle lava, the cliff of which is called Hotokeiwa, but the evidence of this being a fault is ambiguous. Dikes are characteristically lacking in the volcanic edifice of Kurofu volcano; so far, a dike is reported by Aramaki (1963) in the head of Jabori river near the southern end of the caldera floor, which is an exceptional example; there is no radial dikes nor dike swarms in the volcanic edifice of Kurofu volcano.

8 Summary

1. Kurofu volcano is the largest polygenetic composite volcano in the Eboshi-Asama volcano group, and its estimated original volume is approximately 40 km^3.
2. The reconstructed shape of the original volcanic cone is symmetrical with an estimated height of the summit crater of approximately 2800 m above sea level.
3. Nearly half of the volcanic cone, including the summit area, was lost during the large-scale sector collapse of Sennin volcano.
4. Kurofu volcano consists of three lava groups: Kurofu (85 vol%), Kengamine (5 vol%), and Mitsuone (10 vol%), stratigraphically lower to upper. The three lava groups are classified based on the whole-rock MgO content on the SiO_2 variation diagram.
5. The ages of the Gippa and Kengamine lava groups are 0.09–0.075 Ma, and that of the Mitsuone lava group is 0.075–0.055 Ma. Kurofu volcano was constructed during a relatively short time span of less than 35,000 years.
6. The eruption style of Kurofu volcano was effusive, Strombolian to Vulcanian with no associated voluminous fallout pumice and pyroclastic density current deposits.
7. Most cone-forming clastic rocks are not pyroclastic density current deposits or fallout rocks but autoclastic breccias and their secondary volcaniclastic rocks, according to paleomagnetic measurements.
8. The lithofacies comprising the composite cone of Kurofu volcano in cross-section from the near vent area toward the margin near the foot of volcanic edifice is estimated as follows: (1) agglutinate and lavas in the near vent area; (2) abundant autoclastic volcanic breccias and tuff breccias with lower volumes of pyroclastic rocks and intercalated lava flows; (3) mainly lava flows with minor amounts of pyroclastic, autoclastic, and volcaniclastic rocks; (4) pyroclastic density current deposits, lahar deposits and rare lava flows at the foot of the volcanic edifice.
9. The volcanic edifice of Kurofu volcano is cut by several normal faults, the representative of which is the Tomi fault.
10. Dikes are rare in the volcanic edifice of Kurofu volcano; there are no radial dikes or dike swarms.

Acknowledgments We are grateful to Aramaki, S., Takemoto, H., Nakajima, T., Mukai, Y., Kikuchi, M., Kobayashi, T., Ichikawa, H., Orito, K., Miura, M., Ogawa, T., Makishima, K., Horiba, T., Mera, G., and Hagino, R. for their cooperation and help with the study. We also appreciate Martin, A. for critical reading and revising of the manuscript.

References

Aramaki S (1963) Geology of Asama volcano. J Fac Sci Univ Tokyo Sec II XIV:229–443

Grunbensky MJ, Smith GA, Geisman JW (1998) Field and paleomagnetic characterization of lithic and scoriaceous breccia at Pleistocene Broken Top volcano, Oregon Cascades. J Volcanol Geotherm Res 83:93–114

Machida H, Arai F (2003) Atlas of tephra in and around Japan. University of Tokyo Press, Tokyo, 336p (in Japanese)

Nishiki K, Takahashi K, Matsumoto T (2013) Spatial and temporal change of volcanic field in Asama-Eboshi volcanoes, central Japan. J Geol Soc Japan 119:474–487. (in Japanese with English abstract)

Smith GA, Grunbensky MJ, Geismann JW (1999) Nature and origin of cone-forming volcanic breccias in the Te Jerenga formation, Ruapehu, New Zealand. Bull Volcanol 61:64–82

Takahashi M, Nakajima T, Mukai Y, Yasui M, Kanamaru T (2008) Major element chemistry for eruptive products of Asama-Kurofu volcano, central Japan: summary for data of 288 samples. Proc Inst Nat Sci Nihon Univ 43:195–216. (in Japanese with English abstract)

Takahashi M, Kanamaru T, Yasui M (2024) Whole-rock major element chemistry for eruptive products of Asama volcano group: summary of analytical data of 1,922 samples. Proc Inst Nat Sci Nihon Univ 59: 47–108 (in Japanese with English abstract)

Takemoto H (1999) Paleoenvironmental change and volcanic activities in the northwestern part of northern Kanto. Doctoral Thesis, Graduate School of Science and Technology, Ibaraki University. (in Japanese with English abstract)

Geology of Asama-Sennin Volcano

Masaki Takahashi, Maya Yasui, Tatsuo Kanamaru, and Masashi Nagai

1 Introduction

The Sennin volcano was previously thought as a member of Kurofu volcano (Aramaki 1963), but now it is considered to be a separate volcanic edifice and not part of Kurofu volcano (Fig. 1). The style of eruption of Sennin volcano is explosive and contrasts to the predominantly effusive eruption style of Kurofu volcano. The eruption style of the Asama volcano group had thus changed from effusive to explosive since the formation of Sennin volcano. It is difficult to estimate accurately the eruptive volume of Sennin volcano because most of the volcanic edifice was lost by a large-scale sector collapse. The geology of Sennin volcano, the eruption style of which is unique in the Asama volcano group, is overviewed in this chapter.

M. Takahashi (✉)
Institute of Natural Sciences, College of Humanities and Sciences, Nihon University, Tokyo, Japan
e-mail: takahashi.masaki@nihon-u.ac.jp

M. Yasui · T. Kanamaru
Department of Earth and Environmental Sciences, College of Humanities and Sciences, Nihon University, Tokyo, Japan
e-mail: yasui.maya@nihon-u.ac.jp; kanamaru.tatsuo@nihon-u.ac.jp

M. Nagai
Volcano Disaster Resilience Research Division, National Research Institute for Earth Science and Disaster Resilience, Tsukuba, Ibaraki, Japan
e-mail: mnagai@bosai.go.jp

2 History of Volcanic Activity

The onset of volcanic activity of Sennin volcano was ca 30,000 years calBP based on tephra-stratigraphy (Tsuji et al. 2004). This was following a dormant period lasting approximately 25,000 years after the cessation of volcanic activity of Kurofu volcano. The volcanic rocks of Sennin volcano underlain those of Kurofu volcano with a sharp unconformity (Fig. 2). The eruption of Sennin volcano was explosive and characterized by sub-Plinian eruptions; its eruptive products comprised a densely welded pyroclastic cone, lava flows, and fallout pumice deposits. The volcanic edifice of Sennin volcano collapsed together with the edifice of Kurofu volcano at ca 28,000 years calBP (Higuchi 1989). The sector collapse was followed by voluminous sub-Plinian fallout pumice eruptions without any time gaps, resulting in a large-scale debris avalanche, which flowed down along the Agatsuma and Tone rivers and reached the Maebashi city district in the Kanto plain ca 50 km to the east of Sennin volcano (Takemoto and Kubo 2003; Yoshida and Sugai 2006, 2007; Yasui et al. 2019). The main portion of the volcanic edifice of Sennin volcano including its eruption center was completely lost by the sector collapse, and the only remnants of the volcanic edifice can be found on the northwestern slope of Kurofu volcano (Fig. 1). The volcanic activity of Sennin vol-

S. Aramaki et al. (eds.), *The Asama Volcano*, Active Volcanoes of the World,
https://doi.org/10.1007/978-981-95-5799-8_6

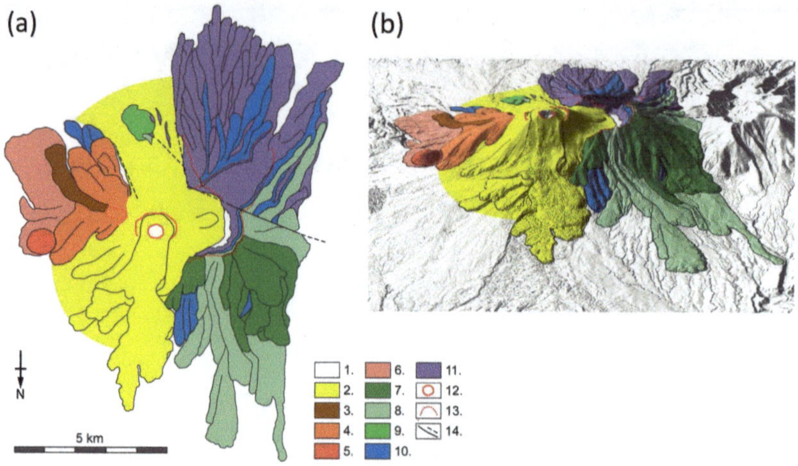

Fig. 1 Geologic sketch map of Sennin volcano. The present volcanic edifice of Sennin volcano is distributed on the northern flank of Kurofu volcano. (**a**) Geologic sketch map. (**b**) Perspective view of the geologic map of Sennin volcano from the north. 1: alluvium and volcanic rocks other than volcanic edifices of Asama volcano group. 2: eruptive products of Maekake volcano. 3: Hotokeiwa upper andesitic lava. 4: Hotokeiwa middle rhyolitic to dacitic lava. 5: Koasama rhyolitic lava dome. 6: Hotokeiwa lower rhyolitic lava. 7: Lavas of Sennin volcano. 8: Mitsuone lava group of Kurofu volcano. 9: Sekisonsan lava. 10: Kengamine lava group. 11: Gippa lava group. 12: craters of Maekake volcano. 13: caldera rim formed by sector collapse. 14: faults (solid line) and inferred faults (dashed line). These figures were created using Kashmir 3D (http://www.kashmir3d.com/) and DEM data provided by the Geospatial Information Authority of Japan (GSI)

Fig. 2 Photographs showing a section of the densely welded pyroclastic cone of Sennin volcano at the northern end of the wall of Yunotaira collapsed caldera. Five densely welded lava-like layers are exposed on the wall, which overlay the volcanic rocks of the Gippa lava group with angular unconformity. (**a**) Densely welded pyroclastic cone of Sennin volcano viewed from the east. (**b**) Close view of the internal structure of a densely welded pyroclastic cone of Sennin volcano, exhibiting five lava-like layers from WP1 to WP5

cano ceased after this large-scale sector collapse producing voluminous debris avalanche deposits and the eruption of large-scale fallout pumice. The duration of volcanic activity of Sennin volcano was short with a time span of less than 2000 years.

3 Densely Welded Pyroclastic Cone

A section of a densely welded pyroclastic cone is well preserved in the wall of Yunotaira caldera formed by the sector collapse (Takahashi et al. 2013). In the caldera wall to the north of Sennin-iwa cliff with a height of nearly 100 m, at least five lava-like sheets of densely welded pyroclastic rocks and intercalated volcanic breccias are exposed (Fig. 2). The thickness of each lava-like sheet is more than 10 m (WP1), 28 m (WP2), 26 m (WP3), 9 m (WP4), and more than 1 m (WP5), respectively (Fig. 3). In the top of the cliff, a fallout pumice deposit with a thickness of 60 cm is emplaced, the upper half of which is flattened and grades to upper densely welded compact rock (a welded fallout pumice bed) with a thickness of more than 1 m (Fig. 4). The densely welded pyroclastic rocks grade from volcanic breccias composed of weakly agglutinated clinker-like blocky spatter, and include black glassy lenses with reddish oxidized matrix (Fig. 5); eutaxitic texture and fragmented crystals are observed under the microscope (Fig. 6).

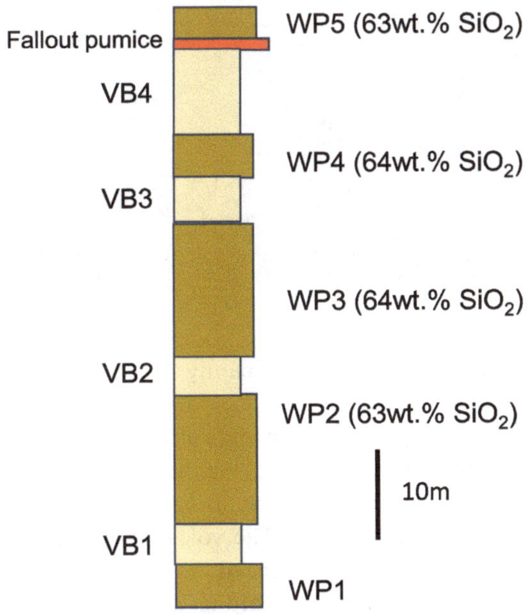

Fig. 3 Columnar section of the densely welded pyroclastic cone on the cliff-wall of Fig. 2. WP: lava-like densely welded pyroclastic rock. VB: volcanic breccia

As there are no erosional gaps or paleosols between the sheet of densely welded pyroclastic rocks and intercalated volcanic breccias, the deposition of volcanic rocks was continuous and the welded pyroclastic cone was constructed during a short time span by a single large-scale eruption just like the Kamayama or Maekake pyroclastic cone of Maekake volcano (Yasui and Takahashi 2015).

The welded pyroclastic rocks are andesite with phenocrystic content of 42–45 vol%; phenocrysts are plagioclase, orthopyroxene, and clinopyroxene. The whole-rock SiO_2 contents are 63–64 wt% (Takahashi et al. 2008) (Fig. 7).

4 Lava Flows

The andesitic lava flows of Sennin volcano are distributed on the slope of Kurofu volcano near Mt. Jakotsudake (2366 m above sea level in altitude) in the rim of the collapsed caldera, which moved down to the north and reached the northern foot of Kurofu volcano up to ca 3 km from the present caldera rim (Fig. 1). The lavas are up to 50 m in thickness. The phenocrysts of andesitic lavas are broken and fragmented, showing characteristic texture of clastogenic lava microscopically. Aramaki (1963) considered the lavas as members of the Mitsuone Group, but the whole-rock chemical compositions are nearly the same as those of the welded pyroclastic cone of Sennin volcano (Fig. 7).

The lavas are andesite with phenocryst content of 42–45 vol%; phenocrysts are plagioclase, orthopyroxene, and clinopyroxene. The whole-rock SiO_2 contents are 61–63 wt% (Takahashi et al. 2008) (Fig. 7).

5 Fallout Pumice Deposits

The fallout pumice deposits erupted from Sennin volcano are called the Itahana brown pumice (As-BP). The columnar section of an outcrop near Karuizawa station ca 16 km southwest from the present summit crater of Maekake volcano is shown in Fig. 8. A tuff with a thick-

Fig. 4 Photographs showing the welded fallout pumice deposit at the top of the caldera wall near the Sennin-iwa cliff. (**a**) Fallout pumice deposit and lava-like densely welded pyroclastic rocks (WP5). (**b**) Flattened pumice layer grades into densely welded pyroclastic rocks. (**c**) Close view of the flattened pumice layer. (**d**) Non-welded fallout pumice deposit with angular pumice clast

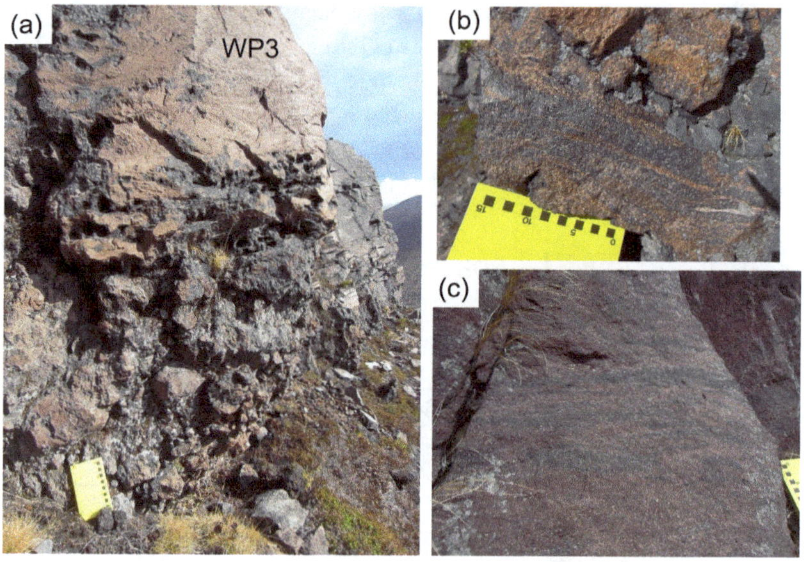

Fig. 5 Photographs showing a volcanic breccia deposit and lava-like densely welded pyroclastic rock (WP3). (**a**) Volcanic breccia deposit grades into the lava-like densely welded pyroclastic rock. (**b**) Black glassy lenses in reddish oxidized matrix of densely welded pyroclastic rock. (**c**) Fiamme of black glass in a reddish matrix of densely welded pyroclastic rock

ness of 10 cm (AT tuff) which erupted at ca 30,000 years calBP from the Aira caldera, southern Kyushu, is present at the bottom of the columnar section. The As-BP fallout pumice comprises six units from As-BP 1 (the Murota Pumice) to As-BP 6 intercalated by thin beds of paleosol. The eruption age of As-BP 6 is estimated to be ca 28,000 years calBP, concurrent with the large-scale sector collapse and debris avalanche deposits, indicating that the sector collapse was similar to the collapse at Mt. St. Helens triggered by a magma eruption but lacking pyroclastic density current deposits. The debris avalanche deposit is overlain by the As-BP 6 tuff member. The volcanic event at ca 28,000 years commenced with large-scale Vulcanian eruptions resulting in deposition of fallout tuff beds, which was followed by the

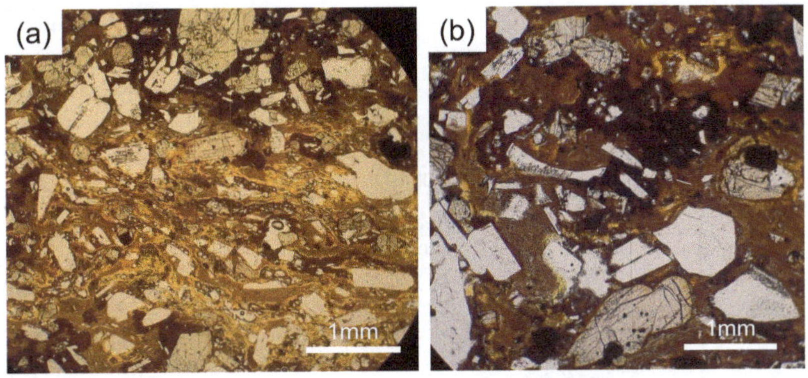

Fig. 6 Photomicrographs of the lava-like densely welded pyroclastic rock. Eutaxitic texture is conspicuous and crystals are fragmented. (**a**) Eutaxitic texture is conspicuous in groundmass. (**b**) Plagioclase phenocrysts are broken to pieces showing fragmented outlines

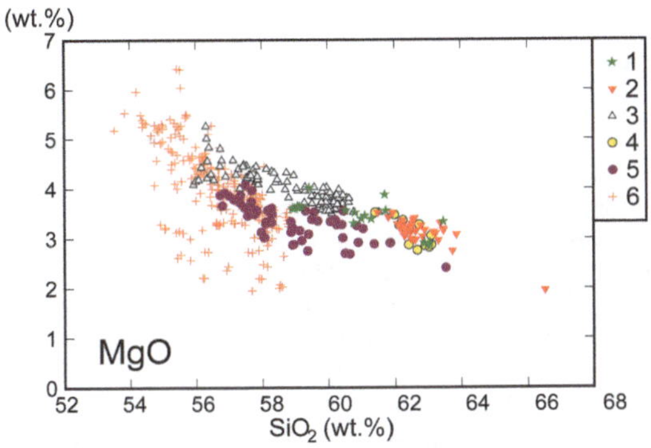

Fig. 7 MgO versus SiO_2 diagram for the eruptive products of Kurofu and Sennin volcanoes. 1: As-BP. 2: Sennin volcano. 3: Mitsuone lava group. 4: Sekisonsan lava. 5: Kengamine lava group. 6: Gippa lava group

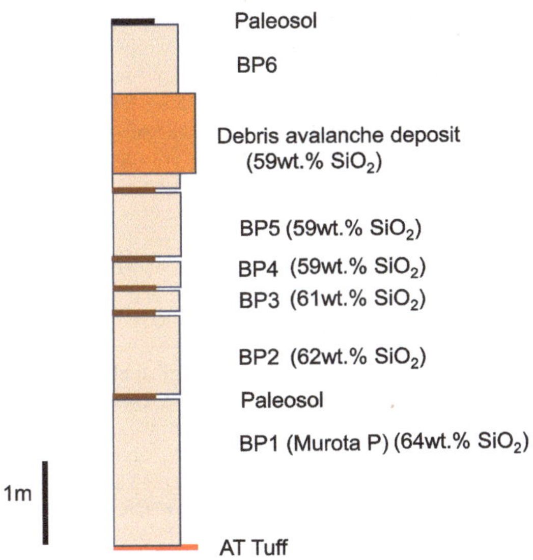

Fig. 8 Columnar section of As-BP at Karuizawa station ca 16 km to the ESE of the summit crater of Maekake volcano

large-scale sector collapse and the deposition of sub-Plinian fallout pumice beds.

The whole-rock SiO_2 contents of pumice clasts decrease from As-BP 1 to As-BP 6; As-BP 1 is 64 wt%, As-BP 2 is 62 wt%, As-BP 3 is 61 wt%, As-BP 4 is 59 wt%, and As-BP 5 is 59 wt% (Fig. 8). The pumice clasts included in the debris avalanche deposit are 59 wt% SiO_2. The formation of remnant densely welded pyroclastic cone and lava flows is thought to be contemporaneous with the eruptions of As-BP 1 to As-BP 3, as their whole-rock chemical compositions are nearly the same and they can be correlated with each other by whole-rock chemistry.

6 Volcanic Activity Associated with Sector Collapse

The details of volcanic activity followed by the large-scale sector collapse are displayed in the core samples obtained by bore-hole drilling at Manzanbo located to the east of Maekake volcano ca 7 km from the summit crater (Yasui et al. 2019). As-BP 6 in the drilling core sample is underlain by the debris avalanche deposit with no evidence of time-gap between the two deposits. As-BP 6 comprises fallout pumice and tuff (Fig. 9); the pumice bed is produced by sub-Plinian eruption, and tuff beds by Vulcanian eruption. The tuff beds are abundant in the early to middle stages. The whole-rock SiO_2 contents of pumice clasts decrease from 61 to 58 wt% from the bottom upward.

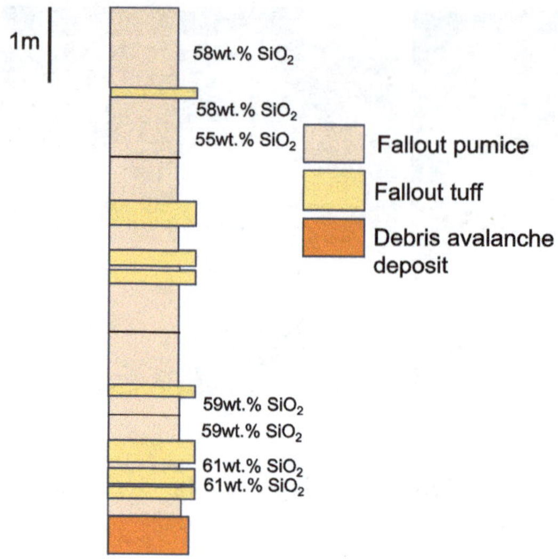

Fig. 9 Columnar section of As-BP6 from bore-hole core drilled at the Manzanbo, Karuizawa town, ca 7 km ESE of the summit crater of Maekake volcano

7 Summary

1. The Sennin volcano commenced its volcanic activity ca 30,000 years BP after a dormant period of approximately 25,000 years.
2. The Sennin volcano was explosive (mainly sub-Plinian), and erupted a densely welded pyroclastic cone, lava flows, and fallout pumice deposits; the fallout pumice deposit (As-BP) comprises at least six units intercalating thin paleosols between each unit.
3. The volcanic rocks of Sennin volcano are andesite with 55–64 wt% SiO_2. The whole-rock chemical compositions are plotted on the same variation trend as those of the Mitsuone lava group of Kurofu volcano in SiO_2 variation diagrams, though they are more silicic.
4. The sector collapse of the volcanic edifice of Sennin volcano occurred at ca 28,000 years calBP, which also involved the volcanic edifice of Kurofu volcano. The large-scale debris avalanche deposits were produced by the collapse. The collapse was associated with an eruption of large-scale sub-Plinian fallout pumice deposits (As-BP 6), similar to that of Mount St. Helens but without formation of pyroclastic density current deposits.
5. The duration of volcanic activity of Sennin volcano is relatively short, being less than 2,000 years.

Acknowledgments We are grateful to Aramaki, S., Takemoto, H., Nakajima, T., Masuda, T., Sakumoto, M., and Kuwano, N. for their cooperation and help with the study. We also appreciate Martin, A. for critical reading and revising of the manuscript.

References

Aramaki S (1963) Geology of Asama volcano. J Fac Sci Univ Tokyo Sec II XIV:229–443

Higuchi K (1989) Study of the history of volcanic activity of Asamayama. Chikuma 63:15–33. (in Japanese)

Takahashi M, Nakajima T, Mukai Y, Yasui M, Kanamaru T (2008) Major element chemistry for eruptive products of Asama-Kurofu volcano, central Japan: summary for data of 288 samples. Proc Inst Nat Sci Nihon Univ 43:195–216. (in Japanese with English abstract)

Takahashi M, Ichikawa H, Kanamaru T, Yasui M, Maseguchi T (2013) Proximal volcaniclastic and pyroclastic rocks of the Asama-Kurofu volcano at the Sennin-iwa located in the wall of Kurofu sector collapsed caldera: volcanic breccia and tuff breccia of the Gippa lava group and welded pyroclastic rocks of the Sennin lava group. Proc Inst Nat Sci Nihon Univ 48:141–168. (in Japanese with English abstract)

Takemoto H, Kubo S (2003) Tephra stratigraphy of the Okuwa debris avalanche deposit of Asama volcano. Proc Inst Nat Sci Nihon Univ 38:55–64. (in Japanese)

Tsuji S, Miyaji N, Arai F (2004) Tephrostratigraphy and chronology of Asama volcano at Minami-Karuizawa as a basis of studies on environmental and disaster history. Bull Natl Mus Jpn Hist 118:165–187. (in Japanese with English abstract)

Yasui M, Takahashi M, Kanamaru T (2019) Rediscovery of the Higuruma debris avalanche deposit in Asama volcano. Proc Inst Nat Sci Nihon Univ 54:123–142. (in Japanese with English abstract)

Yoshida H, Sugai T (2006) Morphological characteristics of hummocks originating from the 24ka sector collapse event of Asama volcano, central Japan. J Geogr 115:638–646. (in Japanese with English abstract)

Yoshida H, Sugai T (2007) Topographical control of large-scale sediment transport by a river valley during the 24ka sector collapse of Asama volcano, Japan. Geomorphol Relief Process Environ 13:217–224

Multiple Sector Collapses of Asama-Sennin and Asama-Kurofu Volcanoes and Associated Large-Scale Debris Avalanche Deposits

Maya Yasui, Masaki Takahashi, Tatsuo Kanamaru, and Masashi Nagai

1 Introduction

An east-facing horseshoe-shaped collapsed topography, known as the Yunotaira Caldera, can be found on the west side of Maekake volcano. The highest point of the cliff of the caldera is the summit of Mt. Kurofu (2404 m) (Fig. 1a). In the caldera wall, the eruptive products of Sennin volcano unconformably overlie that of Kurofu volcano (Aramaki 1963); however, only a few traces of Sennin volcano remain in the caldera walls and flank slopes. Debris avalanche deposits generated by the sector collapse of Sennin volcano are distributed at the foot of Kurofu volcano. The sector collapse was so large that associated volcanic mudflow deposits reached more than 123 km away from Kurofu volcano (Yoshida and Sugai 2006). Aramaki (1963) estimated that the original conical volcanic edifice reached an altitude of more than 2800 m prior to the sector collapse. In the eruptive history shown in chapter "Geology of Asama-Sennin Volcano", the sector collapse is estimated to have occurred around 28 ka ago in the late stage of Sennin volcano which was before the start of volcanic activity at Hotokeiwa volcano. The debris avalanche deposits are thickly covered by the eruptive products of Hotokeiwa and Maekake volcanoes.

After confirming buried debris avalanche deposits at more than 30 locations and plotting these locations on a map in this study (Fig. 2), it was found that the distribution is much wider than previously thought. It is also widely distributed on the west side, opposite the Yunotaira caldera mentioned above. Plural collapsed depressions are found in the volcanic edifice, and many normal faults are also recognized that cut through the edifice. Here, we describe the occurrence and distribution of debris avalanche deposits from the proximal area to far away. And then, we mention the estimated amount of collapse and reconstruction of the volcanic edifices. The trigger for the large-scale sector collapse at Sennin Volcano will also be discussed. Although the

M. Yasui (✉) · T. Kanamaru
Department of Earth and Environmental Sciences, College of Humanities and Sciences, Nihon University, Tokyo, Japan
e-mail: yasui.maya@nihon-u.ac.jp; kanamaru.tatsuo@nihon-u.ac.jp

M. Takahashi
Institute of Natural Sciences, College of Humanities and Sciences, Nihon University, Tokyo, Japan
e-mail: takahashi.masaki@nihon-u.ac.jp

M. Nagai
Volcano Disaster Resilience Research Division, National Research Institute for Earth Science and Disaster Resilience, Tsukuba, Ibaraki, Japan
e-mail: mnagai@bosai.go.jp

S. Aramaki et al. (eds.), *The Asama Volcano*, Active Volcanoes of the World,
https://doi.org/10.1007/978-981-95-5799-8_7

Fig. 1 Topography centered around Yunotaira caldera (**a**) and a photo of the northern caldera wall (**b**). This figure was created using Kashmir 3D (http://www.kashmir3d.com/) and DEM data provided by the Geospatial Information Authority of Japan (GSI). Open star: Summit of Mt. Kurofu (2404 m ASL) Y: Yunotaira caldera

exact duration of period is uncertain, repeated sector collapses occurred prior to the formation of the Hotokeiwa Volcano. The most recent collapse, at approximately 28 ka, was a major event involving the Sennin Volcano and the underlying Kurofu Volcano. The period described above may be regarded as an "era of major collapses" of Asama volcano group.

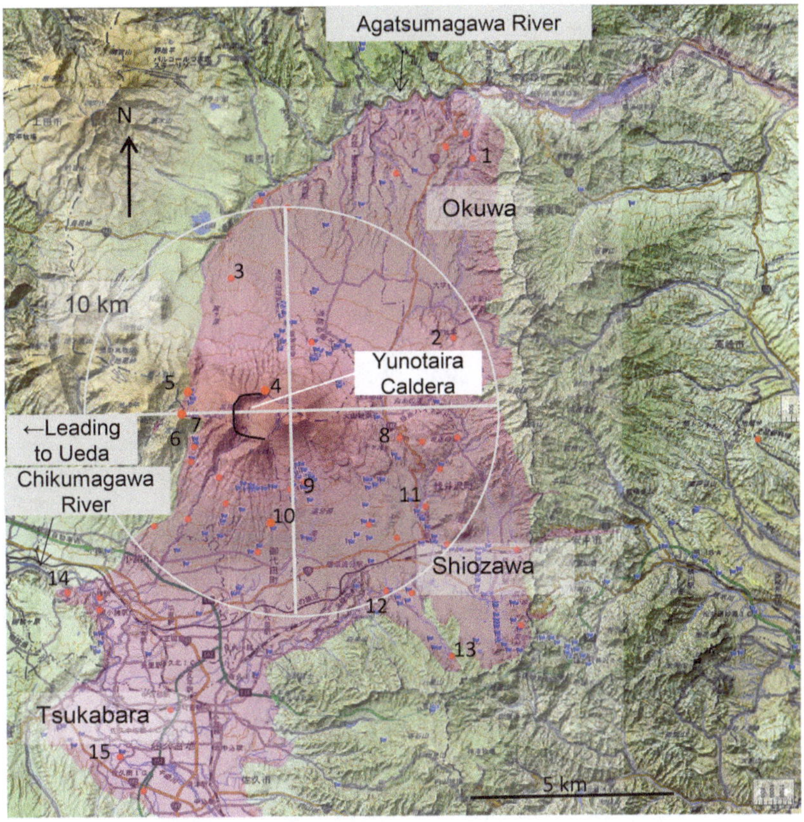

Fig. 2 Confirmed locations of debris avalanche deposits and their estimated distribution areas. The red circles are the localities where debris avalanche deposits were confirmed in this study. Blue symbols indicate other survey localities. Location numbers mentioned in the manuscript are also indicated on the map. This figure was created using Kashmir 3D (http://www.kashmir3d.com/) and DEM data provided by the Geospatial Information Authority of Japan (GSI)

2 Geological Features on the Collapsed Wall of Kurofu and Sennin Volcanoes

The cliffs of Yunotaira Caldera (Fig. 1b) expose sections of Kurofu volcano, which was active until about 50 ka, and Sennin volcano, which was active after 30 ka, with an erosion gap in between (chapter "Geology of Asama-Kurofu Volcano"). Kurofu volcano is mainly composed of an alternation of layers of pyroclastic rocks, predominantly volcanic breccia and relatively thin lava flows, and based on differences in stratigraphy and whole rock chemical composition, it is divided into three groups: the Gippa, Kengamine, and Mitsuone Groups. Remarkably altered rocks can be seen on the NNW crater wall called Shirozore (Fig. 1b). On the other hand, Sennin volcano is composed of pumice fall deposits and lava flows, and the remains of a pyroclastic cone made of strongly welded pyroclastic rock which can be seen at Nokogiri-dake (Fig. 1b). The Itahana Brown pumice (BP) are pumice fall deposits of Sennin volcano. Some of it directly overlie the debris avalanche deposits on the SE foot (e.g., Fig. 3f).

Fig. 3 Photo of debris avalanche deposit. (**a**) Flow mounds topography in the Okuwa area, Loc.1. (**b**) A flow mound at Loc.3. The debris avalanche deposit is covered by YPk pumice fall deposit of Hotokeiwa volcano. KM: Kamayama, M: Maekake volcano, K: Kurofu volcano, (**c**) Jigsaw crack observed in a megablock at nearby Loc.11, (**d**) a megablock with a length of 51 m and a height of 3 m or more at Loc.7, (**e**) Jigsaw crack seen in a megablock at Loc.6, (**f**) matrix of heterogeneous debris avalanche at Loc.11. The deposit is directly covered by the BP pumice fall deposit of Sennin volcano

3 Occurrence and Distribution of Debris Avalanche Deposits

In this study, debris avalanche deposits have been confirmed to be distributed in almost all directions around Kurofu and Sennin volcanoes (Fig. 2). Debris avalanche deposits are buried by younger deposits (e.g., Fig. 4b). However, the thickness of the overlying deposits varies depending on location. Over a wide area on the southern and northern foot of the present Maekake volcano, they are covered by pumice flow deposits from Hotokeiwa volcano, forming a flat plateau when completely buried by the pyroclastic flow deposits (e.g., Fig. 7c). On the SSW foot, according to local inhabitants' memories, the Tsukabara area (Fig. 2) was flattened by excavation since the 1960s, but before that, the original undulated surface was well preserved. This is because the cover of the pumice flow deposits is thinner than in the north. In the Okuwa area on the north-northeast foot and the Shiozawa area on the southeast foot (Fig. 2), black soil and eruptive products from Maekake and Hotokeiwa volcanoes cover the debris avalanche deposits, but this area is generally undulated. Where pyroclastic density current deposits from Maekake volcano are not detected, pyroclastic fall deposits cover the surface with mantle bedding; therefore, the undulations rich in flow mounds remain (e.g., Figs. 3a and 5f Meoto-iwa) (Yasui et al. 2019).

Fig. 4 Photo of debris avalanche deposit at Loc.5 (Takamine highland). (**a**) Whole view of the trench excavation site. Arrow: Excavated meter-sized blocks. (**b**) Outcrop and the trench. X: Black soil of Maekake volcano stage. A pumice fall layer (UB) was detected in the soil. Below it, glass shards of K-Ah were detected in the soil. *K-Ah is one of the useful widespread tephra in Japan that erupted from Kikai caldera, Southern Kyushu in 7.3 ka ago. Y: Pumice flow deposit of ca. 12 ka from Hotokeiwa volcano. **^{14}C, cal YBP, Z: Debris avalanche deposit (DA). (**c**) Heterogeneous matrix of the DA deposit. (**d**) Various blocks excavated from the DA. (**e**) Hand specimen of a phenocryst-rich block. Coarse clinopyroxene phenocrysts are conspicuous

Three main characteristics of the occurrence of the debris avalanche deposits in the field are summarized as follows: (1) Most of the blocks contained in the deposit are subangular to subrounded-shaped, gray, dense lava fragments (Figs. 4d and 6d). There are also lava fragments containing coarse pyroxene phenocrysts several millimeters in size, which are seen in the Gippa-Kengamine Group (Figs. 4e and 5cX). Fragments of reddish gray strongly welded pyroclastic rock with conspicuous welded lenses like the rock observed at Sennindake on the Yunotaira Caldera wall are also characteristically included (Fig. 6f, g). (2) Mega blocks can be over 50 m wide (Fig. 3d) and have characteristic jigsaw cracks (Fig. 3c, e). The core sample mentioned above appears to contain plural mega blocks of several meters thick or more, and some parts are significantly altered (Fig. 7aY). (3) The matrix is heterogeneous with various colors and often contains a patchwork of deformed soft blocks (Figs. 3f, 4c and 6e). As an example, as a result of a trench excavation on the western side of Kurofu volcano (Loc.5 in Fig. 2), a debris avalanche deposit covered by a pumice flow deposit of the 12 ka cal YBP from Hotokeiwa volcano was newly identi-

Fig. 5 Photo of debris avalanche deposit on the southern flank slope and foot. Photos **a**, **b**, **c**, and **e** are related to Mt. Sekisonsan. (**a**) "Zazen-kutsu" cave at Loc.Y in **d**. It was a place for training of the Buddhist priests in the past. Arrow shows the entrance of the cave. Pyroclastic rock can be seen here. (**b**) Lava block rich in jigsaw cracks at Loc.Z in **d**. (**c**) Hand specimens. X: Lava rich in coarse clinopyroxene phenocrysts at Loc.X in **d**. Y: Hand specimen at Loc.Y in **d**. (**d**) Map showing areas with different rock types and chemical compositions. This map was processed with Kashmir 3D as same as in Fig. 1a. (**e**) Welded pyroclastic rock observed in the inner wall of Zazen-kutsu at Loc.Y

fied (Fig. 4b). The debris avalanche is more than 4 m thick with characteristics of (1) and (3). The matrix is heterogeneous, and the deposit contains various blocks up to several meters in diameter (Fig. 4a, d, e).

The confirmed minimum thickness of the debris avalanche deposits is only a few meters (Fig. 6a), but one core sample drilled on the ESE flank slope of Maekake volcano was at least 27 m thick, although the drilling did not reach the base of the debris avalanche deposit (Fig. 7a, Loc.8 in Fig. 2) (Yasui et al. 2019).

One debris avalanche deposit forms a dome-like topography called Mt. Sekisonsan (1668 m) with a relative height of approximately 250 m on the southern flank slope, with a maximum slope of 30°. Sekisonsan is divided into multiple parts within an area of 860 × 1300 m (Fig. 1a). The

Fig. 6 Photo of debris avalanche deposit. (**a**) A section of trench excavation at Loc.10 in Fig. 2. From the lower order, AT volcanic ash, 1.2-m-thick debris avalanche deposits, pumice flow deposit from Hotokeiwa volcano, and black soil that was formed after 10,000 years ago, can be seen. (**b**) An excavated meter-sized large block, (**c**) close-up photo of 12 cm-thick AT ash, a famous widespread tephra erupted from the Aira caldera 30,000 years ago and one of the useful key beds in Japan, (**d**) Rock fragments contained in the debris avalanche deposit, (**e**) heterogeneous matrix of the debris avalanche deposit, (**f**) a hand specimen of densely welded pyroclastic rock (agglutinate) at Sennin-dake (Loc.4), (**g**) Agglutinate block contained in the debris avalanche deposit at Loc.2

volume of Sekisonsan is estimated to be approximately 0.06 km^3 (Aramaki 1963). Volcanic breccia consisting of lava fragments and matrix can be seen in a southern outcrop (Fig. 5a, e). Dense lava-like rocks with remarkable jigsaw cracks are exposed locally on the east and west sides of Sekisonsan (Fig. 5b). The size and amount of phenocrysts vary depending on location, with coarse clinopyroxene phenocrysts noticeable near the summit (Fig. 5cX). A whole-rock chemical analysis shows a wide range of compositions from the Gippa-Kengamine, Mitsuone, and Sennin Groups (Fig. 5d). The origin of Sekisonsan is discussed below.

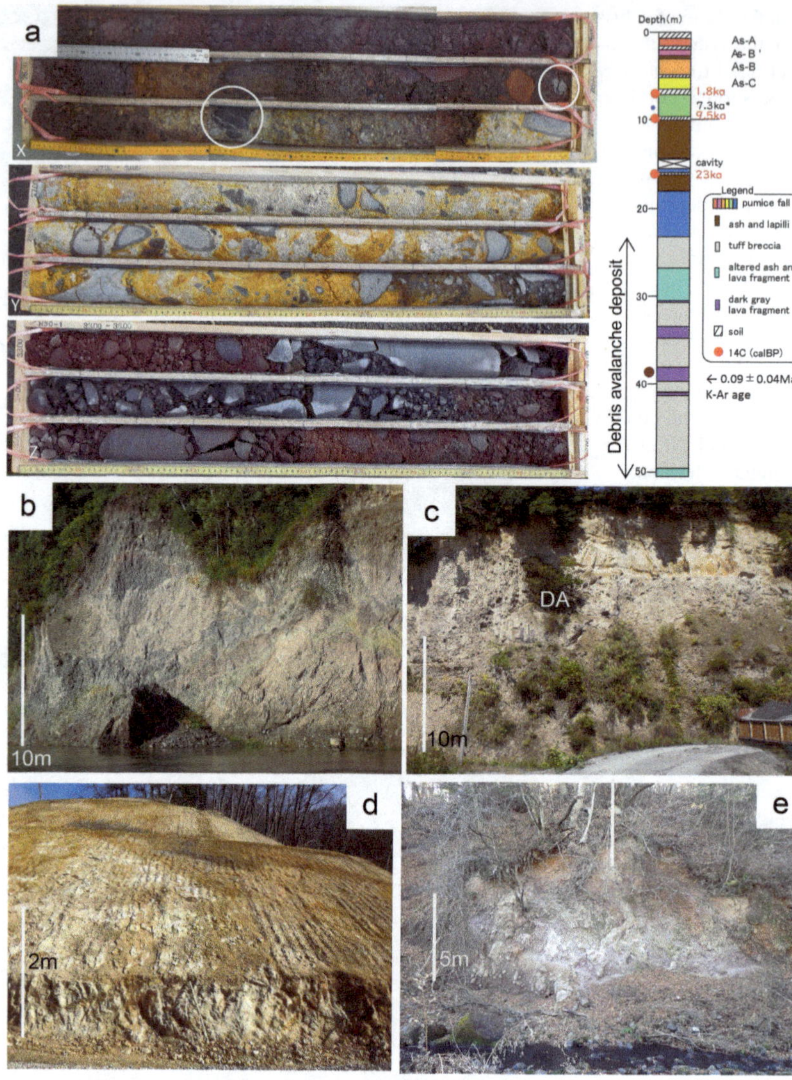

Fig. 7 Photo of debris avalanche deposit. (**a**) Drilling core sample at the eastern foot (Loc.8 in Fig. 2). The columnar section is also shown on the right. *: Detected level of the widespread tephra, K-Ah. For this ash layer, see caption of Fig. 4b. Depth (-m) X: 24–27 m, Y: 27–30 m, Z: 33–36 m. Jigsaw cracks can be seen in blocks in Photo X. Photo Y shows altered volcanic breccia. Dark gray massive lava fragments and reddish-brown oxidized breccia are seen in Photo Z. Photos **b**–**e** show a heterogeneous matrix of debris avalanche deposits observed in distal areas. (**b**) Loc.14, (**c**) Loc.15, (**d**) Loc.13, (**e**) Loc.12. DA: debris avalanche deposit

4 Large-Scale Sector Collapse

Until now, it was thought that the Yunotaira Caldera was the remnants of the collapsed Kurofu volcano. However, it was confirmed that debris avalanche deposits were widely distributed under the younger deposits in all directions around the volcano. It was recently discovered that debris avalanche deposits are also distributed on the west side of Yunotaira Caldera which opens to the East. It has become clear that the scale of the collapse, or the lost volcanic edifice, is larger than previously thought.

4.1 Estimating the Volume of Sector Collapse

The area covered by debris avalanche deposits around the volcanic edifice is estimated to be approximately 412 km^2 from Fig. 2. The base of the debris avalanche deposits around the volcano is barely visible, so its thickness is still unknown. If the average thickness of the debris avalanche deposit is assumed to be 10 m, the estimated volume is approximately 2.3 km^3. The thickness of the debris avalanche deposit was measured at least 27 m in a 50 m-long borehole drilled on the east-southeast flank slope of the volcano (Loc 8 in Fig. 2). However, as the borehole did not penetrate the base of the deposit, the maximum thickness at this locality is probably even thicker. So, it is plausible that the average thickness may exceed 10 m. On the other hand, the accumulation of secondary mudflows has been confirmed even in places far away from the volcano. In the northern foot of the volcano, the debris avalanche deposits entered Agatsumagawa River and flowed eastward. Associated thick mudflow deposits have been confirmed in Maebashi City, more than 120 km away (e.g., Takemoto and Kubo 2003). Yoshida (2004) estimated the volume of this mudflow to be approximately 4 km^3. Even in the southern foot of the volcano, mudflow deposits have been confirmed at Ueda, more than 52 km away, after flowing westward along Chikumagawa River (Togashi and Yokoyama 2015). When these mudflow deposits are added to the estimates from the widely distributed debris flow avalanche deposits identified in this study, the total volume reaches an order of 10 km^3 in total. In other words, it is thought that this was a fairly large-scale sector collapse. The volume of the depressed portion of the Yunotaira Caldera is estimated only to be about 1 km^2 which is insufficient to account for the total volume of debris avalanche deposits identified in this study. This suggests that a much larger volcanic edifice was lost due to the sector collapse, and that in other words, the original summit of the preexisting volcano might have been more than 3000 m ASL (Fig. 8).

4.2 Clues for Reconstructing the Volcanic Edifice

It is challenging to reconstruct the upper part of the volcanic edifices of Senin and Kurofu volcanoes as so much of the original volcanic edifices were lost in the sector collapse. If the distributions of the lava flow on the flank slope of Kurofu volcano (Gippa and Kengamine) are extended to the summit area and the volcanic edifice is reconstructed, a relatively long, east-west oval-shaped crater area is estimated. In whole-rock chemical analyses of 148 lava blocks contained in the debris avalanche deposits distributed in the northern and southern foot of the volcano, more than 80% of the blocks were of Mitsuone, Gippa, Kengamine, and less than 20% of Sennin origin (Takahashi et al. 2018). Most of Sennin volcano, a few remnants of which can only be found on the northwest side of Yunotaira Caldera, has collapsed, and the underlying Kurofu volcano also appears to have collapsed significantly (Fig. 8c).

The remaining Kurofu and Sennin volcanic edifices have been cut by many faults and have complex topography. In addition to the largest Tomi Fault, which extends from west-northwest to east-southeast, other remarkable faults are obvious on the east side of Kengamine and on the east side of Nokogiridake (Fig. 8a). These faults might have been syn-collapse faulting or preexisting active faults cutting through the volcanic edifices that might have facilitated dissection during sector collapse.

At the Socompa volcano in South America, it has been confirmed that a huge toreva block with a short flow distance remains on the slope at the time of collapse (Wadge et al. 1995). After Yagi (1936), Sekisonsan was thought to have been a lava dome, but Yasui et al. (2023) argued against this due to differences in its petrological properties depending on location (Yasui et al. 2023). Its eastern part is composed of rocks from Sennin volcano, whereas the western part and summit area contain rocks from Kurofu volcano (Fig. 5d). This indicates directly that Sekisonsan is a large mass of the volcanic edifice that was dissected out during the sector collapse. In other words, the huge toreva block of Sekisonsan was formed during the dissection of the volcanic edifice. The distance traveled by the toreva block was estimated to be several kilometers, and it appears to have become stuck on the slope of the volcano. Its shape indicates that it was on the way to breaking up into several pieces (Fig. 1a).

At the southeastern foot of the volcano, a pumice fall deposit directly covers the top of the debris avalanche deposits (e.g., Localities 11 and 12, Fig. 3f). This indicates that a Plinian eruption

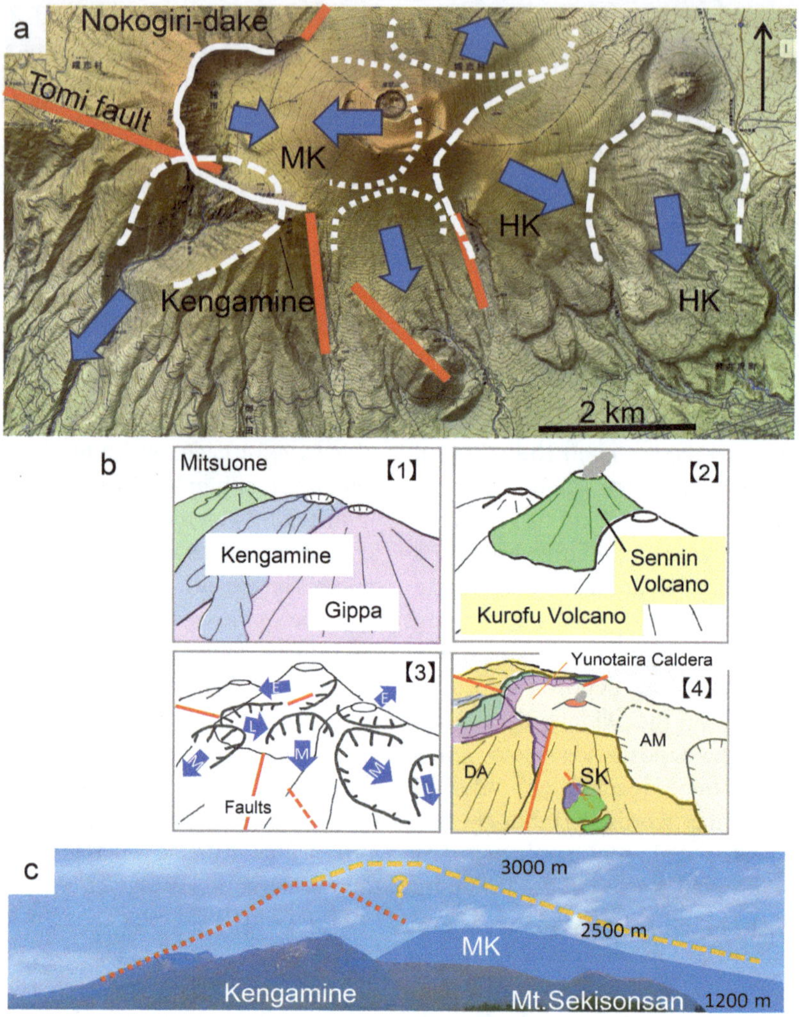

Fig. 8 Reconstruction of the volcanic edifice before sector collapse (illustrative reconstruction). MK: Maekake volcano, HK: Hotokeiwa volcano, (**a**) Faults and collapsed landforms found in volcanic bodies. This figure was created using Kashmir 3D (http://www.kashmir3d.com/) and DEM data provided by the Geospatial Information Authority of Japan (GSI). Arrows indicate the direction in which the debris avalanche, estimated from deposits on the slope and foot of the volcano, flowed downhill. Note that the lava flows from Hotokeiwa Volcano, which became active later, occupy the inside of the collapsed amphitheaters. White dashed line: Collapse landforms inferred from present topography. White dotted line: Collapse parts inferred from the distribution of deposits. Solid line: Amphitheater clearly preserved at present, (**b**) Schematic illustrative diagram showing the stages before and after the collapse (not to scale). [1] Topography after the end of Kurofu volcano activity, [2] Sennin volcano before the sector collapse, [3] Assumed localities and directions of the sequential sector collapse. E: Early stage, M: Middle stage, L: Late stage [4] Immediately after the series of the sector collapse. DA: debris avalanche deposit, SK: Mt. Sekisonsan, AM: amphitheater, (**c**) Remnant of Sennin and Kurofu volcanoes after sector collapse seen from the south. Restored, assumed outline before the collapse and approximate altitudes are also shown

occurred immediately after the collapse, so it is possible that at least one sector collapse triggered a Plinian eruption, like the eruption at Mount St. Helens in 1980 (see chapter "Geology of Asama-Sennin Volcano"). A massive magma intrusion beneath the volcanic edifice may have stimulated a preexisting fault system developed inside the volcanic edifice, creating multiple sectors, and causing a chain reaction of collapses in different directions (Fig. 8b [3]). Therefore, it is assumed that the sector collapses occurred in multiple stages in various directions. The base of the debris avalanche deposits is barely visible, and no outcrop where multiple debris avalanche deposits overlap has been confirmed to date. Although it is suggested that they collapsed in various directions one after another at some time, it is unclear whether there was any time gap between them, or whether these were sequential or not. An image of what Yunotaira Caldera might have looked like immediately after the collapse is shown in Fig. 8b [4], but the actual topography is unclear because the amphitheater spaces are almost occupied by the younger volcanoes of Hotokeiwa and Maekake (Fig. 8c). The timing of collapse was estimated as early, middle, and late from the topography and labeled (Fig. 8b [3]). The amphitheaters further east are thought to have formed during the later stages of collapse.

In directions other than southeast to east, pumice flow deposits from Hotokeiwa volcano after 16 ka ago overlie the top of the debris avalanche deposits without any soil between them. In other words, it seems that the devastation of land around the volcano continued for over 10,000 years after the large-scale sector collapse about 28 ka ago.

5 Summary

The Yunotaira Caldera located on the west of Maekake volcano is an east-facing amphitheater-shaped caldera. The walls of the caldera expose the internal structure of Sennin and Kurofu volcanoes, and were formed after the sector collapse of Sennin volcano around 28 ka ago. At the foot of Maekake volcano, debris avalanche deposits containing rocks from Sennin and Kurofu volcanoes are thickly covered by younger deposits. A geological survey conducted in this study confirmed buried debris avalanche deposits also on the west side of Yunotaira Caldera, indicating that they are distributed over a wider area than previously thought. Based on estimated and known thicknesses of proximal debris avalanche deposits and associated distal mudflow deposits, the volume of the collapse is estimated to be of the order of 10 km^3. The debris avalanche deposit distribution around the volcano implies a series of sector collapses in multiple directions. The collapse event of Sennin volcano also caused major destruction of the underlying Kurofu volcano. It is assumed that at least one of the sector collapses triggered a Mount St. Helens 1980-like Plinian eruption. The large-scale sector collapses may have been caused by a combination of factors, including the development of faults cutting through the volcanic edifices, alteration by volcanic gases, and magma intrusion and accumulation beneath the volcano. It is not known whether the multiple sector collapses were the result of a single event, but based on estimates from faults developing in the volcanic edifice, it is possible that they occurred in a chain reaction on a geological scale within a short period of time.

Acknowledgments The authors would like to express thanks and appreciation to Aramaki S. and Takarada S. who provided helpful discussions. Former Nihon University students Horiba T, Mera G, Hikawa S, and Ogino R helped us with geological survey and laboratory work. Martin A kindly revised our manuscript. We also thank Obayashi H, Hirakawa T, and Okui M for their help in the field.

References

Aramaki S (1963) Geology of Asama Volcano. J Fac Sci Univ Tokyo Sect 2 14:229–443

Takahashi M, Kikuchi M, Orito K, Yasui M, Kanamaru T, Takemoto T (2018) Major element chemistry for clasts of volcanic rocks in the large-scale debris avalanche and volcanic mud flow deposits formed by the sector collapse of Asama-Kurofu and Asama-Sennin volcanoes: summary of 148 analytical data and the origin of clasts of volcanic rocks. Proc Inst Nat Sci Nihon Univ 53:51–75. (in Japanese with English abstract)

Takemoto H, Kubo S (2003) Tephrostratigraphy of the Ohkuwa Debris Avalanche Asama Volcano, Central Japan. Proc Inst Nat Sci Nihon Univ 38:55–64. (in Japanese with English abstract)

Togashi H, Yokoyama H (2015) Origin of the Ueda mud flow deposit inferred from the geomorphic history of the Ueda Basin in Nagano Prefecture, central Japan. Bull Nagano Environ Conser Res Inst 11:1–8. (in Japanese with English abstract)

Wadge G, Francis PW, Ramirez CF (1995) The Socompa collapse and avalanche event. J Volcanol Geotherm Res 66:309–336

Yagi T (1936) Asama Volcano (with geological map of Asama volcano). Shinano Educational Association. 533p. (in Japanese)

Yasui M, Takahashi M, Kanamaru T (2019) Rediscovery of the Higuruma Debris Avalanche Deposit in Asama Volcano. Proc Inst Nat Sci Nihon Univ 54:123–142. (in Japanese with English abstract)

Yasui M, Obayashi H, Ogino R, Hikawa S, Kanamaru K, Sato Y, Okui M, Takahashi M (2023) Large-scale and complex sector collapses at Asama Kurofu and Sennin volcano. Programme and abstracts the Volcanol Soc Japan: A3-02 (in Japanese)

Yoshida H (2004) Mud flow deposits derived from the activity of Asama Volcano and their impacts on the geomorphic development of the northwestern part of the Kanto Plain, Central Japan. Geogr Rev Japan 77:544–562. (in Japanese with English abstract)

Yoshida H, Sugai T (2006) Morphological characteristics of hummocks originating from the24 ka sector collapse event of Asama Volcano, Central Japan. J Geogr 115:638–646. (in Japanese with English abstract)

Geology of Asama-Hanareyama and Hotokeiwa Volcanoes

Masaki Takahashi, Maya Yasui, Masashi Nagai, and Tatsuo Kanamaru

1 Introduction

The Hanareyama monogenetic and Hotokeiwa polygenetic volcanoes are unique in the Eboshi-Asama volcano group, in that they are characterized by thick dacitic to rhyolitic lava flows and voluminous fallout pumice and pyroclastic density current deposits (Figs. 1, 2 and 3). The drastic change of eruption style from Sennin volcano to Hanareyama and Hotokeiwa volcanoes may be closely related to the large-scale sector collapse of the volcanic edifices of Sennin and Kurofu volcanoes at ca 28,000 years calBP. The eruption age of Hanareyama monogenetic volcano is ca 24,000 years calBP and the onset of Hotokeiwa polygenetic volcano is ca 23,000 years calBP. ^{14}C dating of the eruptive products of Hanareyama and Hotokeiwa volcanoes was made by Togashi (1983), Higuchi (1989), Nakamura et al. (1997), and our study (unpublished data). Whole-rock chemical compositions of Hanareyama and Hotokeiwa volcanoes are reported in Takahashi et al. (2008).

The Hotokeiwa polygenetic volcano is a flat lava cone essentially consisting of three thick lava flows: they are the lower rhyolitic lava, middle rhyolitic to dacitic lava, and upper andesitic lava (Fig. 7). Aramaki (1963) estimated that the shape of Hotokeiwa volcano is one-quarter of a cone with an apparent height of 0.4 km, radius of 5 km, and total volume of ca 2.6 km^3, but the total volume of edifice of Hotokeiwa volcano is smaller and approximately 0.78 km^3 according to a recent estimate. The summit crater of Hotokeiwa volcano is inferred to be located on the eastern flank of the Maekake volcano at 2150 m above sea level, but it is now completely buried by the eruptive products of Maekake volcano. Pyroclastic density current deposits are widely distributed on the northern and southern foot of Hotokeiwa volcano and reach up to ca 19 km from the current summit crater of Maekake volcano on the northern foot and ca 21 km on the southern foot of the volcano, respectively (Fig. 8). Aramaki (1963) concluded that the construction of the Hotokeiwa lava cone preceded the eruption of voluminous fallout pumice and pyroclastic density current deposits, but now these eruptions were estimated to be concurrent with the lava flows based on tephra stratigraphy and whole-rock chemistry of erupted products (Fig. 18).

M. Takahashi (✉)
Institute of Natural Sciences, College of Humanities and Sciences, Nihon University, Tokyo, Japan
e-mail: takahashi.masaki@nihon-u.ac.jp

M. Yasui · T. Kanamaru
Department of Earth and Environmental Sciences, College of Humanities and Sciences, Nihon University, Tokyo, Japan
e-mail: yasui.maya@nihon-u.ac.jp; kanamaru.tatsuo@nihon-u.ac.jp

M. Nagai
Volcano Disaster Resilience Research Division, National Research Institute for Earth Science and Disaster Resilience, Tsukuba, Ibaraki, Japan
e-mail: mnagai@bosai.go.jp

S. Aramaki et al. (eds.), *The Asama Volcano*, Active Volcanoes of the World,
https://doi.org/10.1007/978-981-95-5799-8_8

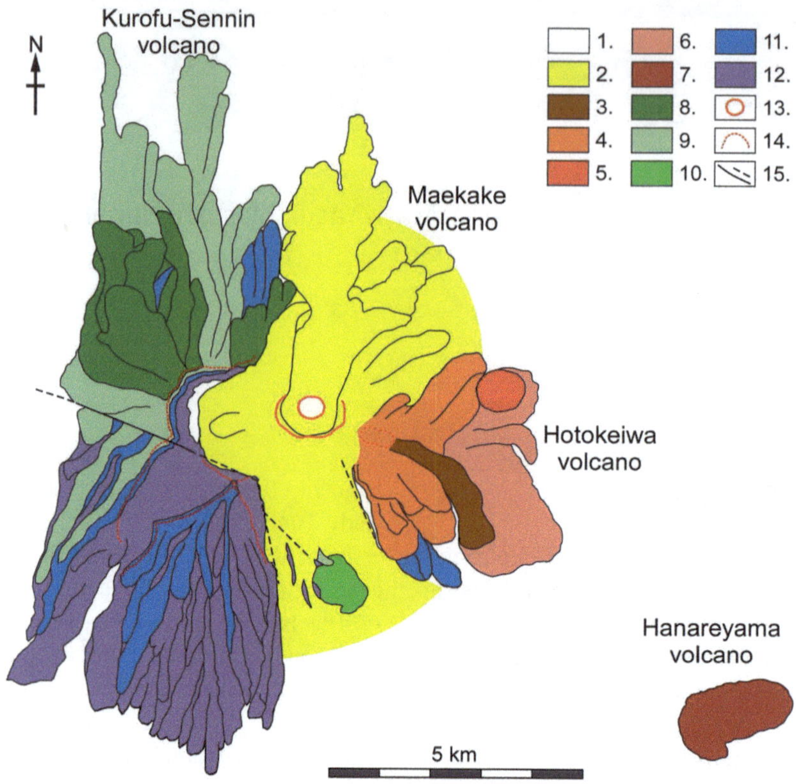

Fig. 1 Geologic sketch map showing the Hanareyama lava dome and Hotokeiwa volcano. 1: volcanic rocks and alluvium younger than Kurofu and Sennin volcanoes. 2: eruptive products of Maekake volcano. 3: Hotokeiwa upper andesitic lava. 4: Hotokeiwa middle rhyolitic to dacitic lava. 5: Koasama rhyolitic lava dome. 6: Hotokeiwa lower rhyolitic lava. 7: Hanareyama lava dome. 8: Lavas and pyroclastic cone of Sennin volcano. 9: Mitsuone lava group of Kurofu volcano. 10: Sekisonsan lava. 11: Kengamine lava group of Kurofu volcano. 12: Gippa lava group of Kurofu volcano. 13: Craters of Maekake volcano. 14: Rim of collapsed Yunotaira caldera. 15: Fault (solid line) and inferred faults (dashed line)

2 Hanareyama Volcano

Hanareyama volcano is a monogenetic volcano which erupted at ca 24,000 years calBP (Nakamura et al. 1997). The volcano is located in Karuizawa town, approximately 10 km to the SE of the summit crater of Maekake volcano (Fig. 1). The volcanic activity of Hanareyama volcano commenced with the eruption and deposition of the Kumoba dacitic to rhyolitic pyroclastic density current deposit, followed by effusive dacitic to rhyolitic lava eruptions which built up the Hanareyama lava dome.

2.1 The Kumoba Pyroclastic Density Current Deposit

The Kumoba pyroclastic density current deposit is distributed to the south of the Hanareyama lava dome over a 3 × 3 km area with an estimated erupted volume of less than 0.01 km^3. The eruption center was beneath the Hanareyama lava dome which formed later. The maximum thickness of the Kumoba pyroclastic density current deposit is approximately 6 m and comprises lower and upper units intercalated by an ash layer (Fig. 4) (Tsuji et al. 2004). The lower unit with a thickness of 5 m

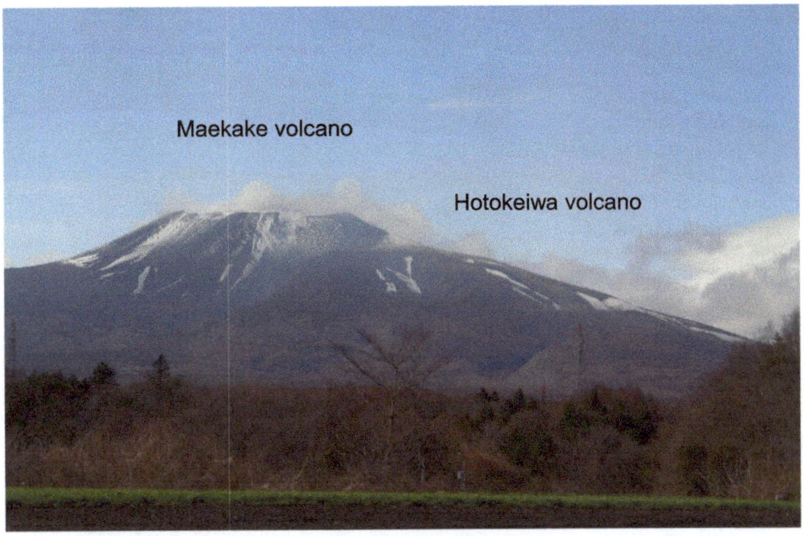

Fig. 2 Photograph showing Hotokeiwa volcano viewed from the south. Hotokeiwa volcano is a flat-shaped lava cone and overlain by the eruptive products of Maekake volcano

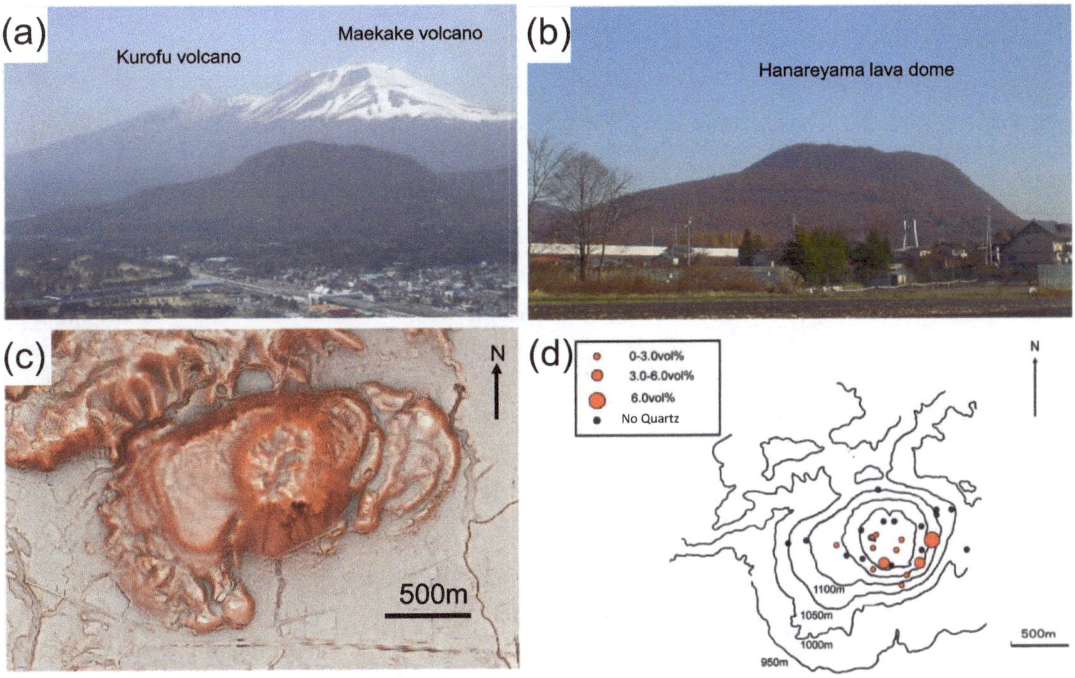

Fig. 3 Photographs and topographic relief of the Hanareyama lava dome. (**a**) Hanareyama lava dome, Maekake and Kurofu-Sennin volcanoes, viewed from the Karuizawa town. (**b**) Hanareyama lava dome viewed from Naka-Karuizawa to the southwest of Mt. Hanareyama. (**c**) Topographic relief map of the Hanareyama lava dome (courtesy of Chiba, T.). (**d**) Map showing the modal contents of phenocrystic quartz associated with hornblende and biotite; the localities of samples with no quartz, hornblende and biotite are also shown

is massive and includes a lot of lens-like layers composed of weakly vesiculated pumice clasts, the thickness of which is a few tens cm (Fig. 4). The stratified upper unit with a thickness of 1 m shows cross-laminations. The dense pyroclastic density current first erupted to form the lower unit, followed by a dilute and turbulent pyroclastic density current which produced the upper unit.

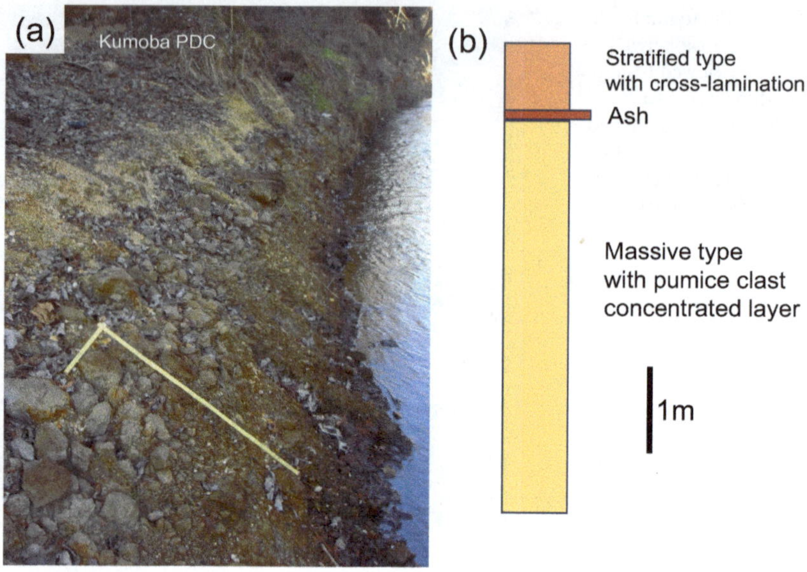

Fig. 4 Photograph and representative columnar section of the Kumoba pyroclastic density current deposit. (**a**) Photograph showing the Kumoba pyroclastic density current deposit with a lens-like layer of pumice clasts concentrated. (**b**) Representative columnar section of the Kumoba pyroclastic density current deposit

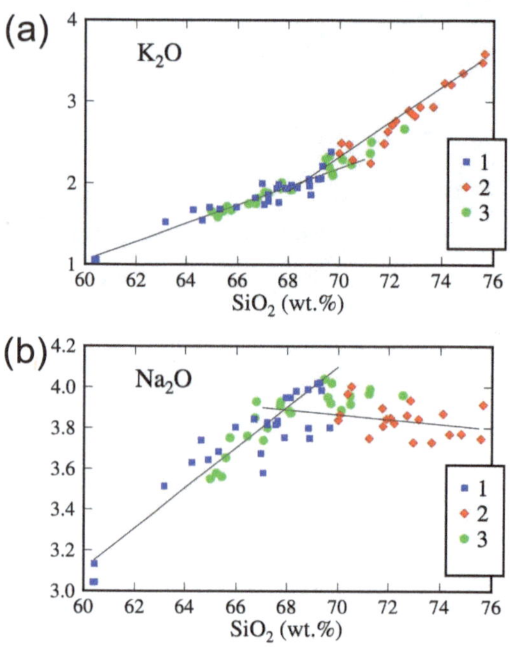

Fig. 5 SiO_2 variation diagram of the Hanareyama lava dome and Kumoba pyroclastic density current deposit for K_2O (**a**) and Na_2O (**b**) (Takahashi et al. 2008). 1: Hanareyama lava dome including no phenocrystic quartz, hornblende, and biotite. 2: Hanaraeyama lava dome including phenocrystic quartz, hornblende, and biotite. 3: Kumoba pyroclastic density current deposit

The whole-rock SiO_2 contents of pumice clast are 60–70 wt% (Fig. 5) with crystals of plagioclase, orthopyroxene, clinopyroxene, ±olivine, ±hornblende, ±biotite, and ±quartz.

2.2 The Hanareyama Lava Dome

The Hanareyama lava dome is an exogenous lava dome with a basal area of 2.3 × 1.0 km and a net height of 0.26 km (Fig. 3). The eruptive volume is estimated to be ca 0.2 km^3. A lot of thick but small lava flows effused at first. The lava flows changed to be larger, finally constructing a lava dome to form Mt. Hanareyama with a summit of 1256 m above sea level in altitude.

The lavas are dacitic to rhyolitic with whole-rock SiO_2 contents of 65–76 wt% (Fig. 5) and phenocrystic content of 8–21 vol%; phenocrysts are plagioclase, orthopyroxene, clinopyroxene, ±olivine, ±hornblende, ±biotite, and ±quartz. The rhyolitic rocks include quartz, hornblende and biotite in addition to plagioclase, orthopyroxene, and clinopyroxene, which erupted in the later stage and comprise the main part of the summit of the effused lava dome (Fig. 3). The hornblende-bearing rhyolitic to dacitic rocks and hornblende-free andesitic to dacitic rocks show distinct linear compositional trends with different inclinations in the K_2O vs. SiO_2 and Na_2O vs. SiO_2 diagrams

(Fig. 5); both magmas may have been derived from different magmatic systems but erupted simultaneously.

3 The Eruption History of Hotokeiwa Volcano

The eruption history of Hotokeiwa volcano is summarized in Fig. 6 (Nakamura et al. 1997; Tsuji et al. 2004; our study). The onset of Hotokeiwa volcano was the eruption of the Shiraitonotaki fallout pumice deposit (As-SP), the age of which is ca 23,000 years calBP. The eruption of Shiraitonotaki fallout pumice deposit was followed by the effusive eruption of the Hotokeiwa lower rhyolitic lava, which was overlain by the Koasama rhyolitic lava dome (Fig. 7). The next large-scale eruptions were the Okubozawa 1 fallout pumice and pyroclastic density current deposits at ca 21,000 years calBP, and the Okubozawa 2 fallout pumice and pyroclastic density current deposits at ca 20,000 years calBP. The climactic large-scale eruptions that occurred at ca 16,000 years calBP are As-YP (the Itahana yellow pumice) and As-YPk (the Kusatsu or Tsumagoi yellow pumice) fallout and Komoro 1 pyroclastic density current deposit (Fig. 8). The Hotokeiwa middle dacitic to rhyolitic lava was associated with the climactic eruption phase (Fig. 7). The As-Sj (the Sojya pumice) fallout and Komoro 2 pyroclastic density current deposits

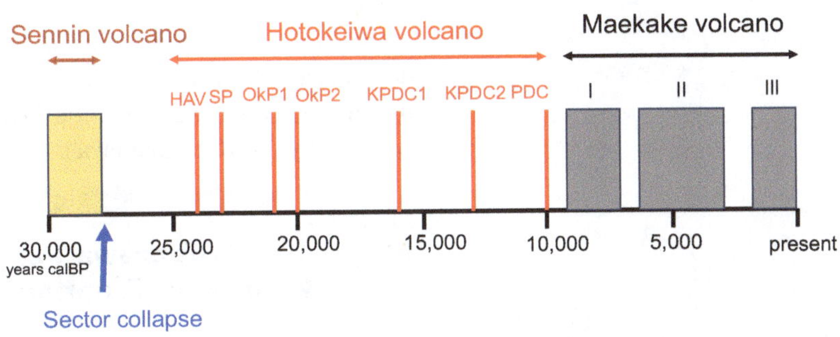

Fig. 6 Diagram showing the history of volcanic activity of Senin, Hanareyama, Hotokeiwa, and Maekake volcanoes. HAV: Hanareyama volcano. SP: Shiraitonotaki fallout pumice eruption. OkP 1: Okubozawa 1 fallout pumice and pyroclastic density current eruption. OkP 2: Okubozawa 2 fallout pumice and pyroclastic density current eruption. KPDC 1: Komoro 1 pyroclastic density current, As-YP and As-YPk fallout pumice eruption. KPDC 2: Komoro 2 pyroclastic density current and As-Sj fallout pumice eruption. PDC: pyroclastic density current deposit ca 10,000 calBP. I: active stage I of Maekake volcano. II: active stage II of Maekake volcano. III: active stage III of Maekake volcano

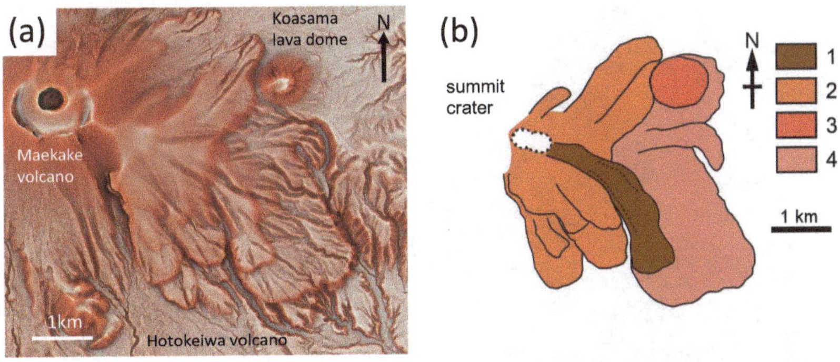

Fig. 7 Topographic relief and geological sketch maps of Hotokeiwa volcano. (**a**) Topographic relief map (courtesy of Chiba, T.). (**b**) Geological sketch map. 1: Hotokeiwa upper andesitic lava. 2: Hotokeiwa middle dacitic to rhyolitic lava. 3: Koasama rhyolitic lava dome. 4: Hotokeiwa lower rhyolitic lava

erupted at ca 13,000 years calBP (Fig. 8). The last large-scale eruption was the pyroclastic density current deposit at ca 10,000 years calBP (< 0.2 km^3 DRE in volume). It is probable that the formation of the Hotokeiwa upper andesitic lava was closely related to the Komoro 2 eruption.

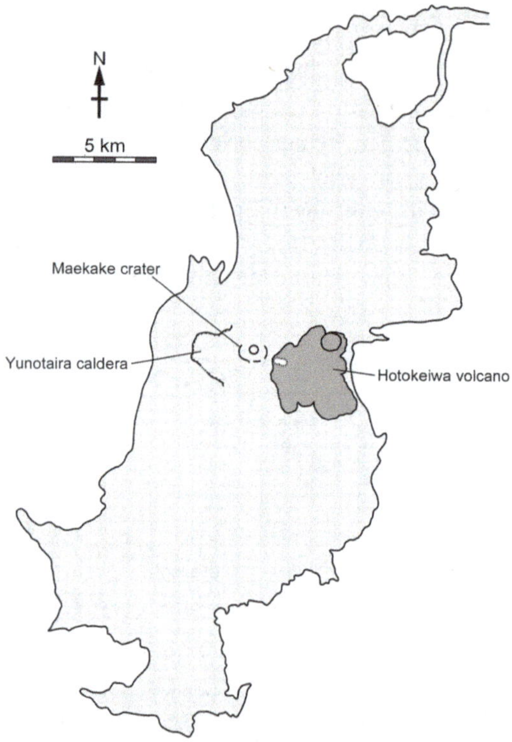

Fig. 8 Sketch map showing the distribution of Komoro pyroclastic density current deposits (including both Komoro 1 and 2 pyroclastic density current deposits) and Hotokeiwa volcano based on Aramaki (1963)

4 The Shiraitonotaki Fallout Pumice Deposit

The Shiraitonotaki fallout pumice deposit (Fig. 9a) is overlain by the Hotokeiwa lower rhyolitic lava evidenced in the bore-hole core drilled at the western foot of Koasama lava dome (Fig. 22). The Shiraitonotaki fallout pumice deposit was generated from a Plinian eruption at ca 23,000 years calBP. The axis of the isopach map of the deposit extends toward ENE (Fig. 9b). The eruptive volume of As-SP is ca 1.21 km^3 DRE. The whole-rock SiO_2 contents of pumice clast of the Shiraitonotaki fallout pumice deposit are 66–72 wt% (Fig. 18) with phenocrystic minerals of plagioclase, orthopyroxene, clinopyroxene, ±hornblende, and ±biotite. The eruption of Shiraitonotaki fallout pumice deposit was not associated with any pyroclastic density current deposits and followed by the effusive eruption of the Hotokeiwa lower rhyolitic lava, which was covered by the Koasama lava dome.

5 The Okubozawa 1 and 2 Pyroclastic Eruptions

The distribution of Okubozawa 1 and 2 pyroclastic density current deposits is limited to near the Okubozawa valley on the southeastern foot of Hotokeiwa volcano. In the Okubozawa valley, the Okubozawa 1 fallout pumice deposit (As-OkP1) with a thickness of ca 15 cm is under-

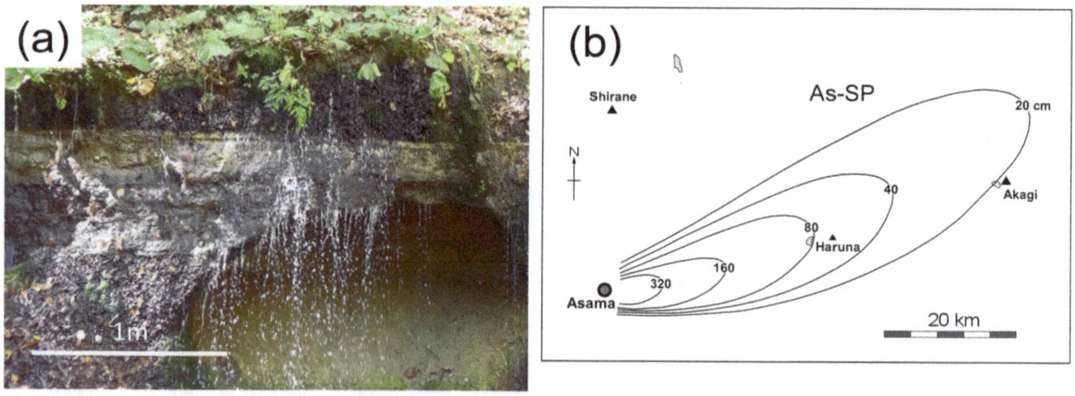

Fig. 9 Photograph and isopach map of the Shiraitonotaki fallout pumice deposit. (**a**) Shiraitonotaki fallout pumice deposit at the Shiraitonotaki fall. (**b**) Isopach map of the Shiraitonotaki fallout pumice deposit (Nakazawa et al. 1984)

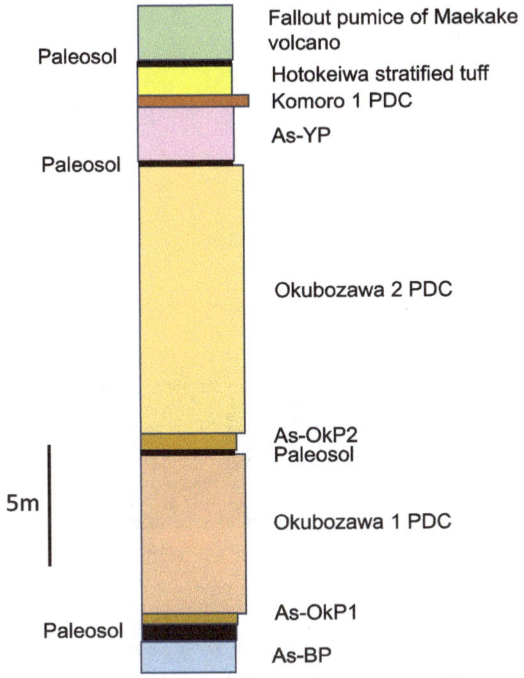

Fig. 10 Columnar section of the Okubozawa 1 and 2 pyroclastic density current deposits in the Okubozawa valley, Karuizawa town

lain by the As-BP fallout pumice of Sennin volcano and intercalates a paleosol layer (Figs. 10 and 11). The As-OkP1 is directly overlain by the Okubozawa 1 pyroclastic density current deposit with a thickness of about 6 m. The Okubozawa 1 pyroclastic density current deposit is overlain by the Okubozawa 2 fallout pumice deposit (As-OkP2) and intercalates a paleosol layer. The As-OkP2 with a thickness of about 1 m is overlain by the Okubozawa 2 pyroclastic density current deposit with a thickness of 12 m. The Okubozawa 2 pyroclastic density current deposit is overlain by the As-YP fallout pumice deposit and intercalates a paleosol layer. The axes of the isopach maps of the As-OkP 1 and 2 are toward southeast from the inferred source vent (Nakazawa et al. 1984) (Fig. 12a). The Okubozawa 1 and 2 pyroclastic eruptions lack associated lava flows. The estimated total eruptive volume of the OkP 1 and 2 fallout pumice and associated pyroclastic density current deposits are ca 0.26 and 0.21 km^3 DRE, respectively.

The ages of Okubozawa pyroclastic eruptions 1 and 2 are ca 21,000 and 20,000 years calBP, respectively. The whole-rock SiO_2 contents of pumice clasts of OkP 1 and 2 fallout and pyroclastic density current deposits are 66–71 wt% and 65–69 wt%, respectively (Fig. 18); their phenocrystic minerals are plagioclase, orthopyroxene, and clinopyroxene.

6 The Komoro 1 and 2 Pyroclastic Eruptions

The climactic eruption of Hotokeiwa volcano was the Komoro 1 pyroclastic eruption, which occurred at ca 16,000 years calBP (Fig. 8). The beginning of the climactic eruption was a Plinian eruption of As-YP fallout pumice (Figs. 10 and 12c); the axis of its isopach map is oriented toward ESE (Nakazawa et al. 1984) (Fig. 12b). The fallout of As-YP was followed by the eruption of Komoro 1 pyroclastic density current deposit, which traveled ca 19 km from the present summit crater of Maekake volcano on the northern foot and ca 21 km on the southern foot of the volcano with a distributed area of 40 × 14 km (Fig. 8); it spread and filled the topographic relief and formed a large pyroclastic density current plateau with an almost flat surface (Fig. 13a, c), the thickness of which is several meters to 30 m (Fig. 13). The valleys dissecting the plateau constructed characteristic box canyons named as the "Tagiri" canyons (Fig. 13b). On the northern foot of the volcano, the As-YPk is underlain by the main unit of the Komoro 1 pyroclastic density current deposit with no evidences of time-gap (Fig. 14a–c); the axis of isopach map is oriented toward NE (Fig. 14d). The As-YPk is the most voluminous Plinian fallout pumice deposit in the Komoro 1 eruptive episode.

The Komoro 1 pyroclastic density current deposit shows no evidence of welding and often includes lenses composed of coarse pumice clasts with a length of 1–20 m and thickness of 0.5–2 m (Fig. 15a, b) and many fragments of charcoal (Fig. 15c), which comprises at least three units intercalating stratified ash beds (Fig. 15d). The

Fig. 11 Photographs of the Okubozawa 1 and 2 pyroclastic density current deposits. (**a**) Okubozawa 2 pyroclastic density current deposit (OK2 PDC) is overlain by a paleosol and As-YP fallout pumice deposit (YP). (**b**) As-OkP 2 fallout pumice deposit (OkP 2) is overlain by the Okubozawa 2 pyroclastic density current deposit (OK2 PDC). (**c**) As-OkP 2 fallout pumice deposit (OkP 2) is underlain by a paleosol. (**d**) Okubozawa 1 pyroclastic density current deposit (OK1 PDC) is underlain by the As-OkP 1 fallout pumice deposit (OkP 1) and a paleosol layer

stratified ash beds consist of various sizes of sandy tuff with lithic and glassy fragments, which are estimated to be fallout tuffs produced by large-scale Vulcanian eruptions. The thickness of a single bed of stratified ash bed is less than 10 cm in general, and their total thickness is several meters to centimeters and is thicker in the eastern foot of the volcano (Fig. 16). The stratified ash beds are labeled UG in areas in the Kanto plain to the east of Asama volcano group (Suzuki 1991). The estimated erupted volume of Komoro 1 pyroclastic density current deposit is 1.32 km^3 DRE (Aramaki 1963) and those of As-YP and As-YPk are 0.41 and 0.83 km^3 DRE, respectively; the total volume of eruptive products of Komoro 1 pyroclastic eruption is ca 2.56 km^3 DRE, and ca 2.81 km^3 DRE including the Hotokeiwa middle dacitic to rhyolitic lava. The whole-rock SiO_2 contents of essential clasts of Komoro 1 fallout pumice and pyroclastic density current deposits are 52–70 wt% (Fig. 18); their phenocrystic minerals are plagioclase, orthopyroxene, clinopyroxene, and rare olivine.

The Komoro 2 pyroclastic density current deposit, which occurred at ca 13,000 calBP, was preceded by the sub-Plinian eruption of As-Sj fallout deposit (Fig. 15d) (Takemoto and Kubo 1995; Tsuji et al. 2004). The Komoro 2 pyroclas-

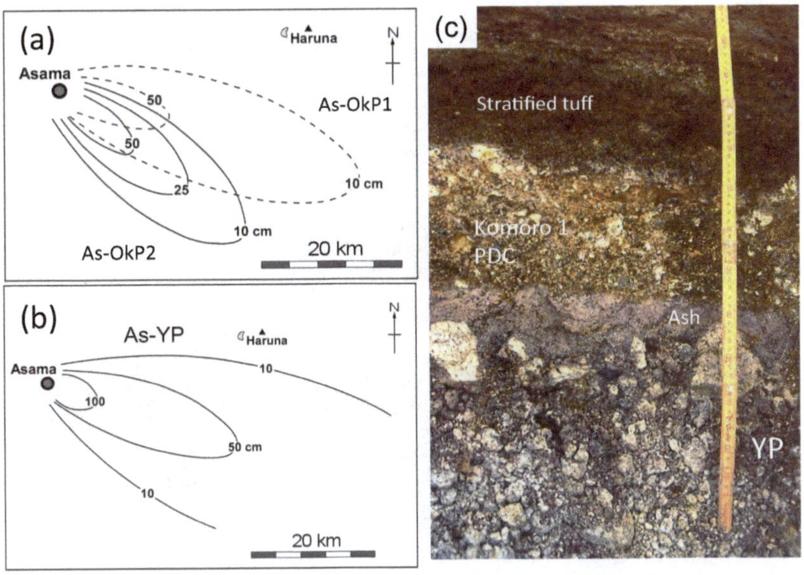

Fig. 12 Isopach maps of the Okubozawa 1–2 and Itahana yellow pumice fallout deposits (As-YP) (Nakazawa et al. 1984). (**a**) Isopach map of the Okubozawa-1 (As-OkP 1) and -2 (As-OkP 2) pumice fallout deposit. (**b**) Isopach map of the Itahana yellow pumice fallout deposit (As-YP) (**c**) Photograph showing the As-YP (YP) and overlying ash, Komoro 1 pyroclastic density current deposit (Komoro 1 PDC), and stratified fallout tuff deposits

Fig. 13 Photographs showing the Komoro 1 pyroclastic density current deposit. (**a**) Plateau consisting of thick pyroclastic density current deposit with a thickness of ca 25 m dissected by a valley in Komoro city. (**b**) "Tagiri" box canyon dissecting a plateau consisting of pyroclastic density current deposits with a thickness of ca 15 m in Miyota town. (**c**) Thick pyroclastic density current deposit with a thickness of ca 15 m

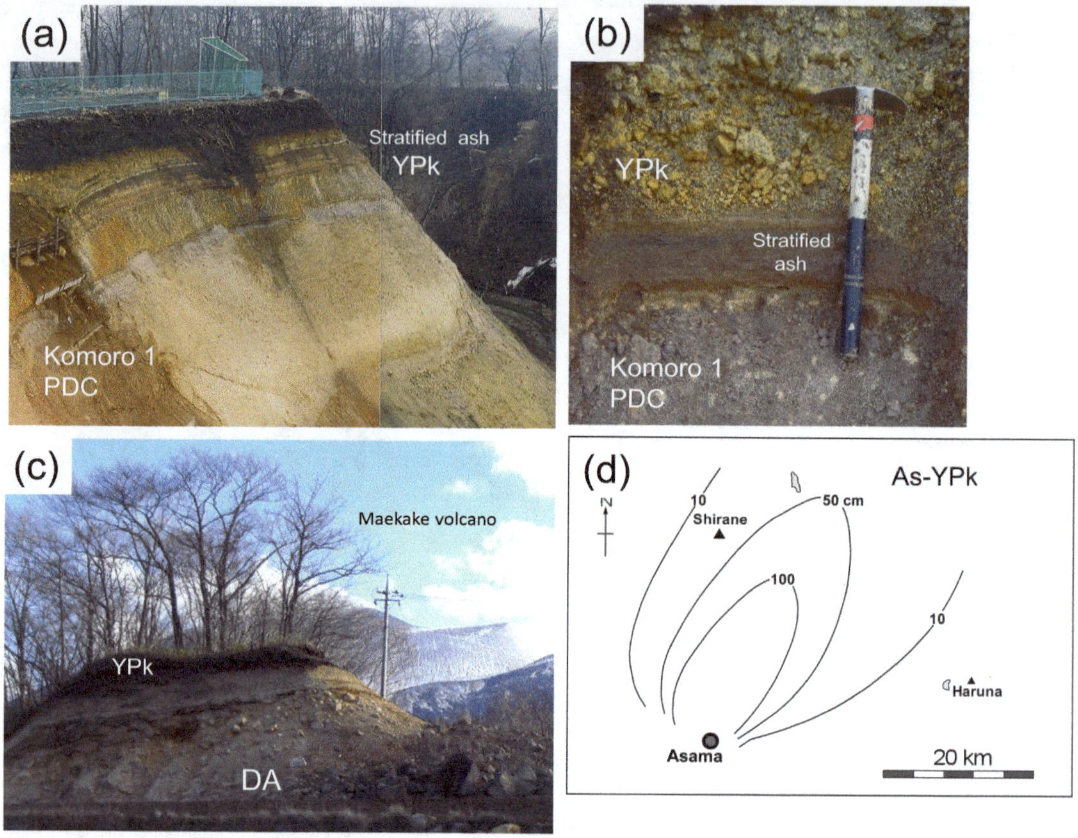

Fig. 14 Photographs and an isopach map of the As-YPk fallout pumice deposit. (**a**) As-YPk (YPk) is underlain by the Komoro 1 pyroclastic density current deposit. The eruptive products comprise the Komoro 1 pyroclastic density current deposit (Komoro 1 PDC), stratified ash beds, As-YPk fallout pumice deposit (YPk), stratified ash beds, paleosol, ash of Komoro 2 pyroclastic density current deposit and black soil. (**b**) Close view of the As-YPk fallout pumice deposit (YPk) overlying the stratified ash bed and Komoro 1 pyroclastic density current deposit. (**c**) As-YPk fallout pumice deposit directly overlays the Okuwa debris avalanche deposit (DA). (**d**) Isopach map of As-YPk, the axis of which is NE (Nakazawa et al. 1984)

tic density current deposits are widely distributed covering the flat surface of the Komoro 1 pyroclastic density current plateau with a paleosol, but their thickness is less than 10 m in general and thinner than those of the Komoro 1 pyroclastic density current deposit (Figs. 15a, b, d and 17a, b); they include a lot of charcoal fragments and sometimes standing tree trunks (Fig. 17a, b). The eruptive volume of Komoro 2 pyroclastic density current deposit estimated by Aramaki (1963) is ca 0.4 km^3 DRE and the total volume of Komoro 2 pyroclastic eruption including fallout deposits is estimated to be ca 0.44km^3 DRE. The Komoro 2 pyroclastic density current deposit often includes a lot of scoria clasts with andesitic composition. The whole-rock SiO_2 contents of pumice and scoria clasts of Komoro 2 pyroclastic density current deposit are 55–69 wt% (Fig. 18); their phenocrystic minerals are plagioclase, orthopyroxene, clinopyroxene, and rare olivine.

In the SiO_2 vs. MgO diagram, three groups with different MgO content are distinguished (Fig. 18)—Group 1 (Hanareyama lava dome and Kumoba pyroclastic density current deposit, Shiraitonotaki fallout pumice deposit, Hotokeiwa lower lava and Koasama lava dome, and Okubozawa 1 and 2 fallout pumice and

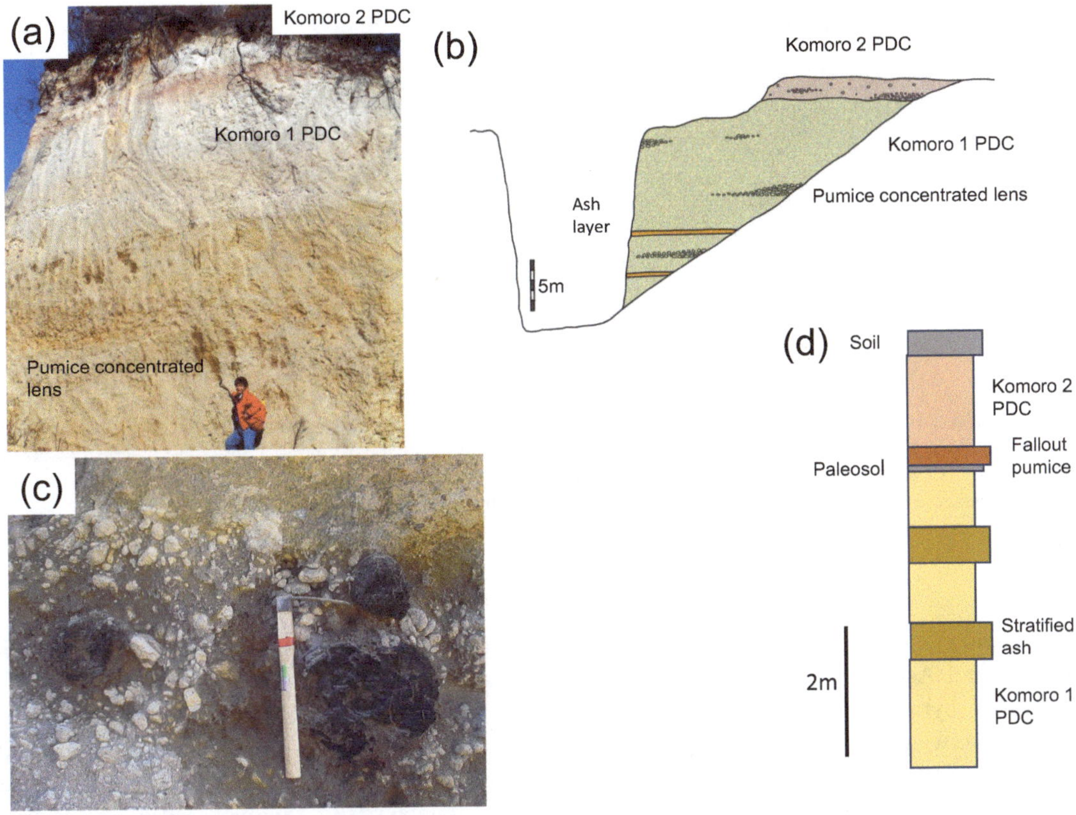

Fig. 15 Internal structure of the Komoro 1 pyroclastic density current deposit. (**a**) Pyroclastic density current deposit, with evidence of pinkish oxidation at the top of the cliff and a pumice clast concentrated lens. (**b**) Schematic cross section of the Komoro 1 and 2 pyroclastic density current deposits, with pumice clasts concentrated lens and thin ash layers (Aramaki, 1963). (**c**) Charcoal woods in the Komoro 1 pyroclastic density current deposit. (**d**) Columnar section of Komoro 1 and 2 pyroclastic density current deposits at Yui, Karuizawa town; Komoro 1 pyroclastic density current deposit is intercalated by two layers of stratified ash; Komoro 2 pyroclastic density current deposit is underlain by a fallout pumice deposit (As-Sj)

Fig. 16 Stratified ash beds associated with in the Komoro 1 pyroclastic density current deposit. (**a**) Stratified ash bed with a thickness of more than 2 m overlying the Komoro 1 pyroclastic density current deposit. (**b**) Close view of the stratified ash beds

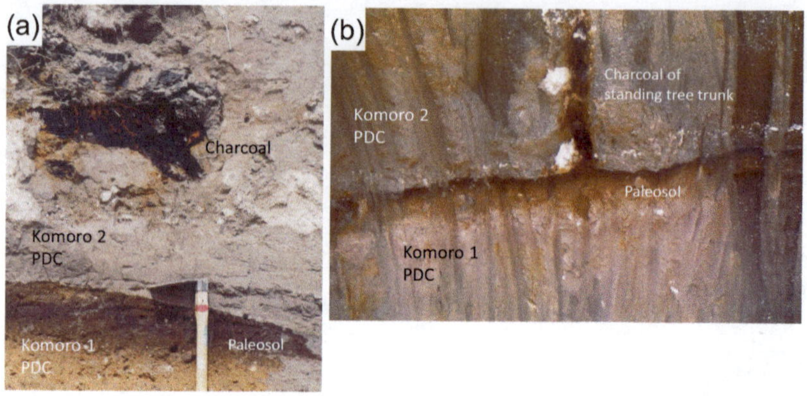

Fig. 17 Paleosol layers at the boundary between the Komoro 1 and 2 pyroclastic density current deposits. (**a**) Komoro 2 pyroclastic density current deposit with a fragment of charcoal wood underlain by a paleosol and Komoro 1 pyroclastic density current deposit. (**b**) Komoro 2 pyroclastic density current deposit is underlain by a paleosol and the Komoro 1 pyroclastic density current deposit; a charcoaled standing tree trunk is included in the Komoro 2 pyroclastic density current deposit

pyroclastic density current deposits), Group 2 (As-YP and As-YPk fallout pumice deposits, Komoro 1 pyroclastic density current deposit, and Hotokeiwa middle dacitic to rhyolitic lava), and Group 3 (Komoro 2 pyroclastic density current deposit). The MgO content of each group increases with time from Group 1 to Group 3 (Fig. 18).

7 The Hotokeiwa Lavas and Koasama Lava Dome

The volcanic edifice of Hotokeiwa volcano comprises the Hotokeiwa lava flows and Koasama lava dome.

7.1 The Hotokeiwa Lower Lava

The Hotokeiwa lower rhyolitic lava is the largest lava flow in Hotokeiwa volcano (Fig. 7); its area is ca 7.3 km^2 and estimated erupted volume is ca 0.51 km^3 assuming the average thickness as 70 m. The thickness of the lava flow measured in cliff exposures of the sides and leading edge of the Hotokeiwa lower lava is 50–80 m and that of the central portion is 100 m which was revealed by a drilling at the western foot of the Koasama lava dome.

The whole-rock SiO_2 content of the lower rhyolitic lava flow is 72–73 wt% and low in MgO (Fig. 18), which are nearly the same as those of the Shiraitonotaki fallout pumice and Koasama lava dome. The results of drilling indicate that they probably erupted successively with no time intervals between each in the order of the Shiraitonotaki fallout pumice deposit, the Hotokeiwa lower rhyolitic lava, and the Koasama lava dome.

7.1.1 Marginal Cliff of Lava

The margins of the Hotokeiwa lower lava are well exposed in cliffs along the Okubozawa valley, where a section of rhyolitic lava more than 20 m in thickness can be seen. In the lava flow section, more than four lens-like layers of volcanic breccia with a height of 2–5 m and length of 10–50 m are included (Fig. 19); they alternate with lavas and dip in the direction of flow. The lavas show flow foliation or banded structure; some of them are spherulite and the foliation is often folded (Fig. 20). The flow foliation is formed by shear stress and crystals are rotated and fractured by shearing (Fig. 20b); the banding is also stretched and fractured to show lens-like structures (Fig. 20c, d).

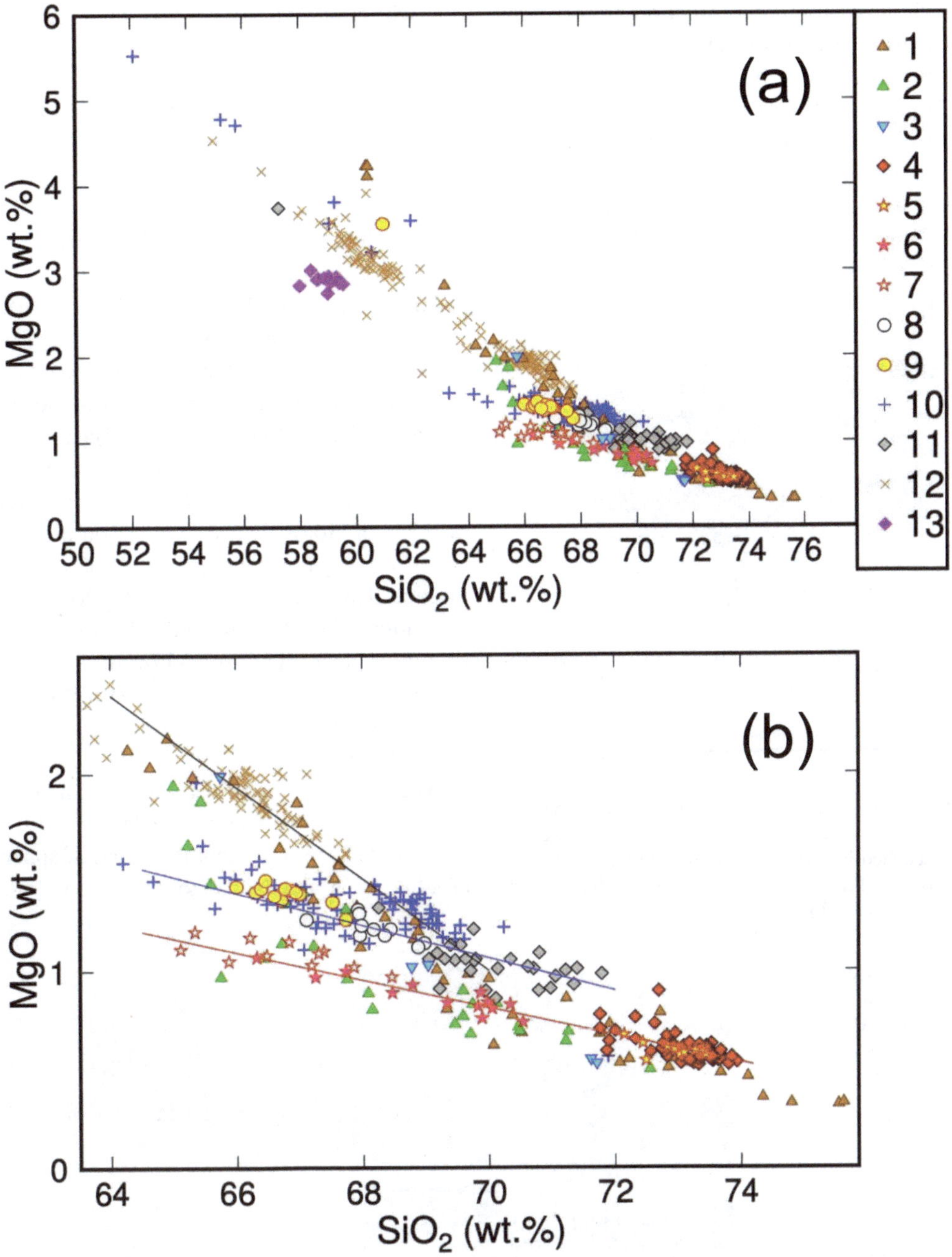

Fig. 18 SiO_2 variation diagram of eruptive products of Hanareyama and Hotokeiwa volcanoes for whole-rock MgO content (Takahashi et al. 2008). (**a**) Diagram showing the SiO_2 content ranging from 50 to 76 wt%. (**b**) Diagram showing the SiO_2 content ranging from 64 to 76 wt%. 1: Hanareyama lava dome. 2: Kumoba pyroclastic density current deposit. 3: Shiraitonotaki fallout pumice deposit. 4: Hotokeiwa lower lava. 5: Koasama lava dome. 6: Okubozawa 1 fallout pumice and pyroclastic density current deposits. 7: Okubozawa 2 fallout pumice and pyroclastic density current deposits. 8: As-YP fallout pumice deposit. 9: YPk fallout deposit. 10: Komoro 1 pyroclastic density current deposit. 11: Hotokeiwa middle lava. 12: Komoro 2 pyroclastic density current deposit. 13: Hotokeiwa upper lava

The flow foliation and lineation are revealed by measuring the anisotropy of magnetic susceptibility (AMS) of the Hotokeiwa lower lava (Fig. 21). The Hotokeiwa lower lava is grouped into three areas based on the results of AMS over a distance of 60 m. In Area 1, both magnetic foliation and lineation dip with inclinations of less than 26° outward from the interior of the lava flow, and the lava is pushed outward. In Area 2, both magnetic foliation and lineation dip with inclinations of less than 23° in the direction of lava flow. In Area 3 near the front of the lava flow, the magnetic foliation dips inward with an inclination of less than 37° but lineation is upward, indicating that it is a ramp structure often developed near the front of lava flows.

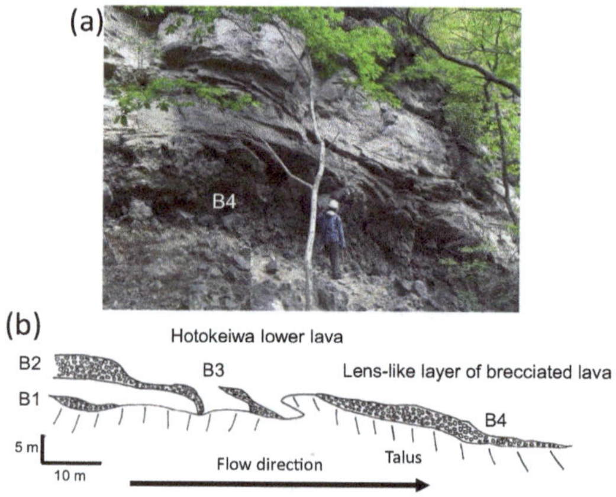

Fig. 19 Photograph and sketch of the marginal side cliff of Hotokeiwa lower rhyolitic lava flow. (**a**) Lens-like layers of brecciated lava (B4) included in the lava flow. (**b**) Four lens-like layers of brecciated lava (B1 to B4) included in the lava flow

7.1.2 Section of Central Portion of Lava

A 200 m deep bore-hole was drilled in the western foot of Koasama lava dome, the site of which is located on the Hotokeiwa lower rhyolitic lava. The drill-core from the surface down to 200 m revealed the following (Fig. 22): the fallout tephra of Maekake volcano (0–13 m in depth), fallout tephra and pyroclastic density current deposits of Hotokeiwa volcano (Komoro 1 pyroclastic density current deposit, YP, or YPk fallout pumice deposits: 13–33 m in depth), Koasama lava dome (33–71 m in depth), Hotokeiwa lower rhyolitic lava flow (71–178 m in depth), Shiraitonotaki fallout pumice deposit (178–192 m in depth), and andesitic lava of Sennin volcano (192–203 m in depth).

The Hotokeiwa lower rhyolitic lava from the top to bottom comprises a chilled marginal zone of obsidian (72–75 m in depth), a zone of spherulite (75–88 m in depth), a zone of flow foliation

Fig. 20 Close view and photomicrograph of the Hotokeiwa lower rhyolitic lava. (**a**) Flow structure (foliation) with spherules. (**b**) Photomicrograph of (**a**) showing a fluidal structure. (**c**) Lenses of fractured lava. (**d**) Photomicrograph of (**c**) showing fragmented lenses of glass included in the matrix with flow structure, indicating that fracturing occurred during lava flowing

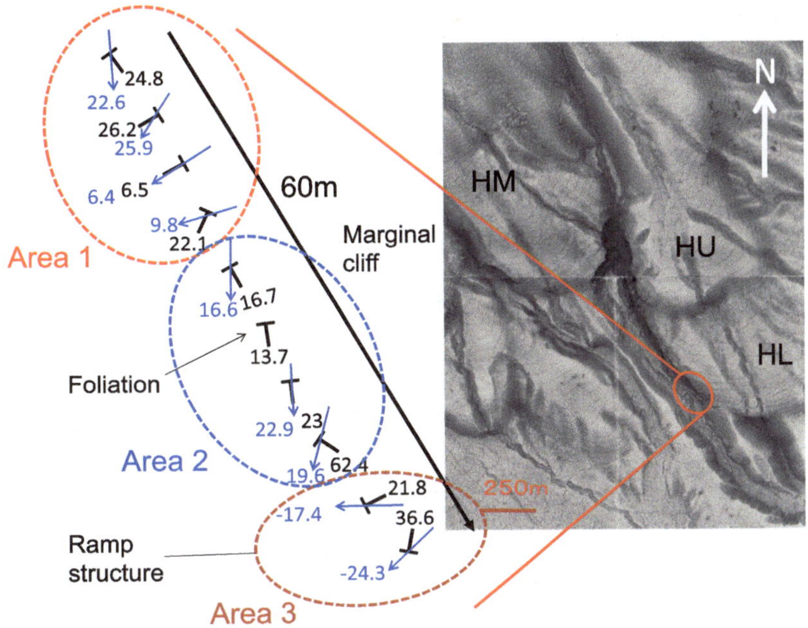

Fig. 21 Anisotropy of magnetic susceptibility (AMS) of the Hotokeiwa lower rhyolitic lava at the margin along the Okubozawa valley showing locally different modes of flow. The magnetic lineation (arrows) and foliation are denoted. The magnetic lineation and foliation are different in each area. Numerals denote dips of foliation and inclinations of lineation in degree. HM: Hotokeiwa middle lava; HU: Hotokeiwa upper lava; HL: Hotokeiwa lower lava. See text for explanation

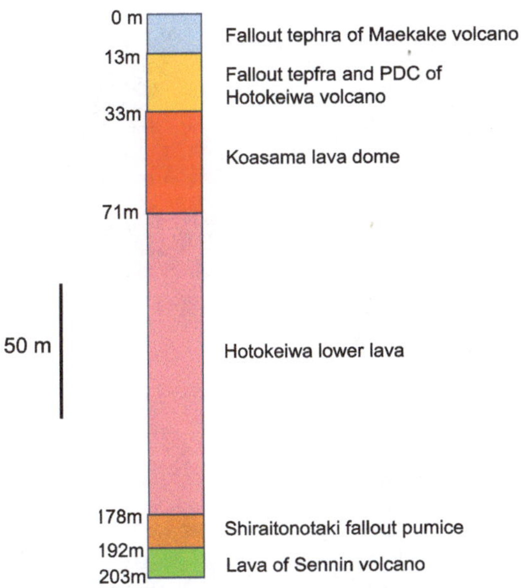

Fig. 22 Complete section of bore-hole core drilled at the western foot of Koasama lava dome. From top to bottom: fallout tephra of Maekake volcano (0–13 m), fallout tephra and pyroclastic density current deposit (PDC) of Hotokeiwa volcano (13–33 m), Koasama lava dome (33–71 m), Hotokeiwa lower lava (71–178 m), Shiraitonotaki fallout pumice deposit (178–192 m), and lava of Sennin volcano (192–203 m)

(88–127 m in depth), and a heterogeneous zone of a mixture of leucocratic and melanocratic domains (127–179 m in depth) (Figs. 23 and 24). The total thickness of the Hotokeiwa lower lava in the drill core is 107 m, indicating that the thickness of the central portion of the Hotokeiwa lower rhyolitic lava exceeds 100 m. The heterogeneous zone with a thickness of 52 m is glassy, and the microlite in the leucocratic domain is larger in grain-size and more abundant in amount than those in the melanocratic domain (Fig. 25); the whole-rock SiO_2 content of leucocratic domains (72–72.5 wt% SiO_2) is slightly lower than that of melanocratic ones (72.5–73 wt% SiO_2).

The phenocrystic contents of Hotokeiwa lower lava are 10–15 vol% with phenocrysts of plagioclase, orthopyroxene, clinopyroxene, and hornblende. The whole-rock SiO_2 content of Hotokeiwa lower rhyolitic lava is 72–73 wt% and rather homogeneous throughout the lava flow; that of the bore-hole core with a thickness of 107 m is also homogeneous with whole-rock SiO_2 content ranging from 72 to 73 wt% (Fig. 18).

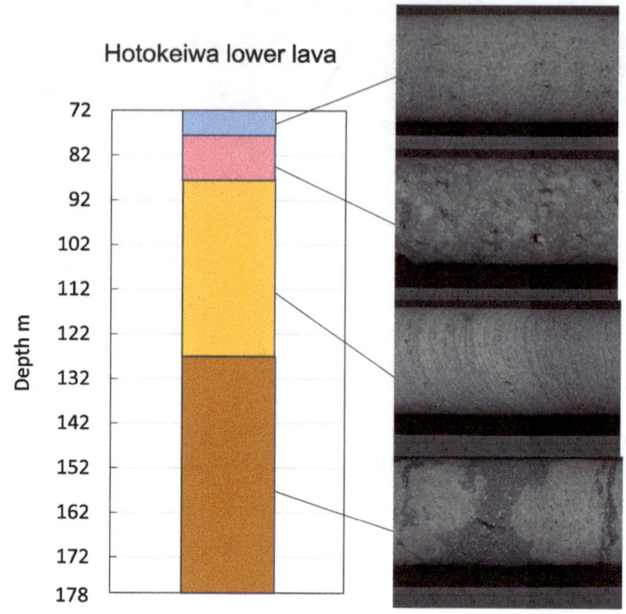

Fig. 23 Bore-hole core section of the Hotokeiwa lower rhyolitic lava. From top to bottom: glassy chilled margin (obsidian) (72–75 m), spherulite (75–88 m), flow structure (88–127 m), and a heterogeneous zone (127–179 m). (**a**) Cross section. (**b**) Photograph of bore-hole core samples

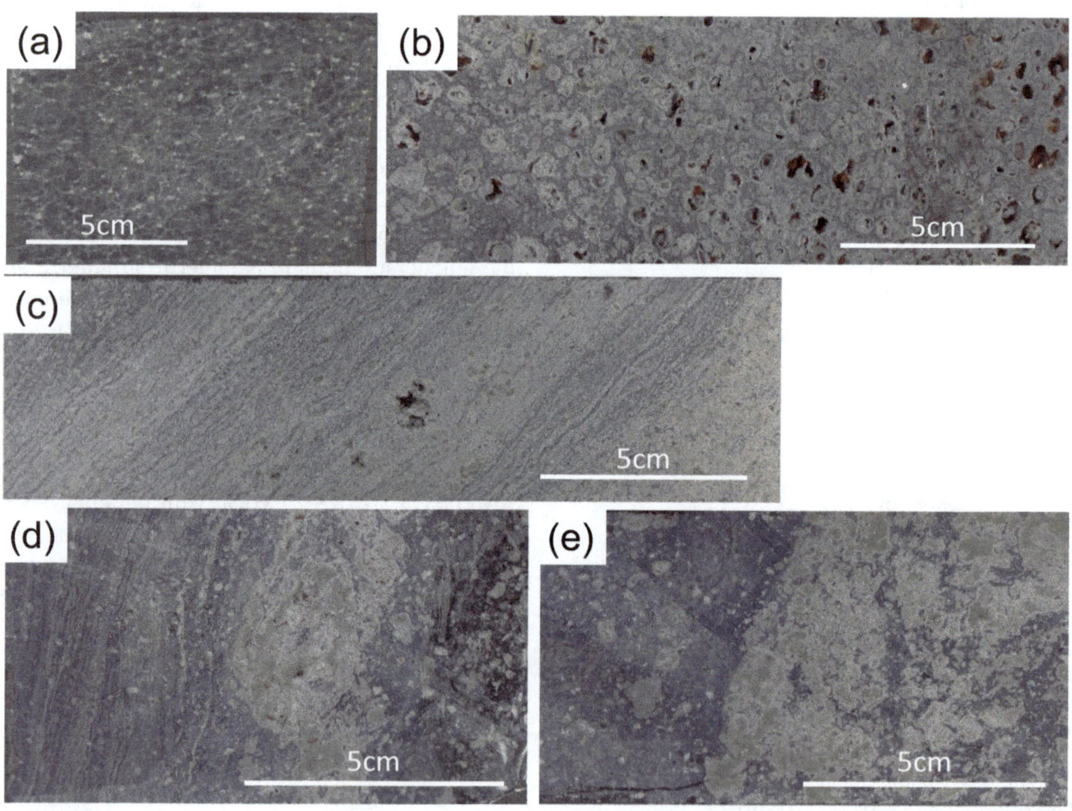

Fig. 24 Photographs showing the bore-hole core samples. (**a**) Glassy chilled margin. (**b**) Spherulite. (**c**) Flow structure (foliation). (**d**) Coexisting flow structure and heterogeneous zone. (**e**) Heterogeneous zone

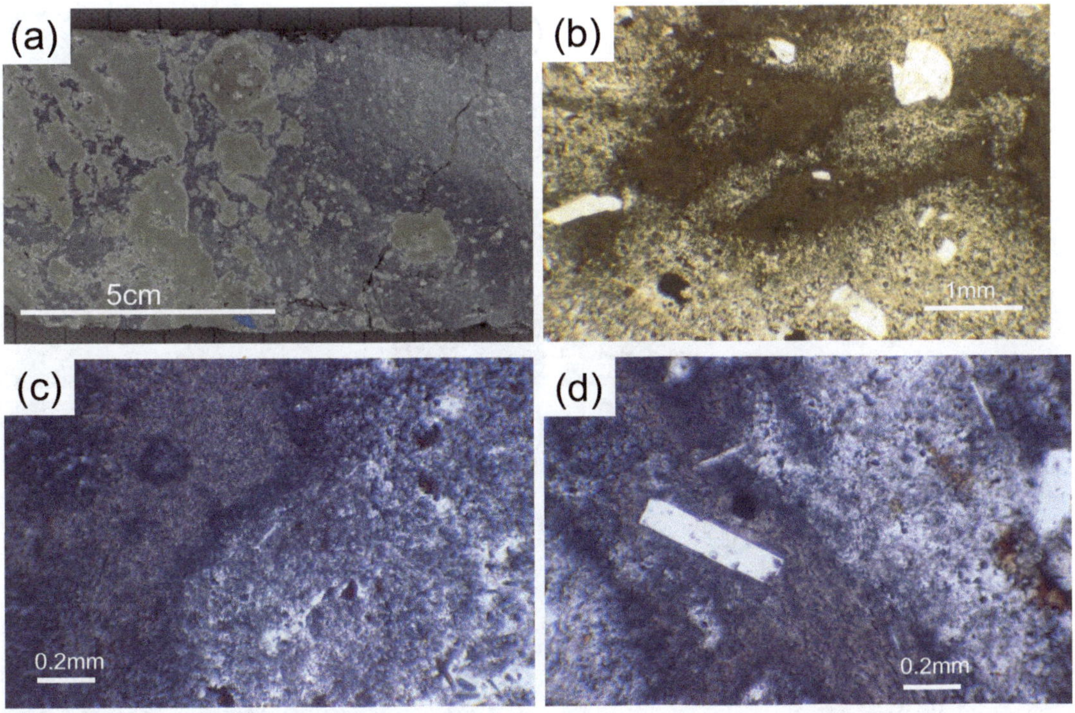

Fig. 25 Photograph and photomicrographs of the heterogeneous zone. (**a**) Photograph of the heterogeneous zone consisting of irregular-shaped leucocratic dmain and matrix of melanocratic domain. (**b**) Photomicrograph of the heterogeneous zone. Leucocratic dimain is coarse-grained with glass and melanocratic domain is fine-grained with glass. (**c**) and (**d**) Magnified photomicrograph of the heterogeneous zone. Leucocratic domain is coarse-grained with glass and melanocratic domain is fine-grained with glass

7.2 The Koasama Lava Dome

The basal diameter of the Koasama rhyolitic lava dome is 0.8 km with a net height of 0.2 km, yielding an estimated volume of ca 0.08 km^3(Aramaki 1963) (Fig. 26a–c). The Koasama lava dome is directly underlain by the Hotokeiwa lower rhyolitic lava with no evidence of time-interval. Flow foliations are conspicuous around the entire circumference of the lava dome with dip inclined inward, indicating that the lava dome is endogenous (Aramaki 1963).

The phenocrystic content of Koasama lava dome is 10–16 vol% with phenocrysts of plagioclase, orthopyroxene, clinopyroxene, and hornblende (Fig. 26d). The whole-rock SiO_2 content of Koasama rhyolitic lava dome is 72–73 wt%, which is the same as those of the Hotokeiwa lower rhyolitic lava (Fig. 18).

7.3 The Hotokeiwa Middle Lava

The Hotokeiwa middle dacitic to rhyolitic lava with an area of ca 4.9 km^2 and estimated total volume of ca 0.25 km^3 consists of two major lobes: the northern and southern lava lobes (Fig. 7). The southern lobe comprises at least two flow units with a total thickness of more than 70 m, which is exposed in the Hotokeiwa cliff (Fig. 27a). Cooling columnar joints and flow foliation or banding are well developed (Fig. 27b).

The phenocrystic contents of Hotokeiwa middle lava are about 30 vol% with phenocrysts of plagioclase, orthopyroxene, and clinopyroxene. The whole-rock SiO_2 content of Hotokeiwa middle dacitic to rhyolitic lava is 68–72 wt% (Fig. 18). There are no direct evidences of tephrastratigraphy; however, the eruption of Hotokeiwa

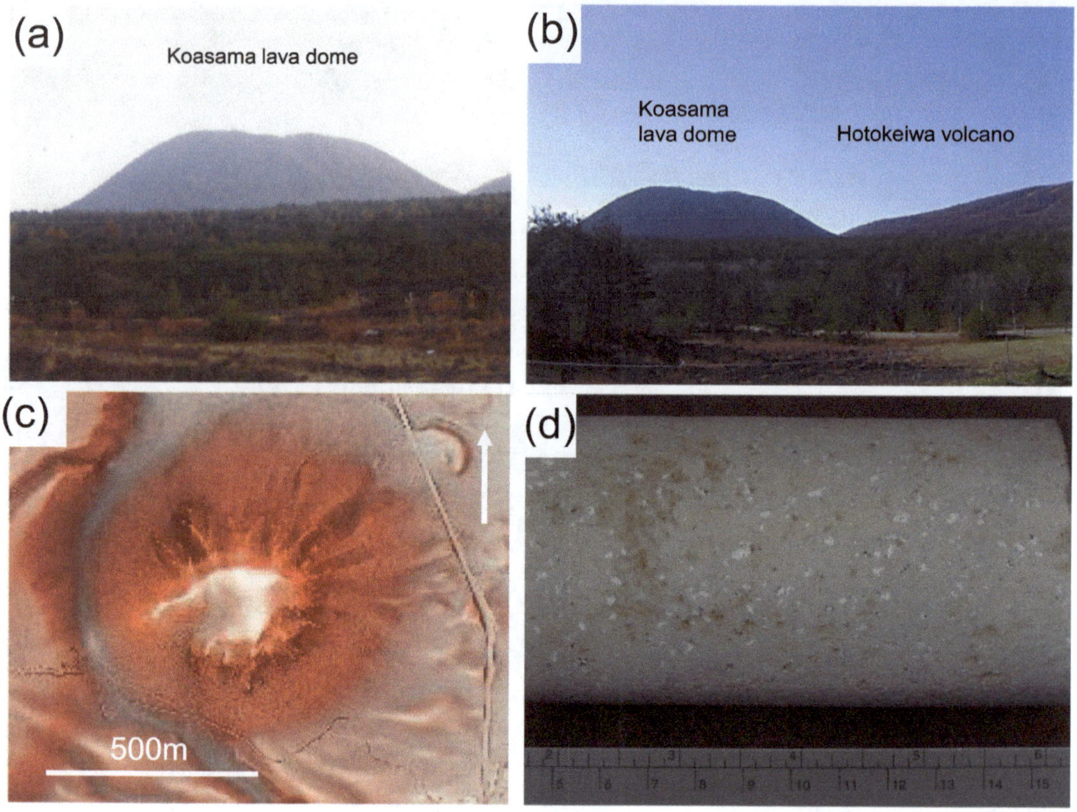

Fig. 26 Photograph, topographic relief map, and lithofacies of the Koasama rhyolitic lava dome. (**a**) Koasama lava dome viewed from the north. (**b**) Koasama lava dome is overlain by the Hotokeiwa middle lava viewed from the north. (**c**) Topographic relief map of the Koasama lava dome. (**d**) Bore-hole core sample of the Koasama lava dome

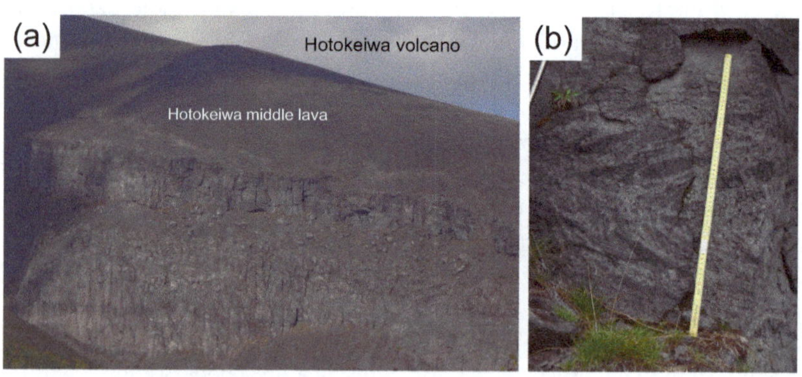

Fig. 27 Photographs showing the Hotokeiwa middle lava flow. (**a**) Hotokeiwa cliff is ca. 70 m high viewed from the summit of Mt. Sekisonzan, which comprises the Hotokeiwa middle lava consisting of two flow units. (**b**) Close view of the Hotokeiwa middle lava with folding flow structure

middle lava is considered to have been concurrent with the As-YP and As-YPk fallouts and the Komoro 1 pyroclastic density current deposits, as the whole-rock MgO compositional trend of Hotokeiwa middle dacitic to rhyolitic lava on the SiO_2 variation diagram is the same as those of such pyroclastic deposits (Fig. 18).

7.4 The Hotokeiwa Upper Lava

The Hotokeiwa upper andesitic lava flow ca 3 km in length and ca 0.7 km in width has a flow front cliff and lava levee (Fig. 7); its average thickness is ca 20 m and its volume is 0.02 km³. The phenocrystic content of the Hotokeiwa upper lava is 38 vol% with phenocrysts of plagioclase, orthopyroxene, and clinopyroxene. The whole-rock SiO_2 content of Hotokeiwa upper andesitic lava is 58–59 wt% (Fig. 18) and the microscopic texture of Hotokeiwa upper lava resembles those of the lavas of Maekake volcano.

8 The Uniqueness of Volcanic Activity of the Hanareyama and Hotokeiwa Volcanoes

The volcanic activities of the Hanareyama and Hotokeiwa volcanoes (5.93 km³ DRE in total eruptive volume) are unique compared with other volcanoes in the Eboshi-Asama volcano group in that they are characterized by voluminous rhyolitic to dacitic fallout pumice and pyroclastic density current deposits, which are the products of Plinian to sub-Plinian eruptions associated with thick silicic lava flows. In contrast, the Sennin and Maekake volcanoes comprise andesitic eruptive materials produced by sub-Plinian to Vulcanian eruptions. The large-scale sector collapse of Sennin and Kurofu volcanoes (Takemoto and Kubo 2003; Yoshida and Sugai 2006; Yasui et al. 2019) probably triggered the more explosive volcanic activity of Hanareyama and Hotokeiwa volcanoes. The large-scale sector collapse may have destroyed the magmatic plumbing system beneath the Sennin volcano and made it difficult to transport magma to the surface from the magma chamber. As a result, silicic magma remained longer in the magma chamber after the large-scale sector collapse. The inflated and pressurized magma chamber fractured the surrounding crust to construct new pathways to the surface, which resulted in the formation of monogenetic lateral volcanoes such as the Hanareyama and Koasama silicic lava domes. The stored voluminous silicic magma occasionally gave rise to the large-scale explosive pumice eruptions; eruptive volume of the largest climactic eruption is 2.81 km³ DRE, yet the time interval of the eruption is longer and the averaged long-term eruption rate (0.00042 km³ DRE/year) is slightly higher than that of Maekake volcano (0.00027 km³ DRE/year) but evidently lower than that of active stage III (A, B-B′ and C) of Maekake volcano (0.0012 km³ DRE/year) (Takahashi et al. 2022) (Fig. 28).

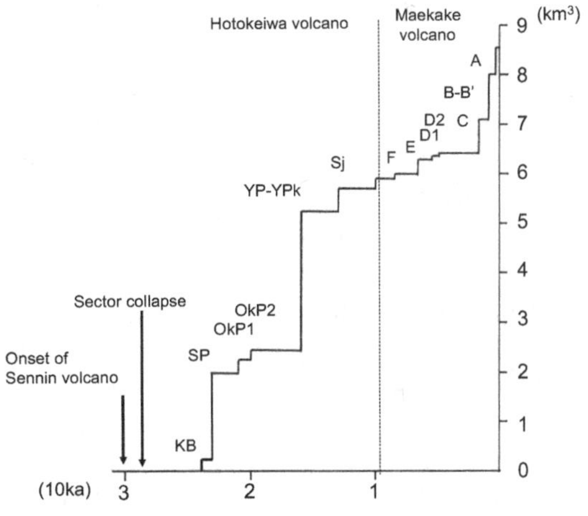

Fig. 28 Diagram showing the relation between eruptive volumes and ages of Hanareyama-Hotokeiwa and Maekake volcanoes (Takahashi et al. 2022). A: 1783 AD eruption (0.51 km³ DRE). B-B′: 1108 AD (0.82 km³ DRE) and 1128 AD (0.10 km³ DRE) eruptions. C: second-century eruption (0.68 km³ DRE). D1: 4800 years calBP eruption (0.06 km³ DRE). D2: 5200 years calBP eruption (0.06 km³ DRE). E: 6500 years calBP eruption (0.27 km³ DRE). F: 8300 years calBP eruption (0.13 km³ DRE). Sj: 13,000 years calBP eruption (0.42 km³ DRE). YP-YPk: 16,000 years calBP eruption (2.81 km³ DRE). OkP1: 21,000 years calBP eruption (0.26 km³ DRE). OkP2: 20,000 years calBP eruption (0.21 km³ DRE). SP: 23,000 years calBP eruption (1.8 km³ DRE). KB: 24,000 years calBP eruption (0.21 km³ DRE)

9 Summary

1. The volcanic activity of the Hanareyama and Hotokeiwa volcanoes is unique in the Asama volcano group and characterized by voluminous rhyolitic to dacitic fallout pumice, pyroclastic density current deposits, and lavas. The thick lava flows constructed a flat shaped lava cone called the Hotokeiwa volcano.
2. The volcanic activity commenced with the eruption of the monogenetic Hanareyama dacitic to rhyolitic lava dome (65–76 wt% SiO_2) and Kumoba pyroclastic density current deposit (60–70 wt% SiO_2) at ca 24,000 years calBP.
3. The volcanic activity of Hotokeiwa polygenetic volcano started with the eruption of dacitic to rhyolitic Shiraitonotaki fallout pumice deposit (As-SP) (66–72 wt% SiO_2), rhyolitic Hotokeiwa lower lava (72–73 wt% SiO_2), and Koasama lava dome (72–73 wt% SiO_2) at ca 23,000 years calBP, which was followed by the dacitic Okubozawa 1 (66–71 wt% SiO_2) and 2 (65–69 wt% SiO_2) fallout pumice and pyroclastic density current deposits at ca 21,000 and ca 20,000 years calBP, respectively.
4. The climactic eruption of Hotokeiwa volcano occurred at ca 16,000 years calBP; the dacitic Itahana yellow pumice fallout deposits (As-YP) (67–68 wt% SiO_2), mainly dacitic Komoro 1 pyroclastic density current deposit (52–70 wt% SiO_2), dacitic Tsumagoi pumice fallout deposit (As-YPk) (66–68 wt% SiO_2), and rhyolitic to dacitic Hotokeiwa middle lava (68–72 wt% SiO_2), erupted. The Sojya pumice fallout deposit (As-Sj), dacitic Komoro 2 pyroclastic density current deposit, and probably Hotokeiwa upper andesitic lava (58–60 wt% SiO_2) erupted at ca 13,000 years calBP. The last eruption of Hotokeiwa volcano was the pyroclastic density current deposit at ca 10,000 years calBP.
5. The Hanareyama lava dome is 0.2 km^3 and Kumoba pyroclastic density current is 0.01 km^3 DRE in volume. The eruptive volumes of the Shiraitonotaki fallout pumice deposit, Hotokeiwa lower lava, and Koasama lava dome are 1.21 km^3 DRE, 0.51 km^3, and 0.08 km^3, respectively. The Okubozawa 1 fallout pumice and pyroclastic density current deposits are 0.16 and 0.10 km^3 DRE in volume. The Okubozawa 2 fallout pumice and pyroclastic density current deposits are 0.11 and 0.10 km^3 DRE in volume.
6. The eruptive volumes of the Itahana yellow pumice fallout, Komoro 1 pyroclastic density currentt, Tsumagoi yellow pumice outfall deposits, and Hotokeiwa middle lava are 0.41, 1.32, 0.83 km^3 DRE, and 0.25 km^3, respectively. The eruptive volumes of the Komoro 2 pyroclastic density current and Sojya pumice fallout deposits, and Hotokeiwa upper lava are 0.4, 0.02 km^3 DRE, and 0.02 km^3, respectively. The volume of pyroclastic density current deposit at 10,000 years calBP is less than 0.2 km^3 DRE.
7. The estimated total eruptive volume of Hanareyama and Hotokeiwa volcanoes is 5.93 km^3 DRE. The eruption rate of Hanareyama-Hotokeiwa volcanoes is 0.00042 km^3 DRE/year, which is higher than that of Maekake volcano (0.00027 km^3 DRE/year) but lower than that of the active stage III of Maekake volcano (0.0012 km^3 DRE/year).
8. The uniqueness of volcanic activity of Hanareyama and Hotokeiwa volcanoes is probably closely related to the large-scale sector collapse of Sennin and Kurofu volcanoes at ca 28,000 years calBP, which may have destroyed the magmatic plumbing system beneath Sennin volcano and made it difficult for magma to migrate to the surface from the magma chamber; resultantly, a large amount of silicic magma was stored in the magma chamber and gave rise to the eruption of voluminous silicic magma.

Acknowledgments We are grateful to Aramaki S., Takemoto H., Mukai Y., Nakajima T., Okada M., Tsurumaki R., Kimoto Y., Abe T., Nagata S., Yajima T., Ishibashi T., Seo N., Sasaki T. for cooperation and help with the study. We also appreciate Martin A. for critical reading and revising of the manuscript.

References

Aramaki S (1963) Geology of Asama volcano. J Fac Sci Univ Tokyo Sec II XIV:229–443

Higuchi K (1989) Study of the history of volcanic activity of Asamayama. Chikuma 63:15–33. (in Japanese)

Nakamura T, Tsuji S, Takemoto H, Ikeda A (1997) ^{14}C age measurements with accelerator mass spectrometry of Asama tephra stratigraphic samples around Minami-Karuizawa, the latest Pleistocene, Nagano prefecture, central Japan. J Geol Soc Japan 103:990–993. (in Japanese with English abstract)

Nakazawa H, Arai F, Endo K (1984) Tephrostratigraphy of the Kurofu-Maekake stages of Asama volcano. In: Abstract of the meeting of the Japan Association for Quaternary Research, 14:69–70. (in Japanese)

Suzuki M (1991) Tephrostratigraphy and petrochemistry of the UG tuff in the uppermost part of Tachikawa loam. Bull Doto Univ 25:87–97. (in Japanese with English abstract)

Takahashi M, Mukai Y, Nakajima T, Yasui M, Kanamaru T (2008) Whole-rock major element chemistry for eruptive products of Asama-Hotokeiwa volcano: summary for data of 307 samples. Proc Inst Nat Sci Nihon Univ 43:167–193 (in Japanese with English abstract)

Takahashi M, Yasui M, Kanamaru T, Nakagawa M (2022) Eruptive volume and age of large-scale pumice eruptions of the Asama-Maekake volcano, central Japan, as revealed by detailed trench surveys: implications for future volcanic hazards. J Disaster Res 17:1080–1089. (in Japanese)

Takemoto H, Kubo S (1995) Tephra of Gunma Miyamabunko, 180p (in Japanese)

Takemoto H, Kubo S (2003) Tephra stratigraphy of the Okuwa debris avalanche deposit of Asama volcano. Proc Inst Nat Sci Nihon Univ 38:55–64. (in Japanese with English abstract)

Togashi S (1983) 14C age of charcoal in the first pumice flow deposits of Asama volcano, Japan. J Vocanol Soc Japan 28:163–165. (in Japanese with English abstract)

Tsuji S, Miyaji N, Arai F (2004) Tephrostratigraphy and chronology of Asama volcano at Minami-Karuizawa as a basis of studies on environmental and disaster history. Bull Natl Mus Jpn Hist 118:165–187. (in Japanese with English abstract)

Yasui M, Takahashi M, Kanamaru T (2019) Rediscovery of the Higuruma debris avalanche deposit in Asama volcano. Proc Inst Nat Sci Nihon Univ 54:123–142. (in Japanese with English abstract)

Yoshida H, Sugai T (2006) Morphological characteristics of hummocks originating from the 24ka sector collapse event of Asama volcano, central Japan. J Geogr 115:638–646. (in Japanese with English abstract)

Geology of Asama-Maekake Volcano

Masaki Takahashi, Maya Yasui, Tatsuo Kanamaru, and Masashi Nagai

1 Introduction

The current volcanic activity of Maekake volcano commenced approximately 9.5 ka (Yasui et al. 2021; Takahashi et al. 2022). The Maekake volcano consists of welded pyroclastic cones, clastogenic lava flows, pyroclastic density current, and fallout pumice deposits (Figs. 1 and 2). The volcanic edifice of Maekake volcano including pyroclastic cones, clastogenic lavas, and pyroclastic density current deposits covers an area of ca 5.5 km in EW and ca 24 km in NS (Fig. 3). The height of the summit of Maekake volcano, consisting of the Kamayama pyroclastic cone which was constructed by the 1783 AD eruption, is 2568 m above sea level. The eruption styles of Maekake volcano are explosive and comprise phreatic, phreatomagmatic, Vulcanian, and sub-Plinian eruptions. In this chapter, the process that constructed the volcanic edifice of Maekake volcano is overviewed.

M. Takahashi (✉)
Institute of Natural Sciences, College of Humanities and Sciences, Nihon University, Tokyo, Japan
e-mail: takahashi.masaki@nihon-u.ac.jp

M. Yasui · T. Kanamaru
Department of Earth and Environmental Sciences, College of Humanities and Sciences, Nihon University, Tokyo, Japan
e-mail: yasui.maya@nihon-u.ac.jp; kanamaru.tatsuo@nihon-u.ac.jp

M. Nagai
Volcano Disaster Resilience Research Division, National Research Institute for Earth Science and Disaster Resilience, Tsukuba, Ibaraki, Japan
e-mail: mnagai@bosai.go.jp

2 Eruption Style

The eruption style of Maekake volcano ranges from small- to intermediate-scale phreatic to Vulcanian eruptions to large-scale sub-Plinian eruptions.

2.1 Small- to Intermediate-Scale Eruptions

The style of the small-scale eruptions is phreatic to phreato-magmatic, the eruptive volumes of which are up to 10^{-5} km^3 DRE, while that of intermediate-scale eruptions are Vulcanian associated with phreatic to phreato-magmatic, Strombolian, and ash eruptions, whose eruptive volumes are 10^{-4} to 10^{-3} km^3 DRE (Nakada et al. 2005; Maeno et al. 2010). Minakami (1935) ranked the small- to intermediate-scale eruptions of Maekake volcano according to explosive energy from A to D (Miyazaki 2003); A: 10^{19} to 10^{20} erg (10^{12} to 10^{13} J); B: 10^{18} to 10^{19} erg (10^{11} to 10^{12} J); C: 10^{17} to 10^{18} erg (10^{10} to 10^{11} J); D: $<10^{17}$ erg ($<10^{10}$ J). However, this classification does not necessarily represent eruptive energy, because it does not take into account eruptive volumes.

S. Aramaki et al. (eds.), *The Asama Volcano*, Active Volcanoes of the World,
https://doi.org/10.1007/978-981-95-5799-8_9

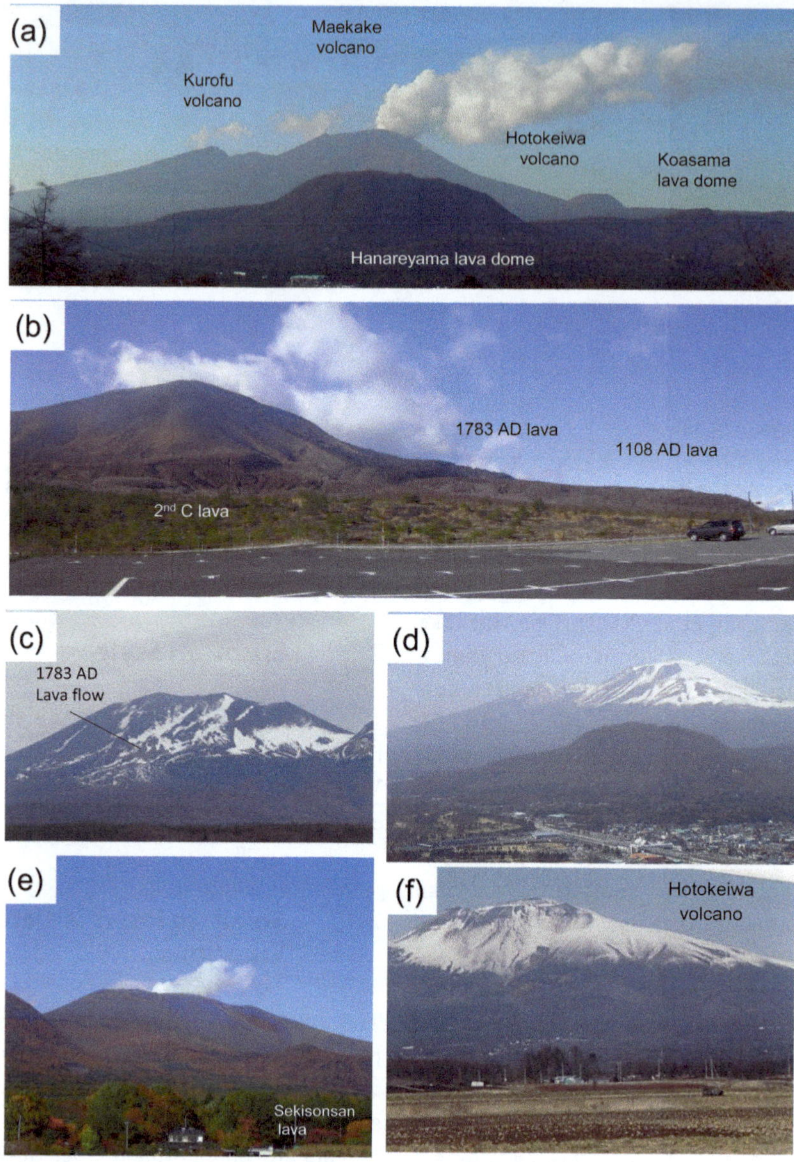

Fig. 1 View of the Maekake volcano. (**a**) Maekake volcano viewed from the east. The central volcanic edifice with an active fumarole is the Maekake volcano. The small hill to the north is the Koasama lava dome. A peak to the south is the rim of the collapsed caldera of Kurofu volcano (Mt. Kengamine). A hill in front is the Hanareyama lava dome. (**b**) Maekake volcano viewed from the northeast. Historical lava flows generated by large-scale eruptions flowed down the northern flank of Maekake volcano—from younger to older, the Onioshidashi (1783 AD), Kamibutai (1108 AD), and Shimobutai and Kuromamegawara (second century) lavas. (**c**) Maekake volcano viewed from the northwest. The slope is composed of the Onioshidashi lava flow from the Kamayama pyroclastic cone in 1783 AD. (**d**) Maekake volcano viewed from the east with Hanareyama lava dome and Karuizawa town in front. (**e**) Maekake volcano viewed from the south. There are no lava flows on the southern flank of Maekake volcano. (**f**) Maekake volcano viewed from the south. The Maekake volcano overlays the Hotokeiwa volcano with a gentle slope to the east

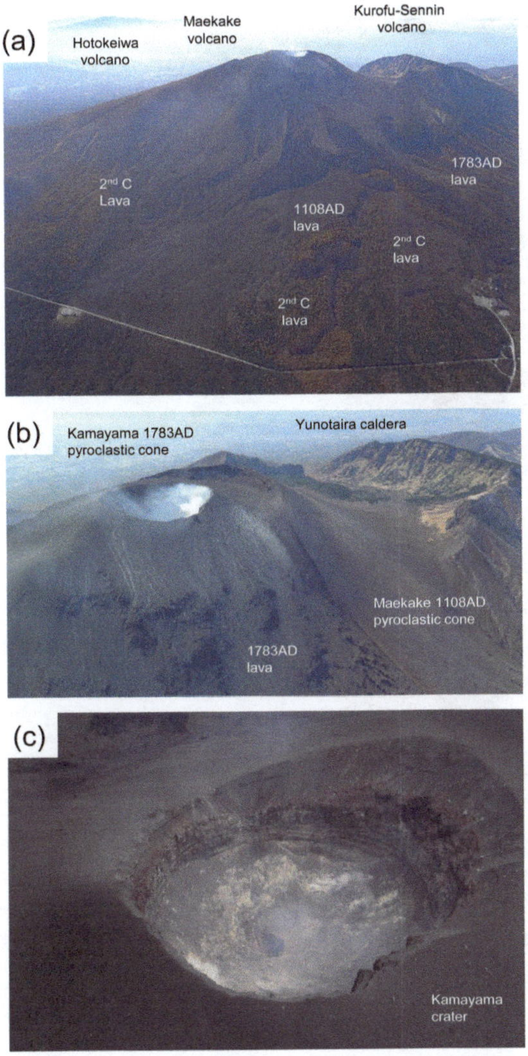

Fig. 2 Aerial photographs showing the Maekake volcano (courtesy of Japan Meteorological Agency and Japan Self-Defense Forces). (**a**) Maekake volcano viewed from the northeast. Recent lavas such as the Onioshidashi (1783 AD), Kamibutai (1108 AD), and Kuromamegawara and Shimobutai (second century) lavas flowed down from the summit crater. (**b**) Craters of Kamayama (1783 AD) and Maekake (1108 AD) pyroclastic cones viewed from the north. (**c**) Crater of Kamayama pyroclastic cone with a diameter of 0.44 km

The main volcanic hazards caused by the intermediate-scale Vulcanian eruptions are fall-out ash, ballistic bombs, pyroclastic density current deposits, and lahars. The ballistic bombs reach up to 4 km of the summit crater, and pyroclastic density current deposits are distributed within 3 km from the summit crater.

The small- to intermediate-scale eruptions of Maekake volcano are grouped into two types: single eruptions lasting up to 4 years and continuous eruptions lasting more than 5 years (Fig. 4; Table 1). Single eruptions are a series of eruptions, in which more than 2 dormant years are present before and after the eruption, and the duration of intermittent eruptive activity is less than 4 years. The continuous eruptions are a series of eruptions, in which more than two dormant years are present before and after the eruption period, but the duration of intermittent activity is long and lasts more than 5 years. The total eruptive volume of continuous eruptions is 10^{-3} km^3 DRE. There were seven periods of continuous eruptions since the sixteenth century (Table 1 and Fig. 4).

Murase et al. (2007) explained based on leveling data that the magma supply rate to the magma chamber was constant during the single small- to intermediate-scale eruptions from 1962 to 2005 AD, while it remarkably increased during the period of continuous small- to intermediate-scale eruptions from 1930 to 1940 AD.

2.2 Eruptive Sequence of Single Intermediate-Scale Eruptions

Single intermediate-scale eruptions occurred in 1973 and 2004 AD. The intermediate-scale eruption in 1973 AD with an eruptive volume of more than 2×10^{-3} km^3 DRE started with an explosive Vulcanian eruption on February 1 associated with small-scale pyroclastic density current (Shimozuru et al. 1975; Murai 1974; Miyazaki 2003) (Figs. 5 and 6). The explosive Vulcanian eruption produced an eruption column 2000–2500 m in height on February 6 and 14. The style of eruption changed from Vulcanian to Strombolian on February 16–18, in which lava probably appeared at the bottom of the crater (Figs. 5 and 6), but did not overflow the rim of the crater. The explosive Vulcanian eruption with small-scale pyroclastic density current occurred again on March 10 after a dormant period of 18 days. The pyroclastic density current reached up to 3 km from the summit crater. After a dormant period of a month, explosive Vulcanian

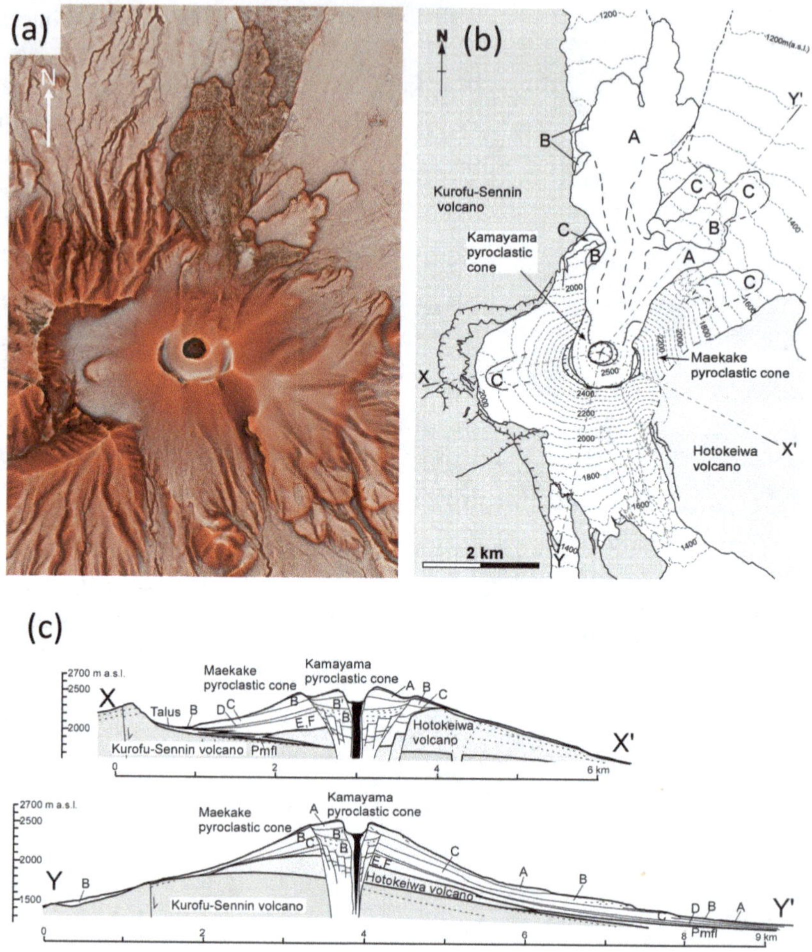

Fig. 3 Map showing the Maekake volcano. (**a**) Topographic relief map of Maekake volcano (courtesy of T. Chiba). (**b**) Geological sketch map with topographic contours. X-X′ and Y-Y′ are cross-section lines. (**c**) Cross-sections of Maekake volcano along the X-X′ and Y-Y′ lines in (b)

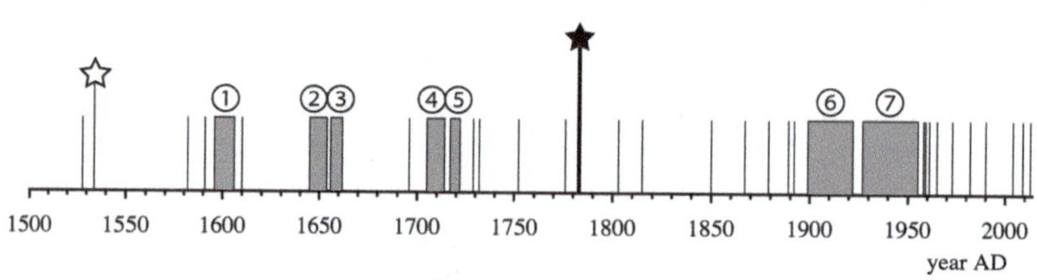

Fig. 4 The eruption history of Maekake volcano since the sixteenth century. Solid star: 1783 AD Class A large-scale eruption (Tenmei eruption); open star: 1532–1534 AD Class B large-scale eruption (As-A′); broad gray band with numeric number: period of continuous eruptions from period 1 (①) to 7 (⑦). Data from Miyazaki (2003)

eruptions with column heights of 4000–4500 m took place on April 18 and 26. The last eruption was a small-scale phreatic to Vulcanian eruption on May 24. The duration of activity of a single intermediate-scale eruption in 1973 AD was approximately 3.8 months.

The intermediate-scale eruption in 2004 AD had an eruptive volume of 2×10^{-3} km^3 DRE and commenced with an explosive Vulcanian eruption with an eruption column height of 3500–5500 m on September 1 (Nakada et al. 2005). After a 2-week dormant period, the volcanic activity

Table 1 Eruptive events (year AD) of the different styles of eruption of Maekake volcano since the sixteenth century. Data from Miyazaki (2003)

Eruption style	Eruptive event (year AD)
Small-scale eruption Phreatic to phreato-magmatic eruption	(1) 1776–1777 (2) 1879 (3) 1965 (4) 1982–1983 (5) 1990 (6) 2009 (7) 2015 (8) 2019
Intermediate-scale eruption A single eruption Vulcanian eruption	(1) 1527–1528 (2) 1582 (3) 1590–1591 (4) 1609 (5) 1695 (6) 1728–1729 (7) 1732-1733 (8) 1752–1754 (9)1803 (10) 1815 (11) 1866–1869 (12) 1889–1890(13) 1894 (14) 1958–1961 (15) 1973 (16) 2004
Intermediate-scale eruption Continuously eruptive period Vulcanian eruption	(1) 1595–1605 (2) 1645–1652 (3) 1655–1661 (4) 1704–1713 (5) 1717–1723 (6) 1899–1924 (7) 1927–1955
Large-scale eruption Sub-Plinian eruption Class A Class B	(1) 1783 (1) 1532–1534

Fig. 5 The eruption of Maekake volcano in 1973 AD. (**a**) Eruption column of a Vulcanian eruption viewed from the east, March 10, 1973 (courtesy of E. Koyama). (**b**) Eruption column of a Vulcanian eruption with a small pyroclastic density current running down the northern slope viewed from the east, February 6, 1973 (courtesy of S. Aramaki). (**c**) Eruption column of a Vulcanian eruption viewed from the east, February 14, 1973 (courtesy of S. Aramaki). (**d**) Lava with a diameter of ca 150 m filling the bottom of Kamayama crater (courtesy of S. Aramaki)

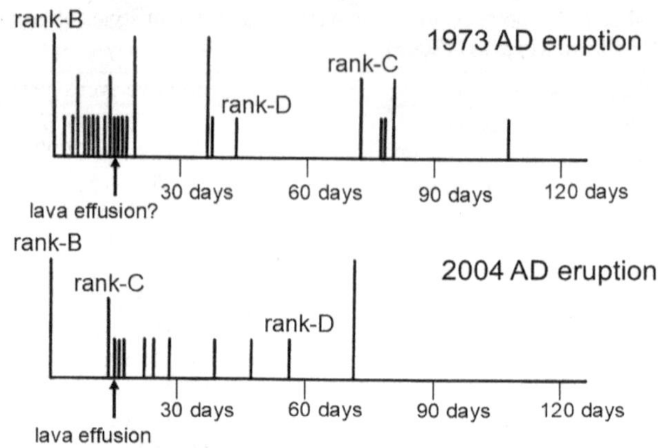

Fig. 6 The eruptive sequence of the intermediate-scale eruptions in 1973 and 2004 AD. The alphabet ranks denote the intensity of explosive energy of eruptions. The time scale is shown in days

resumed on September 14–15 and three Vulcanian eruptions occurred, producing eruption columns of 300, 1000, and 2500 m in succession. The style of eruption changed to Strombolian on September 16–18, and a lava with a diameter of 200 m filled the bottom of the crater, but did not fill the crater and overflow its rim. Following the effusion of lava, small-scale eruptions repeated on September 23, 25, 29 and October 10, 19, and 28. The last explosive Vulcanian eruption with an eruption column height of 3500–5500 m occurred on November 14, and then the eruption in 2004 AD ceased. The duration of activity of the intermediate-scale eruption in 2004 AD was approximately 2.5 months.

The intermediate-scale eruptions thus typically start with explosive Vulcanian eruptions. After repeated Vulcanian eruptions, the eruption style changes to Strombolian with effusion of lava at the bottom of the crater. Vulcanian eruptions resumed, and finally the volcanic activity ceased. The ascent of vesiculated viscous magma occurs during the explosive Vulcanian eruptions. The continuous ascent of viscous magma makes it easy for less viscous magma to rise and the eruption style changes from Vulcanian to Strombolian; ascending magma finally appeared as lava at the bottom of the crater. When the top of the magma column began to descend, the Vulcanian eruptions resumed, blowing off the lava cap filling the conduit. The volcanic activity ceases when the magma head completely descends to a deeper level from the shallow level of the vent.

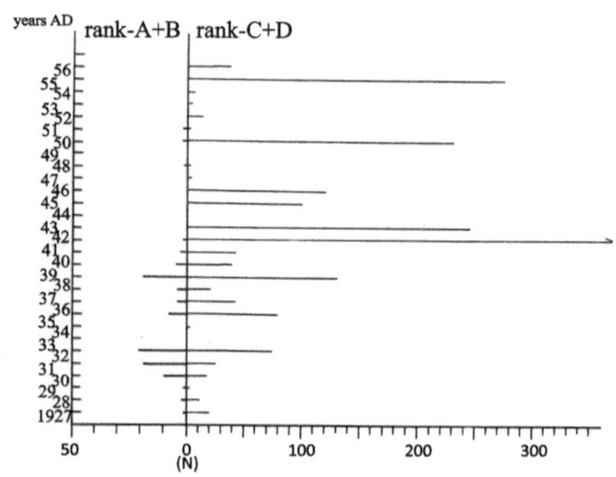

Fig. 7 The eruptive sequence of the continuous small- to intermediate-scale eruptions period 7 from 1927 to 1956 AD. Data from Miyazaki (2003). Vertical line: ages in years (AD); horizontal line: number of eruptions; A, B, C, and D: ranks of explosive energy of eruption

2.3 Eruptive Sequence of Continuous Eruptions

The eruptive sequence of the continuous eruptions in period 7 (1923–1955 AD) is summarized in Fig. 7 (Miyazaki 2003). Explosive Vulcanian eruptions with high explosive energy ranked A and B were conspicuous in the first half of the eruption sequence, whereas intermediate eruptions ranked C and D with low explosive energy including Strombolian and ash eruptions were predominant in the latter half of the sequence. A leveling study revealed that the magma supply rate increased remarkably in the first half of the sequence but decreased in the latter half (Murase et al. 2007).

2.4 Large-Scale Eruption

The large-scale eruptions are sub-Plinian and produce fallout pumice deposits, pyroclastic density current deposits, pyroclastic cones, and clastogenic lava flows. The eruptive volume of the large-scale eruptions which constructed the volcanic edifice of Maekake volcano is considerably higher than those of the small- to intermediate-scale eruptions, which do not alter the volcanic edifice.

The large-scale eruptions are grouped into Class A and B according to eruptive volume. The eruptive volume of Class A large-scale eruptions is 10^{-2} to 10^{-1} km^3 DRE, while that of Class B is 10^{-3} km^3 DRE. Class A corresponds to Class 1 and 2 (Yasui et al. 2021), and Class B to Class 3 and 4 (Yasui et al. 2021), respectively. There were eight Class A large-scale eruptions in the eruptive history of Maekake volcano: As-A (1783 AD), As-B′ (1128 AD), As-B (1108 AD), As-C (second century), As-D1 (4,785 years calBP), As-D2 (5,232 years calBP), As-E (6,447 years calBP), and As-F (8,343 years calBP) eruptions (Takahashi et al. 2022) (Table 2).

2.5 Eruptive Sequence of Class A Large-Scale Eruptions

The eruptive sequence of the Class A large-scale eruptions is estimated from a representative section of fallout pumice and ash deposits 5 km east of the summit crater (Fig. 8). The fallout pumice deposit consists of pumice and ash beds. The pumice beds are products of sub-Plinian eruptions. The ash beds are grouped into two different types: coarse-grained (volcanic sand) and fine-grained (volcanic silt) ash beds. The ash beds are predominantly produced by large-scale Vulcanian eruptions, but some of the fine ash beds are probably derived from ash clouds associated with pyroclastic density current. Figure 9 summarizes the estimated sequence of large-scale eruptions.

The 1783 AD large-scale eruption is described in detail in Yasui and Koyaguchi (2004). The onset of the large-scale eruption of 1783 AD (As-A) was a small-scale eruption on May 9, 1783. Explosive intermediate-scale Vulcanian eruption occurred on June 25 after a dormant period of 46 days, the eruptive product of which is a thin ash bed less than several millimeters in thickness at the bottom of fallout pumice deposit. After a dormant period of 22 days, a sub-Plinian eruption occurred on July 17. After small eruptions continued from July 21 to 26, the sub-Plinian eruption resumed on July 27 and continued until August 3. A violent climactic sub-Plinian eruption occurred on August 4–5. Most of the fallout pumice deposit was formed by this climactic eruption; it is intercalated by some thin beds of pinkish fine ash, which were probably the co-ignimbrite ash of pyroclastic density currents concurrent with the deposition of the fallout pumice. The main eruptive event producing an intermediate-type pyroclastic density current occurred after the pumice fall, but the construction of the pyroclastic cone and formation of clastogenic lava were simultaneous with the eruption of the fallout pumice. The volcanic activity of the climactic eruption had ceased by noon of August 5.

The 1128 AD large-scale eruption (As-B′) started with a large-scale sub-Plinian eruption

Table 2 Eruptive volumes (km^3 DRE) of fallout pumice deposits, pyroclastic cones, pyroclastic density current (PDC) deposits, lavas, and total eruptive products of the Class A large-scale eruptions of Maekake volcano

	As-A	As-B′	As-B	As-C	As-D1 As-D2	As-E	As-F
Fallout pumice deposit	0.03	0.04	0.10	0.05	0.01	0.04	0.01
Pyroclastic cone	0.04	0.05	0.13	0.06	0.01	0.05	0.01
PDC deposit	0.15	0.01	0.36	0.26	0.02	0.08	0.09
Lava	0.29	–	0.23	0.31	0.02	0.10	0.02
Total (km^3 DRE)	0.51	0.10	0.82	0.68	0.06	0.27	0.13

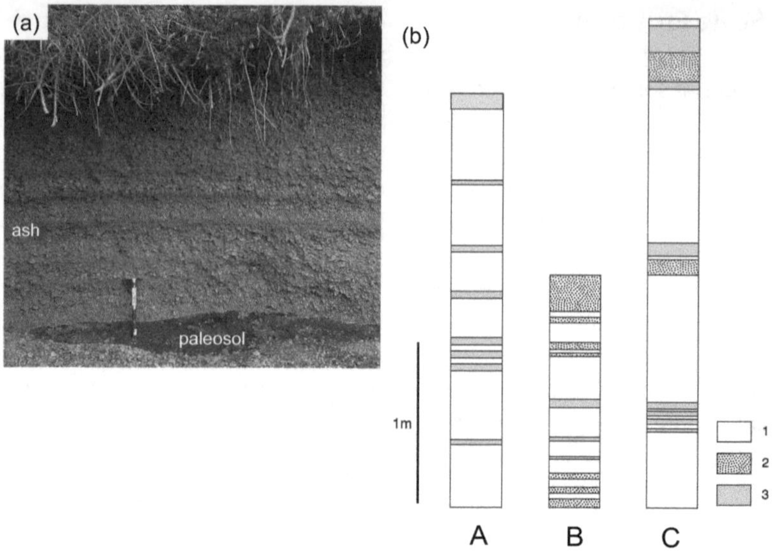

Fig. 8 Columnar sections of fallout pumice deposits produced by a large-scale eruption. (**a**) Photograph of the fallout pumice deposit (As-A) erupted in the 1783 AD large-scale eruption at a locality ca 5 km east from the summit crater. The fallout pumice bed overlays a black paleosol bed. Pinkish ash layers are co-ignimbrite ash. (**b**) Representative columnar sections for fallout pumice deposit of a large-scale eruption in active stage III. A: Fallout pumice deposit of 1783 AD eruption at ca 5 km east from the summit crater. B: A fallout pumice deposit of the 1108 AD eruption at ca 10 km east of the summit crater. C: A fallout pumice deposit of the second-century eruption at ca 4 km east of the summit crater. 1: pumice bed. 2: coarse-grained ash (volcanic sand). 3: fine-grained ash (volcanic silt)

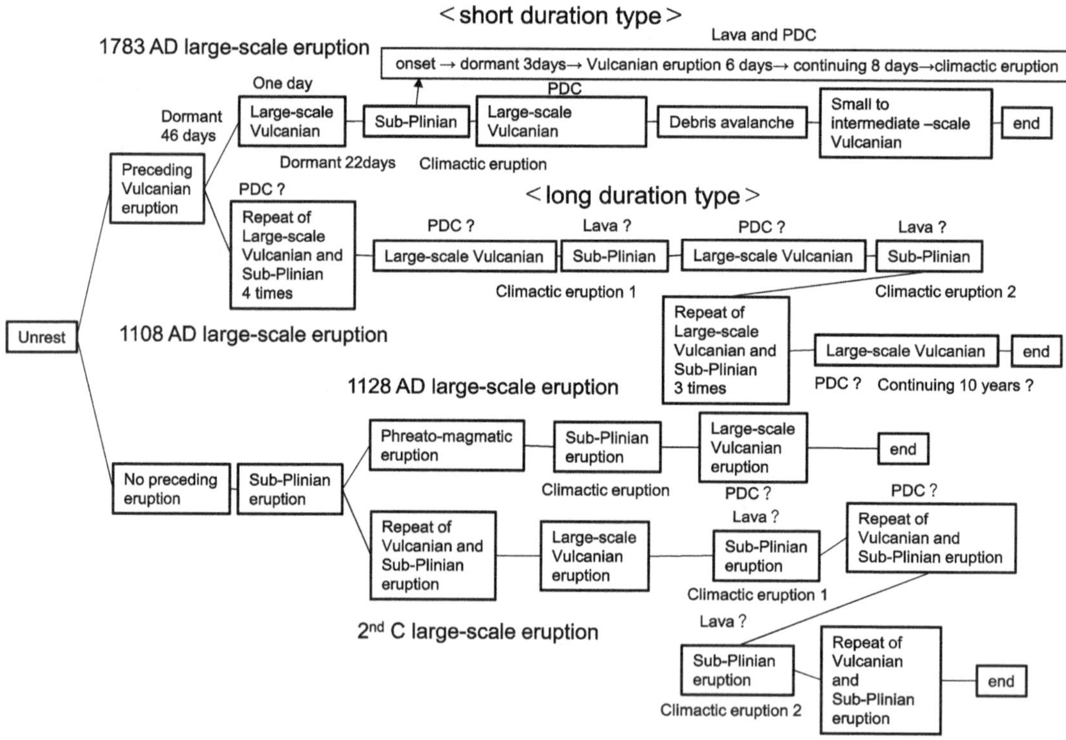

Fig. 9 Estimated sequence of eruption of large-scale eruptions in the active stage III based on the stratigraphy of fallout pumice deposits

without any preceding smaller scale eruption. A phreatomagmatic eruption then occurred, which was followed by a climactic sub-Plinian eruption. The pumice-producing sub-Plinian eruption then changed to large-scale Vulcanian eruptions which may have been associated with pyroclastic density currents. The volcanic activity ended after the large-scale Vulcanian eruptions had ceased.

The 1108 AD large-scale eruption (As-B) commenced with a large-scale Vulcanian eruption, which was followed by four cycles of alternating large-scale Vulcanian and sub-Plinian eruptions. This was then followed by a large-scale Vulcanian eruption, two larger-scale sub-Plinian eruptions intercalated by a large-scale Vulcanian eruption, then by three repeats of alternating large-scale Vulcanian and sub-Plinian eruptions. After a series of large-scale eruptions ceased, the large-scale Vulcanian eruptions were estimated to have continued intermittently for more than 10 years, forming thick stratified red fallout ash deposits. The relationships between the eruptive sequence and the formation of pyroclastic density current, pyroclastic cone, and clastogenic lava are not necessarily clear from outcrop evidences; however, the eruption of pyroclastic density currents was inferred to be related to the Vulcanian eruptions, and the densely welded pyroclastic cone and clastogenic lava were probably formed by lava fountains associated with the sub-Plinian eruption column.

The second-century large-scale eruption (As-C), estimated to be between 1,830 and 1,702 calBP (120–248 AD) with a median age of 1,766 years calBP (184 AD) based on ^{14}C dating of charcoal at the base of pumice deposits, commenced with a large-scale sub-Plinian eruption and was followed by repeated alternating large-scale Vulcanian and sub-Plinian eruptions. After a large-scale Vulcanian eruption, a large-scale sub-Plinian eruption occurred again. Then, a large-scale Vulcanian and sub-Plinian eruptions repeated alternately, followed by a final large-scale sub-Plinian eruption. The second-century large-scale eruption ceased after the large-scale Vulcanian eruptions and the last sub-Plinian eruption. The relationships between the eruptive sequence and the formation of pyroclastic density current, pyroclastic cone, and clastogenic lava in the second-century large-scale eruption are not clear from outcrop evidences; however, the eruption of pyroclastic density currents was inferred to be related to the Vulcanian eruptions, and the densely welded pyroclastic cone and clastogenic lava were probably formed by lava fountains associated with the sub-Plinian eruption column.

The eruptive sequence of the Class A large-scale eruptions is divided into two types: one associated with a preceding large-scale Vulcanian eruption and one without. The former is the 1783 and 1108 AD eruptions, whereas the latter is the 1128 AD and second-century eruptions. Both types can also be divided according to the duration of the eruption sequence. The 1783 and 1128 AD eruptions occurred over a relatively short time period lasting weeks to months, whereas the 1108 AD and second-century eruptions were much longer and continued for years. The eruptive sequence of the Class A large-scale eruptions is rather complex; the eruptive sequence is not the same in every case.

3 Construction of Volcanic Edifice

The formation of pyroclastic cones, lava flows, and pyroclastic density current deposits contribute to the construction of the volcanic edifice. At the climactic stage of the large-scale eruption of 1783 AD, a big lava fountain concurrent with the ascending sub-Plinian eruption column was observed and painted (Fig. 10). The lava fountain probably formed the welded pyroclastic cone and clastogenic lava flows which constructed the volcanic edifice of Maekake volcano.

3.1 Construction of Pyroclastic Cone

Lava fountains and ejecta of intensive Strombolian eruptions construct pyroclastic cones around the volcanic vent. The present pyroclastic cones of Maekake volcano are the Maekake and

Fig. 10 Old painting of the 1783 AD eruption showing the lava fountain concurrent with an ascending eruption column with volcanic lightning (courtesy of H. Misaizu)

Kamayama pyroclastic cones, which were constructed by the large-scale eruptions of 1108 and 1783 AD.

The base diameter and crater of the Kamayama welded pyroclastic cone are 1.34 and 0.44 km, respectively; its height is 170 m and its volume is 0.04 km^3 (Yasui and Koyaguchi 2004). The crater wall of Kamayama crater is composed of unit A (>55 m), unit B (95 m), and unit C (28 m) volcanic deposits (Fig. 11h). Unit A is older pyroclastic material and overlain by unit B with angular unconformity; unit B is densely welded eruptive products of the 1783 AD eruption with columnar joints. Unit C is eruptive products after the 1783 AD eruption. The majority of the Kamayama welded pyroclastic cone was built by lava fountains within 15 hours during the climactic eruption on August 4 (Figs. 12 and 13). The lava fountain effused concurrently with the uprising sub-Plinian eruption column which produced fallout pumice deposits over a wider area.

The diameters of the base and crater of Maekake welded pyroclastic cone are 3.4 and 1.3 km, respectively; its height is 0.5 km and its volume is 0.18 km^3; most parts of the present cone of Maekake volcano are composed of the Maekake welded pyroclastic cone. The wall inside of Maekake crater with a maximum thickness of 70 m consists of 4 units of densely welded pyroclastic fall deposits: Units 1, 2, 3, and 4. Maekake pyroclastic cone is a composite cone made up of two pyroclastic cones (Fig. 11c, d). Units 1 and 2 erupted from the western pyroclastic cone (Fig. 12a,b), and after the collapse of the crater of the western pyroclastic cone, Units 3 and 4 erupted from the eastern crater to form the eastern pyroclastic cone. The original two craters collapsed and amalgamated to form the present large ovoidal crater of Maekake welded composite pyroclastic cones (Yasui and Takahashi 2015).

3.2 Clastogenic Lava Flows

Lava fountains often result in fountain-fed lava flow (clastogenic lava flow). The nature of clastogenic lava flow depends on many complicated factors, such as temperature and viscosity of magma, eruption rate, size and cooling rate of effused spatters or clasts (Wolff and Sumner 2000). The lava fountains with high temperature and low viscosity magma fall down to the surface and are accumulated to form fountain-fed lava just like the ordinary overflow lava as is seen in the Hawaiian volcanoes such as the Kilauea volcano. The lava fountain with low temperature, low effusion rate, and relatively high viscosity magma constructs a weakly welded pyroclastic or spatter cone. Various types of clastogenic lava flows are produced within the variation between the above two end-member types. The lava fountain with relatively high temperature and high effusion rate does not construct a conspicuous pyroclastic cone around the vent but flows down like overflow lava; however, it is not a simple liquid lava but a densely welded rheomorphic clastogenic lava. Some pyroclastic cones with densely welded fluidal interior flow down preserving the surficial shape of the original pyroclastic cone (rheomorphic pyroclastic cone). The Onioshidashi lava of the 1783 AD eruption (Figs. 13, 14, 15 and

Fig. 11 View of the Maekake and Kamayama pyroclastic cones. (**a**) Aerial view of the Kamayama crater from the north (courtesy of T. Koyaguchi). (**b**) Maekake and Kamayama pyroclastic cones viewed from the rim of Yunotaira caldera. (**c**) Western crater wall of the Maekake densely welded pyroclastic cone. The crater wall consists of Unit 1 and 3 welded pyroclastic fall deposits. (**d**) Northwestern crater wall of the Maekake densely welded pyroclastic cone. The crater wall comprises Unit 1 and 2 welded pyroclastic fall deposits. (**e**) Aerial view of the Kamayama crater from the northeast (courtesy of T. Chiba). (**f**) Aerial view of the Maekake and Kamayama pyroclastic cones from the north. The Onioshidashi lava flowed down from the Kamayama pyroclastic cone (courtesy of the Miyota Jomon Museum). (**g**) Aerial view of the Kamayama pyroclastic cone in the crater of the Maekake pyroclastic cone from the west (courtesy of T. Koyaguchi). (**h**) Aerial view of the crater of Kamayama densely welded pyroclastic cone from the southeast. The crater wall is composed of Unit A, B, and C welded pyroclastic fall deposits

16), and the Kuromamegawara, Shimobutai, Higashiizumisawa, and Maruyama lavas from the second-century eruption (Fig. 21) are densely welded clastogenic lavas, and the Kamibutai lava of the 1108 AD eruption is a rheomorphic pyroclastic cone (Figs. 17, 18, 19 and 20). These main clastogenic lavas of Maekake volcano flowed down the north side of the volcano because of the asymmetric shape of the lava fountain which inclined to the north (Fig. 10).

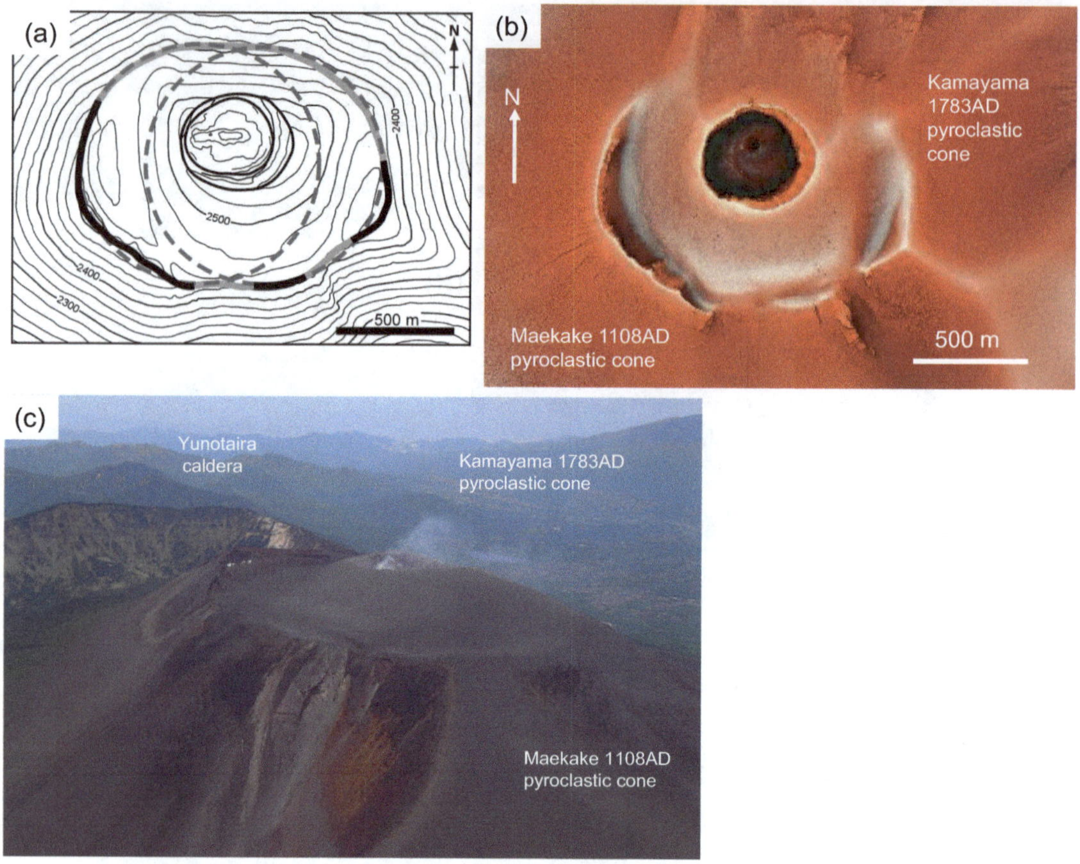

Fig. 12 Photograph, topographic, and topographic relief map showing the Maekake (1108 AD) and Kamayama (1783 AD) pyroclastic cones. (**a**) Topographic map of the Kamayama pyroclastic cone in the ovoidal crater of the Maekake pyroclastic cone (Yasui and Takahashi 2015). The Maekake pyroclastic cone is a composite cone consisting of two craters. (**b**) Topographic relief map of the Kamayama pyroclastic cone in the ovoidal crater of the Maekake pyroclastic cone (courtesy of T. Chiba). (**c**) Aerial view of the Kamayama and Maekake pyroclastic cone from the southeast (courtesy of Japan Meteorological Agency and Japan Self-Defense Forces)

The Onioshidashi lava of the 1783 AD eruption consists of flow units of L1, L2 (a, b and c), and L3 in the order of eruption (Inoue 2002, 2006) (Fig. 14a, b), which arrived up to 5.8 km from the summit crater of the Kamayama pyroclastic cone. The earlier lavas of L1 and L2a are enriched in whole-rock SiO_2 content and similar to those of the As-A fallout pumice deposit, while the later lavas of L2b, L2c, and L3 are less abundant in whole-rock SiO_2 content and correspond to those of the Agatsuma and Kambara pyroclastic density current deposits (Fig. 14c).

The surface morphology of the Onioshidashi lava comprises several topographic units: the flat surface, plateau, extrusion of internal fluidal materials, spines, and blocks (Inoue 2006) (Fig. 15a, b). The original rather flat but rough and ragged surface (Fig. 16c) was broken into plateaus by extension cracks caused by the movement of lava, which further fractured into pieces to form spines or blocks. The interior of the lava is composed of densely welded fluidal rocks with eutaxitic texture (Fig. 16b), which was revealed by a 70 m bore-hole drilled in the central part of the lava (Yasui and Koyaguchi 2004). The internal fluidal materials extruded or squeezed out to the surface of the clastogenic lava through the extensional cracks to form a fluidal lava-like

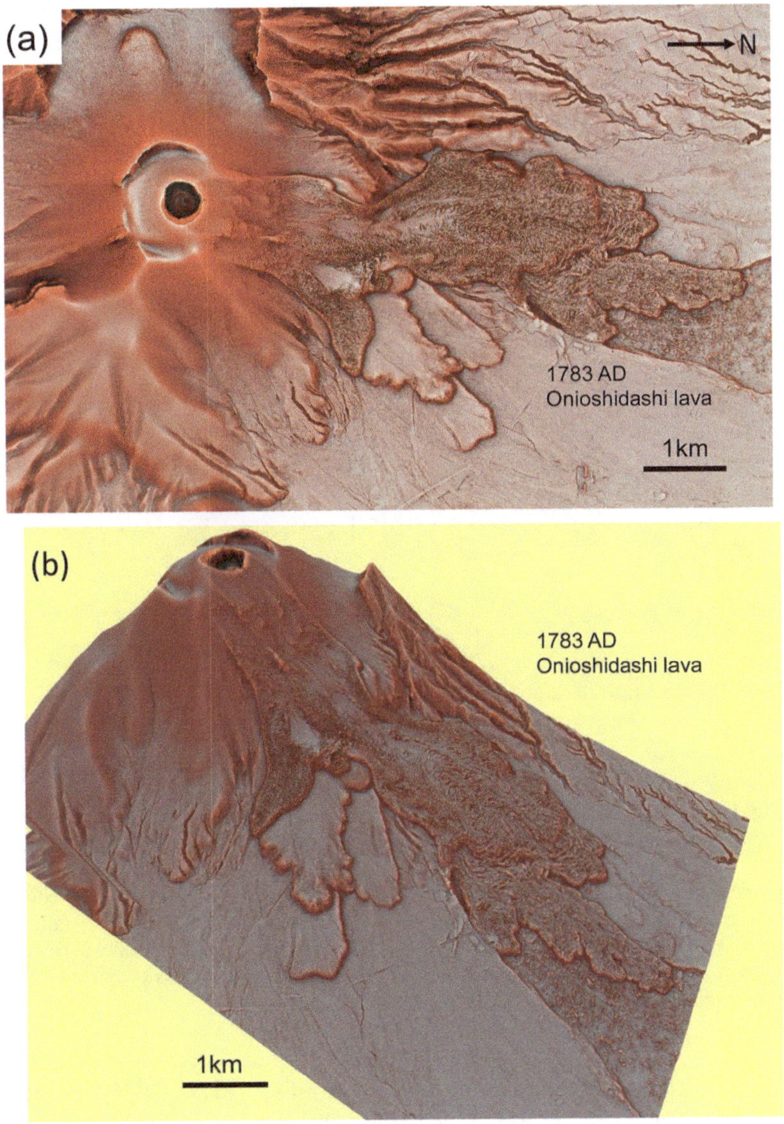

Fig. 13 Topographic relief maps of the Onioshidashi lava flow of 1783 AD. (**a**) Topographic relief map of the Onioshidashi lava flow (courtesy of T. Chiba). (**b**) Rotated view of the topographic relief map of Onioshidashi lava flow (courtesy of T. Chiba)

smooth surface. The extruded internal fluidal materials sometimes form blocky lava consisting of blocks with polygonal smooth surfaces (Fig. 16d).

The interior of Kuromamegawara lava consists of densely welded rocks, the glassy lens of which is folded by the movement of lava (Fig. 22a). The lateral marginal cliff of Kuromamegawara lava comprises a stratified agglutinate of alternation of a thin glassy melanocratic layer and reddish oxidized vesiculated layer with a thickness of several cm (Fig. 22b). The Kuromamegawara and Shimobutai lavas reached up to 4 km from the crater of the Kamayama pyroclastic cone (Fig. 21).

The Kamibutai lava comprises two different units: an alternation of lava and pyroclastic rocks (Fig. 18a, b) and stratified volcanic breccia beds (Fig. 20a), which were part of a welded pyroclastic cone that traveled down slope by rheomorphism and reached up to 4.5 km from the crater of the Kamayama pyroclastic cone (Takahashi and Yasui, 2013). The thickness of the lava is about 1 m and that of intercalated tuff breccia is about 1–4 m (Figs. 18d, e and 19a, b, c, e). A thin layer of volcanic breccia comprising rounded clasts of

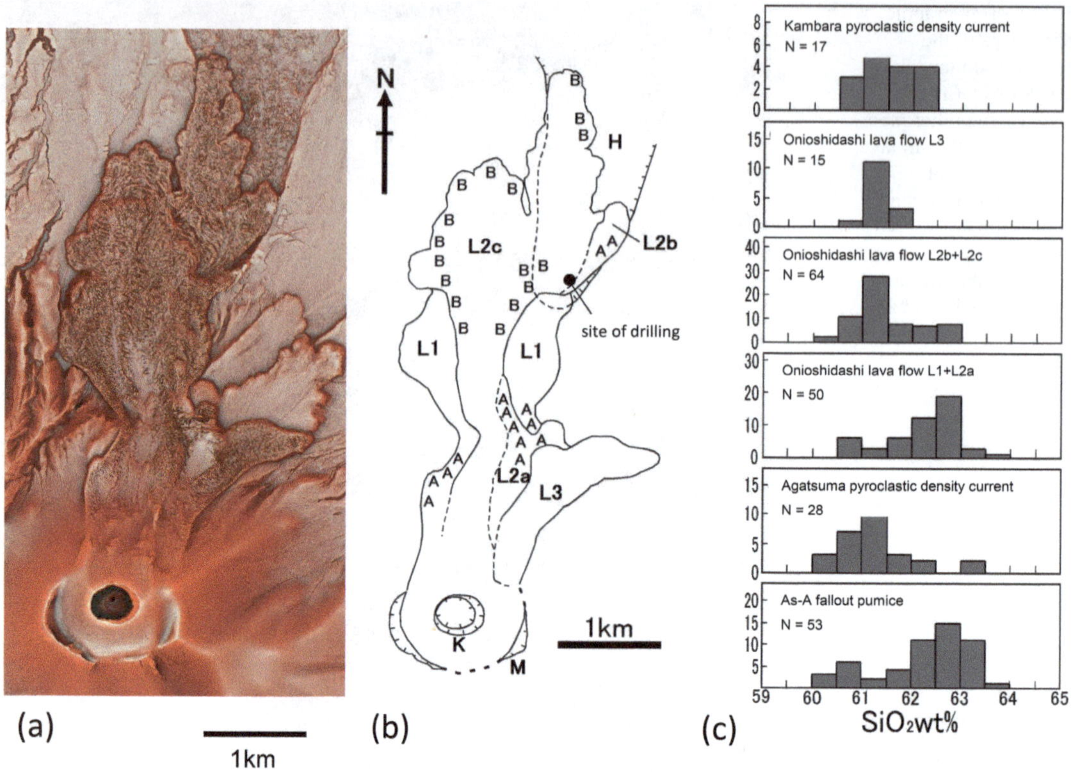

Fig. 14 Flow units of the Onioshidashi lava flow and histograms showing their whole-rock SiO_2 contents together with those of the As-A fallout pumice deposit, Agatsuma pyroclastic density current deposit, and Kambara pyroclastic density current deposit. (**a**) Topographic relief map of the Onioshidashi lava flow. (**b**) Map showing the flow units of the Onioshidashi lava flow (Inoue 2002). The flow units are L1, L2a, L2b, L2c, and L3 in the order of eruption. A: blocks of Agatsuma pyroclastic density current deposit; B: fragments of welded pyroclastic rocks. The site of drilling is shown. (**c**) Histograms showing the whole-rock SiO_2 contents of each unit of Onioshidashi lava flow, As-A fallout pumice deposit, Agatsuma pyroclastic density current deposit, and the Kambara pyroclastic density current deposit

cabbage-like scoria grades to a layer of lava-like densely welded pyroclastic rocks (Fig. 18d, e). The frontal half of the Kamibutai lava comprises stratified volcanic breccias (Fig. 20a–c) and the other half consists of an alternation of lava and tuff breccia; they dip toward the summit crater and are probably a ramp structure (Figs. 18a, b and 20a). Close view of a densely welded pyroclastic rock is shown in Fig. 19d. The alternation of lavas and pyroclastic rocks is developed on the upper slope of Maekake volcano which is a member of the Maekake densely welded pyroclastic cone (Figs. 18c and 19f).

3.3 Pyroclastic Density Current Deposits

The eruptions of Maekake volcano are often associated with pyroclastic density current deposits. Aramaki (1957) classified the pyroclastic density current into three types: pumice, block and ash, and intermediate type pyroclastic density currents. The intermediate type pyroclastic density current deposits include weakly vesiculated volcanic bombs or a cluster of rounded clasts of cabbage-like scoria (Fig. 25a, b); the pyroclastic density current deposits of Class A

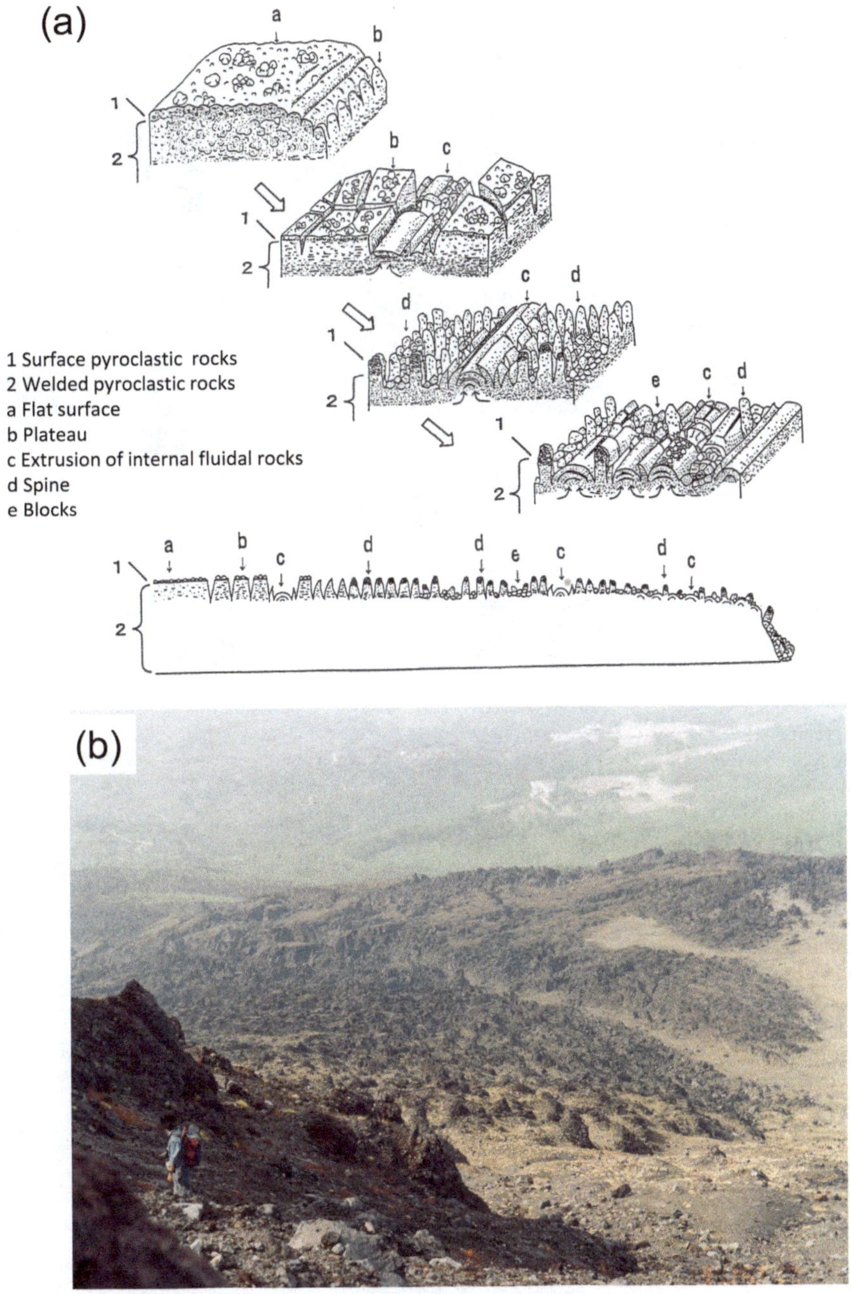

Fig. 15 The surface topography of Onioshidashi lava. (**a**) Units of surface topography of Onioshidashi lava (Inoue 2006). (**b**) Photograph of the surface topography of Onioshidashi lava, showing the various topographic units

large-scale eruptions of Maekae volcano are intermediate type, yet the block and ash flow deposits are rarely included in the Oiwake pyroclastic density current deposit. Intermediate type pyroclastic density currents are rare in Japanese Quaternary volcanoes, and their abundant occurrence is the unique characteristic of Maekake volcano. The style of eruption of intermediate type pyroclastic density current of Maekake volcano is thought to be the same as the fountain

Fig. 16 Photographs of the Onioshidashi lava flow. (**a**) Aerial view of the Onioshidasahi lava on the flank of Maekake volcano denoting its flow units, which shows no evidence of originating from the summit crater of Kamayama pyroclastic cone (courtesy of the Miyota Jomon Museum). (**b**) Densely welded rocks showing eutaxitic texture inside of the Onioshidashi lava flow in the drilling core from Onioshidashi lava flow. The site of drilling is shown in Fig. 14b. (**c**) Surface of the Onioshidasahi lava consisting of fragmented welded pyroclastic rocks. (**d**) Block lava formed by extruding internal fluidal materials of the lava

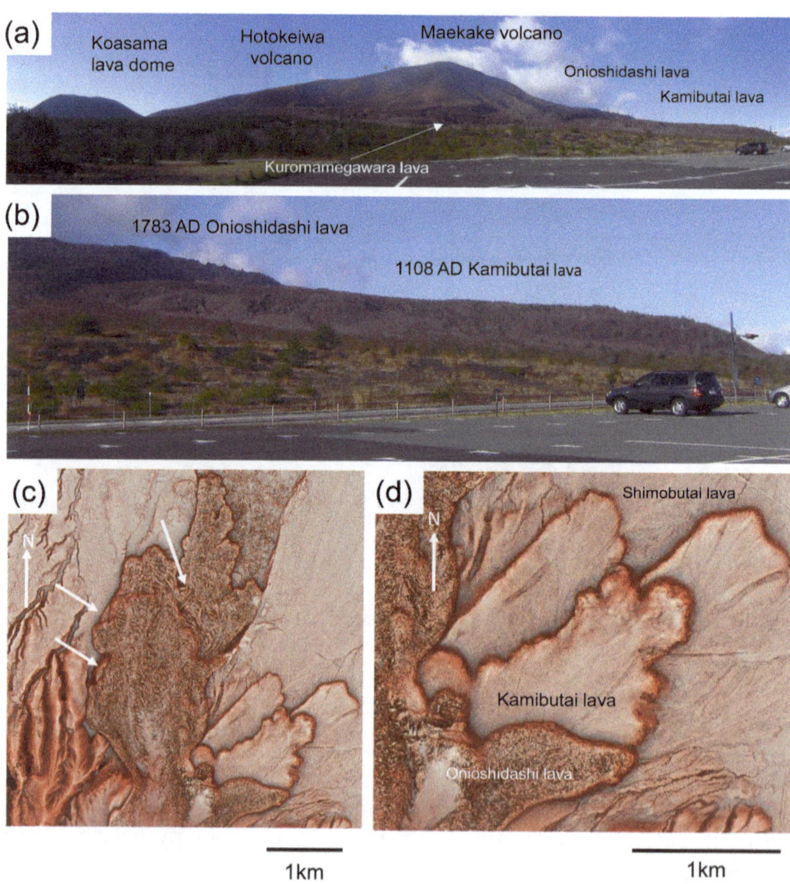

Fig. 17 Photographs and topographic relief maps of the Kamibutai lava (1108 AD). (**a**) Photograph showing recent lavas of Maekake volcano on its northern flank: Onioshidashi (1783 AD), Kamibutai (1108 AD), and Kuromamegawara (second century) lavas. (**b**) Close view of the Kamibutai lava overlain by the Onioshidashi lava. (**c**) Main part of Kamibutai lava is covered by the Onioshidashi lava and appears at several localities beneath the Onioshidashi lava (arrows) (courtesy of T. Chiba). (**d**) Topographic relief map of the Kamibutai lava (courtesy of T. Chiba)

Fig. 18 Photographs showing close views of the Kamibutai lava. (**a**) View of the eastern marginal cliff of Kamibutai lava consisting of an alternation of lava-like densely welded and non-welded to weakly welded pyroclastic rocks. (**b**) Alternation of densely welded and non-welded to weakly welded pyroclastic rocks, which incline toward the summit crater of Maekake volcano. (**c**) Alternation of densely welded and non-welded to weakly welded pyroclastic rocks found on the southern upper slope of Maekake pyroclastic cone. (**d**) Close view of the alternated densely and non-welded to weakly welded pyroclastic rocks. (**e**) Volcanic breccia grades into densely welded pyroclastic rock

collapse type eruption of the Soufriere Hills volcano, Montserrat (Druitt et al. 2002; Cole et al. 2002), because the occurrence of pyroclastic density current deposits of Maekake volcano is very similar to the pyroclastic density current deposits produced by the fountain collapse type eruption. The most intermediate type pyroclastic density currents of Maekake volcano were produced not by the column collapse of the sub-Plinian eruption but by the Vulcanian eruptions, as the fountain collapse type eruption is a kind of Vulcanian eruption; they occurred associated with the Vulcanian eruptions.

The Agatsuma pyroclastic density current deposit of the 1783 AD large-scale eruption and the Oiwake pyroclastic density current deposit of 1108 AD large-scale eruption spread widely on the volcanic apron of Maekake volcano and arrived up to 8 and 15 km from the summit crater, respectively (Fig. 23). They comprise a lot of flow units (Fig. 24) and their total eruptive volume is 0.15 and 0.36 km^3 DRE, respectively. The eruptive volume of Kotaki pyroclastic density current deposit of the second-century large-scale eruption is 0.26 km^3 DRE and intermediate between those of the Agatsuma and Oiwake pyroclastic density current deposits. Many flow units of the Agatsuma and Kotaki pyroclastic density current deposits are densely welded rocks with a reddish oxidized top including flattened dense scoria clasts; they also show cooling columnar joints (Fig. 25c), but those of the Oiwake pyroclastic density current deposit are mostly non-welded to weakly welded (Fig. 25a). The thickness of each flow unit is less than a few meters in general (Fig. 25); however, those set-

Fig. 19 Photographs showing alternated beds of Kamibutai lava. (**a**) Columnar section of Kamibutai lava consisting of the alternation of densely and non-welded to weakly welded pyroclastic rocks. (**b**) Lava-like densely welded pyroclastic rock in columnar section (a). (**c**) Non-welded to weakly welded pyroclastic rocks in columnar section (a). (**d**) Black glassy lenses in the reddish oxidized matrix of densely welded pyroclastic rocks. (**e**) Reddish oxidized weakly welded tuff breccia in the alternated bed. (**f**) Alternated bed of reddish oxidized tuff breccia and densely welded pyroclastic rocks at the southern upper slope of Maekake pyroclastic cone

tled in valleys are thick occasionally up to several tens of meters in thickness.

3.4 Construction of Volcanic Edifice: Relation Between Eruptive Volume and Age

The products of the Class A large-scale eruption comprise fallout pumice deposit, pyroclastic cone, clastogenic lava, and pyroclastic density current deposit. The estimated volumes of these eruptive products of Class A large-scale eruption are summarized in Table 2. The total eruptive volume (DRE) of each Class A large-scale eruption is as follows: 0.51 km^3 (As-A eruption), 0.10 km^3 (As-B′ eruption), 0.82 km^3 (As-B eruption), 0.68 km^3 (As-C eruption), 0.06 km^3 (As-D1 and As-D2 eruptions), 0.27 km^3 (As-E eruption), and 0.13 km^3 (As-F eruption).

The relation between the volume of Class A large-scale eruption and its age is shown in

Fig. 20 Photographs showing the bedded volcanic breccias and tuff breccias of Kamibutai lava. (**a**) Volcanic breccia and tuff breccia beds incline toward the summit crater. (**b**) Clast-supported volcanic breccia with angular clasts of andesite. (**c**) Matrix poor tuff breccia

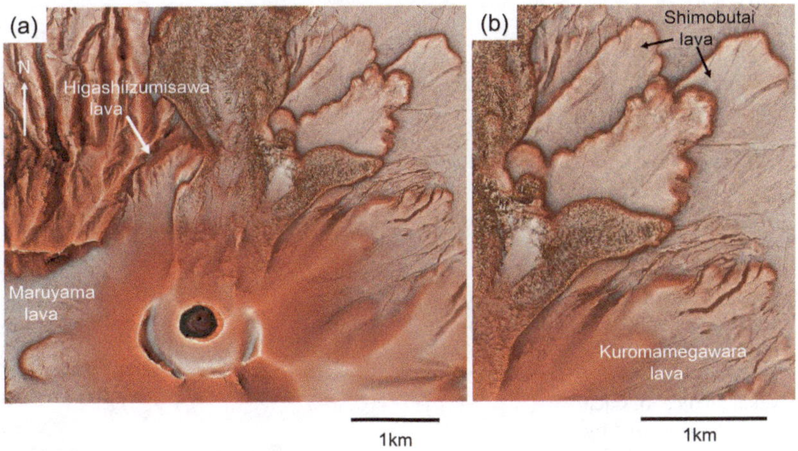

Fig. 21 Topographic relief map of the lavas effused during the second-century eruption (courtesy of T. Chiba). (**a**) Topographic relief map of Higashiizumisawa and Maruyama lavas. (**b**) Topographic relief map of Kuromamegawara and Shimobutai lavas

Fig. 22 Photographs of the lavas effused in second century. (**a**) Folded lenses in the densely welded Kuromamegawara lava, the matrix of which is reddish due to oxidation. (**b**) Outcrop of Kuromamegawara lava comprising stratified agglutinate of alternation of a thin vesiculated glassy melanocratic layer and reddish oxidized layer with a thickness of several cm

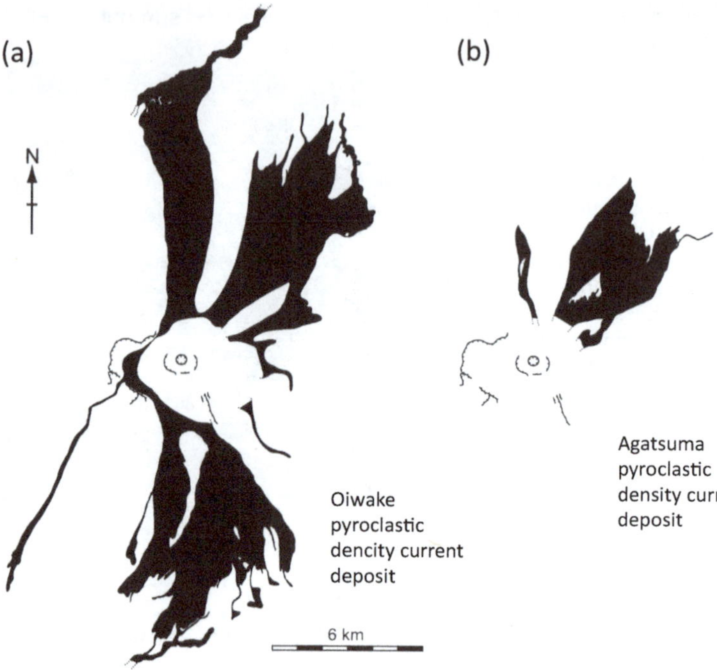

Fig. 23 Maps showing distribution of the Oiwake (1108 AD) and Agatsuma (1783 AD) pyroclastic density current deposits (Aramaki 1963). (**a**) Oiwake pyroclastic density current deposit. (**b**) Agatsuma pyroclastic density current deposit

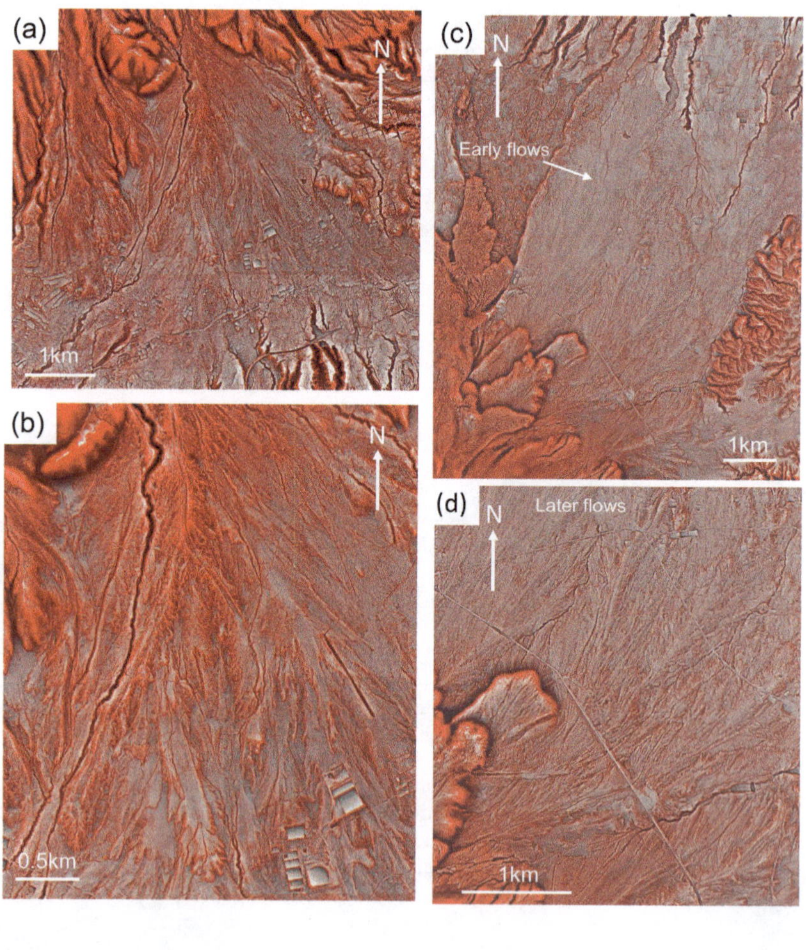

Fig. 24 Topographic relief maps of the Agatsuma and Oiwake pyroclastic density current deposits (courtesy of T. Chiba). The pyroclastic density current deposit comprises many flow units. (**a**, **b**) Oiwake pyroclastic density current deposit in the southern foot of Maekake volcano. Each flow is characterized by a levee and front cliff. (**c**) Agatsuma pyroclastic density current deposit in the northern foot of Maekake volcano. The early flows with low aspect ratio (thickness vs. area) are widely distributed, while the later flows with high aspect ratio (thickness vs. area) show a levee and front cliff. (**d**) Later flows with levees and front cliffs

Fig. 25 Photographs showing a close view of a pyroclastic density current. (**a**) Non-welded Oiwake pyroclastic density current deposit including a lot of weakly vesiculated cabbage-like rounded scoria clasts. (**b**) Non-welded Agatsuma pyroclastic density current deposit in which cabbage-like rounded scoria clasts are concentrated. (**c**) Densely welded Agatsuma pyroclastic density current deposit with a thickness of ca 1 m. The dark colored scoria clasts are flattened by compaction and the top of the deposit is reddish in color due to oxidation. The cooling columnar joints are conspicuous

Fig. 26. The eruptive history of Maekake volcano consists of active stages of high volcanic activity I (9,400 to 7,300 years calBP), II (6,500 to 3,200 years calBP), and III (<1,766 years calBP), and relatively active stages (without large-scale eruptions) of lower volcanic activity 1 (7,300 to 6,500 years calBP) and 2 (3,200 to 1,765 years calBP). The eruption rate of active stages I, II, and III is 0.6×10^{-4}, 0.1×10^{-3}, and 0.11×10^{-2} km^3/year, respectively; the eruption rate is largest in active stage III.

A total of 80% of the volume of the eruptive products of Maekake volcano were produced in active stage III, in particular, most of the volcanic edifice of Maekake volcano was constructed in the last ca 2,000 years by the four Class A large-scale eruptions during a relatively short time span. And 89 vol% of the eruptive products of Maekake volcano is proximal (pyroclastic cones, clastogenic lavas, and pyroclastic density current deposits) and comprise the volcanic edifice, while the fallout deposit is only 11 vol%.

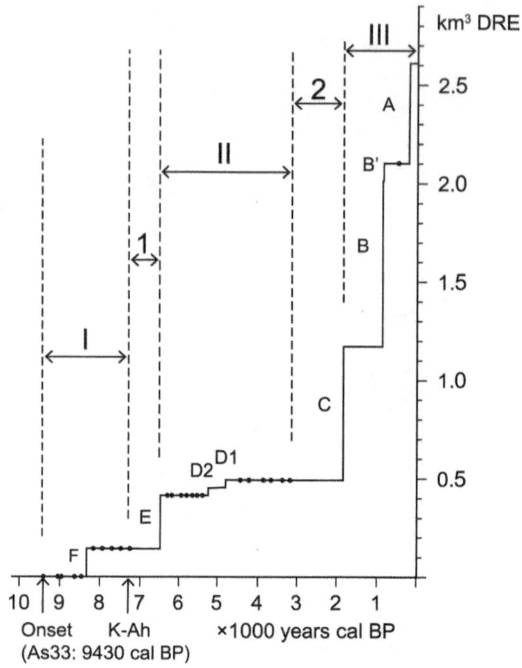

Fig. 26 Diagram showing the relation between cumulative volume of large-scale eruptions and their eruptive ages of Maekake volcano (Takahashi et al. 2022). A: As-A (1783 AD). B′: As-B′ (1128 AD). B: As-B (1108 AD). C: As-C (second century). D1: As-D1 (4,785 years calBP). D2: As-D2 (5,232 years calBP). E: As-E (6,447 years calBP). F: As-F (8,343 years calBP). Solid circle: Class B large-scale eruption. I: active stage I. II: active stage II. III: active stage III. 1: moderately active stage 1. 2: moderately active stage 2. K-Ah: Kikai-Akahoya tuff erupted from the Kikai caldera, southern Kyushu (7,300 calBP)

4 Summary

1. The eruption style of Maekake volcano consists of small- to intermediate-scale and large-scale eruptions. The small- to intermediate-scale eruptions are phreatic, phreatomagmatic, and Vulcanian, while the large-scale eruptions are mainly sub-Plinian.
2. The small- to intermediate-scale eruptions are grouped into single and continuous eruptions. There have been seven periods of continuous eruptions since the sixteenth century. The magma supply rate to the magma chamber is estimated to be constant during the period of single small- to intermediate-scale eruption, while it is estimated to have increased during the period of continuous small- to intermediate-scale eruptions.
3. The eruption of single small- to intermediate-scale eruptions start with Vulcanian eruptions, which shifts to Strombolian and ash eruptions and effusion of a lava in the crater; the eruption sequence ceased after a final Vulcanian eruption. The Vulcanian eruptions are predominant in the first half of the period of continuous eruptions, whereas the Strombolian and ash eruptions are the main eruption style in the latter half.
4. The large-scale eruptions are classified into Class A and B depending on eruption volume. There have been eight Class A large-scale eruptions since the onset of volcanic activity of Maekake volcano (9,430 years calBP): A (1783 AD; 0.51 km^3 DRE), B′ (1128 AD; 0.10 km^3 DRE), B (1108 AD; 0.82 km^3 DRE), C (184 AD; 0.68 km^3 DRE), D1 (4,785 years calBP; 0.06 km^3 DRE), D2 (5,232 years calBP; 0.06 km^3 DRE), E (6,447 years calBP; 0.27 km^3 DRE), and F (8,343 years calBP; 0.13 km^3 DRE). The eruption sequence of class A large-scale eruptions comprises Vulcanian and sub-Plinian eruptions, the order of which is not the same for each eruption.
5. The volcanic edifice of Maekake volcano comprises pyroclastic cones, clastogenic lava flows, and pyroclastic density current deposits which were the products of large-scale eruptions.
6. The pyroclastic cones and clastogenic lava flows are produced by lava fountains associated with the sub-Plinian eruption columns; presently exposed pyroclastic cones are the Kamayama (As-A; 0.04 km^3 DRE) and Maekake (As-B; 0.18 km^3 DRE) and the clastogenic lava flows are the Onioshidashi (As-A; 0.29 km^3 DRE), Kamibutai (As-B; 0.23 km^3 DRE), and Shimobutai-Kuromamegawara-Higashiizumisawa-Maruyama (As-C; 0.31 km^3 DRE).
7. The intermediate-type pyroclastic density current deposits were probably produced by

eruptions following fountain collapse; presently exposed pyroclastic density current deposits are the Agatsuma (As-A; 0.15 km^3 DRE), Oiwake (As-B; 0.36 km^3 DRE), and Kotaki (As-C; 0.26 km^3 DRE).

8. The history of volcanic activity of Maekake volcano comprises active stage I (9,430 to 7,260 years calBP), II (6,447 to 3,160 years cal BP), and III (1,760 years calBP) which is intercalated by relatively active stage 1 (7,261 to 6,446 years calBP) and 2 (3161 to 1,759 years calBP) based on eruptive ages and estimated volumes of the large-scale eruptions. The eruption volume and rate are the largest in active stage III. The total eruptive volume of Maekake volcano is at least 2.63 km^3 DRE.
9. 80 vol% of the volcanic edifice of Maekake volcano was constructed by the large-scale eruptions of the active stage III, namely during the second century to present.
10. 89 vol% of the eruptive products of Maekake volcano is proximal and comprise the volcanic edifice, while the fallout deposit is only 11 vol%.

Acknowledgments We are grateful to Aramaki, S., Takemoto, H., Shimotomai, T., Sakagami, M., Tsuchihashi, H., Nakamura, Y., Kurosawa, T., Saita, M., Shimizu, Y., Hongo, Y., Niijima, M., Sato, K., Akimoto, N., Iwasaki, T., Shimizu, T., Tamura, H., Ishikawa, S., Yamashita, D., Moro, K., Mizota, M., Orihara, K., and Osaka, I. for their cooperation and help with the study. We also appreciate Martin A. for critical reading and revising of the manuscript.

References

Aramaki S (1957) Classification of pyroclastic flows. J Volcanol Soc Japan Ser 2 1:47–57. (in Japanese with English abstract)

Aramaki S (1963) Geology of Asama volcano. J Fac Sci Univ Tokyo Sec II XIV:229–443

Cole PD, Calder ES, Sparks RSJ, Clark AB, Druitt TH, Young SR, Herd RA, Harford CL, Norton GE (2002) Deposit from dome-collapse and fountain-collapse pyroclastic flow at Soufriere Hill volcano, Montserrat. In: The eruption of Soufriere Hill volcano, Montserrat, from 1995 to 1999. Geol Soc London, Memoirs, vol 21, pp 231–262.

Druitt TH, Young SR, Bapite B, Bonadonna C, Calder ES, Clark AB, Cole PD, Harford CL, Herd RE, Luckett R, Ryan G, Voigt B (2002) Episodes of cyclic Vulcanian explosive activity with fountain collapse at Soufriere Hill volcano, Montserrat. In: The eruption of Soufriere Hill volcano, Montserrat, from 1995 to 1999. Geol Soc London, Memoirs, vol 21, pp 281–306.

Inoue M (2002) Sequence of bulk-rock chemistry of the Onioshidashi lava flow, the 1783 eruption of Asama volcano. Rep Geogr Fac Liter Kanazawa Univ 10:17–23. (in Japanese with English abstract)

Inoue M (2006) The Onioshidashi lava flow of Asama volcano as clastogenic lava flow. The Earth monthly, vol 28, pp 223–230 (in Japanese)

Maeno F, Suzuki Y, Nakada S, Koyama E, Kaneko T, Fujii T, Miyamura J, Onizawa S, Nagai M (2010) Course and ejecta of the eruption of Asama volcano on 2 February 2009. J Volcanol Soc Japan 55:147–154. (in Japanese with English abstract)

Minakami T (1935) Recent activities of Asama volcano. Jishin 1:319–339. (in Japanese with English abstract)

Miyazaki T (2003) Re-examination of the reports of activities of Asama volcano. Bull Earthq Res Inst Univ Tokyo 78:283–463. (in Japanese with English abstract)

Murai I (1974) The eruptive activity of Asama volcano in 1973 and the associated minor pyroclastic flows. Pro Rep Earthq Res Inst Univ Tokyo 13: 127–162. (in Japanese with English abstract)

Murase M, Ono K, Ito T, Miyajima R, Mori H, Aoyama H, Oshima H, Yoshida Y, Terada A, Koyama E, Takeda T, Watanabe H, Kimata F, Fujii N (2007) Time-dependent model for volume changes in pressure sources at Asama volcano, central Japan due to vertical deformations detected by precise leveling during 1902-2005. J Volcanol Geotherm Res 164:54–75

Nakada S, Yoshimoto M, Koyama E, Tsuji H, Urabe T (2005) Comparative study of the 2004 eruption with old eruptions at Asama volcano and the activity evaluation. J Volcanol Soc Japan 50:303–313. (in Japanese with English abstract)

Shimozuru D, Utibori S, Gyoda N, Koyama E, Miyazaki T, Matsumoto T, Osada N, Terao H (1975) The 1973 explosive activity of Asama volcano: general description of volcanic and seismic events. Bull Eartq Res Inst Univ Tokyo 50:115–151. (in Japanese with English abstract)

Takahashi M, Yasui M (2013) Proximal volcanic geology and field excursion guide to the Asama-Maekake volcano: historical eruptive products of the Asama-Maekake volcano distributed around the Kuromamegawara area. J Volcanol Soc Japan 58:311–328. (in Japanese with English abstract)

Takahashi M, Yasui M, Kanamaru T, Nakagawa M (2022) Eruptive volume and age of large-scale pumice eruptions of the Asama-Maekake volcano, Central Japan, as revealed by detailed trench surveys: implications for future volcanic hazards. J Disaster Res 17:1080–1089

Wolff JA, Sumner JM (2000) Lava fountain and their products. In: Sigurdsson H, Houghton B, McNutt SR,

Rymer H, Stix J (eds) Encyclopedia of volcanoes. Academic Press, New York, pp 321–329

Yasui M, Koyaguchi T (2004) Sequence and eruptive style of the 1783 eruption of Asama volcano, Central Japan. Bull Volcanol 66:243–262

Yasui M, Takahashi M (2015) Formation of the andesitic welded pyroclastic cones by pyroclastic eruption recorded in the summit area of Asama-Maekake volcano and the collapsed caldera wall of Kurofu volcano. J Volcanol Soc Japan 60:109–123. (in Japanese with English abstract)

Yasui M, Takahashi M, Kanamaru T, Nagai M (2021) High-resolution reconstruction of the history of large-scale eruptions of Asama-Maekake volcano based on the stratigraphy of pyroclastic fall deposits. J Volcnol Soc Japan 66:293–325. (in Japanese with English abstract)

High-Resolution Reconstruction of the Eruptive History of Asama-Maekake Volcano Based on the Stratigraphy of Pyroclastic Fall Deposits: Example of Geological Survey Utilizing Extensive and Numerous Trench Excavations

Maya Yasui, Masaki Takahashi, Tatsuo Kanamaru, and Masashi Nagai

1 Introduction

What types of volcanic activity has the Asama-Maekake Volcano exhibited over the past 10,000 years? Activity of the Maekake volcano can be divided into two types of eruptions. They are large-scale eruptions such as the 1783 eruption, which generates obvious geological units, and small and medium-sized eruptions (e.g., vulcanian eruption, continuous ash emission, and phreatic eruption). To study large-scale eruption, we conducted trench excavation surveys in a wide area around the volcano and numerous ^{14}C dating measurements. And then, we determined the stratigraphy and estimated the scale of the eruption. Results enabled us to reconstruct the eruptive history of the large-scale eruptions of Maekake volcano in detail (Yasui et al. 2021). Small and medium-sized eruptions occurred frequently in the first half of the twentieth century and are well documented, but they become less and less recorded as they get older, and little is known before the sixteenth century. The individual eruption volumes of small and medium-sized eruptions are too small to form clear geological units. Therefore, we investigated the traces of small and medium-sized eruptions by analyzing volcanic ash particles contained in the soil between the products of large-scale eruptions. Here, we will introduce the method and results that were tried and reconstructed detailed 10,000-year eruption history of Maekake Volcano.

M. Yasui (✉) · T. Kanamaru
Department of Earth and Environmental Sciences, College of Humanities and Sciences, Nihon University, Tokyo, Japan
e-mail: yasui.maya@nihon-u.ac.jp; kanamaru.tatsuo@nihon-u.ac.jp

M. Takahashi
Institute of Natural Sciences, College of Humanities and Sciences, Nihon University, Tokyo, Japan
e-mail: takahashi.masaki@nihon-u.ac.jp

M. Nagai
Volcano Disaster Resilience Research Division, National Research Institute for Earth Science and Disaster Resilience, Tsukuba, Ibaraki, Japan
e-mail: mnagai@bosai.go.jp

S. Aramaki et al. (eds.), *The Asama Volcano*, Active Volcanoes of the World,
https://doi.org/10.1007/978-981-95-5799-8_10

2 Process of Reconstructing Eruptive History of Large-Scale Eruptions

Generally speaking, eruptive products from older times are often buried by the overlying younger deposits, and the older the eruption, the more difficult it becomes to identify a single eruptive event. In addition, the distribution of pyroclastic fall deposits is affected by the wind direction above the summit crater at the time of the eruption, and pyroclastic density currents and lava flows avoid topographical heights on their way, so the deposits that appear vary depending on the location. Therefore, the stratigraphy at any given locality does not represent the entire eruption history. In general, the closer you are to the crater, the thicker the younger deposits are, and the chances of observing older deposits are extremely reduced. In the distal areas where flow deposits, such as pyroclastic density currents, do not reach, pyroclastic fall deposits have the advantage of being able to trace back the stratigraphy to older times. On the other hand, there is also the problem that the eruption clouds of the smaller eruptions do not reach distant places. Unless we conduct detailed geological surveys over a wide area from near the volcanic edifice to far away, and even in the vertical direction at individual locations, we will not be able to see the full picture of the eruption history. In other words, in order to comprehensively determine the stratigraphy and reconstruct the history of eruptions, it is essential to understand the horizontal spread of numerous deposits and determine their stratigraphy.

An intensive geological survey of Maekake Volcano has been conducted since 2016 under the Integrated Program for Next-Generation Volcano Research and Human Resource Development promoted by the Ministry of Education, Culture, Sports, Science, and Technology. We conducted a wide-ranging geological survey of more than 180° around the crater by excavating trenches at 36 locations using heavy machinery, drilling 50 m of boring on the east-southeast slope (Loc. M101), as well as normal geological survey (Fig. 1a, c). This method involves observing and describing the deposits buried underground, performing multiple dating measurements, and comparing different locations. This is a new method for reconstructing the history of large-scale eruptions such as the 1783 eruption. For instance, from the ground surface to the red line in Fig. 2b–d is the black soil that has accumulated over the past 10,000 years and the pyroclastic fall deposits. Fall units of pumice and volcanic ash that fell during large-scale eruptions are interbedded with the black soil. The depth to the red line is a little over 1 m in the north (Loc. 13) and south (Loc. 19) of the crater, but it is about 5 m in the east (Loc. 8). At Loc. 6 in the east-southeast direction, since the thickness of As-A and As-B are large, the bottom of the deposits about 2000 years ago can be seen at a depth of 4 m at Loc.6 (Fig. 2a). Drilling of boring excavation at 5 km east-southeast of the summit crater revealed 10,000-year deposits 10 m thick. The above survey results show that the plumes of individual sub-Plinian eruptions were often dispersed to the east-southeast of the crater by westerly winds. Figure 3 shows columnar sections of the deposits of the representative localities. In order to correlate the deposits between the different localities, extensive ^{14}C dating more than 100 were obtained. In most cases, black soil immediately beneath the pyroclastic fall deposits was measured; however, charcoals contained in the pyroclastic density current deposits are also used for age determination. The level at which volcanic glass shards of K-Ah were detected is marked in the columnar sections of each location (Fig. 3). It is a co-ignimbrite ash derived from Kikai caldera in southern Kyushu about 7300 years ago. It is one of famous widespread tephra in Japan and is useful as a key bed. Isopach maps of the pyroclastic fall deposits were made as many as possible to reveal the distribution of the deposits (Fig. 4). These data were used for reconstructing the total stratigraphy of the Asama-Maekake Volcano.

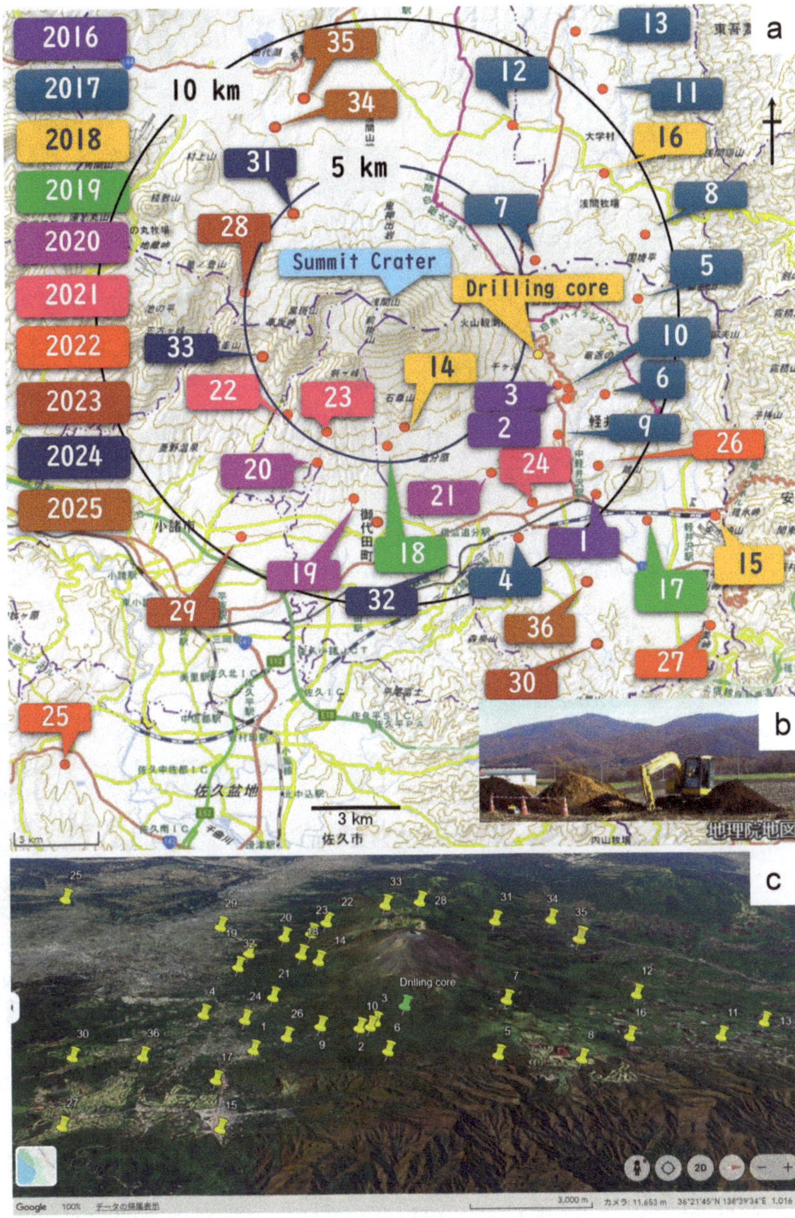

Fig. 1 (**a**) Map showing trench locations 1–36 based on GSI Maps (electronic National Land Web Map) provided by the Geospatial Information Authority of Japan (GSI). (**b**) Excavation scene in trench 13. (**c**) Bird's-eye view of Asama Volcano and surrounding area from the east created using Google Earth (© Google), with 36 trench excavation sites and one boring site added by the authors

3 Classification of the Large-Scale Eruption and Its Eruptive History

Pyroclastic fall deposits are given alphabetical names from youngest to oldest, such as As-A, B, and C (Aramaki 1963). For example, As-A corresponds to the deposits of the 1783 AD eruption and As-B to the 1108. By comparing the distribution of pyroclastic fall deposits, we were able to classify the scale of large-scale eruptions into four classes. Quantitative volumetry is not easy for older deposits due to the difficulty of making isopach maps in more detail than younger deposits. Instead of that, we compared the area surrounded by the 64 and 16 cm isopach lines (Fig. 4a). An eruption like the 1783 eruption with a thick layer of pumice over a wide area is classified as Class 1, an eruption with only a few pumice particles scattered in the soil is classified as

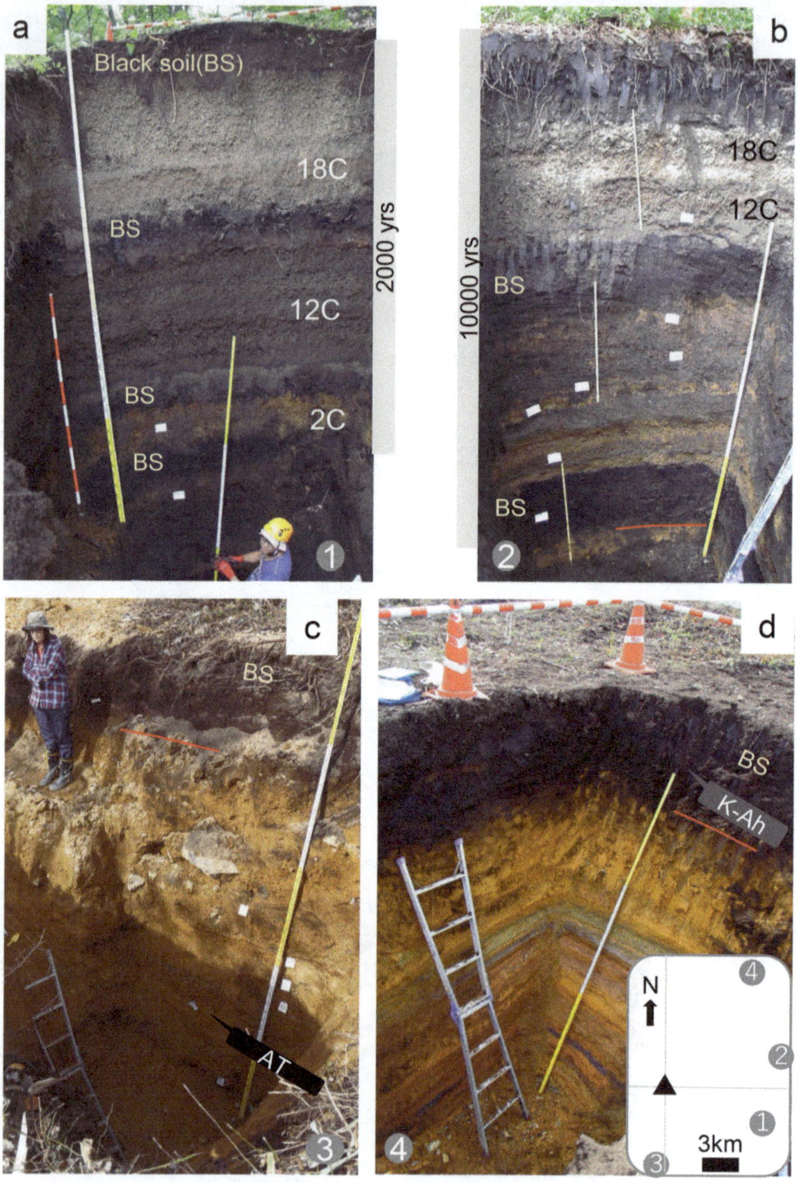

Fig. 2 Photographs of the representative trench excavations. (**a**) Thick pyroclastic fall deposits of As-A and As-B exposed on the trench wall at Trench 6. The longest scale is 5 m long. (**b**) At trench 8, the entire sequence of the deposits of the Asama-Maekake Volcano could be observed above the red line on the photo. The scale on the right side is 3 m long. (**c**) At Trench 19, a debris avalanche deposit is excavated under the pyroclastic flow deposit of the Hotokeiwa volcano. A 12 cm-thick ash fall called "AT" can be seen close to the bottom of the trench. It is one of the famous widespread tephra fall in Japan that was derived from Aira caldera southern Kyushu in 30 ka. (**d**) A 90-cm-thick black soil layer at Trench 13. A speech bubble shows the horizon in which K-Ah was found. Beneath the soil, stratified ash fall layers of the Asama-Hotokeiwa Volcano are remarkable. The stepladder is 2.44 m long. The map shows the distance and direction of the four trenches in this figure from the summit crater. BS: Black soil

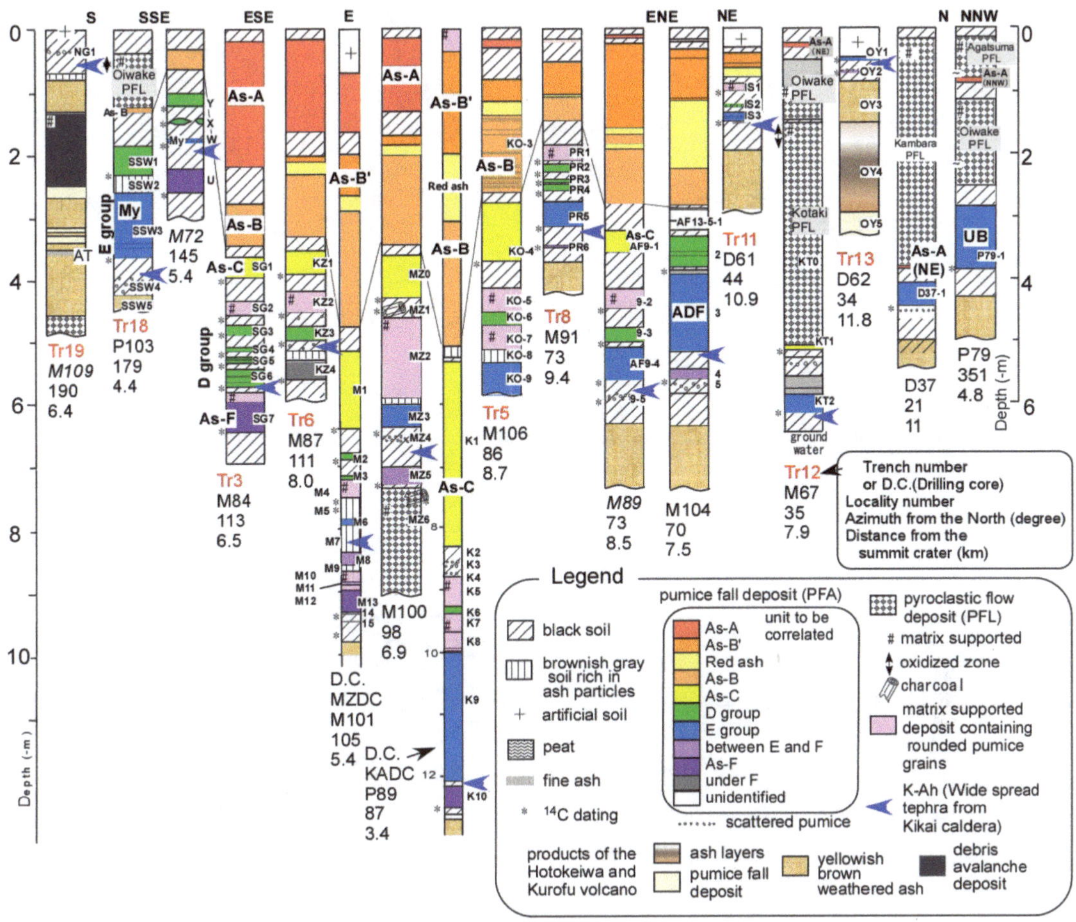

Fig. 3 Columnar sections of the trench excavations, the major outcrops, and the drilling core samples. The trench excavation numbers are the same as Fig. 1a, c

Class 4, and an intermediate level is classified as 2 or 3. Specifically, based on the difference in the area surrounded by the 64 and 16 cm isopach lines of the isopach map, the pyroclastic fall deposits were classified into four classes as shown in Fig. 4c.

Class 1: As-A, As-B′, As-B, As-C, E-Group (E-ADF); Class 2: an example of D-Group (D-SSW), E-Group (E-My and E-UB), and As-F; Class 3: As-A′, part of D-Group and E-Group, and between E-Group and As-F, and below As-F; Class 4: part of D-Group and E-Group, and between E-Group and As-F, and deposits overlain by As-F. In the Class 1 eruptions that brought about As-A, B, and C, a layer of ash fall was formed over 50 km from the summit crater. On the other hand, Class 3 and Class 4 deposits are rarely found in remote areas.

Looking at the past 10,000 years (Fig. 5), the first serious Class 1 eruption that resulted in pumice fall (As-F) was about 8300 years ago, and the next one was around 6300 years ago. Around 3000–6000 years ago, the scale of the eruptions was not as large as the 1783 eruption, but pumice fell to the foot of the volcano intermittently to form the D-Group. In other words, from the present to 2000 years ago, a Class 1 eruption has occurred on average about once every 700 years (represented by four arrows in Fig. 5). This means that the pattern of activity has changed significantly in terms of the size and frequency of eruptions. This is also discussed in chapter "Petrology

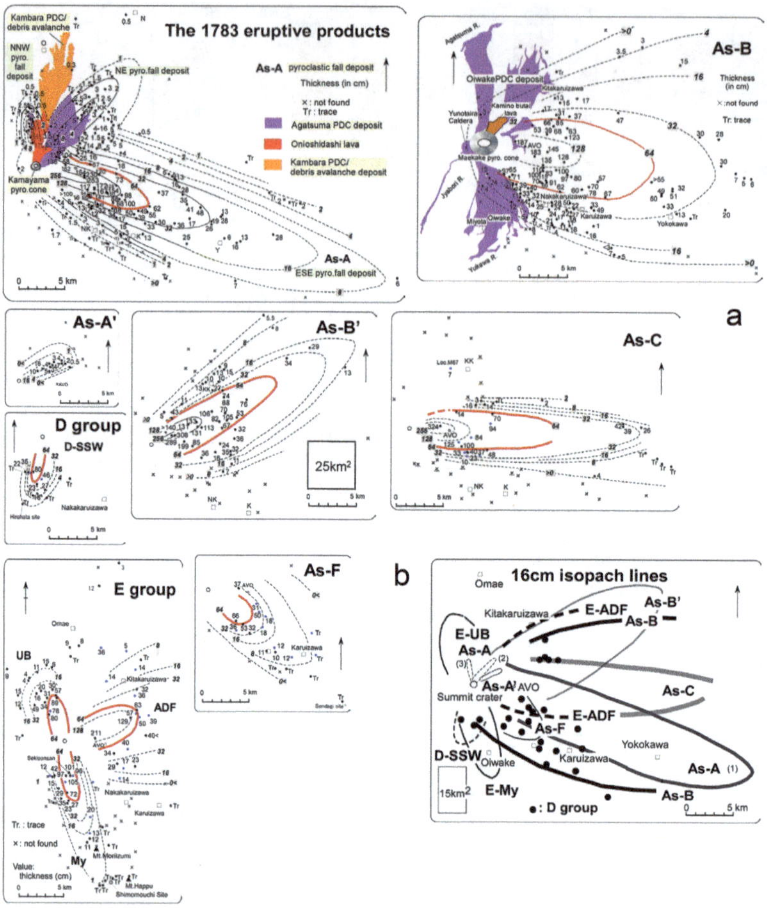

Unit	Subunit	Area(km²)	Class
64cm-line			
D group	D-SSW	4.9	
E group	UB	8.7	2
E group	My	9.2	
As-F		11.8	
E group	ADF	26.4	
As-A	ESE	35.1	
As-B'		46.9	1
As-C		54.7	
As-B		110.6	

Unit	Subunit	Area(km²)	Class
16cm-line			
As-A	NNW	1	
As-A'		1.4	3
As-A	NE	2.1	
D group	D-SSW	17.1	
E group	UB	28	2
E group	My	28.5	
As-F		32.7	
As-B'		133.6	1
As-A	ESE	162.3	

Fig. 4 (**a**) Isopach maps of the pyroclastic fall deposits of the Asama-Maekake Volcano. This was created based on Figure 5 and Table 3 of Yasui et al. (2021). The 64 cm isopach line is colored in red. (a) Pyroclastic fall deposit A (As-A). Distribution of pyroclastic flow deposits (APF: Agatsuma), lava flow (ONI: Onioshidashi) (Aramaki 1963), and a summit pyroclastic cone (KM: Kamayama) (Yasui and Koyaguchi 2004) are also shown. (b) Pyroclastic fall deposit A′ (As-A′). (c) Pyroclastic fall deposit B′ (As-B′). (d) Pyroclastic fall deposit B (As-B). Distribution of pyroclastic flow deposits (Oiwake), lava flow (Kaminobutai) (Aramaki 1963), and a summit cone (Maekake) (Yasui and Takahashi 2015) are also shown. (e) Pyroclastic fall deposit C (As-C). (f) Pyroclastic fall deposit D-SSW (D group). (g) Pyroclastic fall deposit of the E group: UB, My, and ADF. (h) Pyroclastic fall deposit F (As-F). The blue circles in the figure below are where ^{14}C dates were measured: As-C, D-SSW, E-group, and As-F. × in map for As-C: Point where the soil immediately below the pumice layer below As-B shows an older age than As-C. AVO: Asama Volcano Observatory, The University of Tokyo, Other place name abbreviations K: Karuizawa, NK: Nakakaruizawa, KK: Kitakaruizawa, N: Naganohara, O: Omae, Y: Yokokawa. (**c**) Table showing the areas enclosed by the isopach lines of 64 and 16 cm of the pyroclastic fall deposits of the Asama-Maekake Volcano. Individual 64 cm lines are indicated by red lines in (**a**), and the 16 cm lines are shown in (**b**)

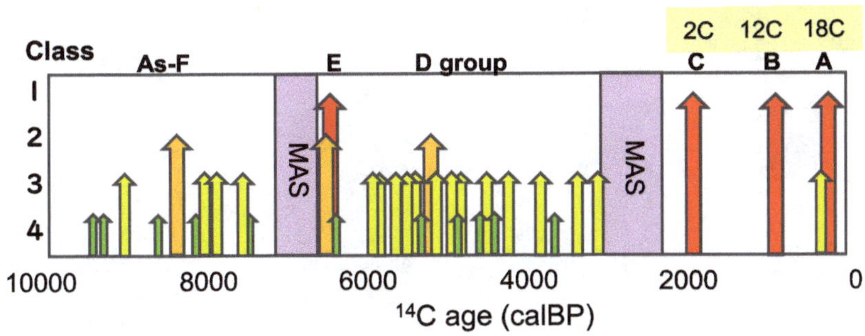

Fig. 5 10,000-year history of large-scale eruptions of the Asama-Maekake Volcano. Diagram showing the relationship between the measured ^{14}C ages (cal BP) and the class given to each deposit. For As-A and As-B, ^{14}C ages were measured mainly using charcoals contained in the pyroclastic flow deposits. The other ^{14}C ages are from soil samples immediately under the pyroclastic fall deposits. MAS: moderately active stage

of the Magmatic Plumbing System of Asama-Maekake Volcano".

4 10,000 Years History of the Small and Medium-Sized Eruptions

Breadcrust bombs are sometimes found on flank slopes (Fig. 6a). The surface layer of Kamayama, at the top of the Maekake volcano, is covered with ballistic materials including volcanic bombs from the 2004 and 1973 eruptions (Fig. 6b). In the first half of the twentieth century, continuous ash falls and Vulcanian eruptions were frequent until around 1961, but after that, the only times that magma rose to the crater floor were during the 1973 and 2004 eruptions. At the top of the Kamayama crater wall, Unit-C, which was deposited by small and medium-sized eruptions after the 1783 eruption, can be seen (Fig. 6c). Since there are many documents dating after the end of the nineteenth century (e.g., Miyazaki 2003; Japan Meteorological Agency n.d., website), the layered structure seen in Unit C can be thought to have been formed by repetitive eruptions in the first half of the twentieth century. The number of records prior to the 1783 eruption decreases over time, but before the fifteenth century, many records are ambiguous as to whether or not an eruption occurred. The oldest record is of ash fall in the seventh century, but it is unclear whether it originated from Asama Volcano (Miyazaki 2003). Although there are many records of eruptive activity since the end of the nineteenth century, human-written records are hardly useful for reconstructing the history of small and medium-sized eruptions that occurred before then.

In small and medium-sized eruptions, individual plumes do not leave clear ash fall deposits as geological units, but traces of the eruption can be recognized as volcanic ash particles of lava fragments contained in the soil (e.g., Fig. 6f–i). In other words, we can evaluate the activity of small and medium-sized eruptions from vertical changes in the features of ash particles in soil at a given location. We evaluated the temporal changes in the activity of small and medium-sized eruptions using the boring core samples at locality M101 at the east-southeast foot of the volcano (Fig. 7). The soil sandwiched between pyroclastic fall deposits (As-A, B, etc.) from large-scale eruptions are seen in eight horizons ([1]–[8] in Fig. 7a), and they were collected in 2 cm increments. If volcanic soil is sampled at equal intervals, it is expected that the content of fine-grained soil will increase compared to ash particles during the period of volcanic inactivity, and the content of coarse-grained volcanic ash will increase during the active period. And if the sampling location is below the dispersion axis of the eruption cloud, angular coarse particles will be deposited thickly. In other words, the deposition rate can vary with depth. However, by focusing on the lee of prevailing winds, this

Fig. 6 Photos related to Vulcanian eruptions in the twentieth century. (**a**) Breadcrust bomb found on the eastern flank slope. Distance from the summit crater is 3.5 km. (**b**) Numerous volcanic blocks on the flank slope of Kamayama pyroclastic cone can be seen. You can see the volcanic shelter that was dented by impact of the volcanic blocks during the 2004 eruption. (**c**) Unit-C seen just below the rim of the Kamayama crater. * Huge volcanic block erupted during the eruption on September 23, 1950. (**d**) Microscopic photo of a volcanic bomb from an eruption in the twentieth century. Phenocryst minerals (plagioclase and pyroxene) are displaced by faults. The particle sizes in the micrographs (e,f,g,h,and i) are 1–2 mm. (**e**) Microphotograph of volcanic ash particles from the October 10, 2004, eruption. There are many angular-shaped, massive lava fragments. Microphotographs (**f–i**) are volcanic ash particles contained in the boring core sample. (**f**) Angular-shaped, massive lava fragment volcanic ash particles contained in the black soil below As-F in the boring core. (**g**) Subrounded volcanic ash particles contained in the soil below As-C. (**h**) Particles with visible faults contained in the black soil below As-F. (**i**) Volcanic ash particles contained in the black soil below As-F. Pyroclastic veins rich in crystal fragments can be seen

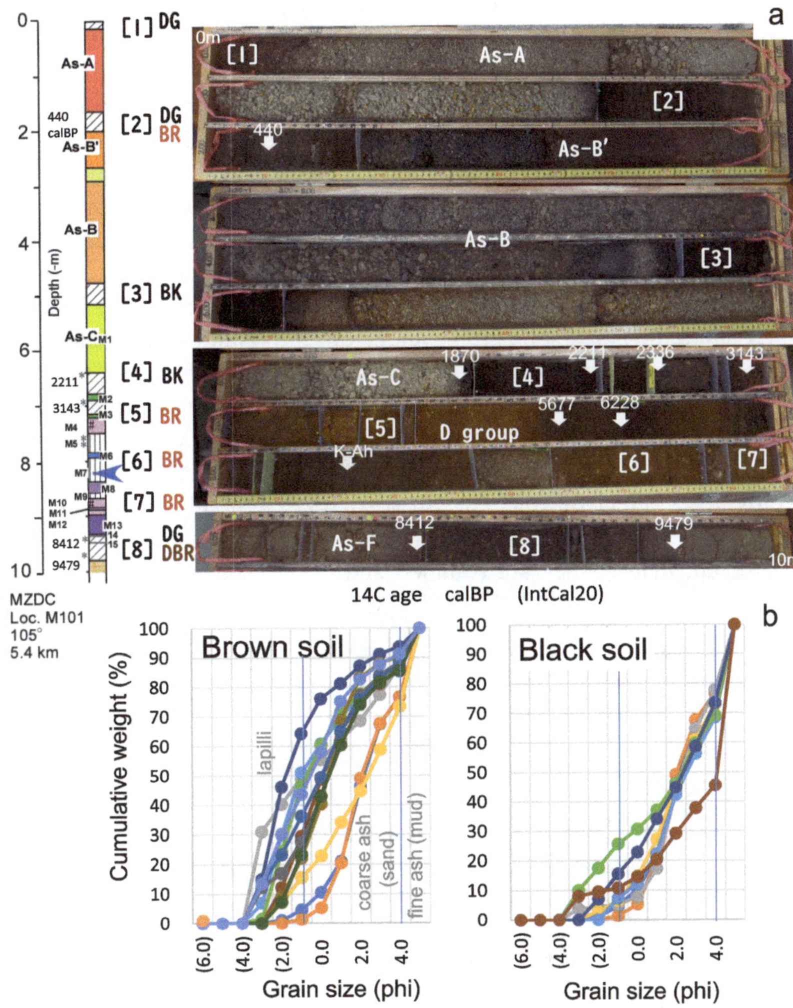

Fig. 7 Columnar section and photo of the drilling core sample at Loc. M101. Values on the photo are 14C ages (calBP). The interval corresponding to depths between 4 and 10 m includes the lower part of As-B, As-C, D group, E-ADF, and As-F. For the detail, see the columnar section in Fig. 3. [1]~[8] show the soil. Note the difference in color of these soil. Grain-size distributions of the brown soil and black soil are shown in the diagram below. Colors of soil are indicated on the right of the columnar section in the following; BK: Black, DG: Dark gray, BR: Brown, DBR: Dark brown

methodology may be useful in reconstructing the rise and fall of small and medium-sized eruptions.

What is notable about the borehole core sample from site M101 is the color of the soil. Brown soil rich in coarse-grained volcanic ash particles is recognized in some depths. For instance, angular lava fragments are detected in the brown soil at a depth of 9.7 m between As-F and the weathered volcanic soil of Hotokeiwa Volcano (Fig. 7a [8]). Fragmented crystals are observed in such grains under the microscope (Fig. 6h–i). Breadcrust bombs from twentieth-century eruptions are characterized by fractured crystals under microscope (Fig. 6d) (Yasui et al. 2005). They are characterized by in situ cracked crystals and shears (faults) and sometimes exhibit vein-like structures filled with crystal fragments. These textures suggest the shear failure of magma around the explosion point (Yasui et al. 2005). Similar fracture textures were found in volcanic ash particles at multiple depths in the core sample (Fig. 6h–i). This suggests that these are the remains of a past explosive eruption. Therefore, the volcanic ash detected at a depth of 9.7 m probably originated from the intense type A or type B vulcanian eruptions according to Minakami's (1935) classification. The ^{14}C age of the soil at a depth of 9.83 m is 9479 calBP. Therefore, the earliest activity of Vulcanian eruptions of Maekake Volcano that left traces in the soil is thought to have occurred approximately 9400 years ago (Fig. 7a[8]). Up until about 11,000 years ago, Hotokeiwa Volcano produced large-scale pyroclastic flows several times during its late-stage eruptions. The entire area at the foot of the volcano would have been completely devastated and covered with thick pyroclastic flow deposits and related materials. After about 1500 years, it appears that vegetation has recovered and soil has begun to form, at least around the boring site. On the other hand, the black soils are developed at the level below As-A, between As-B and As-C, below As-C, and below As-F (Fig. 7a [2][3][4][7]). These depths are rich in fine-grained soil material less than 1/16 mm (Fig. 7b); in other words, the content of volcanic ash particles is low. Many of the volcanic ash particles are weathered, that is, they are rounded and often altered (Fig. 6g). This means that the volcano had little activity or was almost dormant.

However, there is a 400-to-500-year age gap directly above As-C and As-B′ (Arrows in Fig. 8). It is thought that it took a long time for vegetation to recover due to the devastation of the land, which was covered with a thick layer of pumice. In other words, when there is no soil, records of information on small and medium-sized eruptions are missing, so information on the rise and fall of small and medium-sized eruptions cannot be obtained. To address such issues, it may be necessary to conduct additional surveys at locations with different distances and directions from the crater. Current estimates suggest that the small and medium-sized eruptions were almost dormant, at least immediately prior to the large class 1 eruptions of As-B and As-C.

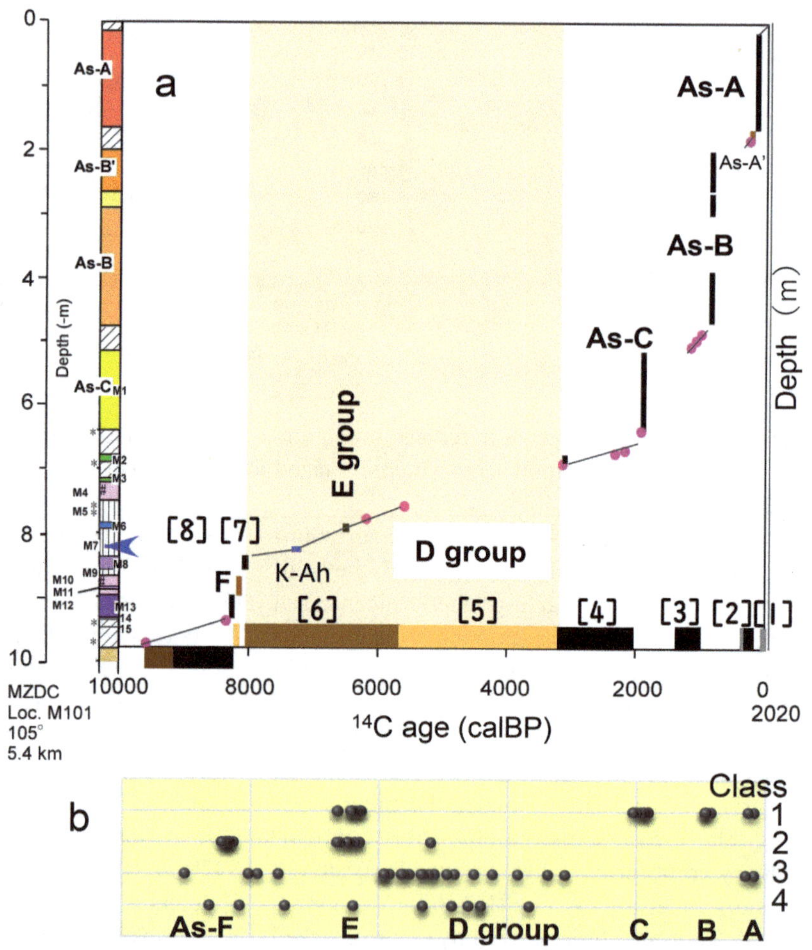

Fig. 8 A: Relation between depth of the drilling core sample of Loc. M101 and ^{14}C age (cal BP) of the soil. [1]~[8] and colors of the horizontal bars corresponds to the numbers and the color of the soil on the photo in Fig. 7. For the details of the columnar secition, see the legend in Fig. 3. K-Ah: 7300 years old widespread tephra derived from Kikai caldera in the southern Kyushu. B: Relation between the classification of the large-scale eruption (Class 1~4) and ^{14}C age (cal BP) of soil immediately below the pyroclastic fall deposits. Individual data are for the soil samples taken in the trench excavations at many localities. For the Class 1~4, see in the text 3

5 Relationships Between Activities at Different Scales

In the long term, active periods of small and medium-sized eruptions appear between large-scale eruptions. However, it is thought that there are variations in the period and scale of small and medium-sized eruptions. For example, in the first half of the twentieth century, magma was often confirmed inside the summit crater, and there were periods of continuous activity, but the number of eruptions has decreased abruptly since 1961. The only single eruption periods in which magma rose to the crater floor and was active for several months occurred in 1973 and 2004. Continuous active eruption periods and single eruption periods in the past 100 years are explained by differences in magma supply from deep (Murase et al. 2007). Takahashi et al. (2022) considered that there were multiple continuous eruption periods after the fifteenth century based on the historical records. From the perspective of the soil analysis as mentioned in Sect. 4, soil is useful as a medium for recording traces of small and medium-sized eruptions. The drilling site targeted in this study is in a direction where sub-Plinian plumes from large-scale eruptions are likely to expand, so Fig. 8, which shows layer thickness and age, can be regarded to approximate the long-term eruptive rate. The following can be considered. During the E-Group and D-Group periods from 3000 to 6000 years ago, class 3 and class 4 small-scale Subplinian eruptions occurred intermittently over a long period. In this period, brown soil developed suggesting that small and medium-sized eruptions also occurred intermittently. In other words, although the scale of eruption was not large, the release of magma as pyroclastic material has been extremely active during this period. In contrast, below As-B, As-C, and As-F, black soils with a few ash particles developed. It is noteworthy that before the major eruptions occurred approximately 8300 years ago, in the second century, and in the twelfth century, small and medium-sized eruptive activity is thought to have been almost dormant (Fig. 7a[8][4][3][2]). In other words, large eruptions appear to occur almost suddenly during periods of calm, which is important for understanding long-term trends in volcanic activity. In other words, the rise and fall of small and medium-sized eruptions may reflect the conditions of the underground magma supply system during the preparation period for large-scale eruptions. The above is an experimental study using a boring core at the east-southeast foot of the mountain. Since the expansion of plumes from small and medium-sized eruptions is expected to be influenced by the direction of tropospheric winds, it will be necessary to confirm the same for deposits distributed in other directions.

6 Summary

To reconstruct in detail the 10,000-year eruptive history of Maekake Volcano, we investigated the following two types of eruptions: large-scale eruptions that form distinct geological units (e.g., 1783 AD eruption), and small- to medium-sized eruptions, such as the Vulcanian eruptions that were active in the first half of the twentieth century. To reconstruct history of the large-scale eruptions, extensive surveys of trench excavations and numerous ^{14}C dating were performed to determine the overall stratigraphy. The scale of the large-scale eruption was classified into four classes based on the distribution map of the pyroclastic fall deposits. The rise and fall of small and medium-sized eruptions were studied experimentally using analysis of the volcanic ash-bearing soil contained in drill core samples on the ESE flank slope.

The earliest Vulcanian eruptions occurred around 9400 years ago, and the first major eruption occurred around 8300 years ago, causing pumice fall. After that, there were intermittent small and medium-sized eruptions until about 3000 years ago, with occasional small-scale sub-Plinian eruptions (Classes 3 and 4). In the last 2000 years, there have been three large-scale eruptions (Class 1) that erupted large amounts of magma, but the intervals between eruptions are long, about 700 years, and the frequency is low. Immediately before the large-scale eruptions around 8300 years ago, the second century, and the twelfth century, there was apparently a quiet period about small and medium-sized eruptions. Long-term trends in eruptive activity of various sizes may be linked to the evolution of the underground magma plumbing system and will be a clue for future long-term predictions.

Acknowledgments The authors would like to express thanks and appreciation to Aramaki S and Nakagawa M for their warm encouragement. This research was supported by the Integrated Program for Next-Generation Volcano Research and Human Resource Development by the Ministry of Education, Culture, Sports, Science and Technology (MEXT), Japan. We are grateful to the Chubu Regional Forest Office, Forestry Agency, Nagano Prefecture Saku Regional Development Bureau, SEIBU REAL ESTATE INC., Asama Resort, Hoshino Resorts, Karuizawa Kazakoshi School, Nagisan Service Co., Ltd., Nihon University, Karuizawa Town, Tsumagoi Village, the town and village staff, and the many residents around the foot of Mt. Asama, and other private companies and organizations, for their cooperation in land use for trench excavation. Former Nihon University students Sakagami M, Yamashita D, Sakumoto S, Orihara K, and Osaka I helped us with geological survey and laboratory work. We also thank Obayashi H, Sato Y, and Okui M for their tremendous help in the field.

References

Aramaki S (1963) Geology of Asama volcano. J Fac Sci Univ Tokyo 14 sec. 2, 229–443

Japan Meteorological Agency (n.d.) Historical activity in Asamayama in active volcanoes in Kanto and Chubu Region, National Catalogue of The Active Volcanoes in Japan (THE FOURTH EDITION). https://www.data.jma.go.jp/vois/data/tokyo/STOCK/souran_eng/volcanoes/045_asamayama.pdf

Minakami T (1935) The explosive activities of volcano Asama in 1935 (part 2). Bull Earthq Res Inst Univ Tokyo 13:790–800

Miyazaki T (2003) Re-examination of the records of activities of Asama volcano. Bull Earthquake Res Inst Univ Tokyo 78:283–463. https://repository.dl.itc.u--tokyo.ac.jp/record/32551/files/IHO78401.pdf

Murase M, Ono K, Ito T, Miyajima M, Mori H, Aoyama H, Oshima H, Yoshida Y, Terada A, Koyama E, Takeda T, Watanabe H, Kimata F, Fujii N (2007) Time- dependent model for volume changes in pressure sources at Asama volcano, Central Japan due to vertical deformations detected by precise leveling during 1902-2005. J Volcanol Geotherm Res 164:54–75

Takahashi M, Yasui M, Kanamaru T, Nakagawa M (2022) Eruptive volume and age of large-scale pumice eruptions of the Asama-Maekake Volcano, Central Japan, as revealed by detailed trench surveys: implications for future volcanic hazards. J Disaster Res 17:1080–1089

Yasui M, Koyaguchi T (2004) Sequence and eruptive style of the 1783 eruption of Asama volcano, Central Japan: a case study of an andesitic explosive eruption generating fountain-fed lava flow, pumice fall, scoria flow and forming a cone. Bull Volcanol 66:243–262. https://doi.org/10.1007/s00445-003-0308-8

Yasui M, Takahashi M (2015) Formation of the Andesitic Welded Pyroclastic Cones by Pyroclastic Eruption recorded in the Summit area of Asama-Maekake Volcano and the Collapsed Caldera Wall of Kurofu Volcano (<Special Section> Determination of the Construction of an Outcrop Database to Reveal Eruptive History (2))Bull Volcanol Soc Jpn 60:109–123

Yasui M, Takahashi M, Kanamaru T, Nagai M (2005) Textural variety in the eruptive products of vulcanian eruptions between 1108 A.D. and 2004 A.D. on Asama-Maekake volcano (<special section> the 2004 eruption of Asama volcano (2)). Bull Volcanol Soc Jpn 50:501–517

Yasui M, Takahashi M, Kanamaru T, Nagai M (2021) High-resolution reconstruction of history of large-scale eruptions of the Asama-Maekake volcano: inferred from the pyroclastic fall deposits. Bull Volcanol Soc Japan 66:293–325

AD1783 Large-Scale Eruption of Asama-Maekake Volcano

Maya Yasui, Masaki Takahashi, Tatsuo Kanamaru, and Masashi Nagai

1 Introduction

The 1783 eruption was the latest large-scale eruption of Maekake Volcano, erupting approximately 0.5 km^3 DRE of andesitic magma. Not only did lava flows and pyroclastic flows flow down to the foot of the volcano, but ash from the sub-plinian eruption's plume reached in the east as far as the Kanto Plain (Fig. 1). This eruption was so shocking that it was recorded by many people at the time. A famous old drawing of the eruption at its climax on the night of August 4 (Fig. 4b and chapter "Volcanic Disasters of the Great Eruptions in AD1108 and AD1783 of Asama Volcano Revealed by Archaeological and Historical Research") shows pumice falling from an eruption cloud, volcanic lightning, and numerous spatter falls from the eruption cloud. It shows the magnitude of its impact. A local historian collected old records and published modern language translations and explanations in five books (Hagiwara 1985). Yagi (1936) conducted pioneering geological research. Minakami (1942) conducted a detailed distribution survey of pyroclastic fall deposits distributed in the southeast direction. Aramaki (1956) gave a detailed description of pyroclastic flows and lava flow. Aramaki (1957) classified pyroclastic flows based on differences in volume and occurrence and positioned the Agatsuma pyroclastic flow as an intermediate-type pyroclastic flow. Yasui and Koyaguchi (2004) reconstructed the sequence of eruptions (Fig. 2) from an analysis of historical records and the correspondence between geological stratigraphy and discussed the eruption style. According to them, intermittent eruptions escalated during the 3 months of activity starting on May 9, 1783. It reached its peak around 15 h from the night of August 4. A sub-plinian eruption occurred with violent lava fountains, and clastogenic lava was discharged at a high rate. The image is that a Hawaiian high fountaining and a sub-plinian eruption coexisted (Fig. 3). This can be said to be a characteristic case of one of the eruption styles of andesitic magma.

The 1783 activity did not end with a climactic phase that erupted a large amount of magma. At around 9:30 a.m. on August 5, huge detonations could be heard in distant places more than 500 km. Immediately after the explosion, debris avalanche rushed through the northern foot of the volcano directly hit Kambara Village and then entered the Agatsuma River. The flow that gained water became a volcanic mudflow (lahar) that reached from the Kanto Plain to the Pacific Ocean, causing an unprecedented disaster. This disaster killed more than 1500 people, making it one of the worst in the history of volcanic

M. Yasui (✉) · T. Kanamaru
Department of Earth and Environmental Sciences, College of Humanities and Sciences, Nihon University, Tokyo, Japan
e-mail: yasui.maya@nihon-u.ac.jp

M. Takahashi
Institute of Natural Sciences, College of Humanities and Sciences, Nihon University, Tokyo, Japan

M. Nagai
Volcano Disaster Resilience Research Division, National Research Institute for Earth Science and Disaster Resilience, Ibaraki, Japan

S. Aramaki et al. (eds.), *The Asama Volcano*, Active Volcanoes of the World,
https://doi.org/10.1007/978-981-95-5799-8_11

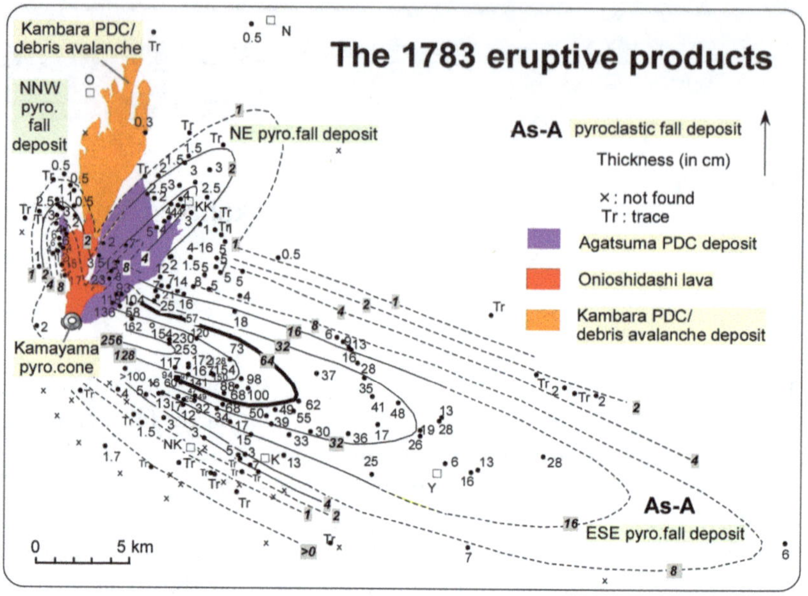

Fig. 1 Map showing the distribution of the eruptive products of the 1783 eruption. Some information is added to Fig. 5a of Yasui and Koyaguchi (2004)

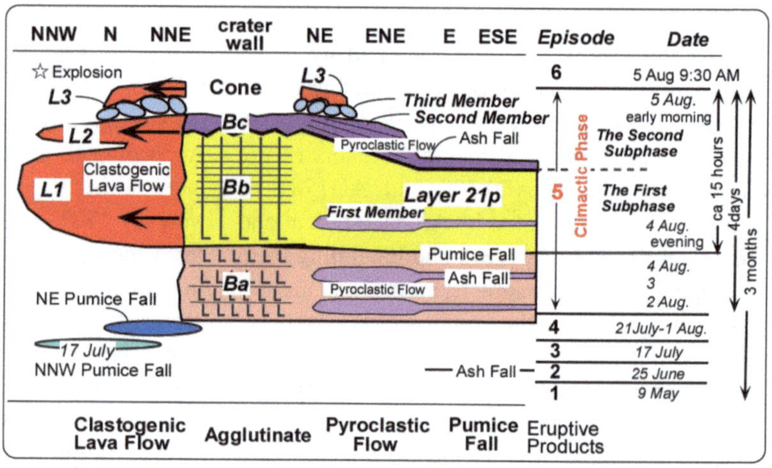

Fig. 2 Schematic illustration of the eruptive sequence of the 1783 eruption. Information on Episode 6 is added to Fig. 16 of Yasui and Koyaguchi (2004). For the explanation of the first, second and third members, see the descriptions of pyroclastic flow deposits in Yasui and Koyaguchi (2004)

disasters in Japan. However, the causes of disasters caused by a series of flow phenomena are not well understood. Some historical records, including paintings from before the 1783 eruption, show a pond on the northern flank slope of the volcano with the place name Yanai. Pitch-black smoke and streams spew out from there, and several paintings depict a mudflow that continues from the northern foot of the volcano to the Agatsuma River. On the other hand, there are also paintings depicting a stream gushing out from the summit crater. Officials at the time who investigated immediately after the eruption also reported that they were unsure whether the eruption came from the summit or the mountainside. At the northern foot of the volcano, a strange sight spreads out, with huge lava blocks sometimes more than 20 m scattered within a box-shaped valley. From the beginning of the twentieth century, geologists drew attention to this phenomenon (e.g., Yagi 1936). Since the 1950s, detailed geological surveys and archaeological excavations, and discussions have started by Dr. Aramaki and others. Here, we will summarize the geological characteristics of the deposits left between the northern flank slope and the Agatsuma River, and present one interpretation of their origin for future discussions. And then, we add other recent

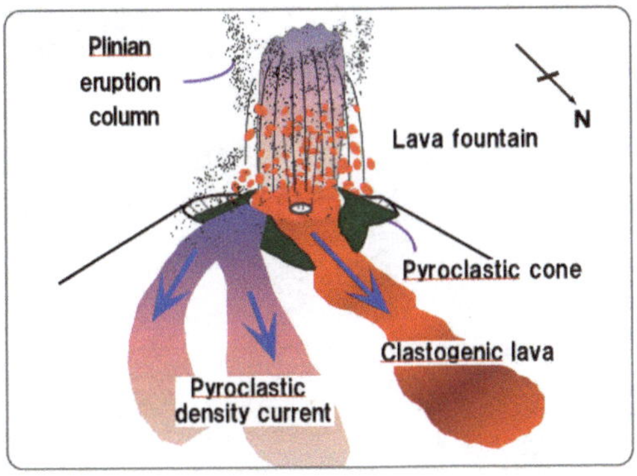

Fig. 3 Schematic illustration showing the image of eruptive style of the climactic eruption. Modified from Fig. 17 of Yasui and Koyaguchi (2004)

developments. Pyroclastic flows have many flow units, and recent surveys have confirmed 2117 vertical tree molds. These can be said to be traces of a large-scale buried forest caused by pyroclastic flows, which is rare even in the world. It will be introduced here because this is important as it can constrain the nature of pyroclastic flows like velocity and temperature.

2 Eruptive Sequence and Eruptive Style Based on the Eruptive Products and Old Documents

Yasui and Koyaguchi (2004) reconstructed the eruptive sequence of the three-month activity in detail by linking many old documents with geological stratigraphy (Fig. 2). The sequence and eruptive style are summarized in the following. Pyroclastic fall deposits are geologically detected in three directions (Fig. 1), each of which was dated from analysis of more than 900 descriptions of the old documents. Old records indicating the start of the eruption date back to May 9 (Episode 1 in Fig. 2), but the eruption was not large enough to leave any geological evidence. The eruption that left volcanic ash (ESE-1a) as a first geological unit occurred on June 25 (Episode 2) and was also correlated in a painting (Fig. 4a). Subsequent eruptions that produced the pumice layers were dated 17 July (Episode 3) to the northwest direction, late July to the northeast, and 2–4 August to the southeast (Episode 5), respectively. The intervals between eruptions that can be read from the old records have shortened with time. The number of eruptions and the eruption intensity increased exponentially at the end of the period.

The climactic eruption lasted about 15 h from the night of August 4. A famous old painting seen from the southern foot sketched the scene of the sub-plinian eruption (Fig. 4b). The effects of intense ash fall in the east-southeast direction due to this eruption were recorded in many towns along the old Nakasendo highway.

The upper half of the ESE pyroclastic fall deposit, layer 21p, is coarse-grained and thick pumice layer, while the lower half consists of many fall units (Fig. 5c). From this, the upper half can be compared to the climactic eruption from the night of August 4, and the lower half to the intermittent eruptions from August 2 to the afternoon of August 4 (Episode 5). Agglutinates with well-developed columnar joints can be seen on the crater wall of Kamayama on the summit (Fig. 5d). The upper half of the crater wall is thick and strongly welded agglutinates (Bb, Bc), whereas the lower half shows stratified multiple layers (unit Ba). Because their vertical patterns are the same, the Ba of the crater wall can be correlated with the pumice layer in the lower half of the foot. In other words, unit Ba was thought to have been formed during the eruptions from August 2 to 4.

The painting after the eruption shows the Kamayama pyroclastic cone on the summit, and there is an annotation that a mountain was created through the eruption (Fig. 4d). If you look at the old paintings sequentially, you will see that no pyroclastic cone was painted before the climactic eruption. This suggests that most of the pyroclastic cone (Bb, Bc) was formed during the climactic sub-plinian eruption. Onioshidashi lava is distributed on the northern slope of the Kamayama crater (Figs. 1 and 5a, b). Since the cross-section of Onioshishi lava also matches the unit Bb and Bc on the northern crater wall, it is thought that the lava flowed out at the same time as the formation of the pyroclastic cone, too.

Fig. 4 Old paintings of the 1783 eruption. These paintings are owned by Misaizu H. (**a**) June 25 eruption, (**b**) The climactic eruption in the night of August 4, (**c**) Pyroclastic density current and its ash cloud in the evening of August 4. (**d**) The situation after the eruption. A pyroclastic cone can be seen at the top of the volcano. There is an annotation that a mountain was created through the eruption

On the other hand, plural ash layers are interbedded in the ESE pyroclastic fall deposits (Fig. 5c). They are characteristically fine-grained less than 1/16 mm mostly and pink to purple colored. Since there are some outcrops where pyroclastic flow deposits are directly covered with an ash layer with the same feature, these are considered to be oxidized co-ignimbrite ash. These facts indicate that pyroclastic flows occurred many times after August 2. The 20a ash layer (Fig. 4c) implies pyroclastic flow immediately before the Plinian eruption at its peak and corresponds to a painting in Fig. 4c. Pyroclastic flow deposits have many flow units. On the flank slope of the ENE, the 21p pumice fall layer of the climactic eruption is overlain by pyroclastic flow deposits (Fig. 5e). The bulk rock SiO_2 content of the juvenile materials ranges from 60 and 64 wt.% and shows a bimodal distribution, with pumice fall tending to be felsic and pyroclastic flows to be more mafic. The 22a ash layer overlying the 21p pumice layer of the ESE pyroclastic fall deposit is thick and rich in dark brown mafic glass. From the above, most of the pyroclastic flows are thought to have occurred after the sub-plinian eruption at its peak, which brought about 21p. Concerning the eruptive style, the following must be a clue. A vigorous lava fountain was drawn in the paintings of the climactic eruption (Fig. 4b). Syn-eruptive lava toward the northern flank slope (Fig. 5a, b) shows clastogenic features, including oxidized fiamme and abundant broken crystals (Yasui and Koyaguchi 2004). Therefore, fountain-fed clastogenic lava generation was considered to be dominant in the climactic eruption. That is, about 15 h climactic eruption was characterized by the simultaneous progress of a vigorous lava fountain and sub-plinian eruption generating a vast clastogenic lava flow onto the flank slope, pumice fall in the wide area, and

Fig. 5 Field photos of eruptive products from the 1783 eruption. (**a**) Onioshidashi lava on the northern slope of Maekake Volcano seen from the NNW foot. (**b**) Blocky surface of Onioshidashi lava. Toeizan Kan'ei-ji Branch Temple, where memorial services are held for the eruption victims, can be seen on the lava field. (**c**) The ESE pyroclastic fall deposit. The upper half (Layer 21p) corresponds to the climactic eruption. (**d**) Northwestern crater wall of Kamayama in the summit. For the units Ba, Bb, Bc, see Sect. 2. (**e**) An outcrop in the proximal area where Agatsuma pyroclastic density current deposit overlies Layer 21p. The distance from the summit crater is about 2.5 km. Note that abundant coarse pumice grains are contained in the 21p pumice fall deposit

cone growth in the summit. Abundant pyroclastic flow generations onto the ENE flank slope also occurred in the latter half of the climactic eruption (Fig. 5e). Schematic image of the eruption style of the climactic eruption is in Fig. 3. It gives the image that a Hawaiian high-fountaining and a Plinian eruption coexist, but 94% of the pyroclastic material is thought to have fallen near the crater. This may also be one of the characteristics of Asama-Maekake Volcano.

3 Additional Explosions That Caused Serious Disasters

As first mentioned in Sect. 1, the activity did not end with the eruption peak. At around 9:30 a.m. on August 5, a huge detonation was heard and imme diately after that, the flow rushed down the northern foot of the volcano directly hit Kambara Village and flowed into the Agatsuma River (Episode 6). Here, we will consider what happened based on the geological features and old documents.

There is a semicircular depression approximately 220 m in diameter on the Onioshidashi lava that is about 3.5 km NNE from the summit crater. Aramaki (1963) called this semicircular depression an amphitheater. From there, a box-shaped valley continues in a narrow fan-shaped area that opens approximately 30° northward, and deposits of the Kambara debris avalanche can be seen inside the valley. The box-shaped valley is 1.1–2 km wide and up to 40 m deep, and the height of the cliffs on both sides decreases toward the north, reaching almost zero around 4.5 km north of the depression (Aramaki 1956). For convenience, here, we will define Area X as the area from the depression to the tip of the Onioshidashi lava, Area Y as far as the tip of the box-shaped valley to the north, and Area Z as the area that extends north to the Agatsuma River, including the Kambara area (Fig. 6). Near the boundary between Areas Y and Z, the number and size of mounds rapidly decrease (Fig. 6). The deposits in Areas X to Z are matrix-supported, consisting of huge blocks and matrix facies (Fig. 7). The deposit is characterized by containing various rocks and loose materials from the surface layer at the foot of the volcano including fragments of black soil. The matrix facies also contain many "soft blocks" of loose materials such as black soil and weathered volcanic ash (Fig. 7b). It is thought that the soft surface layer at the foot of the volcano was stripped off and transported by the flow. The flow is thought to be a debris avalanche. In Area Z, the deposit basically fills in the surface where the ground at that time was eroded. This suggests that the flow had strong ablation ability throughout Areas X to Z.

On the surface of Areas X and Y, there are many huge blocks of lava with several meters in size or larger. The farther south there are, the more gigantic lava blocks are found on the surface in Area Y (Fig. 7a). Most of the lava blocks are less than 6 m high and are rather flat. It is estimated that there are 3400 blocks over 5 m in height in these areas (Inoue 2009). Chilled margins and cooling joints several centimeters wide develop within a few centimeters of the surface

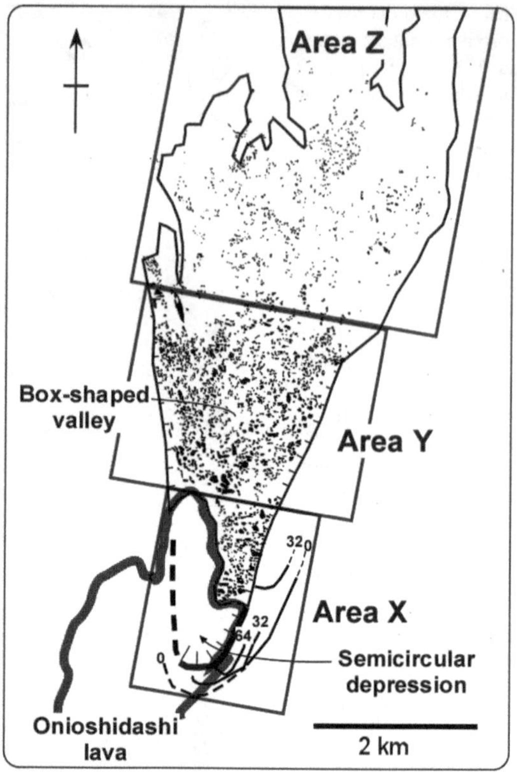

Fig. 6 Topography of the flank slope north of the downstream area of the Onioshidashi lava and distribution map of the Kambara pyroclastic flow/debris avalanche deposit. Blocks observed in the box-shaped valley are mapped. The isopach map (thickness in cm) of the blast deposit that distributes around the semicircular depression is also shown. Modified from Fig. 4 of Tanaka et al. (2012)

of block. These blocks show the magma composition of the 1783 eruption and Aramaki (1956) considered emplacement at high temperatures based on paleomagnetic studies. Their petrological characteristics and density are very similar to Onioshidashi lava.

Blast deposits are distributed around the depression in Area X (Fig. 6). The deposit consists of dense, angular-shaped dark gray lava fragments (Figs. 7c, d). It contains carbonized wood chips, indicating that it was deposited under high-temperature conditions (Aramaki 1956). The layer tends to be thicker near the southern edge of the depression (Fig. 6) and consists of three units (Fig. 7d). Unit-1 is thick in the southeastern part of the depression but is not seen above the Onioshidashi lava. On the

Fig. 7 Field photos of the Kambara pyroclastic density current/debris avalanche deposit and blast deposit. (**a**) A huge block approximately 19 m wide and 9 m high. (**b**) Matrix of debris avalanche deposit in Area Z. A soft block of alternating layers of volcanic ash. (**c**) Angular essential lava fragments that contained in the blast deposits. (**d**) Units 1 and 2 of blast deposits covering NE pyroclastic fall deposits (pumice fall layer)

other hand, units 2 and 3 cover Onioshidashi lava. The above circumstantial evidence points to plural explosions on the flank slope of the volcano and a strong northward flow immediately after.

The reasons for the strong northward flow are unclear, but for future discussions, we would like to present one possible scenario below. We will illustrate a series of events in chronological order as A to E (Fig. 8). Here, we assume the starting point is when the Onioshidashi lava, generated during the peak of the eruption, covers Yanai Swamp.

(a) The violent eruption that continued for 15 h starting in the evening of August 4 and continued all night stopped in the early morning of August 5. It seems that the Onioshidashi lava was still progressing slowly, covering the original Yanai swamp along the way.
(b) The first phreatic explosion occurred when the moisture in Yanai swamp and the groundwater directly below evaporated. An infrasound occurred, and a detonation was heard at remote locations. Immediately after the explosion, part of the flank slope of the volcano collapsed, and a flow of lava fragments from the explosion and materials from the collapse rushed northward. At this time, the powerful current rushed forward while digging up the ground, engulfing Kambara Village on the way. It appears that after entering the Agatsuma River, it continued to flow as a debris avalanche for a while, then took in river water and turned into a debris flow.
(c) Immediately after the first explosion, blast deposits (Unit-1) are observed around the

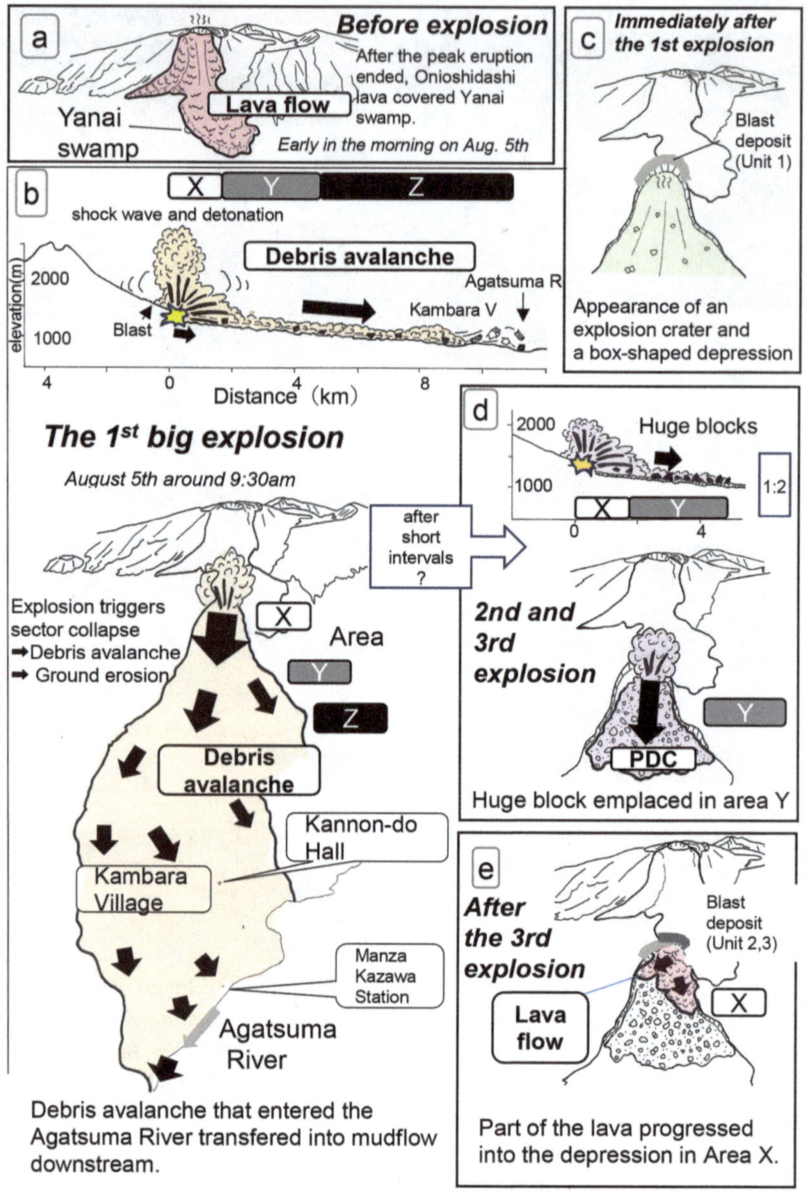

Fig. 8 Schematic diagram of the explosion and flow phenomena on the north flank of Maekake Volcano on the morning of August 5, 1783. For (**a**)–(**e**), see Sect. 3

explosion crater. A box-shaped depression can be seen connected to the north of the explosion crater.

(d) In the second and third explosions, the explosion point shifted slightly to the west, and blast deposits (Units 2 and 3) were deposited around the crater. The explosion produced a coarse mass of lava. The huge blocks of lava fragment flowed a short distance northward and became emplaced in the depression (Area Y).

(e) After the third explosion, lava was exposed in the explosion crater. Suddenly losing its gravitational support, the lava flowed into the depression and traveled about 2 km north.

From the above image, we can see the possibility that it can be simply understood as a series of phenomena that occurred in a chain sequence starting with a secondary explosion of lava. One can imagine that the leading edge of the flow tore away the surface layer of the northern foothills, which includes most of

Kambara Village, and pushed it forward while taking in the surface layer, and the continuing flow filled the entire area. In the above scenario, the idea is that the initial explosion would generate a powerful debris avalanche, and subsequent explosions would generate a pyroclastic flow-like flow consisting mainly of giant lava fragments. This is an attempt to explain field observations in multiple stages. However, this idea cannot explain the strong thrust of the explosion northward, which easily reached the Agatsuma River, more than 10 kilometers away. Aramaki (1992) believed that the origin was at the top of the volcano, and that the flow progressed by converting potential energy into kinetic energy and scraping away at the ground at the time, but this possibility cannot be denied. In the future, it is necessary to further investigate the causes of this phenomenon, including complex factors.

According to the 1980 distribution map of debris avalanches and volcanic mudflows from Mount St. Helens (Glicken 1996), the debris avalanche likely traveled about 20 km after entering the water system. The Kambara debris avalanche is thought to have advanced considerably westward after entering the Agatsuma River. Drawings from the time show that the huge lava mass was swept over 100 km by Lahar. The distance of migration from the debris avalanche to Lahar and the actual behavior of the coarse lava masses are also issues that remain to be addressed in the future.

4 Other Recent Developments: Vertical Tree Molds Caused by Pyroclastic Flows

Many tree molds were discovered on the northern slope of Maekake Volcano more than 90 years ago (Tsumagoi Village 2022). A part of the tree mold distribution area is protected by Tsumagoi Village and is open for walking (Fig. 9a–c). Some areas were designated as a special natural monument of the country in 1952 as a lava tree mold of Asama Volcano. However, they are in the distribution area of pyroclastic flow deposits from the 1783 eruption, and they were formed not by lava flow as follows.

A recent survey conducted by Tsumagoi Village, Gunma Prefecture, confirmed at least 2117 tree molds in areas north-northwest and northeast of the summit crater (Tsumagoi Village 2022). The diameter of tree molds is 0.15–1.2 m, and the depth is 0.3–7 m. The deposit of pyroclastic flow composed of blocks and matrix can be seen in the inner wall of individual tree molds (Fig. 9d, e). Characteristics include that the tree molds are in the form of standing trees. Therefore, although they are tree-shaped cavities, it can also be said to be the kind of large-scale buried forest.

A lobe-like topography is remarkable on the surface of the entire area of the 1783 pyroclastic density currents, suggesting the flows were generated one after another and piled up. When observing the inner wall of a tree mold, the coarse-grained blocks are often deformed to follow the shape of the tree trunk (Fig. 9c). The oxidized zone bounds of multiple flow units can be seen on the inner wall of the tree mold (e.g., Fig. 9e). These features suggest that the pyroclastic flow flowed down many times, and the trees were able to hold on without falling until the pyroclastic flow stopped depositing.

Although some are slightly slanted, most of the tree molds are upright, indicating that the trees were not mown down by the pyroclastic flow. Regarding the diameter of the tree mold, the smallest one is about 15 cm. By analogy with the relationship between the maximum instantaneous wind speed of a typhoon and fallen trees, the existence of a tree mold with a diameter of 15 cm suggests that the flow is slower than several tens of meters per second. This means that the velocity of the flow that formed the tree molds was in the slow category as pyroclastic density currents.

Comparison with the properties of the Oiwake pyroclastic flow deposit from the twelfth-century eruption provides constraints on the temperature of the pyroclastic flow, too. There are many carbonized pieces of driftwood in the Oiwake pyroclastic flow deposits. It is thought that the Oiwake pyroclastic flow that was mowing down trees was

Fig. 9 Field photos of the tree molds formed by pyroclastic flows in the 1783 eruption. (**a**) Distribution of tree molds in the current forest. They are found on the deposition surface of the 1783 pyroclastic flow. The arrows indicate the position of the tree molds. (**b**) Adjacent three tree molds, their diameters are 61–72 cm and depths are 1.57–2.15 m. (**c**) Example of a large tree mold with a depth 5.7 m and a diameter 85 cm. Note that the coarse blocks concentrated in the upper part of the deposits have been deformed along the shape of the trunk and that no blocks can be seen on the ground surface in the shadow of the tree downstream of the pyroclastic flow. (**d**) Inside view of a large tree mold as shown in figure (**c**). The welded inner wall consists of oxidized blocks and a small amount of matrix ash. Its bottom is buried by fallen leaves. (**e**) Inside view of another large tree mold (depth is 6.85 m and diameter is 82 cm). On the inner wall of the "natural chimney", reddish-brown blocks due to high-temperature oxidation are concentrated and welded with matrix ash. (**f**) Trench excavation survey site of pyroclastic flow deposits containing a tree mold. *TM* tree mold, *OXZ* oxidized zone, (**g**) Carbonized trunk that appeared near the base of the trench. White arrow shows the location of a charred core of the tree shown in figure (**g**)

faster than the Agatsuma pyroclastic flow. The Oiwake pyroclastic flow deposits were not welded and carbonized wood remained in the deposits, so it appears that the temperature was lower than that of the Agatsuma pyroclastic flow. In the Agatsuma pyroclastic flow deposits, at least four cases of carbonized wood chips remaining at the base of tree molds were confirmed. The diameter of the carbonized tree found 2.4 m below the ground in the trench excavation survey was 30 cm, and the diameter of the surrounding tree mold cavity was 78 cm (Figs. 9e, f). Looking at the cross-sectional area ratio, it appears that 85% of the tree's surface has been burned away at this depth. After the pyroclastic flow was emplaced, a chimney-shaped space was formed as the trees burned, and while the rising air current emitted soot, there was a shortage of oxygen below. The temperature near the former ground surface was relatively low. Therefore, the base of the trees did not burn completely, and a small amount of char remained. As the trees burned, the surrounding deposits also welded together, and the hollows where the trees had burned were held together by welded inner walls, providing a picture of the tree mold formation through this series of processes.

The confirmation of large numbers of pit tree molds in welded pyroclastic flow deposits, such as the case in 1783, is probably a rare case anywhere in the world. Information obtained from the tree molds is important for understanding the actual state of "intermediate pyroclastic flows" at Asama-Maekake Volcano.

5 Summary

The 1783 eruption was the most recent major eruption of Maekake Volcano and had a major impact on the surrounding area. The eruption was extensively recorded by contemporaries, and historical records were compiled by local historians. Research combining geological surveys and historical records provided a valuable opportunity to reconstruct the eruption sequence. Approximately 0.5 km^3 DRE of andesitic magma erupted over 3 months with intermittent expansion, and 96% of the magma was ejected during his 15 h of maximum eruption. Much of the pyroclastic material fell to the proximal area around and inside the summit crater and became lava and pyroclastic flows, forming pyroclastic cones. Syn-eruptive sub-plinian eruption column caused heavy ash fall in the east-southeast direction, and the eruption cloud reached as far as Edo (present-day Tokyo). This highlights the magnitude of its impact. Pyroclastic flows also occurred frequently during the latter half of the climactic eruption. Furthermore, as many as 2117 vertical tree molds were discovered in the welded pyroclastic flow deposits. This is a type of large-scale buried forest caused by pyroclastic flows, which is rare in the world. It appears that pyroclastic flows occurred one after another, but it was also found that there were units of slow flow that did not knock down trees.

It was the explosion at around 9:30 a.m. on August 5 and the series of events that followed that caused widespread and serious disaster. The explosion could be heard 500–800 km to the west. Debris avalanche and debris flows occurred, killing more than 1500 people. Although the possibility of a secondary explosion of the lava that covered the marsh was shown here, the cause of the strong northward flow phenomenon remains unclear. The eruption of Maekake Volcano in 1783 is a stark reminder of the destructive power of volcanic activity and the importance of understanding such events to prepare for future disasters.

Acknowledgments The authors acknowledge Aramaki S, Koyaguchi T, and Nakamura M for valuable suggestions and discussion. Former Nihon University students Tanaka E, Kato F helped us with geological survey and laboratory work. We extend our sincere acknowledgement to the Seki T, Noto K, Toi M, and Asama North Geopark members for their help in the field.

References

Aramaki S (1956) The 1783 activity of Asama volcano, part 1. Jpn J Geol Geogr 27:189–229

Aramaki S (1957) The 1783 activity of Asama volcano, part 2. Jpn J Geol Geogr 28:11–33

Aramaki S (1963) Geology of Asama volcano. Geology of Asama volcano. J Fac Sci Univ Tokyo 14:229–443

Aramaki S (1992) Eruption of 1783AD (Tenmei eruption). In: Guide book for field workshop at Asama and Kusatsu-Shirane volcanoes, Japan. IAVCEI Commission on Explosive Volcanism, Geneva, pp 11–18

Glicken H (1996) Rockslide-debris avalanche of May 18, 1980, Mount St. Helens volcano. U.S. Geological Survey, Washington, p 90. https://pubs.usgs.gov/of/1996/0677/

Hagiwara S (1985) Mt. Asama Tenmei eruption historical materials collection I~V. Gunma Prefectural Cultural Activities Promotion Association, Maebashi, p 372 (in Japanese)

Inoue K (2009) Landslide disaster caused by eruption: Tenmei Asama eruption and Kambara mud and stone avalanche, Kokon shoin, p 204 (in Japanese)

Minakami T (1942) On the distribution of volcanic ejecta. (part II.) the distribution of Mt. Asama pumice in 1783. Bull Earthq Res Inst Univ Tokyo 20:93–106

Tanaka E, Yasui M, Aramaki S (2012) Nature of the kambara pyroclastic flow/debris avalanche deposit of the Asama 1783 eruption. Proc Inst Nat Sci Nihon Univ 47:271–286 (in Japanese)

Tsumagoi Village (2022) Nationally designated special natural monument Mt. Asama lava tree mold preservation and utilization plan (in Japanese). https://www.vill.tsumagoi.gunma.jp/www/contents/1663119048894/index.html

Yagi T (1936) Asama volcano (with geological map of Asama volcano). Shinano Educational Association, Tokyo, p 533 (in Japanese)

Yasui M, Koyaguchi T (2004) Sequence and eruptive style of the 1783 eruption of Asama volcano, Central Japan: a case study of an andesitic explosive eruption generating fountain-fed lava flow, pumice fall, scoria flow and forming a cone. Bull Volcanol 66:243–262. https://doi.org/10.1007/s00445-003-0308-8

Large-Scale Eruptions of Asama-Maekake Volcano in the Twelfth Century and Asama-type eruption

Maya Yasui, Masaki Takahashi, Tatsuo Kanamaru, and Masashi Nagai

1 Introduction

The 1783 eruption was the latest large-scale eruption and has an exceptionally large number of historical records. Therefore, when reconstructing the eruption sequence, it is possible to include a time axis in the stratigraphy of the eruptive products from the analysis of old documents. On the other hand, regarding the eruptions in the twelfth century, historical records do not help for reconstructing the eruptive sequence because only one record says that there was a large-scale eruption in the fall of 1108 AD generating serious disasters up to the distant place in the East (see chapter "Volcanic Disasters of the Great Eruptions in AD1108 and AD1783 of Asama Volcano Revealed by Archaeological and Historical Research"). On the other hand, the eruptive products of numerous eruption units are distributed in almost all directions around the crater. Although it is extremely difficult to understand stratigraphical relationships, a lot of geological facts obtained help estimate the eruption sequence and style of the twelfth century eruption. In particular, the repeated eruption pattern in which large-scale Vulcanian eruptions and intermediate type pyroclastic density currents occur one after another is thought to be a characteristic of the activity in the twelfth century and will be introduced as "Asama-type eruption" style here.

M. Yasui (✉) · T. Kanamaru
Department of Earth and Environmental Sciences, College of Humanities and Sciences, Nihon University, Tokyo, Japan
e-mail: yasui.maya@nihon-u.ac.jp

M. Takahashi
Institute of Natural Sciences, College of Humanities and Sciences, Nihon University, Tokyo, Japan

M. Nagai
Volcano Disaster Resilience Research Division, National Research Institute for Earth Science and Disaster Resilience, Ibaraki, Japan

2 Distribution and Characters of the Eruptive Products

The types of eruptive products from the twelfth century eruption are pyroclastic density current deposits, pyroclastic fall deposits (As-B, Red ash, As-B′), Maekake pyroclastic cone, and Kaminobutai lava.

The distribution of eruptive products from the twelfth century eruption and a diagram showing stratigraphy by direction are shown in Figs. 1 and 2. In the proximal area, there are at least three pyroclastic cones (αβγ) (Yasui and Takahashi 2015). Those are larger than the Kamayama pyroclastic cone formed in the 1783 eruption and are arranged in an east-west direction at the summit of Maekake volcano (Fig. 3). Topographically,

S. Aramaki et al. (eds.), *The Asama Volcano*, Active Volcanoes of the World,
https://doi.org/10.1007/978-981-95-5799-8_12

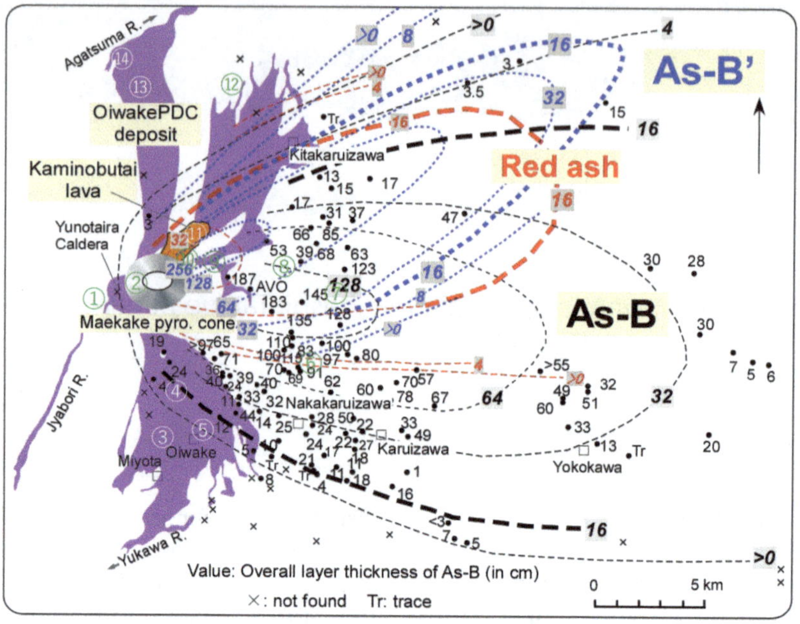

Fig. 1 Distribution of the eruptive products of the twelfth-century eruptions. Distribution of the Oiwake pyroclastic density current deposits (PDC) is from Aramaki (1963). For the pyroclastic fall deposits, information on the Red ash in Fig. 11a of Yasui (2015) and the As-B′ and As-B in Fig. 5 of Yasui et al. (2021) is integrated into this one figure. Isopach map of As-B (Total) is shown with individual measured values in cm. Isopach lines without data of individual locality are shown for the Red ash (Total) and As-B′ (Total)

it is estimated that two depressed craters with a diameter of 1 km are connected in the east and west to form a large crater (Yasui and Takahashi 2015). Kaminobutai lava flowed 3.5 km down the east-northeast slope of the volcano. This lava is very similar in appearance to the pyroclastic cone at the summit. In addition, blocks with the same characteristics as those contained in pyroclastic flows are locally concentrated (Fig. 4). These will be discussed later in Sect. 4.

The types of eruptive products that reached the foot of the volcano are mainly pyroclastic fall deposit (As-B) and pyroclastic density current deposit (Oiwake pyroclastic density current deposit, hereafter).

Oiwake pyroclastic density current deposits are distributed in a wide area ranging from north-northwest to east, south, and west. To the west, it fills the Yunotaira caldera, and some of it also flows into the large valley of the Jabori River on the southwest side (Fig. 1). It consists of many flow units in each direction. The lobe-like surface topography is particularly noticeable in the east and south flank slope, where deposits from the 1783 eruption thinly cover them or not distribute. Very rarely, there are outcrops directly covered by fine-grained, light brown co-ignimbrite ash (Fig. 5a). Oiwake pyroclastic density current deposits form a plateau on the south and northern foot. Stratified mudflow deposits (lahar) often cover the deposit. On the outcrop, the deposit is rarely welded and consists of matrix ash and characteristic uneven blocks (Fig. 5b). The block is unique in that multiple blocks compose a single composite block and are welded together, including matrix ash (Fig. 5c). Brown glass shards and crystal fragments consist of the matrix (Fig. 6d). The fine-grained co-ignimbrite ash that directly covers the lobe-like deposit is also composed of brown glass and is like the matrix of the pyroclastic density current (Fig. 6e).

The overall distribution of the pyroclastic fall deposits (As-B) is shown in Fig. 1. Especially, many fall units were seen from the southeast to the east, indicating that the eruptive plume often expanded in this direction due to the influence of westerly winds. At a locality (M106, Fig. 1(7))

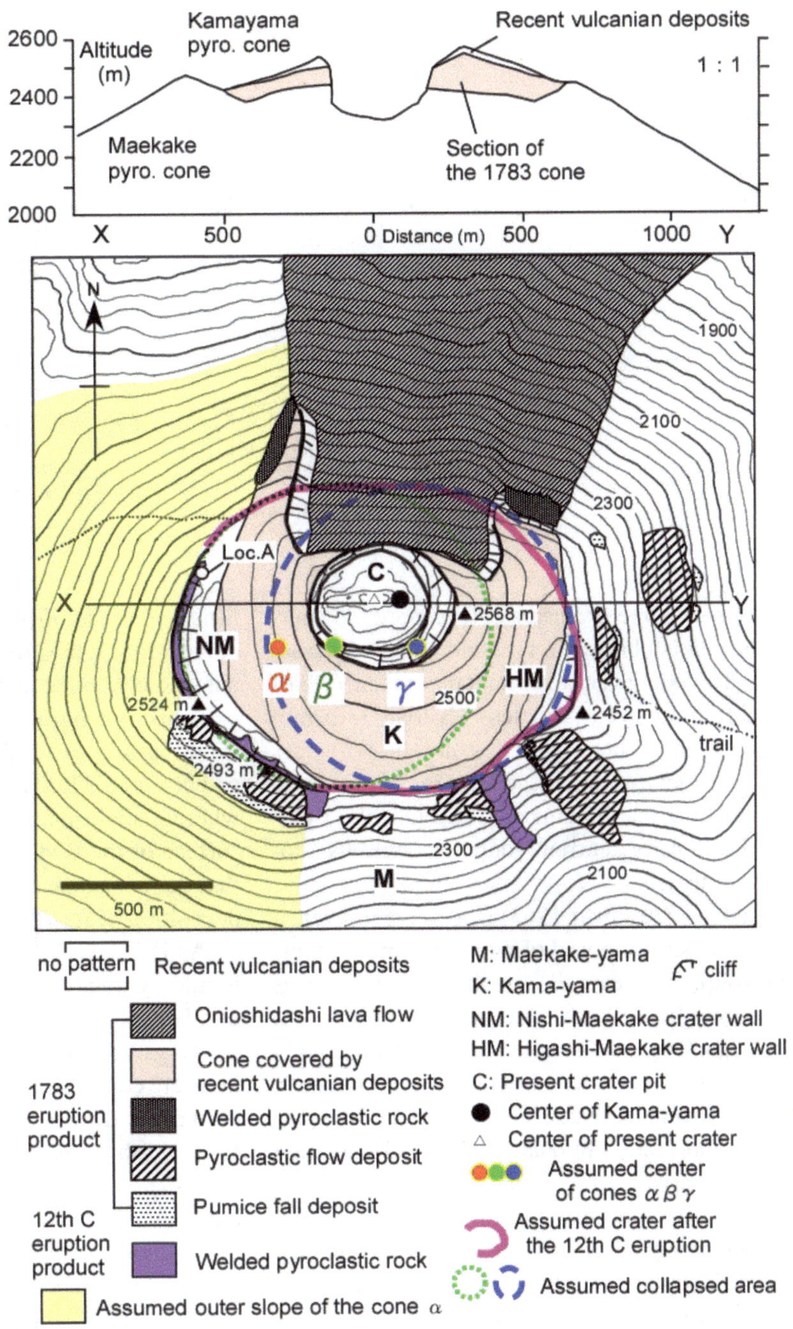

Fig. 2 Geologic map of the summit area of Asama-Maekake volcano. Information on the twelfth-century eruptions is added to Fig. 1 of Yasui and Takahashi (2015). The contour lines on this map were extracted from the 1:5000 Fundamental Map for Volcanoes of Mt. Asama published by the Geospatial Information Authority of Japan (GSI)

that is 8.7 km to the east of the summit crater, 23 fall units of pumice layer, lithic fragment layer, and fine ash layer can be seen with a total thickness of 123 cm (Fig. 7a).

As-B consists of many fall units composed of the pumice layer made of brown pumice (1), the ash/volcanic lapilli layer made of dense lithic fragments (2), and fine-grained ash layers (3). Layers of (2) consist of angular lava fragments like deposits from the twentieth-century Vulcanian eruptions (Fig. 6f). Because the layers remain as recognizable geological units, the scale of the Vulcanian eruption that produced them is much larger than those formed in the twentieth

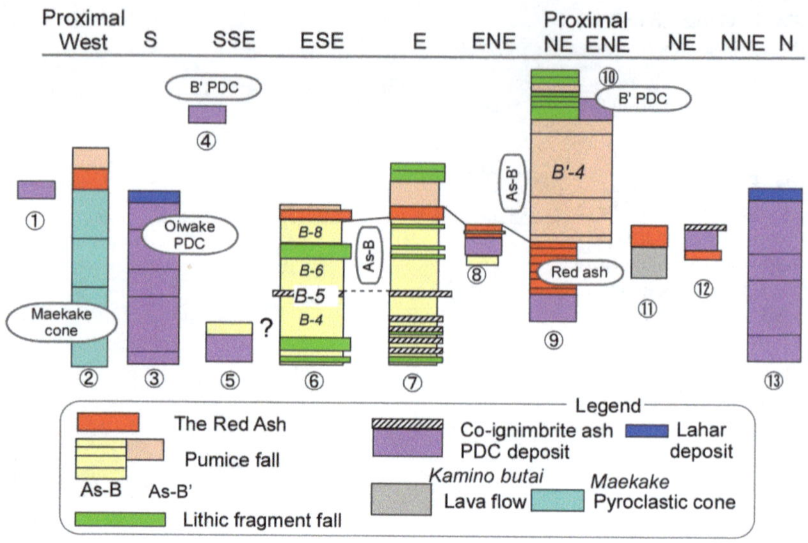

Fig. 3 Schematic columnar sections (not to scale) showing the stratigraphic relationship of the eruptive products of the twelfth-century eruptions. The locations are shown in Fig. 1

century. The fine-grained ash layers are pale purple or pinkish brown in the field and are characterized by being very fine, with more than 70% of it being mud-sized particles less than 1/16 mm (Fig. 7d, e). These particles are brown glass shards under the microscope (Fig. 6c) and are similar to those contained in the Oiwake pyroclastic density current deposit and its co-ignimbrite ash.

Red ash (Aramaki 1963) is a colorful layer of volcanic ash composed of more than 60 fall units (Fig. 8c). The overall distribution of As-B is wide, but the Red ash is distributed from east to northeast (Fig. 1). Coarse fraction of each layer is composed of particles of dense lithic fragments, but fine-grained ash can be seen between the particles. Therefore, bimodal grain-size distribution is characteristic of each layer of the Red ash (Fig. 7b). In the field, the fine particles exhibit various colors such as light purple and pinkish brown. They are mainly mud-sized particles less than 1/16 mm. In contrast, the coarse grains are dark gray and angular (Fig. 7c). Many particles are dense lithic fragments and sometimes show a strongly welded structure under a microscope. Small amounts of pumice grains those chemical compositions are identical to the 1108 magma are also contained in some layers of the Red ash.

As-B′ is a pyroclastic fall deposit whose distribution axis is northeast of the summit crater and covers the Red ash. There are some localities where a slight unconformity can be seen on the surface of the Red ash (Fig. 5d), so it is thought that there is some time gap of erosion before the deposition of As-B′.

As-B′ consists of fall units of at least 7. As-B′ is mainly composed of brown pumice. One fall unit B′-4 characteristically contains banded pumice (Fig. 8b) (Yasui 1994). The uppermost two units are composed of dense lithic fragments. But unlike the Red ash, it does not contain fine-grained ash between grains. In the proximal area, one ash unit of As-B′ contains accretionary lapilli indicating phreatomagmatic origin. On the NE flank slope, small-scale welded pyroclastic density current deposit (B′ scoria flow, Aramaki 1963) is also recognized (Fig. 2). On the south flank slope, lobe-like pyroclastic density current deposits that contain banded pumice distributes (Fig. 3④).

Fig. 4 Photographs of the Kaminobutai lava. (**a**) Topography of the northeastern flank slope of Maekake volcano. A square shows the lateral cliff of the Kaminobutai lava shown in Photo B. This map was created using Kashmir 3D (http://www.kashmir3d.com/) with Digital Map 1:50,000 (map image) and Digital Map 50 m mesh (elevation) published by the Geospatial Information Authority of Japan (GSI). The approval of the Director-General of the Geospatial Information Authority of Japan has been obtained for the creation of this map. (**b**) Southern lateral cliff of the Kaminobutai lava. A square shows the locality (10). The location is shown in Fig. 1. (**c**) A photo showing an occurrence where densely welded parts gradually change into welded blocks. Photographed on a 1 cm grid. (**d**) Close up of the lower part of a welded pyroclastic rock layer. Note that the individual block is similar to the composite block contained in the Oiwake pyroclastic density current deposit. This layer directly overlies a non-welded layer composed of lithic fragments. A dashed orange line marks the boundary. These photos were taken at locality 10

Fig. 5 Photographs of the Oiwake pyroclastic density current deposits. (**a**) A lobe-like Oiwake pyroclastic density current deposit at the near tip of the distribution on the Northern foot of the volcano. Note that light brown colored co-ignimbrite ash directly covers the deposit. This photo was taken at locality 12. The location is shown in Fig. 1. (**b**) A composite block contained in the Oiwake pyroclastic density current deposit. Scale: 5 cm. (**c**) A section of a composite block. Scale: 10 cm. Note that blocks (B) and matrix (M) are welded together. (**d**) Weakly welded Oiwake pyroclastic density current deposit observed on the valley wall of the Agatsuma river on the NNW foot of the volcano. This photo was taken at locality 14. The location is shown in Fig. 1. (**e**) Coarse matrix ash grains of the Oiwake pyroclastic density current deposit. Grains between 1 mm and 2 mm (0 phi). Photographed on 1 mm graph paper

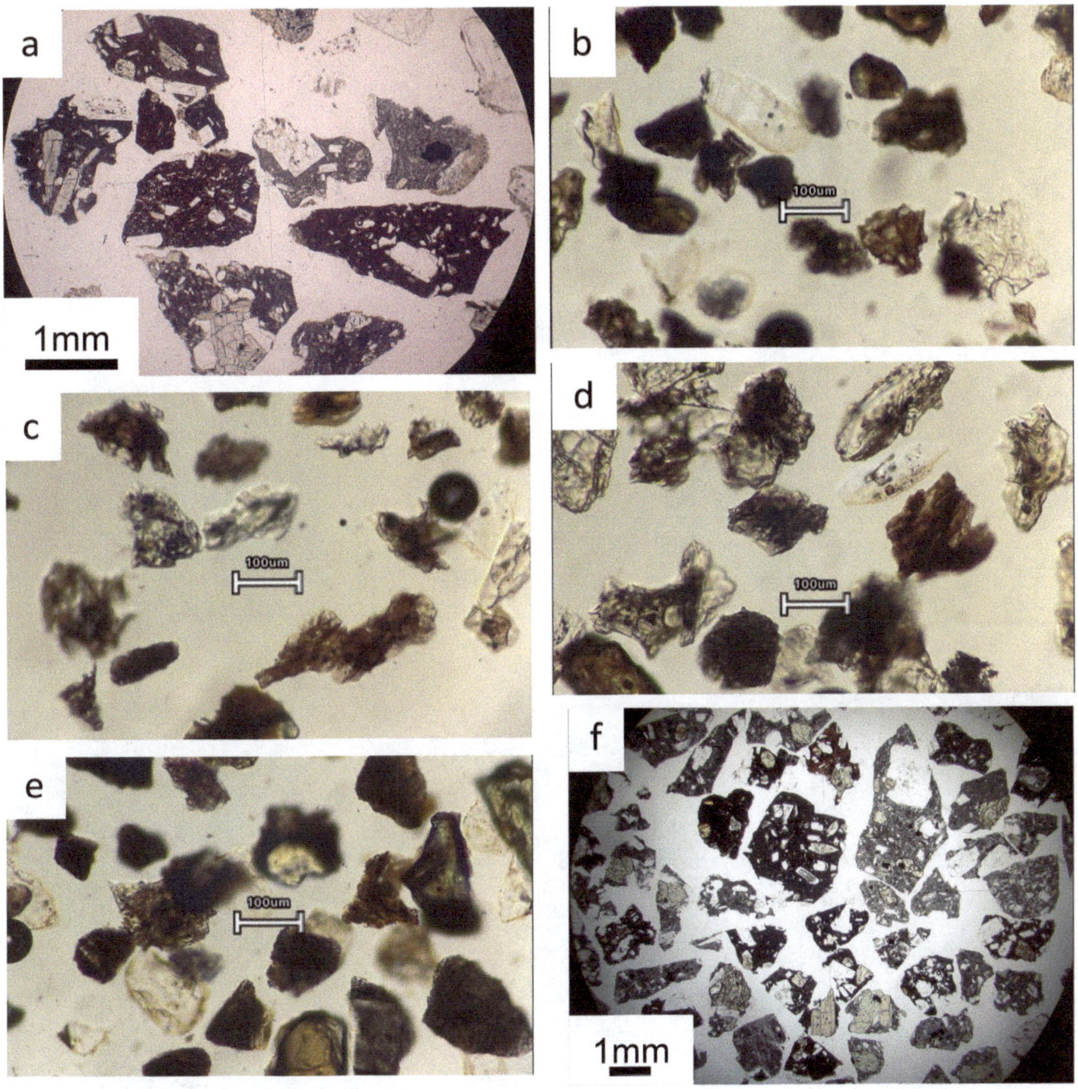

Fig. 6 Photomicrographs of ash particles. All photos were taken with Open nicol. (**a**) Coarse fraction consisting of angular lithic fragments of the Red ash. Grains between 1 mm and 2 mm (0 phi). (**b**) Fine fraction consisting of brown glass shards of the Red ash. Grains between 1/8 mm and 1/16 mm (+4 phi). (**c**) Fall unit (B-5) of As-B consisting of brown glass shards. Grains between 1/8 mm and 1/16 mm (+4 phi). (**d**) Matrix of the Oiwake pyroclastic density current deposit consisting of brown glass shards and crystal fragments. Grains between 1/8 mm and 1/16 mm (+4 phi). (**e**) Co-ignimbrite ash consisting of brown glass shards and crystal fragments. The ash layer directly covers one unit of the Oiwake pyroclastic density currents deposit on the northern foot. Grains between 1/8 mm and 1/16 mm (+4 phi). (**f**) Ash particles from the Vulcanian eruption of November 14, 2004, consisting of angular lithic fragments (0 phi)

Fig. 7 Occurrence on the field, grain-size distribution, and photographs of the representative samples of ash layer of As-B and the Red ash. (**a**) As-B on the Eastern foot of the volcano. This photo was taken at locality 7 (Trench 5). The location is shown in Fig. 1. A square shows the area in photo F. Scale: 5 m. (**b**) Grain-size distribution of a representative sample of the Red ash. (**c**) Photograph of sieved grains. It is the same sample shown by diagram (**b**). Solid star: +2 phi fraction (between 1/2 and 1/4 mm), Open star: fine ash less than 1/16 mm. (**d**) Grain-size distribution of a representative sample of the fall unit (B-5) of As-B. (**e**) Photograph of sieved grains. It is the same sample shown by diagram D. Solid star: fine ash less than 1/16 mm. (**f**) Close-up of stratified layers of As-B

Fig. 8 Photographs of the pyroclastic fall deposits of the As-B′ in the proximal area. (**a**) Proximal facies of the As-B′ where the distance from the summit crater is around 3 km in the ENE direction. This photo was taken at locality 9. The location is shown in Fig. 1. (**b**) Close up of fall unit (B′-4). The circle in the photo shows a typical banded pumice. Note that dark gray massive angular lithic fragments are contained. (**c**) Close up of the Red ash at Loc. 9 in Fig. 1. Note the slight erosional surface on the top of the Red ash here. Scale: 0.26 m

3 Eruptive Sequence of the Twelfth-Century Eruptions

There are many differences between the twelfth-century eruption and the eighteenth-century eruption in terms of eruptive volume, eruption style, eruptive sequence, and duration of the activity. The eruption volume of the 1108 eruption is estimated to be 0.82 km^{3}DRE (Takahashi et al. 2022), which was larger than the 1783 eruption. The eighteenth-century eruption was the intermittent activity for about 3 months and escalated to the climactic activity in the last 15 h (see chapter "AD1783 Large-scale Eruption of Asama-Maekake Volcano"). Although the duration of activity is unknown for the twelfth century, there were many eruptive units, and it appeared to have continued for a long time. Figure 2 shows a schematic columnar section of an outcrop where important stratigraphy can be seen, arranged by direction.

In the proximal area, at least three welded pyroclastic cones (αβγ) were recognized. They are considered to have formed in stages from west to east (Fig. 3). Since the γ cone on the summit and Kaminobutai lava are covered by the Red ash, and their formation was before the Red ash deposition. From the analogy with the eighteenth-century eruption, cones seem to grow by subplinian eruption. On the Nishimaekake crater wall, no pumice fall layers are seen, so the pyroclastic cone and pumice fall layer are likely of the same phase. The Kaminobutai lava was caused by the partial collapse of the easternmost of the γ cone and is thought to have flowed during the latter half of As-B. Because the distribution areas of Oiwake pyroclastic density current and As-B hardly overlap, it is difficult to obtain stratigraphic information directly related to the time of deposition. However, ash fall layers that were interbedded with pumice fall layers (e.g., B-5) are composed of fine-grained brown glass and are understood to be co-ignimbrite ash from pyroclastic flows. In other words, although the main body of the pyroclastic flow has not reached the location where the ash layer was deposited, its stratigraphic level indicates the time when the pyroclastic flow appeared. For example, the Oiwake pyroclastic density current deposits covered with pumice fall, as shown in Loc. (5) in Figs. 1 and 3, may have brought the fine-grained ash layer of B-5 or lower as co-ignimbrite ash. On the other hand, the Red ash consists of coarse-grained lithic fragments of coarse sand size and fine-grained brown glass mostly less than 1/16 mm, the latter of which appears to be co-ignimbrite ash from pyroclastic flows. The coarser grains are composed of angular lithic fragments, similar to the lithic fallout units from the Vulcanian eruption (e.g., 1 September 2004). The possible mechanism will be discussed in 12–4, but the Oiwake pyroclastic density current likely erupted many times during the period that brought about the Red ash. There are many outcrops in which the Red ash covers Oiwake pyroclastic density current deposits of the northeast side, and As-B of the southeast side. In the north, there is an outcrop where the Oiwake pyroclastic density current deposit covers the Red ash (Fig. 1(12)). To summarize the above, subplinian pumice fall layers (e.g., B-4, B-6) were deposited in the early stages of the activity, but pyroclastic flows were also generated in between. Part of pyroclastic cones flowed towards the NE secondarily. After that, there was a period when ash fall and pyroclastic flows erupted for many times generating the Red ash and Oiwake pyroclastic density current deposits. Because at two localities, As-B′ were confirmed to have covered the slightly eroded surface on the top of the Red ash, it seems that there was a short time gap after the series of eruptions of the Red ash, and then As-B′ erupted. Schematic eruptive sequence is shown in Fig. 9.

As As-B′ and As-B show different petrological properties (chemical compositions and petrographic features), they are thought to have been caused by eruptions of the different magmas (see chapter "Petrology of the Magmatic Plumbing System of Asama-Maekake Volcano"). The Red ash is overlain by As-B′ and the chemical compo-

Fig. 9 Diagram showing the eruptive sequence of the twelfth-century eruptions

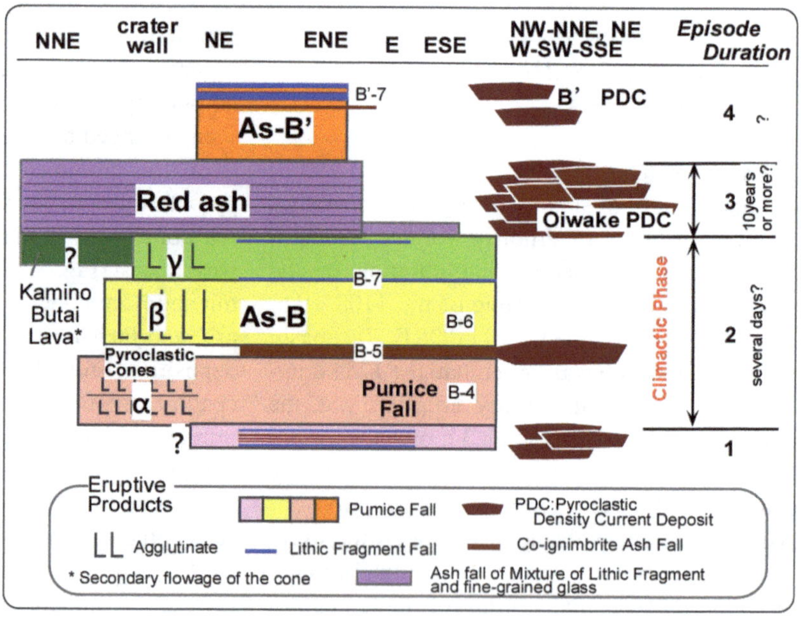

sition of the pumice contained in the Red ash was that of the magma of the 1108 eruption. According to pollen analysis of the Red ash, more than 11 peaks that indicate the spring season were observed, suggesting that the Red ash took a long time to accumulate (Miyano et al. 1994). Since there is a historical record of an eruption in 1128, this may be the eruption that caused As-B′. In other words, it is highly likely that the large-scale eruption that started in 1108 continued for a long time and was followed by an eruption that generated As-B′. Consequently, the estimated sequence of eruptions in the twelfth century, as shown in Fig. 9, it is much more complex than that in the eighteenth-century eruption.

4 Asama-Type Eruption: Large-Scale Vulcanian Explosion Preceded by Generation of the Intermediate-Type Pyroclastic Density Current

Following observations should be a clue for understanding the eruption style in the twelfth-century eruption. Characteristically, the twelfth-century eruption has many eruption units of pyroclastic fall and pyroclastic density current deposits. The individual layer of the Red ash shows bimodal particle size distribution that is composed of coarse lithic fragments and fine glass shards. These glass shards are indistinguishably similar to the matrix ash of Oiwake pyroclastic density current deposits, indicating co-ignimbrite ash. On the crater wall, the inner structure of the pyroclastic cones is observed. They are composed of pyroclastic rocks of various degrees of welding. The Kaminobutai lava on the NE flank slope shows similar features to the pyroclastic cones in the summit area. Therefore, it appears to be a clastogenic lava that has been generated by secondary flowage of the cone. Accumulated composite blocks as same as those contained in the Oiwake pyroclastic density current deposits are observed in Kaminobutai lava.

Here we present one interpretation of the eruption style of the twelfth-century eruption that could explain the characteristics of the eruptive products described above. Figure 10 shows the schematic illustration of the eruption style. We would like to refer to this as an Asama-type eruption here. This image is somewhat similar to the episodes of periodic Vulcanian explosive activity with fountain collapse observed in the Soufrière Hills volcano in 1997 (Druitt et al. 2002). Our image is in the following.

One cycle eruption begins with a Vulcanian explosion (Phase 1) and angular lithic fragments deposit from the eruption cloud onto the foot of the volcano. Immediately after the explosion, sudden decompression occurs to cause vesiculation and fragmentation of magma. For some reason, ascending fragments of magma begin to coalesce within the misty flow in the conduit. As the diameter of the composite block increases, terminal velocity also increases, causing coarse particles to fall back compared with finer grains. A sort of separation may have been at work in which coarse-grained blocks were selectively deposited near or inside the crater when a fountain collapsed to generate a pyroclastic density current (Phase 2 in Fig. 10). After that, in the short quiescence time, fallback blocks that filled the crater continued to weld, forming a kind of plug. When the new magma comes up, pressure builds up under this plug, triggering the next explosion (Phase 1).

Phase 1 progresses to Phase 2 continuously, here. If, immediately after a Vulcanian eruption, a pyroclastic density current was generated and co-ignimbrite ash fell on top of the coarse-grained ash fall layer, then the mixed feature of different ash particles in the individual layer of the Red ash can be explained. Since there are more than 60 layers of the Red ash, the same cycle is considered to have repeated for many times.

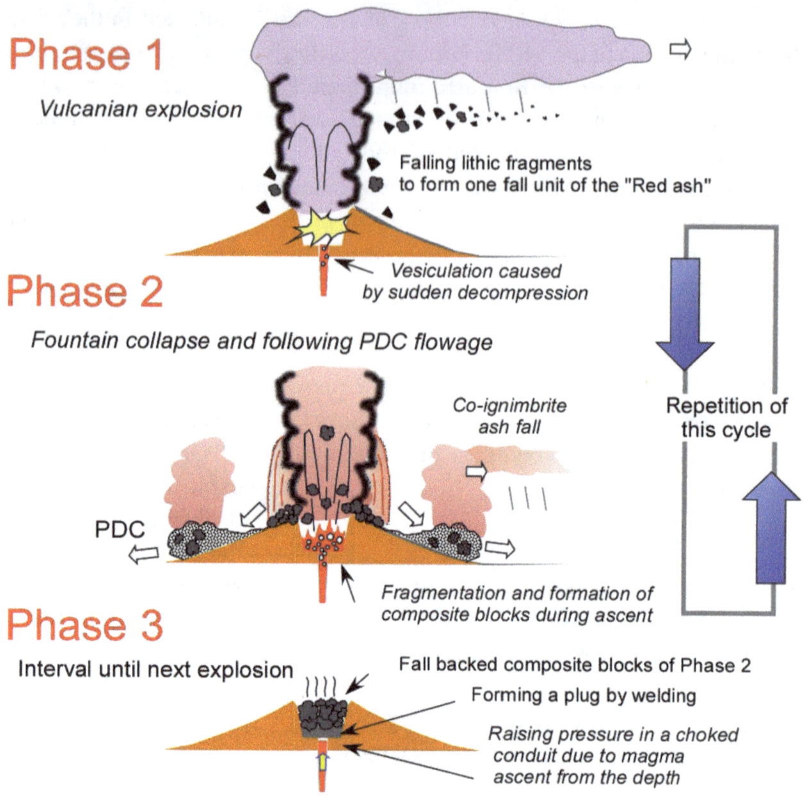

Fig. 10 Schematic illustration of the intermediate-type pyroclastic density current generation in the twelfth-century eruptions. This is the image of the Asama-type eruption proposed here. *PDC* pyroclastic density current

5 Summary

Characteristically, the twelfth-century eruption has many eruption units of pyroclastic fall and pyroclastic density current deposits. Deposits classified as intermediate-type pyroclastic flows by Aramaki (1963) are characterized by a composite block consisting of multiple blocks. The presence of ash fall layers called the Red Ash is also distinctive. Each layer of the Red ash consists of coarse angular lithic fragments and fine brown glass shards. The glass shards are indistinguishable from the matrix ash of Oiwake pyroclastic density current deposits and are believed to be co-ignimbrite ash. These features are clues for understanding the eruptive style. As an interpretation, we considered an eruption style that repeated the cycle of Vulcanian explosions and fountain collapses to generate pyroclastic density currents and called this eruption style "Asama-type eruption." After the Phase 1 Vulcanian explosion, it causes vesiculation due to sudden decompression and magma fragmentation. For some reason, magma fragments begin to coalesce within the misty flow within the conduit to form composite blocks. In Phase 2, fountain collapse, coarse composite blocks with a high terminal velocity selectively deposit inside the crater and separate from the pyroclastic density current. During the short rest period of Phase 3, the blocks that fill the crater will weld together, forming a kind of plug. As new magma rises, it builds up pressure under this plug, causing another explosion (Phase 1). Since there are over 60 layers of the Red ash, it is considered that the same cycle has been repeated over and over again. Based on this image, if co-ignimbrite ash derived from pyroclastic density current was deposited immediately after the deposition of a coarse lithic fragment layer due to a Vulcanian eruption, then its mixed nature of coarse lithic fragments and fine brown glass particles can also be explained.

Finally, we will consider the differences between eighteenth-century eruptions and twelfth-century eruptions. In eighteenth-century eruptions, most pyroclastic density currents occurred within half a day, whereas in the twelfth century, pyroclastic density currents with large vulcanian eruptions appear to have occurred intermittently over long periods of several years or more. Compared to 1783, the total amount of magma erupted was large, the eruption style and sequence were different, and complex activity appears to have continued for a long time. It is also noteworthy that it seems to have triggered an eruption that produced As-B′ with a different magma composition.

Acknowledgments We would like to thank the following people. Aramaki S and Nakamura M provided helpful discussions on the eruptive style of the twelfth-century eruption. Former Nihon University students Miyano Y, Shimotomai T, Akimoto N, Iwasaki T, Moro K, and Mr. Shimizu T helped us with geological survey and laboratory work. We also thank Miyahara T, Obayashi H, and Okui M for their help in the field.

References

Aramaki S (1963) Geology of Asama volcano. Geology of Asama volcano. J Fac Sci Univ Tokyo 14:229–443

Druitt T, Young S, Baptie B, Bonadonna C, Calder E, Clarke A, Cole P, Harford C, Herd R, Luckett R, Ryan G, Voight B (2002) Episodes of cyclic Vulcanian explosive activity with fountain collapse at Soufrière Hills volcano, Montserrat, the eruption of Soufrière Hills volcano, Montserrat from 1995 to 1999. Geol Soc Lond Mem 21:281–306. https://doi.org/10.1144/GSL.MEM.2002.021.01.13

Miyano Y, Yasui M, Endo K, Aramaki S (1994) Sequence and time interval of the eruptions between B and B', Asama volcano, based on pollen analysis. Programme and Abstracts, the Volcanological Society of Japan: 198p (in Japanese with English abstract)

Takahashi M, Yasui M, Kanamaru T, Nakagawa M (2022) Eruptive volume and age of large-scale pumice eruptions of the Asama-Maekake volcano, Central Japan, as revealed by detailed trench surveys: implications for future volcanic hazards. J Disaster Res 17:1080–1089

Yasui M (1994) Magma mingling recorded in "B′-scoria fall deposit" from Maekake stage, Asama volcano. J Mineral Petrol Econ Geol 89(12):439–453

Yasui M, Takahashi M (2015) Formation of the andesitic welded pyroclastic cones by pyroclastic eruption recorded in the summit area of Asama-Maekake volcano and the collapsed Caldera Wall of Kurofu volcano. Bull Volcanol Soc Jpn 60:109–123

Yasui M, Takahashi M, Kanamaru T, Nagai M (2021) High-resolution reconstruction of history of large-scale eruptions of the Asama-Maekake volcano: inferred from the pyroclastic fall deposits. Bull Volcanol Soc Jpn 66:293–325

Petrology of the Magmatic Plumbing System of Asama-Maekake Volcano

Masaki Takahashi, Maya Yasui, and Tatsuo Kanamaru

1 Introduction

The eruptive products of Maekake volcano are andesite with phenocryst content of 18–24 vol% consisting of phenocrysts of plagioclase (12–17 vol%), clinopyroxene (2–4 vol%), orthopyroxene (2–3 vol%), olivine (0.1 vol%), and magnetite (0.2–0.9 vol%). Whole-rock SiO_2 content ranges from 58 to 65 wt%. In this chapter, the origin of the erupted andesitic magma and petrologic aspects of the magmatic plumbing system of Maekake volcano are overviewed.

2 Evidence of Magma Mixing and Mingling

The andesitic rocks of Maekake volcano show several lines of evidence of magma mixing and mingling.

M. Takahashi (✉)
Institute of Natural Sciences, College of Humanities and Sciences, Nihon University, Tokyo, Japan
e-mail: takahashi.masaki@nihon-u.ac.jp

M. Yasui · T. Kanamaru
Department of Earth and Environmental Sciences, College of Humanities and Sciences, Nihon University, Tokyo, Japan

2.1 Plagioclase

Petrographically, the phenocrysts of plagioclase are composed of three types: clear core types with small amounts of melt inclusions, honeycomb texture types including a lot of melt inclusions, and sieve texture types (Fig. 1). The core compositions of honeycomb texture types are high in An (anorthite: $CaAl_2Si_2O_8$) contents, while the sieve texture types are low in An contents (Fig. 2a). The sieve texture is evidence of magma mixing; the rim of low temperature type plagioclase with low An content is dissolved to form a sieve texture, as a result of contact with the relatively high temperature mafic magma (Tsuchiyama 1985). The honeycomb texture is a result of rapid crystal growth from the magma (Kawamoto 1992). The histogram for the core composition of plagioclase phenocrysts shows that those of the 1783 AD large-scale eruption are trimodal (An57-58, An70-71, and An79-80) and those of the 1128 AD large-scale eruption are bimodal (An59-60 and An86-87), but those of the 1108 AD large-scale eruption and the 2004 AD intermediate-scale eruption are rather unimodal (An57-60). The core composition of plagioclase phenocrysts is thus different for each large-scale eruption (Fig. 2a) (Miura et al. 2007). The mode of An contents of low An type plagioclase in the frequency diagram shows that the core compositions of plagioclase phenocryst are nearly the same (An57-60), suggesting that the compositions

S. Aramaki et al. (eds.), *The Asama Volcano*, Active Volcanoes of the World,
https://doi.org/10.1007/978-981-95-5799-8_13

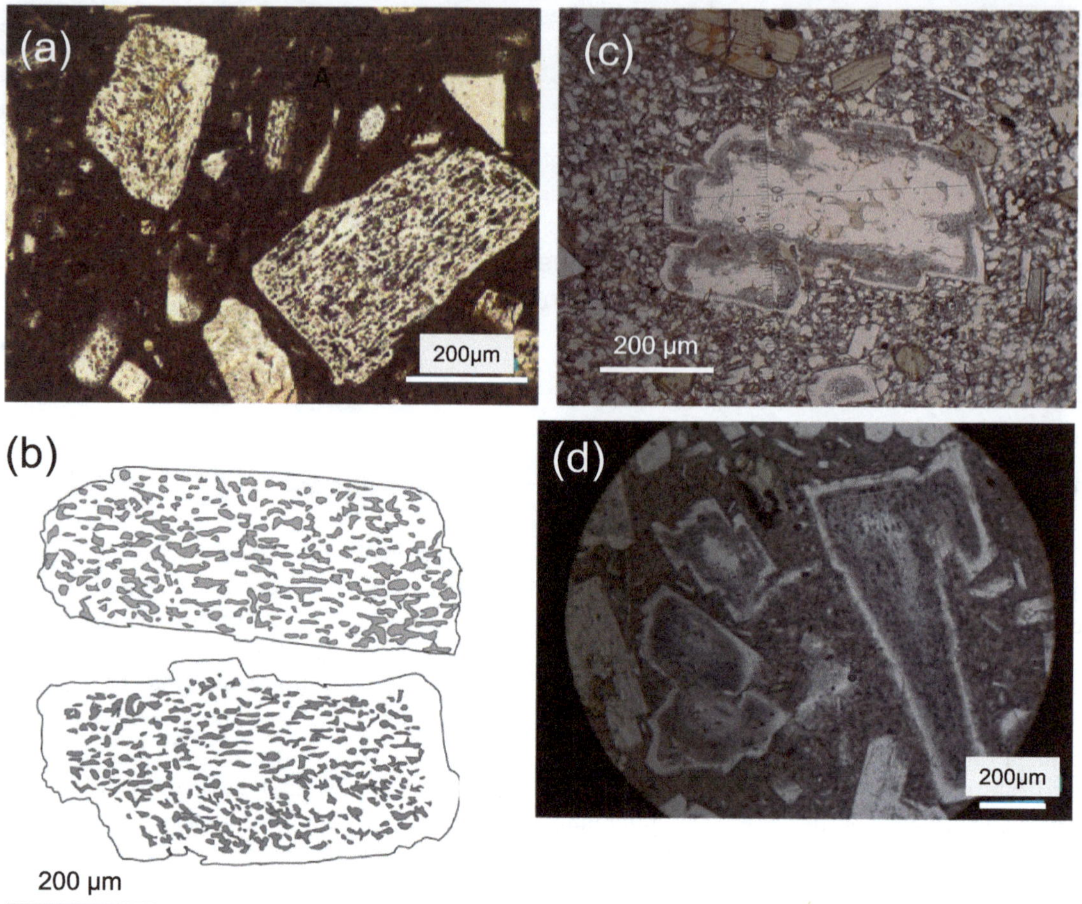

Fig. 1 Types of plagioclase phenocryst. (**a**, **b**) Honeycomb texture. (**c**, **d**) Sieve texture

of silicic end member magmas with low temperature are similar to each other in every large-scale eruption; the compositions of felsic end member magma are nearly the same and constant throughout the eruptive history of Maekake volcano, at least in active-stage III.

2.2 Orthopyroxene

Histograms show the core mode composition of orthopyroxene phenocrysts are nearly the same in the eruptive products of 1783 and 1108 AD; Mg# (Mg/ (Mg + Fe^{2+} mol ratio)) is 62–65 (Fig. 2b). Another small magnesian mode with Mg# 70–78 is present in the histogram; the histograms are bimodal rather than unimodal. The orthopyroxenes show three types compositional zoning: no reverse zoning, single reverse zoning, and double reverse zoning (Fig. 3a, b) (Ichikawa et al. 2009). The ratios of these three orthopyroxene types in the eruptive products of 1783 and 1108 AD eruptions are 15:80:5 and 5:75:20, respectively (Fig. 3c). The reverse zoning at the rim of orthopyroxene is considered to be evidence of magma mixing. No reverse zoning indicates no magma mixing and single reverse zoning indicates a single event of magma mixing. Double reverse zoning indicates that magma mixing occurred twice.

2.3 Olivine

Olivine phenocryst comprises two types: magnesian olivine (Fo80) and ferrous olivine (Fo70)

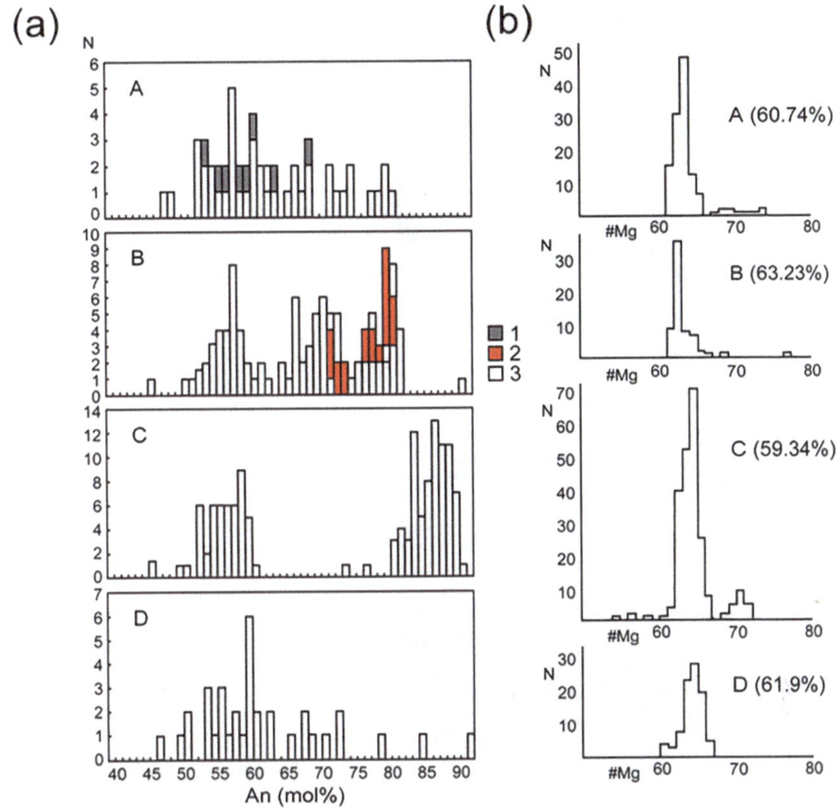

Fig. 2 Histograms of core compositions of plagioclase (**a**) and orthopyroxene (**b**). (a–A) 2004 AD intermediate-scale eruption. (a–B) 1783 AD large-scale eruption (As-A). (a–C) 1128 AD large-scale eruption (As-B′). (a–D) 1108 AD large-scale eruption (As-B). (b–A) 1783 AD large-scale eruption (60.74 wt% whole-rock SiO_2). (b–B) 1783 AD large-scale eruption (63.23 wt% whole-rock SiO_2). (b–C) 1108 AD large-scale eruption (59.34 wt% whole-rock SiO_2). (b–D) 1108 AD large-scale eruption (61.90 wt% whole-rock SiO_2). 1: sieve texture in A. 2: honeycomb texture in B. 3: other textures in A and B

(Yasui 2016). Magnesian olivine is a single crystal but ferrous olivine forms crystal clots with magnesian orthopyroxene (Mg#72) and clinopyroxene (Mg# 80–85). Equilibrium between olivine and pyroxenes can be examined by looking at the Fe-Mg distribution in both minerals (Sakuyama 1981). Magnesian olivine is not in equilibrium with most pyroxenes, because the Mg# number of magnesian olivine is larger than that of pyroxenes; the Mg# number of pyroxenes except for those forming crystal clots with ferrous olivine ranges from 54 to 70 and less than that of magnesian olivine. Magnesian olivine derives from the mafic end-member magma and ferrous pyroxenes originate from the felsic end-member magma. Ferrous olivine and coexisting magnesian pyroxenes may have crystallized from mixed magma or the products of fractional crystallization of mafic end-member magmas. Olivine phenocrysts are often enclosed by reaction-rims consisting of orthopyroxene and Fe-Ti oxide.

2.4 Magnetite

The compositional zoning at the rim of magnetite phenocrysts can be used for diffusion chronometry (Costa et al. 2020). The magnetite phenocrysts in chilled eruptive products of Unzen-Fugendake volcano in eruptions from 1991 to 1995 AD, and those of Kirishima-Shinmoedake volcano from the 2011 AD eruption, show compositional zoning of Ti, Mg, and Al at the rims of phenocrysts which indicate that magma mixing occurred in a short time prior to eruption (Nakamura 1995; Tomiya et al. 2013); magma mixing occurred in less than a few months to several years before the Unzen-Fugendake eruption and just several days before the Kirishima-Shinmoedake eruption.

However, no compositional zonings were observed in the magnetite phenocrysts of chilled eruptive products of Maekake volcano (Kanamaru et al. 2022), regardless of intermediate (2004 AD eruption) or large-scale type eruptions (1783 AD,

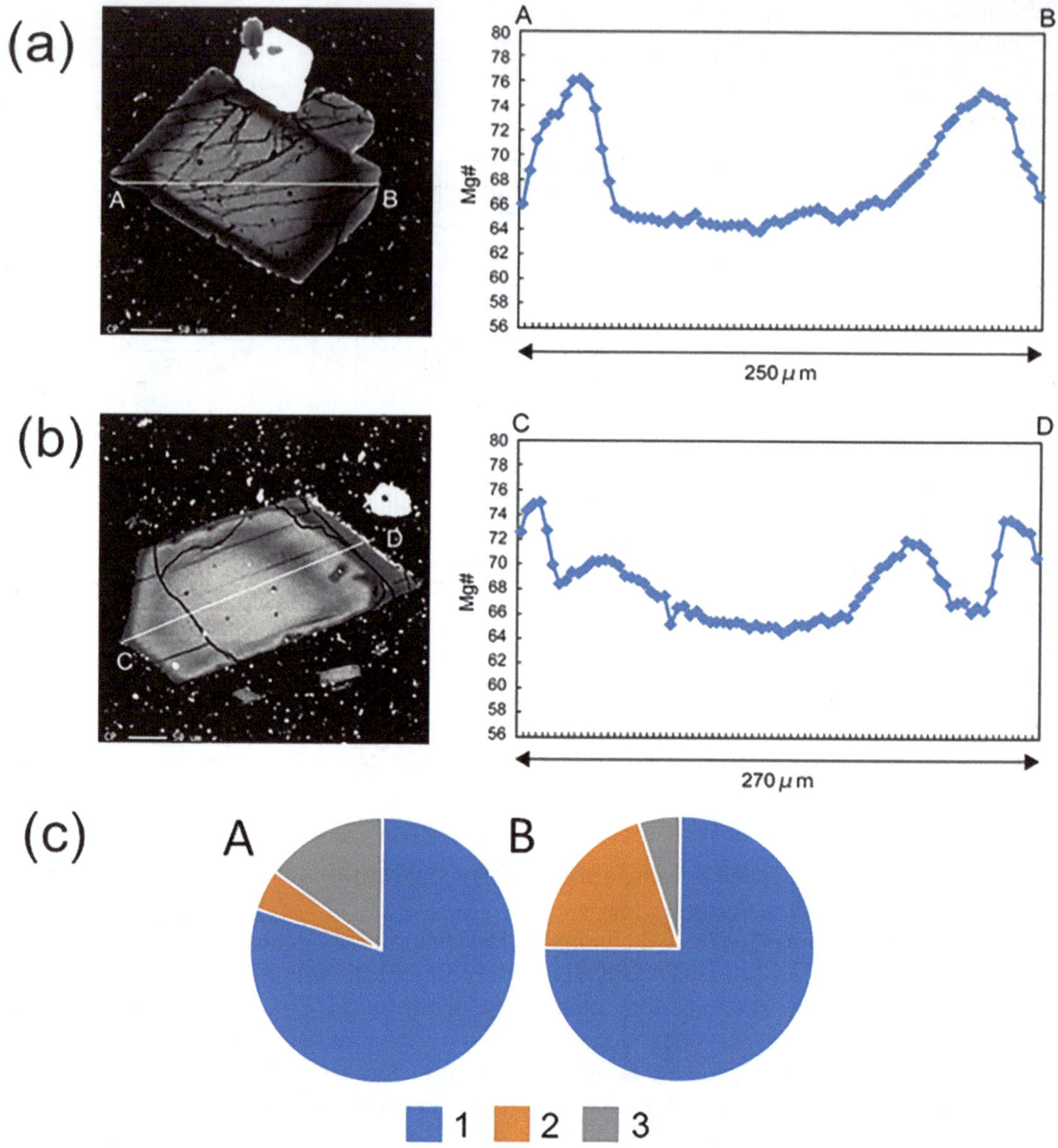

Fig. 3 Reverse compositional zoning of an orthopyroxene phenocryst. The phenocryst in the upper photo (**a**) shows single reverse zoning, and the lower photo (**b**) shows double reverse zoning. The pie chart (**c**) shows the ratio of zoning in orthopyroxenes. 1: single reverse zoning. 2: double reverse zoning. 3: no reverse zoning. (**c-A**) 1783 AD large-scale eruption. (**c-B**) 1108 AD large-scale eruption

1128 AD, and second-century eruptions) (Fig. 4). This indicates that sufficient time had passed to erase compositional zoning by diffusion after magma mixing. There is evidence of magma mixing in the phenocrystic minerals such as plagioclase and pyroxenes excluding magnetite; hence, magma mixing must have occurred for more than several to tens of years before erupting at Maekake volcano; magma mixing had not occurred during the relatively short time period prior to eruption as Unzen-Fugendake or Kirishima-Shinmoedake volcanoes, indicating that magma mixing was not the trigger of eruption at Maekake volcano.

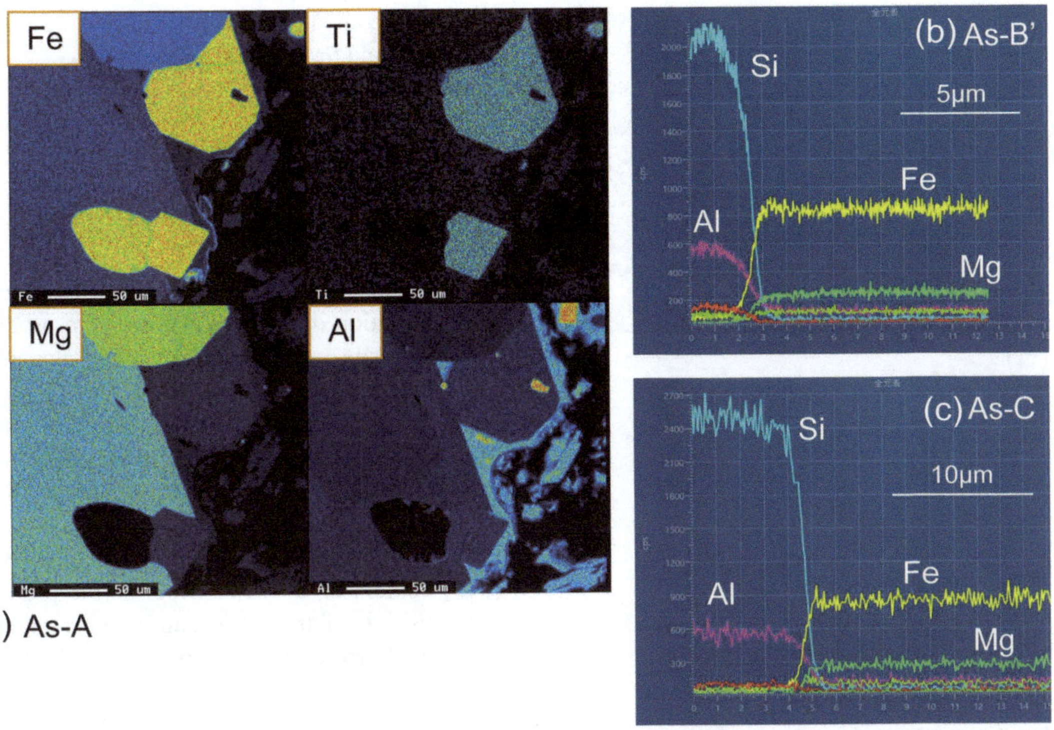

Fig. 4 Compositional zoning of magnetite phenocryst in chilled eruptive products. The left diagram (**a**) shows an elemental map of a magnetite phenocryst from the 1783 AD (As-A) large-scale eruption. The right upper diagram (**b**) is the compositional zoning pattern (line analysis using EPMA) of a magnetite phenocryst from the 1128 AD (As-B′) large-scale eruption. The right lower diagram (**c**) is the compositional zoning pattern (line analysis using EPMA) of a magnetite phenocryst from the second-century large-scale eruption. None of the magnetite phenocrysts show any compositional zoning

2.5 Crystal Clots (Glomeroporphyritic Crystals)

Many crystal clots or glomeroporphyritic crystals are included in eruptive products of Maekake volcano; most of which are composed of plagioclase, orthopyroxene, clinopyroxene, and magnetite with interstitial glass of less than 10 vol% (Fig. 5a, b) (Miura et al. 2007). Crystal clots are considered to be antecrysts (Hildreth 2004) derived from almost crystallized crystal mush in the magma reservoir that was entrained in the erupted magma (Troch et al. 2017). Crystal clots, including plagioclase, comprise two types: one with low An (An 58–68) and one with high An (An76–91) plagioclase. The low An plagioclase type consists of ferrous orthopyroxene (Mg# 63–66) and clinopyroxene (Mg# 65–70), while the high An plagioclase type is composed of orthopyroxene with Mg# 66–75 and clinopyroxene with Mg# 70 to Mg#77 (Yasui and Koyaguchi 2004; Yasui 2016). The low An plagioclase type is derived from felsic end-member magma, while the high An plagioclase type originates from mixed magma. The crystal clots in eruptive products of Maekake volcano are probably the constituents of almost crystallized heterogeneous crystal mush of magma reservoir with liquid of less than 10 vol%, and are entrained in the magma chamber.

2.6 Chemical Composition of Groundmass Glass

The glassy groundmass of eruptive products of the 1783 AD large-scale eruption is heteroge-

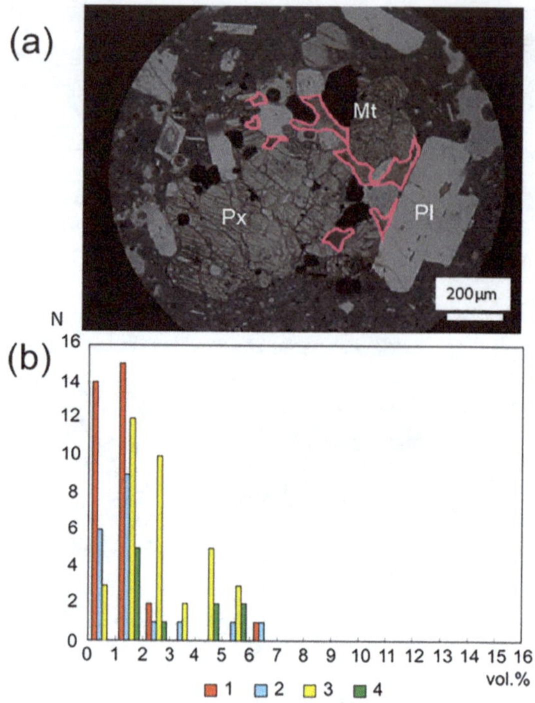

Fig. 5 Crystal clot glass interpreted to be interstitial liquid of low-melt fraction of crystal mush from the magma reservoir. (**a**) Grain-boundary glass of crystal clot (enclosed by a pink line). (**b**) Histogram showing the modal composition of grain-boundary glass. 1: 1783 AD large-scale eruption. 2: 1108 AD large-scale eruption. 3: second-century eruption large-scale eruption. 4: 2004 AD intermediate-scale erption. Px: pyroxene. Pl: plagioclase. Mt: magnetite

neous, consisting of dark to pale brown and clear glasses (Fig. 6c); the brown glass has low SiO_2, and the clear glass has high SiO_2 contents. The SiO_2 contents of groundmass glass range from 64 to 75 wt.% (Yasui 2016) (Fig. 6a). The SiO_2 contents of groundmass glass are bimodal (Fig. 6b). The heterogeneous composition of groundmass glass is evidence of magma mixing and mingling.

3 Estimation for End-Member of Magma Mixing

The andesites of Maekake volcano are produced by the magma mixing of mafic and felsic end-member magmas. The nature of the end-member magmas is estimated by whole-rock major elements, incompatible trace elements, rare earth elements and Sr isotopic compositions.

3.1 Whole-Rock Chemical Composition of Major Elements

The compositional trends of MgO plotted on a SiO_2 variation diagram of the eruptive products of large-scale eruptions at Maekake volcano are different for each large-scale eruption (Takahashi et al. 2007, 2024) (Fig. 7a). Eruptive products of the 1783 AD eruption have the highest MgO content, whereas eruptive products from the 1128 AD eruption have the lowest content. The linear trends of each eruption intersect at a point with SiO_2 content of 67 wt%, which is the probable composition of the felsic end-member magma. The whole-rock MgO compositions of eruptive products of the Asama volcano group are shown in Fig. 7b. The eruptive products of Asama volcano group are grouped into three compositional trends: the Kurofu (Gippa), Maekake-Sennin, and Hotokeiwa trends. The eruptive products of Kurofu volcano (Gippa lava group) show a trend that is low in SiO_2 contents, and MgO content decreases remarkably as SiO_2 content increases, while those of the Maekake-Sennin volcano have a trend that is intermediate in SiO_2 content, and a MgO content that decreases gently as SiO_2 content increases. The eruptive products of the Hotokeiwa volcano have the highest SiO_2 content. Samples from Kurofu and Hotokeiwa volcanoes have trends indicating that the eruptive products were produced by fractional crystallization, whereas samples from Maekake-Sennin volcanoes have linear trends indicating mixing between the Kurofu mafic end-member magmas and the Hotokeiwa felsic end-member magmas.

3.2 Whole-Rock Chemical Composition of Incompatible Trace Elements

The whole-rock chemical compositions of incompatible trace elements of eruptive prod-

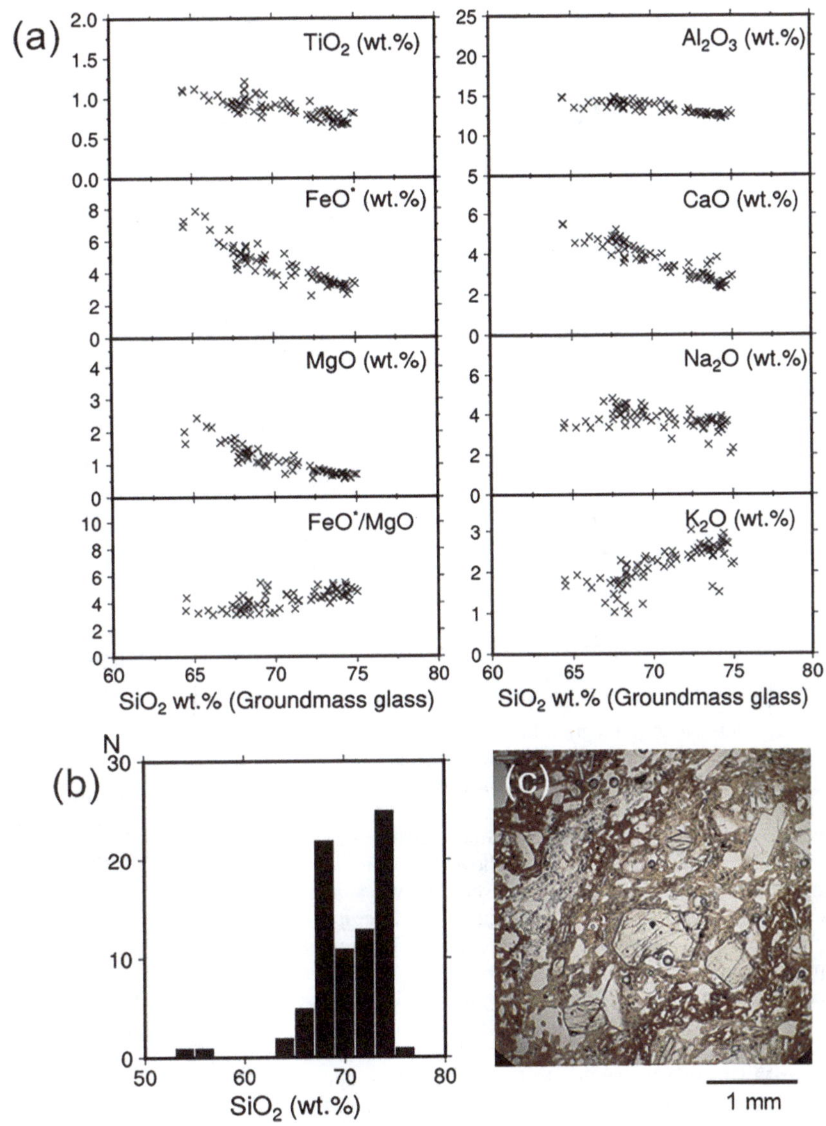

Fig. 6 Chemical compositions of groundmass glass. (**a**) Chemical composition of groundmass glass of eruptive products of the 1783 AD large-scale eruption; the SiO_2 content ranges from 64 to 75 wt%. (**b**) Histogram showing the bimodal distribution of SiO_2 content of groundmass glass of eruptive products of the 1783 AD large-scale eruption. (**c**) Microphotograph of heterogeneous groundmass glass of a volcanic bomb from the 2004 AD eruption; the brown glass is low, and clear glass is high in SiO_2 content

ucts of the Asama volcano group are described in detail in Takahashi et al. (2023). Ba/Y, Rb/Y, Zr/Y, and Rb/Zr ratios of the eruptive products are low in Kurofu volcano (Gippa lava group) and high in Hotokeiwa volcano (Fig. 8). It is difficult to produce the Hotokeiwa magma from the Kurofu magma by fractional crystallization. Those of Maekake-Sennin volcano are a mixture of ratios of eruptive products of both volcanoes, indicating that magma mixing of magmas of both volcanoes produced the magmas of the Maekake-Sennin volcanoes (Fig. 11a).

3.3 Whole-Rock Chemical Composition of Rare Earth Elements

The whole-rock chemical compositions of rare earth elements are published in Takahashi et al. (2023). The chondrite-normalized patterns of rare earth elements for eruptive products of Kurofu (Gippa lava group), Hotokeiwa, and Maekake volcanoes are shown in Fig. 9a. The ratios of heavy rare earth elements (HREE) to light rare earth elements (LREE) of eruptive products of

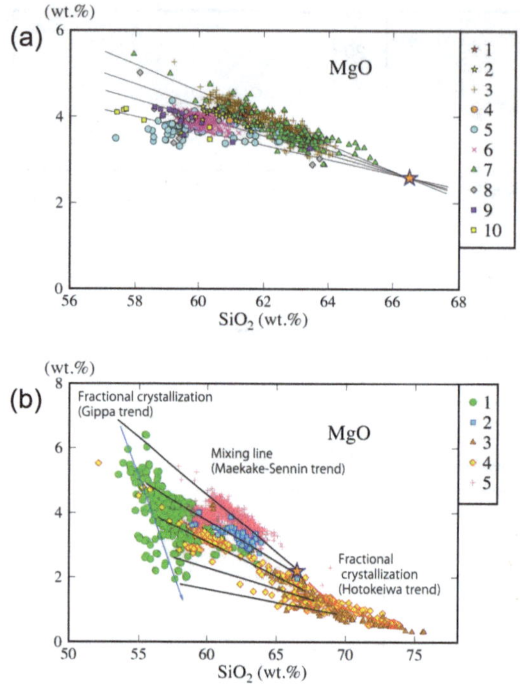

Fig. 7 SiO_2 variation diagram for whole-rock MgO content of the eruptive products of Maekake volcano (**a**) and those of the Asama volcano group (**b**). The large star indicates the projected felsic end-member magma of Maekake volcano. The blue line denotes the Gippa fractional crystallization trend, and the red line shows the Hotokeiwa fractional crystallization trend. (**a**) Maekake volcano. 1: eruptive products of the 2004 AD eruption. 2: eruptive products of the 1940–1960 AD eruptions. 3: eruptive products of the As-A (1783 AD) eruption. 4: eruptive products of the As-A' eruption. 5: eruptive products of the As-B'eruption (1128 AD). 6: eruptive products of the As-B eruption. 7: eruptive products of the As-C eruption. 8: eruptive products of the As-D eruption. 9: eruptive products of the As-E eruption. 10: eruptive products of the As-F eruption. (**b**) Asama volcano group. 1: Kurofu volcano. 2: Sennin volcano. 3: Hanareyama volcano. 4: Hotokeiwa volcano. 5: Maekake volcano

Kurofu volcano (Gippa lava group) are low with no negative europium anomaly; the REE patterns of eruptive products of Kurofu volcano are parallel, and their REE contents increase as the whole-rock SiO_2 contents increase and MgO contents decrease. The variation of patterns of REE contents of eruptive products of Kurofu volcano (Gippa lava group) can be explained by fractional crystallization of olivine and pyroxenes without plagioclase. The HREE/LREE ratios are high in the eruptive products of Hotokeiwa volcano, and those of Maekake volcano are intermediate between the Kurofu (Gippa lava group) and Hotokeiwa volcanoes. The magma mixing of the least and less differentiated eruptive product of Kurofu volcano (Gippa lava group) and that of Hotokeiwa volcano gives rise to the REE patterns of eruptive products of Maekake volcano.

The Ce/Yb ratio of eruptive products of Hotokeiwa volcano cannot be derived from those of Kurofu volcano (Gippa lava group) by fractional crystallization of plagioclase, olivine, clinopyroxene, orthopyroxene, and magnetite (Fig. 9b); the origin of magmas of Kurofu (Gippa lava group) and Hotokeiwa volcanoes is different without any mother-daughter relation. Those of Maekake volcano can be explained by magma mixing of magmas of both volcanoes (Fig. 11b).

3.4 Whole-Rock Chemical Composition of Sr Isotopes

The $^{87}Sr/^{86}Sr$ isotopic ratios for eruptive products of Kurofu volcano (Gippa lava group) are low (0.7039–0.7040), whereas those of Hotokeiwa volcano are high (0.7042) (Takahashi et al. 2023) (Fig. 10). The $^{87}Sr/^{86}Sr$ isotopic ratios for eruptive products of Maekake volcano are intermediate and lie between Kurofu (Gippa lava group) and Hotokeiwa volcanoes. The origin of magmas of Kurofu volcano (Gippa lava group) and those of Hotokeiwa volcano are different, whereas magmas of Maekake volcano can be explained by the magma mixing of magmas of Kurofu (Gippa lava group) and Hotokeiwa volcanoes (Fig. 11c).

4 Petrologic Model of Magmatic Plumbing System of Maekake Volcano

Figure 12 shows the schematic magmatic plumbing system of Maekake volcano, estimated from seismological and geodetic datasets and models (Takeo et al. 2006; Nagaoka et al. 2012; Aoki et al. 2013; Takeo et al. 2022). The magma reservoir of Maekake volcano is estimated to exist at a depth of 6–10 km beneath sea level with a width

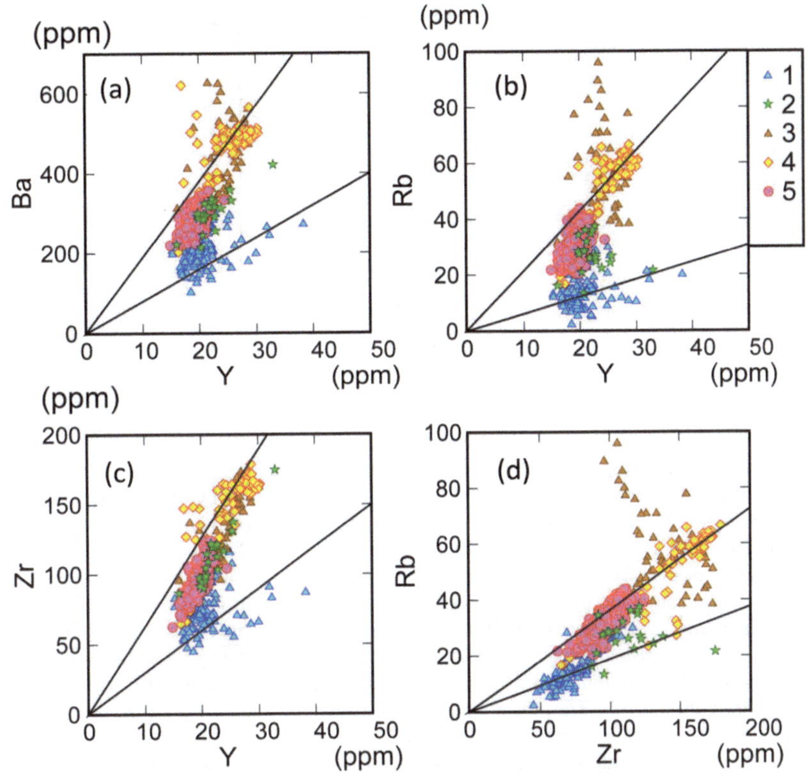

Fig. 8 Diagrams showing incompatible trace element ratios of eruptive products of the Asama volcano group. (**a**) Ba/Y. (**b**) Rb/Y. (**c**) Zr/Y. (**d**) Rb/Zr. 1: Kurofu volcano. 2: Sennin volcano. 3: Hanareyama volcano. 4: Hotokeiwa volcano. 5: Maekake volcano

of ca 5 km (low velocity region less than Vp = 2.5 km/s) and ca 10 km to the west of the volcano vent, based on seismic interpretations and models indicating a region of low seismic velocity (Fig. 12). The magma ascends from the magma reservoir and migrates horizontally eastward through a dike to reach a location approximately 3 km beneath the crater of Maekake volcano; the dike formation was detected by volcanic seismicity and crustal deformation (Murakami 2005). The magma ascends upward through the pathway producing volcanic earthquakes and tremors to erupt from the summit crater, but sometimes the unrest results from a failed eruption. From long-term deformation measured using GNSS, the magma chamber of a sheet-like oblate spheroid (axis ratio is >100) is estimated at a depth of approximately 12 km, which is near the bottom of the modeled low seismic velocity region (Munekane 2021).

The magma mixing producing the erupted andesite of Maekake volcano occurred in the magma reservoir, where both end-member magmas are tapped by independent magma supply systems and sources; the felsic end-member magma cannot be derived from the mafic end-member magma by fractional crystallization but the sources of both end-member magmas are different (Fig. 13).

5 Multiple Magma Chambers and Magma Reservoir

A magma reservoir is composed of liquid and crystal mush (more than 25 vol% crystals), some of which may be a rigid crystal body with low melt fraction exceeding 55 vol% crystals, while the magma chamber indicates a domain mainly filled with liquid within the magma reservoir (Cooper 2017).

The magma chamber system in the magma reservoir is estimated from several lines of evidence: (1) Compositional trends of the whole-rock major element chemistry such as MgO plotted on the SiO_2 variation diagram, which are

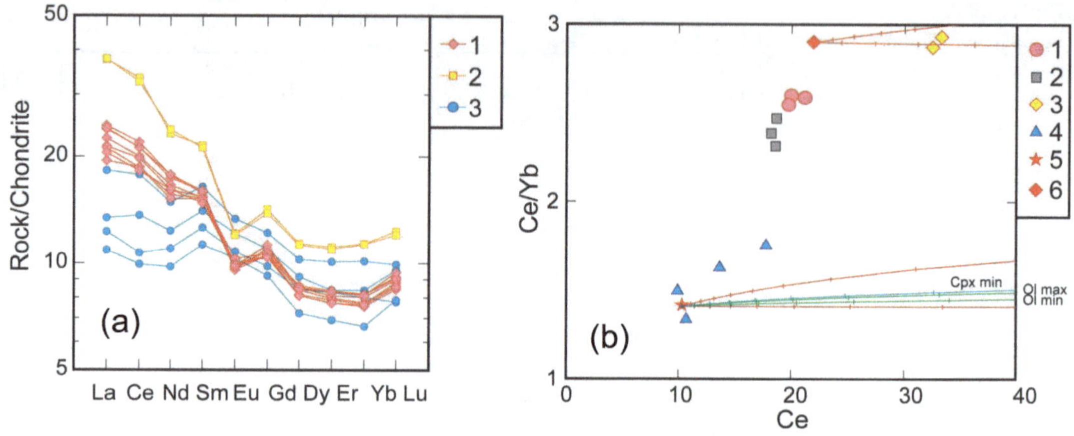

Fig. 9 Plots showing rare earth element patterns and ratios of eruptive products of the Asama volcano group. (**a**) Chondrite normalized pattern. 1: Maekake volcano (andesite). 2: Hotokeiwa volcano (rhyolite). 3: Kurofu volcano (andesite of Gippa lava group). (**b**) Ce/Yb vs. Ce diagram. The blue and green lines denote the fractional crystallization trend based on the Rayleigh fractionation model using the distribution coefficient of Ce and Yb for clinopyroxene (minimum value) and olivine (minimum and maximum values). The lower red line indicates the fractional crystallization trend, assuming Rayleigh fractionation starting with plagioclase (79 vol%), olivine (3 vol%), orthopyroxene (15 vol%), clinopyroxene (2 vol%), and magnetite (1 vol%) (the representative phenocrystic modal composition of andesite of the Gippa lava group); the upper trend uses maximum distribution coefficient values and the lower trend uses minimum distribution coefficient values. It is difficult to derive the eruptive products of Hotokeiwa volcano from the eruptive products of Kurofu volcano by fractional crystallization. The upper red line indicates the fractional crystallization trend (Rayleigh fractionation model) starting with plagioclase (84 vol%), quartz (5 vol%), orthopyroxene (4 vol%), clinopyroxene (2 vol%), and magnetite (5 vol%) (the representative phenocrystic modal composition of rhyolite of Hotokeiwa volcano) (the Hotokeiwa compositional trend). The chemical composition of eruptive products of Maekake volcano is explained by magma mixing between the magma of Gippa lava group of Kurofu volcano (the Gippa compositional trend) and that of the Hotokeiwa volcano (the Hotokeiwa compositional trend). 1: Maekake volcano (1783 AD lava). 2: Maekake volcano (1108 AD lava). 3: Hotokeiwa volcano (Koasama lava dome). 4: Kurofu volcano (Gippa lava group). 5: supposed starting bulk and mineral composition representing the Gippa magmatic trend. 6: supposed starting bulk and mineral composition representing the Hotokeiwa magmatic trend

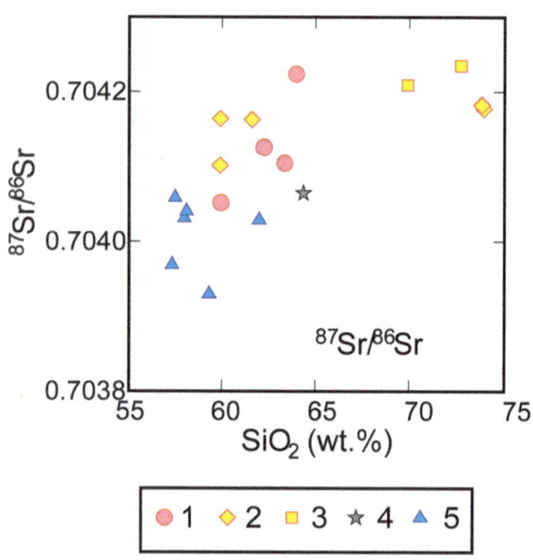

Fig. 10 Diagram showing the relationship between $^{87}Sr/^{86}Sr$ isotopic ratio and SiO_2 content of eruptive products of the Asama volcano group. 1: Maekake volcano. 2: Hotokeiwa volcano. 3: Hanareyama volcano. 4: Sennin volcano. 5: Kurofu volcano

different for each large-scale eruption, suggesting that the magma chamber of each large-scale eruption is independent. (2) Histograms of the core composition of plagioclase phenocrysts which vary for each large-scale eruption, suggesting that the magma chamber of each large-scale eruption is separated. (3) The abundance of several types of reverse-zoned orthopyroxene which also varies for each large-scale eruption, suggesting that the magma chamber of each large-scale eruption is separated. (4) Most phenocrystic minerals not being autocrysts but rather antecrysts (Hildreth 2004) entrained in the magma chamber from the surrounding heterogeneous and low melt fraction crystal mush. (5) Magma mixing and mingling that occurred in the magma chamber between mafic end-member magmas of the Kurofu (Gippa) compositional trend with phenocrysts of magnesian olivine and felsic end-member magma of the Hotokeiwa

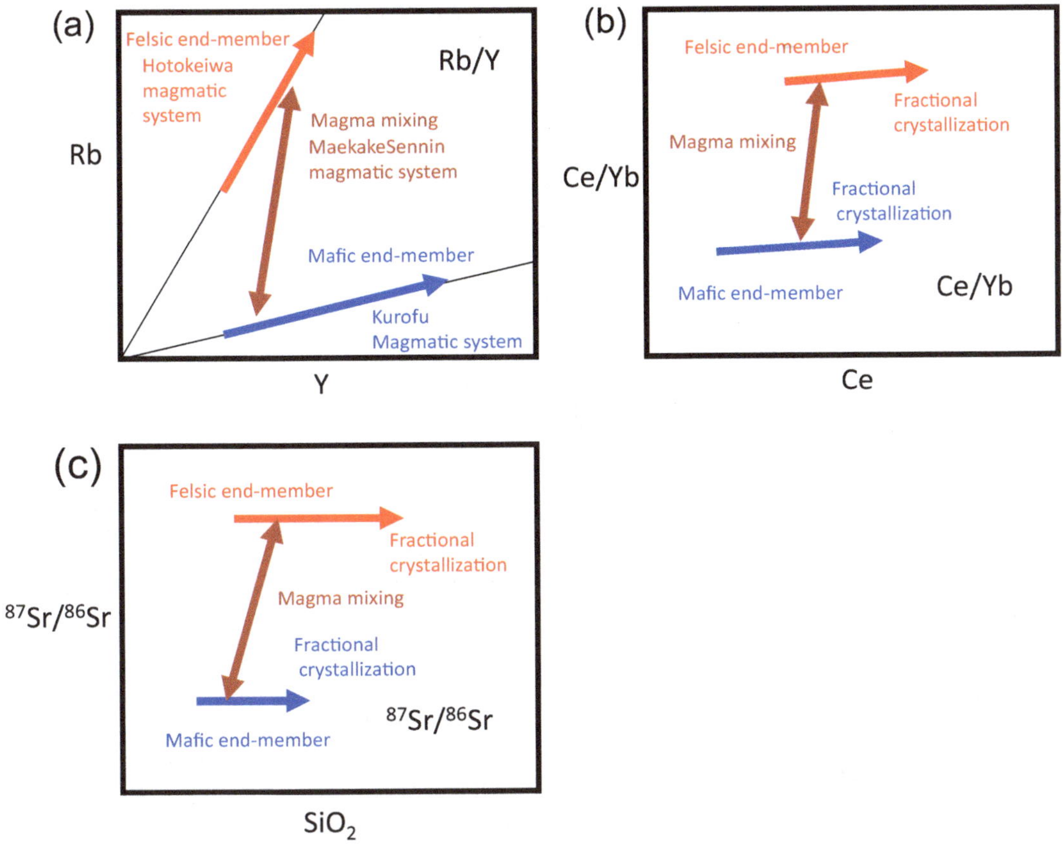

Fig. 11 Schematic diagrams summarizing relationships between the Gippa, Hotokeiwa and Maekake-Sennin compositional trends. (**a**) Incompatible trace element ratio. (**b**) Ce/Yb vs. Ce ratio. (**c**) $^{87}Sr/^{86}Sr$ isotopic ratio vs. SiO_2

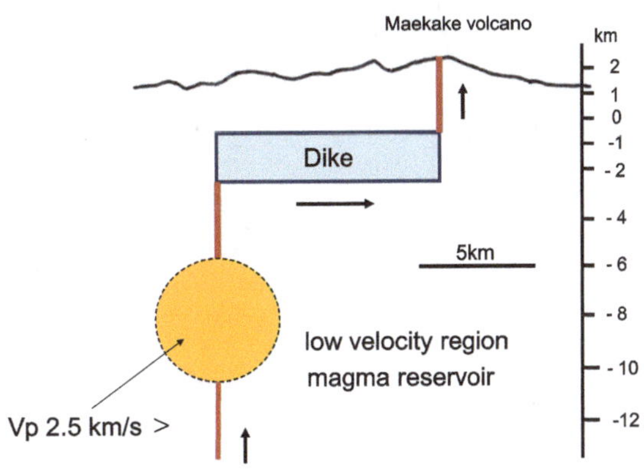

Fig. 12 Conceptualization of the magmatic plumbing system beneath Maekake volcano based on crustal deformation and volcanic seismological monitoring results of Takeo et al. (2022)

compositional trend with phenocrysts of plagioclase, pyroxenes, and magnetite; the Kurofu and Hotokeiwa compositional trends are formed by fractional crystallization of olivine ± pyroxenes and plagioclase + pyroxenes ± magnetite, respectively. Some phenocrysts are crystallized from the mixed magma and/or end-member magmas in the magma reservoir. (6) The mafic and felsic end member magmas are derived from different sources; they are tapped into the magma chamber via separated supply systems and mixed in the magma chamber. (7) The cumulative volume ver-

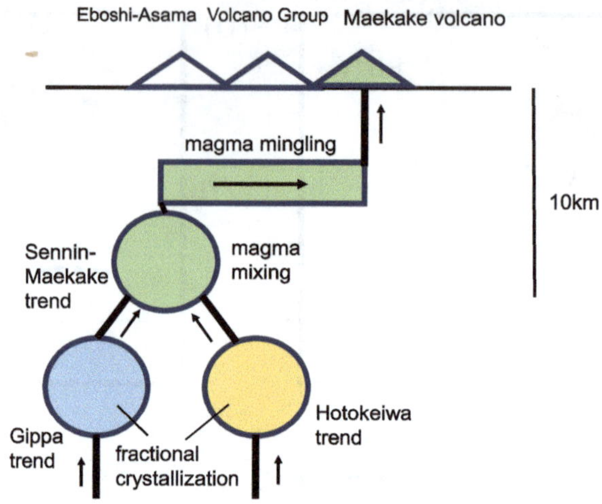

Fig. 13 Schematic diagram showing the petrology of the magmatic plumbing system of Maekake volcano. Mafic end-member magma with Gippa fractional crystallization trend and felsic end-member magma with Hotokeiwa fractional crystallization trend independently recharge the magma reservoir, which are mixed to form andesitic magma with Sennin-Maekake magma mixing trend that subsequently erupts at Maekake volcano after magma mingling during traversing along an EW trending dike

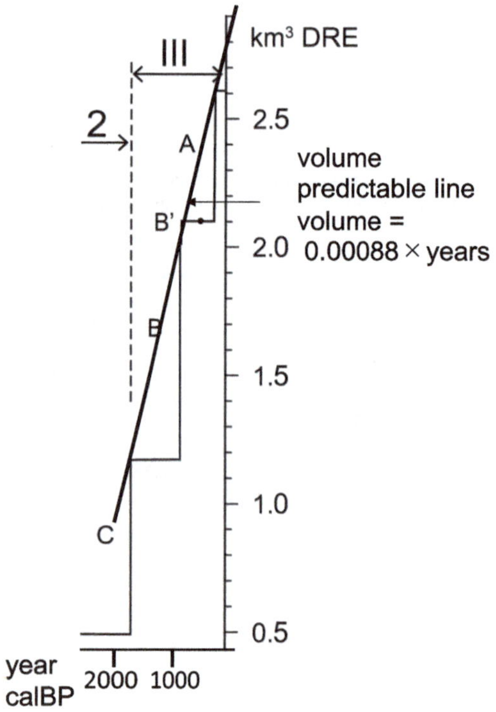

Fig. 14 Cumulative volume versus time of the eruptive products of active stage III of Maekake volcano. The straight line indicates a steady-state output rate of 0.88 km^3 DRE per 1000 years. III: active stage III. 2: relatively active stage 2

sus time diagram for the active stage III shows that the relation among large-scale eruptions is volume-predictable (Takahashi et al. 2022) (Fig. 14), which can be explained if the magma chamber is closed after each large-scale eruption and a new magma chamber starts to form with the same supply rate, which will be the source of next large-scale eruption.

The estimated multiple magma chamber system of Maekake volcano is illustrated schematically in Fig. 15. A small sheet-like lenticular magma chamber filled with a high melt fraction magma formed within the heterogeneous and low melt fraction crystal mush of the magma reservoir. The mafic and felsic end-member magmas are tapped into this magma chamber where magma mixing occurs; the average magma supply rates to the magma chamber are nearly constant. Various types of phenocrysts of the magma reservoir are entrained into the magma chamber. The growing magma chamber finally supplies magma to the surface, which results in a large-scale eruption. After the large-scale eruption, the magma chamber closes, and a new magma chamber starts to store magma in another separated part of the magma reservoir. Some resulting

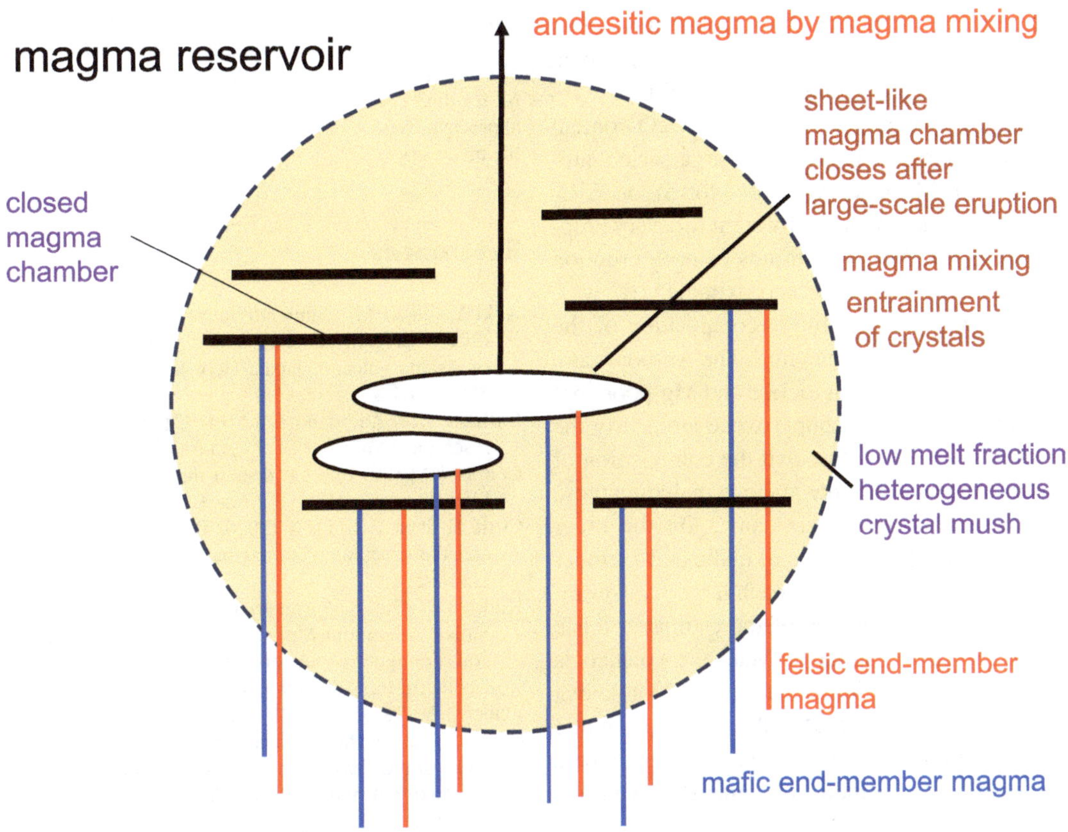

Fig. 15 Conceptualization of a multiple magma chamber system within the magma reservoir of Maekake volcano. See text for explanation

sheet-like lenticular magma chambers may not be able to erupt and crystallize without erupting. The magma reservoir thus formed is a multiple magma chamber system consisting of a sheet and dike complex crystallized to various degrees within a heterogeneous magma reservoir ranging from a high-melt fraction crystal mush to rigid crystal bodies (Cashman and Giordano 2014; Putirka 2017; Sparks et al. 2018).

6 Summary

1. A bimodal distribution of plagioclase core composition, sieve texture of plagioclase, and reverse zoning of orthopyroxene are evidence of magma mixing. Mg-rich olivine, which is in disequilibrium with pyroxenes, is included, which is also evidence of magma mixing between mafic and felsic end-member magmas. The felsic end-member magma includes phenocrysts of plagioclase, ferrous pyroxenes, and magnetite, while the phenocrysts of the mafic end-member magma are Mg-rich olivine. Some phenocrystic minerals crystallized from mixed magmas. Phenocrysts also crystallized in the magma reservoir. A lack of compositional zoning of magnetite indicates that magma mixing did not occur immediately prior to the eruption.
2. The core composition of plagioclase varies in each large-scale eruption. The ratio of orthopyroxenes showing single-reverse compositional zoning versus double-reverse compositional zoning is different in the eruptive products of the 1783 and 1108 AD large-scale eruptions. This suggests that the magma chambers of both large-scale eruptions are separate.

3. Chemical compositions of groundmass glass are heterogeneous (64–75 wt.% SiO_2), which indicates magma mixing and mingling.
4. Variation trends for whole-rock MgO content of eruptive products of the large-scale eruptions of Maekake volcano plotted on SiO_2 variation diagrams are different for every eruption, while the linear trends of each eruption intersect a point with approximately 67 wt% SiO_2, which is the likely composition of the felsic end-member magma. The An-poor core compositions of plagioclase and Mg-poor core compositions of orthopyroxene are nearly the same, also suggesting that the composition of the felsic end-member magma, including these minerals, is nearly the same. On the other hand, the MgO contents of mafic end-member magma are different for each large-scale eruption. The composition of the estimated felsic end-member magma of Maekake volcano is the same as the felsic magmas of Hotokeiwa volcano, while those of the estimated mafic end-member magma can be considered to be a member of the mafic magmas of the Gippa lava group of Kurofu volcano.
5. The incompatible trace element ratios such as Rb/Y, Ba/Y, Zr/Y, and Rb/Zr, rare earth element compositions, and $^{87}Sr/^{86}Sr$ isotopic ratio indicate that the felsic end-member magmas of Hotokeiwa magmatic trends cannot be derived by fractional crystallization of the mafic end-member magmas of the Gippa lava group; the sources of felsic and mafic end-member magmas are different.
6. Magma mixing and mingling occurred in the magma chamber, which is located in the magma reservoir filled with a heterogeneous crystal mush. The felsic and mafic end-member magmas were supplied into the magma chamber from different pathways and sources.
7. A magma chamber empties and closes after each large-scale eruption, then a new magma chamber starts to form in another part of the magma reservoir. Several magma chambers may coexist in the magma reservoir at the same time, although not all will necessarily result in an eruption at the surface.

Acknowledgements We are grateful to Aramaki, S., Miura, K., Kato, F., Nakamura, Y., Ichikawa, M., Ueda, M., Uekusa, T., Osugi, Y., Okada, E., Uematsu, R., Ozawa, M. for their cooperation and help with the study. We also appreciate Martin, A. for critical reading and revising of the manuscript.

References

Aoki Y, Takeo M, Ohminato T, Nagaoka Y, Nishida K (2013) Magma pathway and its structural controls of Asama volcano, Japan. Geol Soc Lond Spec Pub 380:67–84

Cashman KV, Giordano G (2014) Caldera and magma reservoir. J Volcanol Geotherm Res 288:28–45

Cooper KM (2017) What does a magma reservoir look like? The "crystal's-eye" view. Elements 13:23–28

Costa F, Shea T, Ubide T (2020) Diffusion chronometry and the timescales of magmatic processes. Nat Rev 1:201–214

Hildreth W (2004) Volcanological perspectives on Long Valley, Mammoth Mountain, and mono craters: several contiguous but discrete systems. J Volcanol Geotherm Res 136:169–198

Ichikawa M, Takahashi M, Yasui M (2009) Compositional zoning of orthopyroxene phenocrysts in volcanic rocks of the Ten-nin and Tenmei eruptions of Asama-Maekake volcano. Fall Meet Volanol Soc Japan 1:18 (in Japanese)

Kanamaru T, Furukawa K, Zhao X, Suganuma Y (2022) Manetic petrology of pumice fall deposits of the 1783 eruption of Asama volcano, Japan. Earth Planets Space 74(53):1–9

Kawamoto T (1992) Dusty and honeycomb plagioclase: indicators of processes in the Uchino stratified magma chamber, Izu peninsula, Japan. J Volcanol Geotherm Res 49:191–208

Miura K, Takahashi M, Yasui M (2007) Comparative petrography for phenocrysts in volcanic rocks of historical large-scale eruptions of Asama-Maekake volcano, Central Japan. Proc Inst Nat Sci Nihon Univ 42:117–128 (in Japanese with English abstract)

Munekane H (2021) Modeling long-term volcanic deformation at Kusatsu-Shirane and Asama volcanoes, Japan, using the GNSS coordinate time series. Earth Planets Space 73(192):1–15

Murakami M (2005) Magma plumbing system of the Asama volcano inferred from continuous measurements of GPS. Bull Volcanol Soc Japan 50: 347–361

Nagaoka Y, Nishida K, Aoki Y, Takeo M, Ohminato T (2012) Seismic imaging of magma chamber beneath an active volcano. Earth Planet Sci Lett 333:1–8

Nakamura M (1995) Continuous mixing of crystal mush and replenished magma in the ongoing Unzen eruption. Geology 23:807–810

Putirka KD (2017) Down the crater: where magmas are stored and why they erupt. Elements 13:11–16

Sakuyama S (1981) Petrological study of the Myoko and Kurohime volcanoes, Japan: crystallization sequence and evidence for magma mixing. J Petrol 22:553–583
Sparks RSJA, Blundy C, Cashman JD, Rust KV, Jackson AC, M.D. (2018) Formation and dynamics of magma reservoir. Rhil Trans R Soc 377:20180019
Takahashi M, Yasui M, Ichikawa Y, Kamioka Y, Asaka N, Sakagami M, Tanaka E (2007) Whole-rock major element chemistry for eruptive products of Asama-Maekake volcano, Central Japan. Proc Inst Nat Sci Nihon Univ 42:55–70 (in Japanese with English abstract)
Takahashi M, Yasui M, Kanamaru T, Nakagawa M (2022) Eruptive volume and age of large-scale pumice eruptions of the Asama-Maekake volcano, Central Japan, as revealed by detailed trench surveys: implications for future volcanic hazards. J Disaster Res 17: 1080–1089
Takahashi M, Kanamaru T, Okada E, Yasui M (2023) Whole-rock trace element, rare earth element and Sr, Nd, Pb isotopic compositions for eruptive products of the Asama volcano group, Central Japan. Proc Inst Nat Sci Nihon Univ 58:103–152 (in Japanese with English abstract)
Takahashi M, Kanamaru T, Yasui M (2024) Whole-rock major element chemistry for eruptive products of Asama volcano group: summary of analytical data of 1,922 samples. Proc Inst Nat Sci Nihon Univ 59:47–108 (in Japanese with English abstract)
Takeo M, Aoki Y, Ohminato T, Yamamoto M (2006) Magma supply path beneath Mt. Asama volcano, Japan. Geophys Res Lett 33:L15310
Takeo M, Aoki Y, Koyama T (2022) Recent volcanic activity at the Asama volcano and long-period seismic signals. Proc Jpn Acad Ser B 98:416–438
Tomiya A, Miyagi I, Saito G, Geshi N (2013) Short time scales of magma-mixing processes prior to the 2011 eruption of Shinmoedake volcano, Kirishima volcanic group. Japan Bull Volcanol 75(750):3–19
Troch J, Ellis BS, Mark DF, Bindeman IN, Kent AJ, Guillong M, Bachmann O (2017) Rhyolite generation prior to Yellowstone supereruption: insights from the Island Park-Mount Jackson rhyolite series. J Petrol 58:29–52
Tsuchiyama A (1985) Dissolution kinetics of plagioclase in the melt of the system diopside-albite-anorthite, and origin of dusty plagioclase in andesite. Contrib Mineral Petrol 89:1–16
Yasui M (2016) Petrography of the Asama 1783 eruptive products. Pro Inst Nat Sci Nihon Univ 51:231–255 (in Japanese with English abstract)
Yasui M, Koyaguchi T (2004) Sequence and eruptive style of the 1783 eruption of Asama volcano, Central Japan. Bull Volcanol 66:243–262

Geophysics and Geochemistry

History of Seismic Observation and Traditional Classifications of Seismic Signals in Asama Volcano

Minoru Takeo

1 Introduction

Asama-yama seismological observatory, which was established in 1911, is one of the earliest permanently installed observation facilities near an active volcano in the world. This section summarizes the history of seismic observation at Asama volcano and a pioneering work in the classification of seismic signals.

2 History of Seismic Observation

The eruptions and seismic swarms in 1909 motivated seismic observations of Asama volcano in 1910 (Omori 1910, 1912, 1914a, b, c). Omori established the Asama-yama seismological observatory in 1911 at Nagamine (Yuno-taira), which is about 2.3 km west-southwest of the summit crater (Omori 1914a). A photo of an early building of the observatory is shown in Fig. 1. During the spring to autumn season annually from 1911 to 1916, he operated a portable two-component horizontal tremor recorder with a pendulum period of 4.2 s (Fig. 2) and a Gray-Ewing vertical component seismometer with a pendulum period of 10 s (Omori 1914b). This is an epoch-making study for that time.

This observation had finished once, but another seismological observatory was established by the local government of Nagano Prefecture in 1923 at Oiwak, which is about 8 km south-southeast of the summit crater. After that, the Central Meteorological Observatory took over the seismic observations and conducted a steady year-round observation until 1945. Asama Volcano Observatory was established in 1933 by the Karuizawa town, which was later donated to the Imperial University of Tokyo in 1934. Figure 3 shows a photo of an early building of Asama Volcano Observatory. The Asama Volcano Observatory lies about 4 km east of the summit crater and started volcanological observations, including volcanic earthquakes, crustal deformation, geomagnetic, and gravity fields on and near the volcano. After World War II, the seismic observations were summarized by the University of Tokyo Earthquake Research Institute (ERI)

This chapter is a revised version of Takeo et al. (2022) with the inclusion of a reference in the References section.

M. Takeo (✉)
Earthquake Research Institute, The University of Tokyo, Tokyo, Japan

Japan Meteorological Agency, Tokyo, Japan
e-mail: takeo@eri.u-tokyo.ac.jp

S. Aramaki et al. (eds.), *The Asama Volcano*, Active Volcanoes of the World,
https://doi.org/10.1007/978-981-95-5799-8_14

Fig. 1 Photo of the Asama-yama seismological observatory established in 1911

Fig. 2 Photo of a portable two-component horizontal tremor recorder installed at the Asama-yama seismological observatory

and the Japan Meteorological Agency. To date, these entities have continued observations.

During the period from 1933 to 1953, short-period seismometers were installed at the Asama

Fig. 3 Photo of the Asama Volcano Observatory, the Imperial University of Tokyo

Volcano Observatory and at other sub-stations situated on and near the volcano. In 1954, the telemetering system using transmission wires was constructed for data transmission of seismic and geothermal observations. A permanent seismic network of the Asama Volcano Observatory as of 1966 is shown in Fig. 4 (Minakami et al. 1970). A high volcanic activity was observed at the turn of the twentieth century until the 1950s. However, almost all the seismometers installed at Asama volcano prior to the 1960s were short-period sensors, and the main recording systems were smoked-paper recorders. Although seismic observations in Asama volcano have been collected for more than a century, there were only six seismometers as of 1998. The monitoring network was modernized in 2003. Since then, it has rapidly grown. In 2017, it comprised 30 seismometers, 19 of which were equipped with broadband sensors (Fig. 5a). Continuous global navigation satellite system (GNSS) observations have been conducted in the Asama volcano region since 1996. As of 2017, there were 22 continuous GNSS sites within 20 km of the summit crater (Fig. 5b). In addition to seismic and geodetic sensors, other kinds of sensors such as microphones, video and thermal cameras, and a detector of chemical components in volcanic gas have been installed in the small caves located on the west and east rims of the summit crater (Takeo et al. 2022).

Fig. 4 Permanent seismic network of the Asama Volcano Observatory as of 1966

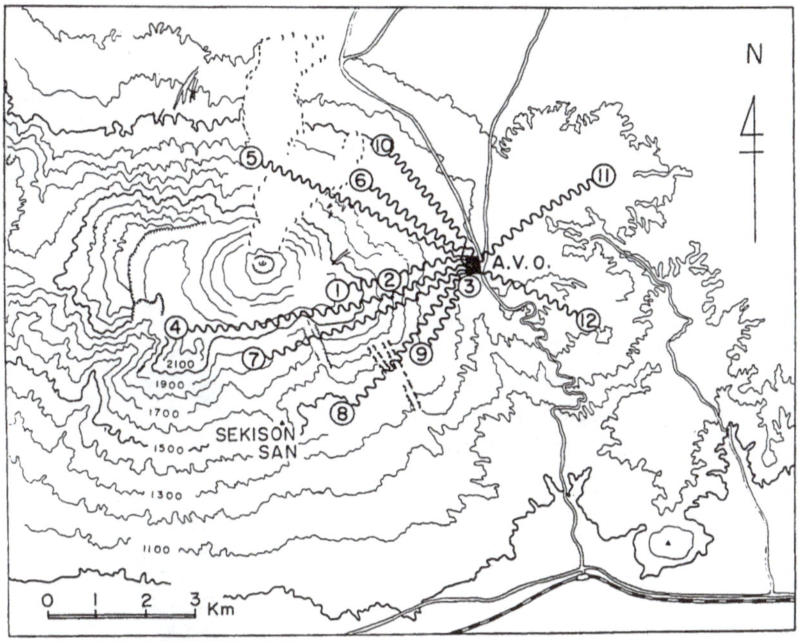

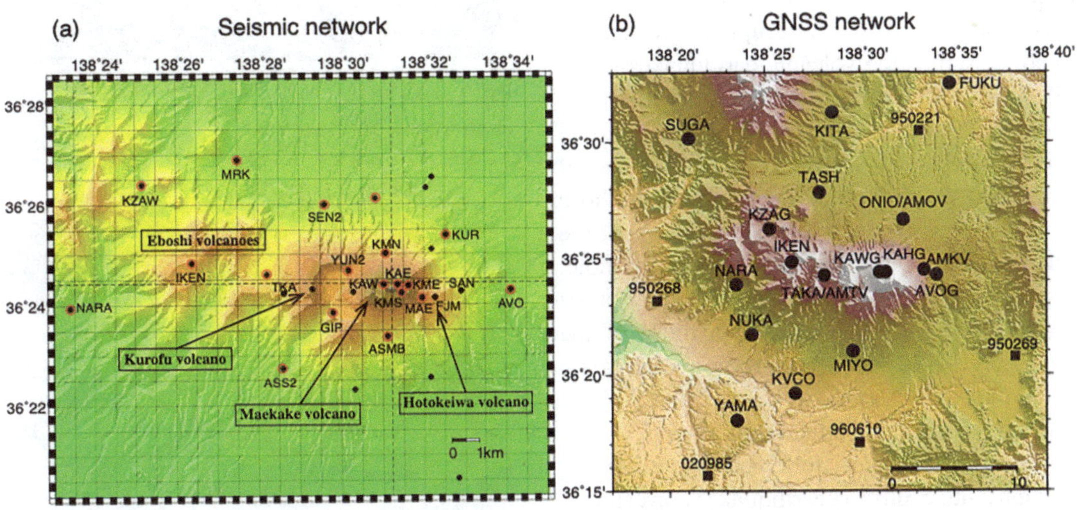

Fig. 5 Topographic map of Eboshi volcanoes, three volcanic edifices (Kurofu, Hotokeiwa, and Maekake) of Asama volcano, and the seismic network (**a**) and GNSS network (**b**). (**a**) Seismic stations with labels are maintained by the University of Tokyo Earthquake Research Institute (ERI). Seismic stations without labels are maintained by the Japan Meteorological Agency and the National Research Institute for Earth Science and Disaster Resilience. Solid circles with red circuits are broadband seismic stations, and those without red circuits are short-period seismic stations. (**b**) GNSS stations marked by circles and small rectangles are maintained by ERI and the Geospatial Information Agency of Japan, respectively

3 Traditional Classification of Seismic Signals

A major achievement during the early days of seismic observations was the classification of volcanic earthquakes, because almost all seismograms were recorded on smoked paper and digital data were not available at that time. Omori (1912) classified the volcanic earthquake into two categories: A category is a volcanic earthquake not accompanying an eruption, and B category is one that is followed by an eruption. Figure 6a, b are examples of volcanic earthquake in A category and in B category, respectively, which were observed by the portable horizontal tremor recorder. The earthquake in B category is dominated by a long-period signal compared with that in A category.

According to the location of hypocenter, the nature of seismic waveform, especially P and S phases, and the relation to eruptions, Minakami (1960) classified the volcanic earthquakes in the following four types. A-type volcanic earthquakes are originating from the depths of about 1–10 km beneath volcanoes. The natures of waveforms are difficult to distinguish from those of tectonic earthquake: the P-phase and the S-phase of seismic waves are clearly discernible. Hypocenters of B-type volcanic earthquakes are limited to an area shallower than those of the A-type events. Consequently, the waveforms are predominated by surface waves, and the S-phase cannot be discerned distinctly. Minakami (1960) classified earthquakes followed by explosive eruptions into explosion earthquakes corresponding to the B category events defined by Omori (1912). Continuous vibrations that originate from volcanoes are classified into volcanic tremors. Typical examples of seismograms corresponding to Minakami's classification are exhibited in Fig. 7.

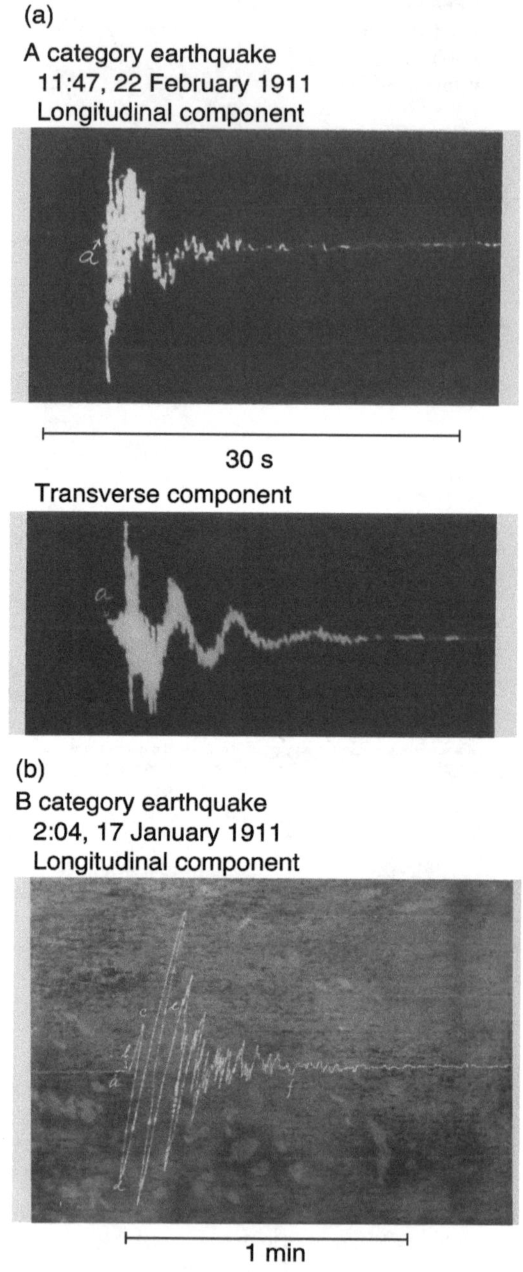

Fig. 6 Examples of volcanic earthquakes in A category and in B category, which were recorded by the portable horizontal tremor recorder. (**a**) A category earthquake at 11:47 on February 22, 1911. (**b**) B category earthquake at 2:04 on January 17, 1911

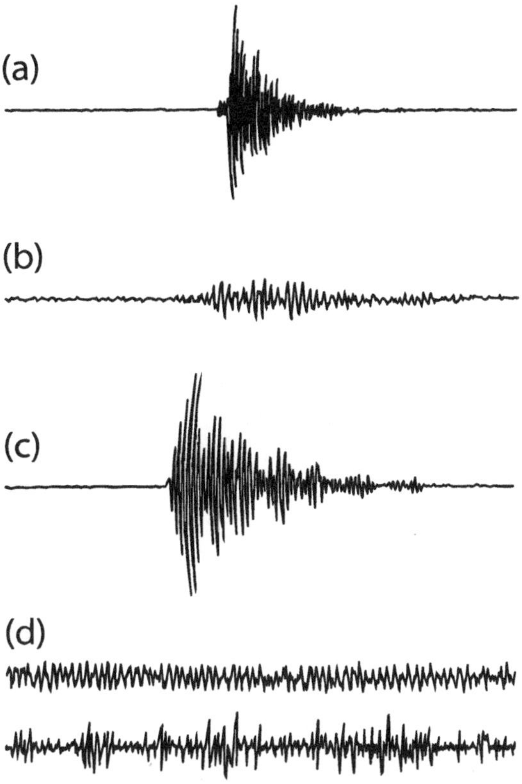

Fig. 7 Typical examples of seismograms on the basis of the traditional classification defined by Minakami (1960). (**a**) A-type volcanic earthquake, (**b**) B-type volcanic earthquake, (**c**) explosion earthquake, (**d**) volcanic tremor

Minakami (1960) discussed the relation between B-type earthquakes and volcanic eruptions over the period from 1930s to 1950s at Asama volcano and exhibits that premonitory phenomena preceding eruption are represented in the remarkable increase in the frequency of B-type earthquakes. The classification of volcanic earthquakes defined by Minakami has been used in other volcanoes for a certain length of time before monitoring networks of volcanic earthquakes were modernized.

4 Summary

Asama volcano is one of the best monitored and scientifically investigated volcanoes in the world. The history of seismic observation at Asama volcano dates back over 100 years. The traditional classification of volcanic earthquakes was established by Minakami (1960) based on the seismic records at Asama volcano.

Acknowledgments We thank the Japan Academy for permission to use several figures and contents from the paper entitled, "Recent volcanic activity at the Asama volcano and long-period seismic signals" by Takeo et al. in Proc. Jpn. Acad., Ser. B 98, 2022.

References

Minakami T (1960) Fundamental research for predicting volcanic eruptions (part 1). Bull Earthq Res Inst Univ Tokyo 38:497–544

Minakami T, Utibori S, Hiraga S, Miyazaki T, Noriya G, Utsunomiya T (1970) Seismometrical studies of volcano Asama (part 1). Bull Earthq Res Inst Univ Tokyo 48:235–301

Omori F (1910) The eruptions of Asama-Yama. Rep Imperial Earthq Invest Committee 67:1–26

Omori F (1912) The eruptions and earthquakes of Asama-Yama. Bull Imperial Earthq Invest Committee 6:1–147

Omori F (1914a) The eruptions and earthquakes of Asama-Yama II: list of the volcanic earthquakes instrumentally registered at the Asama-Yama observatory in 1911 and 1912. Bull Imperial Earthq Invest Committee 6:149–226

Omori F (1914b) The eruptions and earthquakes of Asama-Yama III: remarks on the seismographical observations at Yuno-taira in 1911 and 1912. Bull Imperial Earthq Invest Committee 6:227–257

Omori F (1914c) The eruptions and earthquakes of Asama-Yama IV: strong Asama-Yama outbursts, Dec. 1912 to May 1914. Bull Imperial Earthq Invest Committee 7:1–215

Takeo M, Aoki Y, Koyama T (2022) Recent volcanic activity at the Asama volcano and long-period seismic signals. Proc Jpn Acad Ser B 98:416–438

Eruption Activity of Asama Volcano from 2004 to 2015 and Volcanic Earthquake

Minoru Takeo

1 Introduction

Recent volcanic activity at Asama volcano provides us a lot of high-quality data in seismic, geodetic, and other geophysical and geochemical fields. Based on these data, we reveal the eruption activity from 2004 to 2015. Several unique long-period seismic signals were recorded by the proximal broad-band seismic network and were compared with smoked-paper short-period seismograms recorded before a Vulcanian eruption on November 10, 1958.

2 Eruption Activity from 2004 to 2015

The first moderate-sized eruption since 1983 occurred in 2004. It was preceded by elevated thermal activity at the summit crater, increments of SO_2 emission beginning around 2000, and a few minor eruptions between February and April 2003 (Nakada et al. 2005; Takeo et al. 2006). Figure 1 compares the temporal evolution of the seismic activity with the changes in the GNSS baseline length for sites 950,221 and 950,268 (see Fig. 5) between 1996 and 2009. The activity of the volcanic earthquakes was low between September 1997 and March 2000, except for the last several months. However, the activity of volcanic earthquakes remained high after 2001, and the maximum temperature in the summit crater exceeded 200 °C in autumn 2002 (Takeo et al. 2006). Continuous GNSS data showed that the north-south extension of the volcano started in late July 2004 and lasted about 5 weeks until the first eruption in 2004. Volcanic glows were observed in the last 10 days of July 2004, and the maximum temperature at the bottom of the crater exceeded 500 °C after the sudden extension of the GNSS baseline length.

A seismic swarm started at 15:10 (JST) on August 31, 2004. More than 600 events occurred before the first eruption. Figure 2 displays seismograms of an earthquake swarm which recorded by the vertical component of the broadband seismometer at KAW. The hypocenters of the swarm, which were determined by Yamamoto et al. (2005) using a double-difference algorithm (Waldhauser and Ellsworth 2000), were just below the crater and ranged from 300 to 800 m deep (see Fig. 3), suggesting that the shallowest part of the vent was broken during the swarm activity. A tiltmeter located about 3 km north

This chapter is a revised version of Takeo et al. (2022) with the inclusion of a reference in the References section.

M. Takeo (✉)
Earthquake Research Institute, The University of Tokyo, Tokyo, Japan

Japan Meteorological Agency, Tokyo, Japan
e-mail: takeo@eri.u-tokyo.ac.jp

S. Aramaki et al. (eds.), *The Asama Volcano*, Active Volcanoes of the World,
https://doi.org/10.1007/978-981-95-5799-8_15

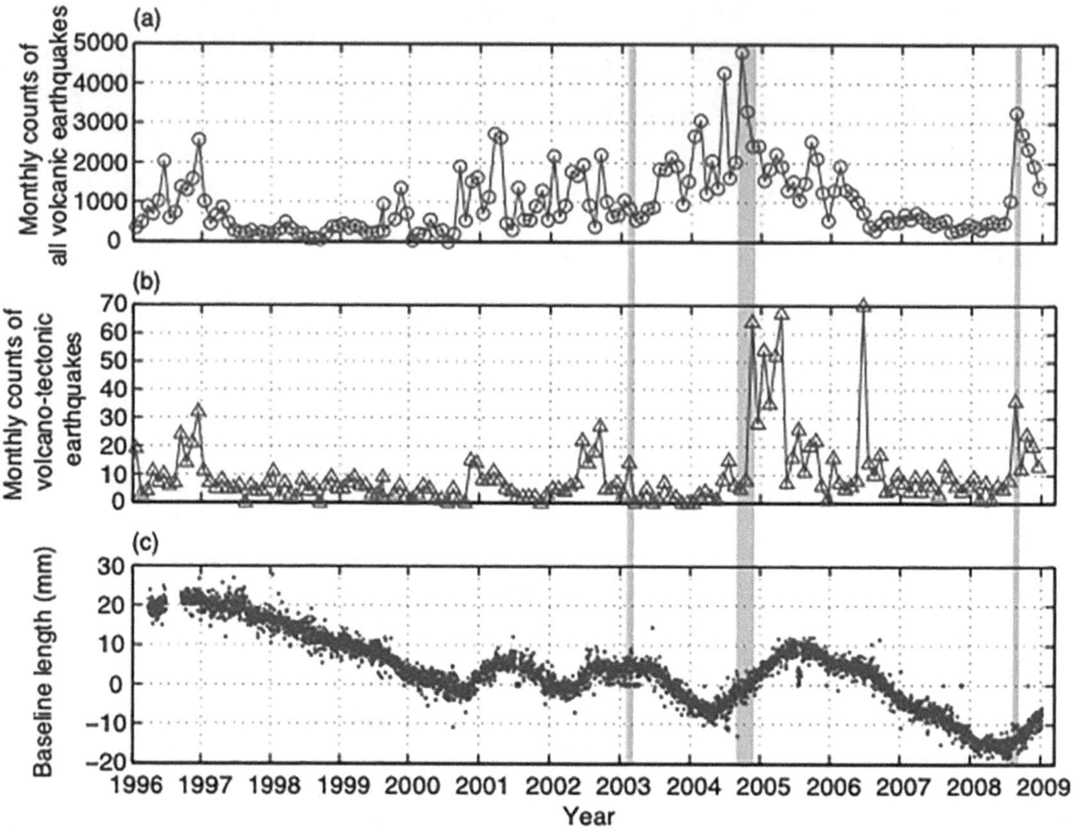

Fig. 1 Temporal evolution of seismic activity and ground deformation between 1996 and 2009. Gray shaded areas indicate eruptive periods from February 6 to March 30, 2003, September 1 to November 14, 2004, and August 10–14, 2008. (**a**) Monthly counts of all volcanic earthquakes triggered by the data acquisition system. (**b**) Monthly counts of volcano-tectonic earthquakes triggered by the data acquisition system. (**c**) Relative changes in the baseline lengths for sites 950,221 and 950,268

from the summit started to tilt toward the east about 29 h before the first eruption, implying an inflation source offset to the west (Churei and Katayama 2006).

At 20:02 (JST) on September 1, 2004, a Vulcanian eruption with a VEI of 2 occurred for the first time in 21 years. The ash plume reached 3.5–5.5 km above the summit or 6–8 km above sea level. After a short period of quiescence, Strombolian eruptions started on 14 and 16–18 September before Vulcanian eruptions occurred on 23 and 29 September, 10 October, 14 November, and 9 December.

The estimated total amount of ejected tephra was about 1.6×10^8 kg (Nakada et al. 2005). The magma in the 2004 eruption was andesite with a composition similar to magmas erupted over the past 10,000 years. A lava dome emerged at the summit by 16 September when continuous Strombolian eruptions began. The change in the magma composition and the increasing amounts of juvenile magma in the eruptive products by that time suggest that juvenile magma may have ascended to the summit crater. The lava dome deflated afterward at its center, indicating that part of the ascended magma descended to the depth. Figure 4 exhibits the hypocenter distribution of volcanic earthquakes between January 1, 2004, and May 24, 2016, using the double-difference algorithm (Waldhauser and Ellsworth 2000).

After 4 years of repose (see Figs. 1 and 4), the seismic activity started to elevate in July 2008, which was 1 month before the first eruption. The earthquakes were mainly located approximately 1 km to the west of the summit and about 1 km

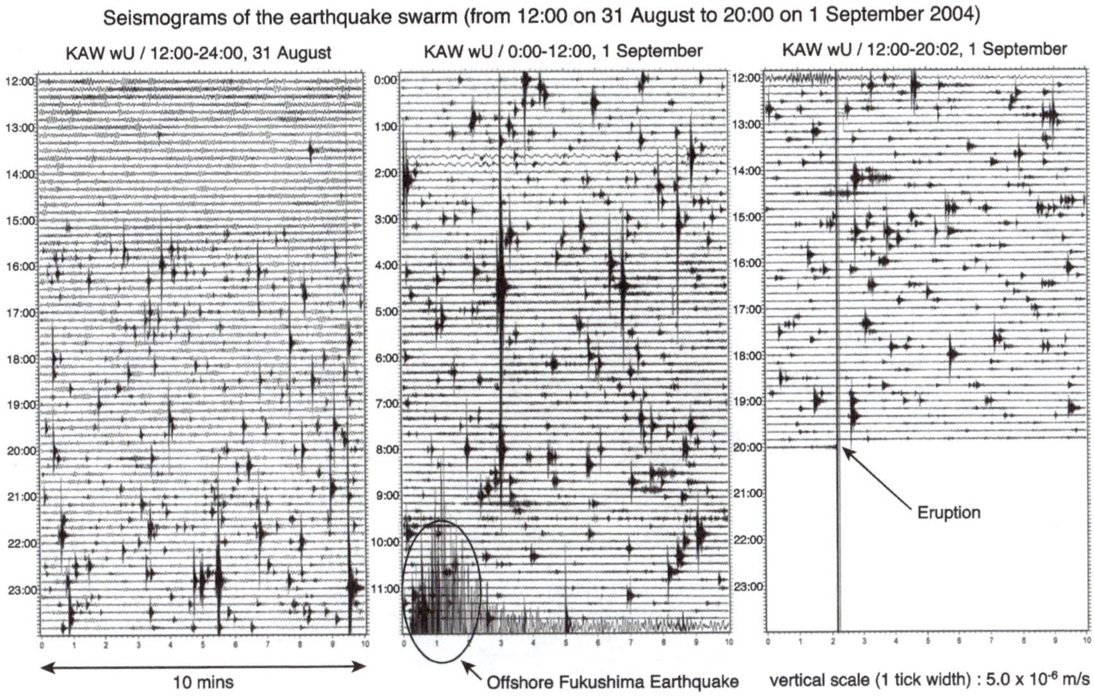

Fig. 2 Seismograms of an earthquake swarm which started at 15:10 (JST) on August 31, 2004

Fig. 3 Distribution of relocated hypocenters of the swarm which occurred between 15:10 on August 31 and 20:02 on September 1, 2004. All hypocenters are relocated using the double-difference algorithm (Waldhauser and Ellsworth 2000). (**a**) Distribution of epicenters with the permanent and temporal stations shown by black crosses. (**b**) North-south cross-section of hypocenters. (**c**) East-west cross-section of hypocenters

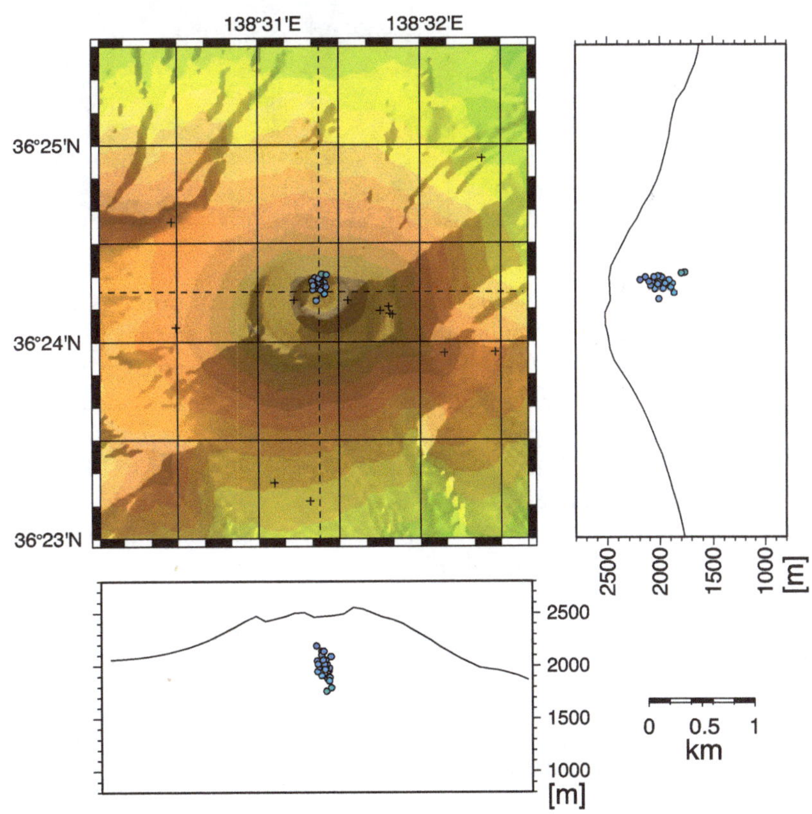

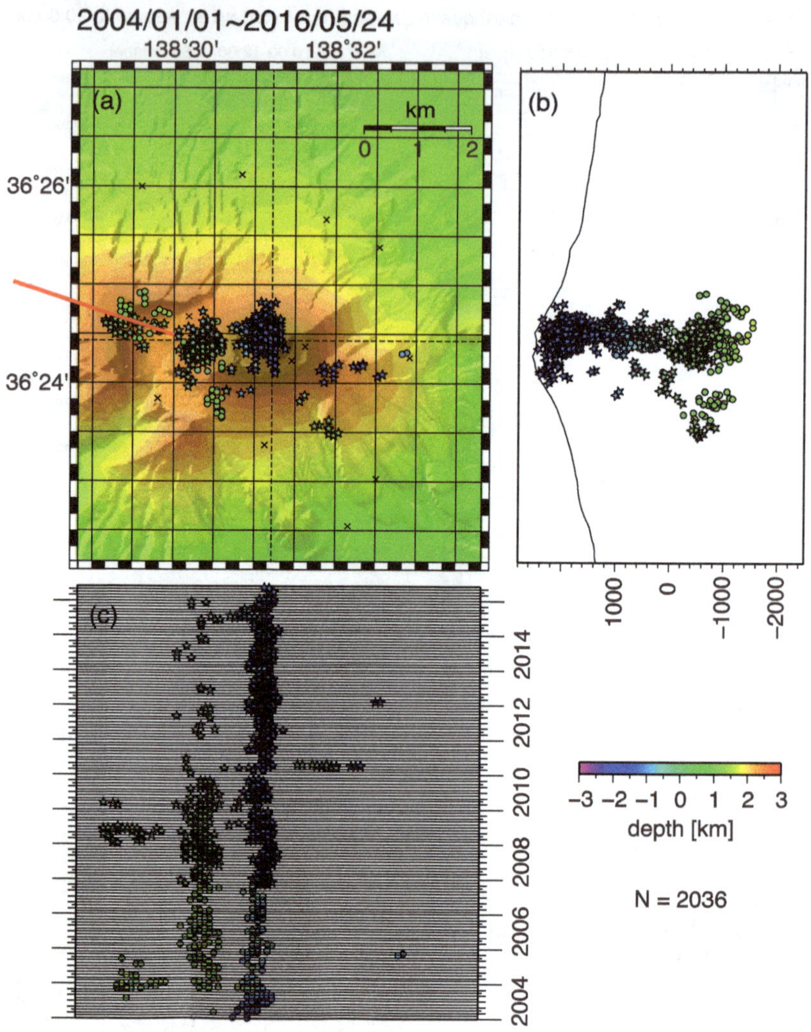

Fig. 4 Distribution of relocated hypocenters around Asama volcano between January 2004 and May 2016. Each hypocenter is colored by its depth. Circles and stars indicate the hypocenters before 2007 and after 2008, respectively. (**a**) Distribution of epicenters with the approximate location of dyke intrusions during crises (red line). (**b**) North–south cross-section of hypocenters. (**c**) Temporal evolution of hypocenters. Horizontal axis represents the longitude of the hypocenters

below sea level. The seismic activity further increased on 8 August, leading to eruptions on 10, 11, and 14 August. These were minor with ash plumes reaching heights of 200–400 m above the summit. The eruptive products did not include juvenile materials. From this, we inferred that the 2008 eruption was the ejection of the remnant magma associated with the 2004 eruption. Since September 2008, an elevated seismic activity has been observed approximately 2 km to the west of the summit and about 1 km below sea level (see Fig. 4).

On February 1, 2009, a day before the largest eruption in 2009, the seismicity just below the summit increased and the tiltmeter about 3 km north from the summit inclined toward the east. This signaled inflation beneath the western flank of the volcano, which led to a summit eruption on the next day. Subsequently, minor eruptions from the summit occurred intermittently on 9–16 February, 15 March, 30 April, and 3 and 27 May, with plume heights less than 500 m. The largest eruption during the 2008–2009 crisis occurred on February 2, 2009, but was smaller than that on September 1, 2004. The ejected mass due to the eruption on 2 February was 2.0–2.4 × 10^7 kg, about 20% of that associated with the September 1, 2004, eruption. The ejected materials included only trace amounts of juvenile materials, suggesting that the eruption was merely an ejection of near-surface materials rather than a magma migration from the depth (Maeno et al. 2010).

After more than 6 years of repose, the seismic activity gradually started to elevate in late April 2015, and the western side of the volcano turned to a north-south extension in May 2015. Tilt motions indicating inflation beneath the western side of the Kurofu mountain ridge started in the beginning of June, and the emission of SO_2 rapidly increased from 200 tons/day on 1 June to 1700 tons/day on 11 June, leading to eruptions on 16 and 19 June. Both eruptions emitted only trace amounts of ash. However, the SO_2 emissions exceeded 300 tons/day and lasted until February 2016.

3 Long-Period (LP) Seismic Signals

Permanent broadband seismic observations near the summit crater started in October 2003. Two broadband seismic stations were installed on the west and east rims of the summit crater before the first eruption in 2004. This proximal broadband seismic network revealed that not only the traditional volcanic earthquakes but also several kinds of unique LP seismic signals occurred intermittently in Asama volcano, but their amplitudes were too feeble to be recorded outside of the Maekake crater.

Figure 5 displays a half-day monitoring record (12:00–24:00 (JST) on June 11, 2004) of the

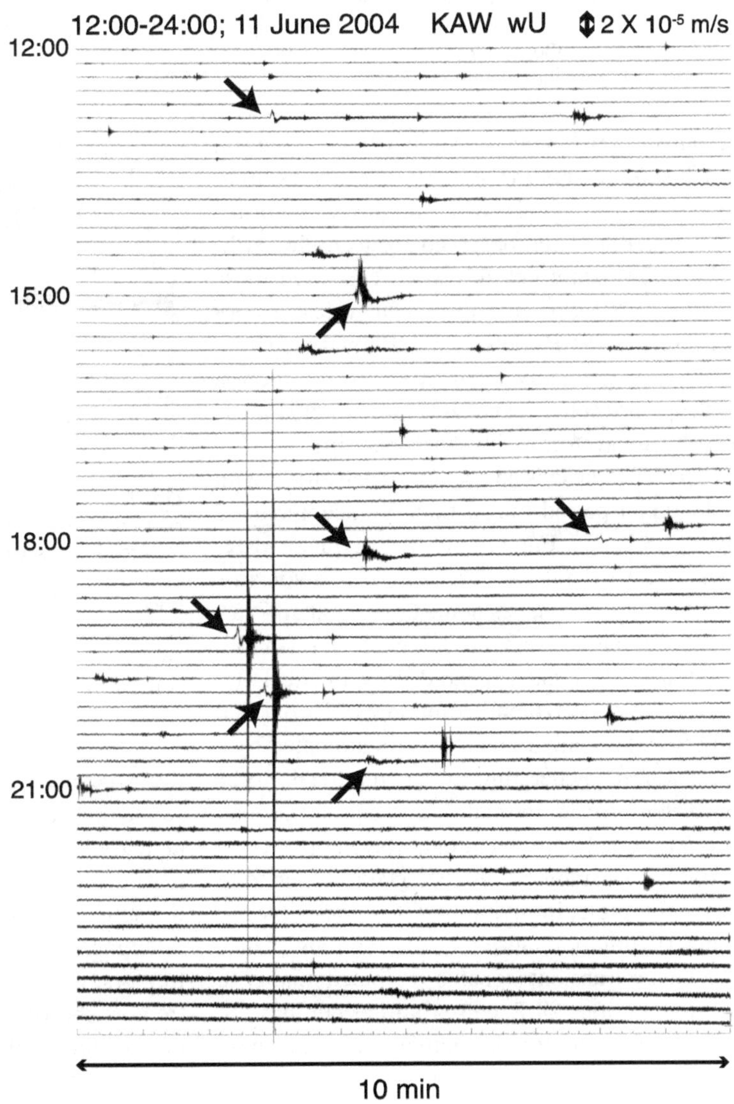

Fig. 5 Half-day record (12:00–24:00) of the broadband seismometer at KAW on June 11, 2004. Arrows indicate several kinds of unique long-period (LP) seismic signals

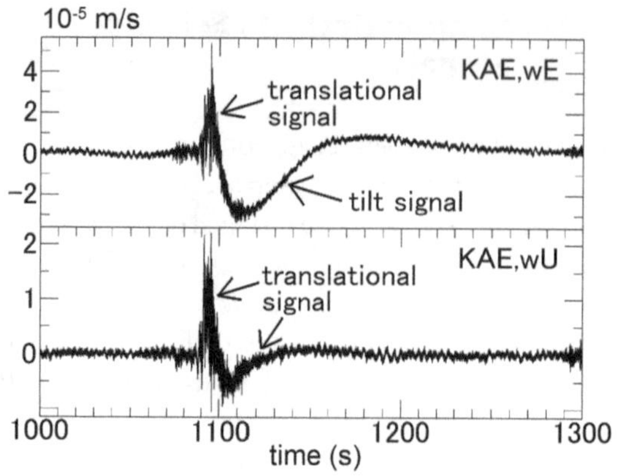

Fig. 6 Very long-period (VLP) pulse observed at Asama volcano. Upper and lower panels represent an east–west and the vertical components of the broadband seismometer at KAE (Maeda et al., 2011). Waveform from $t = 1090$ to 1100 s can be interpreted as a translational motion. Only the horizontal component has a much longer signal, from $t = 1100$ to 1300 s, indicating that this signal is the seismometer's response to a tilt motion

broadband seismometer installed on the west rim of the summit crater (KAW), where arrows highlight seismic signals with unique waveforms, which had not been observed for tectonic earthquakes. Yamamoto et al. (2005) classified these unique waveforms into three categories: a very long-period (VLP) pulse, an earthquake characterized by gradually attenuating an LP motion with upward convex, and an LP tremor characterized by a waveform with sharpened tips.

The VLP pulse is an impulsive waveform with a dominant period longer than 5 s accompanied by a tilt change. Figure 6 shows a typical waveform of a VLP pulse observed at Asama volcano. The initial pulse shapes in the horizontal and vertical components were similar, while a much longer transient signal appeared only in the latter part of the horizontal component. Horizontal broadband seismograms were contaminated by a tilt motion, generating a gravitational force equivalent to an inertial force due to ground acceleration. This gravitational force appeared in a vertical seismogram as the second-order term of the tilt angle, implying that it is negligible for ordinary microradian-order tilt motions. Thus, this longer transient signal is attributed mainly to the tilt motion. VLP pulses still occur but the change of the VLP pulse activity coincides with

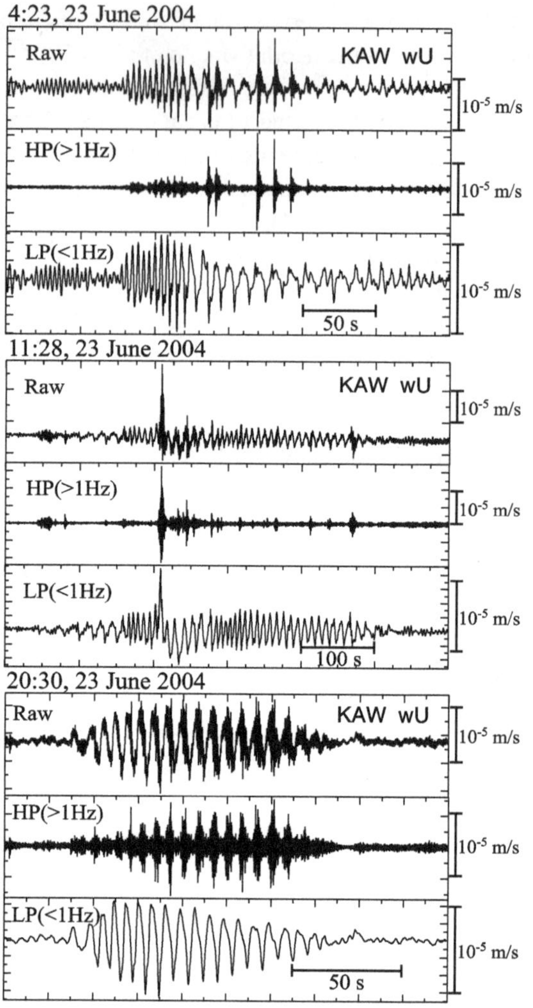

Fig. 7 Unique long-period (LP) volcanic tremors occurring on June 23, 2004. All seismograms are recorded by the broadband seismometer at KAW. Upper traces in each panel are the vertical component of raw records. Middle traces are filtered with a high-pass two-pole Butterworth IIR filter with a corner frequency of 1 Hz. Bottom traces are filtered with a low-pass two-pole Butterworth IIR filter with a corner frequency of 1 Hz

the volcanic activity. Yamamoto et al. (2005) found that similar impulsive pulses have occurred at least since September 2002 based on short-period seismograms recorded on the west rim of the summit crater.

Figure 7 shows three unique LP volcanic tremors, which occurred on June 23, 2004. These tremors showed tips or sawtooth waveforms with periods longer than 5 s and durations longer than

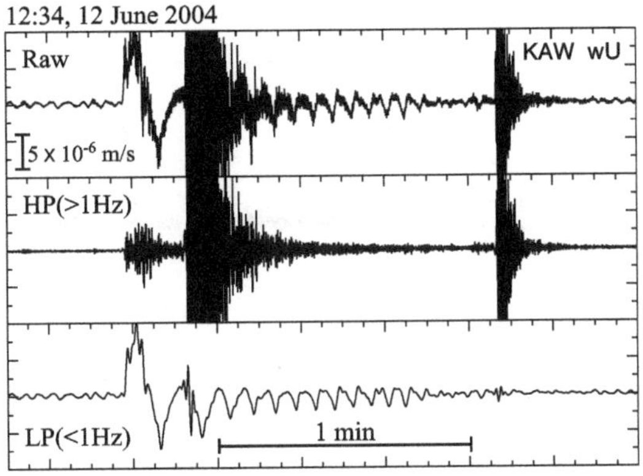

Fig. 8 Magnified view of the long-period (LP) earthquake occurring at 12:34 (JST) on June 12, 2004. Upper trace is the vertical component of the raw record. Middle and the bottom traces are filtered with a high-pass two-pole Butterworth IIR filter and with a low-pass two-pole Butterworth IIR filter, respectively. Both filters have a corner frequency of 1 Hz

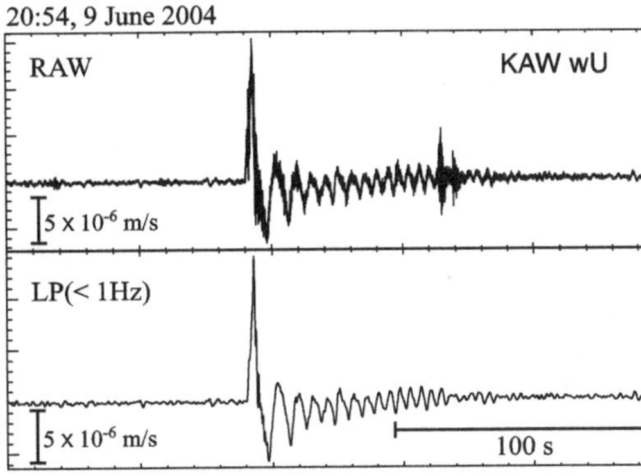

Fig. 9 Quasi-LP earthquake occurring at 20:54 (JST) on June 9, 2004. Upper trace is the vertical component of raw record. Bottom trace is filtered with a low-pass two-pole Butterworth IIR filter with a corner frequency of 1 Hz

100 s. Before the 2004 eruption, unique tremors occurred in June 2004 intensively. The summit broadband stations were destroyed by the first eruption in 2004 but were restored in October 2007. After the restart of the summit broadband seismic observations, these types of tremors were rarely detected until June 2018.

Figure 8 shows a closeup of an LP earthquake characterized by a gradually attenuating motion with an upward convex followed by two large high-frequency events. The signal was filtered with a low-pass IIR filter in the frequency range below 1 Hz to remove the high-frequency component (bottom panel in Fig. 8) as it clearly showed the features of the LP motion. Hereafter, this is referred to as an LP earthquake. Sometimes, an LP earthquake occurs without large high-frequency events (Fig. 9), which is referred to as a quasi-LP earthquake. LP and quasi-LP earthquakes occurred intensively between late-May 2004 and mid-June 2004, and they were rarely detected until May 2018. It is plausible that a fluid flowing through channels in a shallow region of volcanic conduit is analogous to a hydraulic circuit. A hydraulic circuit with a control valve is a candidate for the nonlinear system of the LP tremor or the LP earthquake. Based on a control valve model, Takeo et al. (2022) demonstrated that the occurrence of LP tremors and LP earthquakes qualitatively suggests a progression of blockage in the conduit.

A remarkable earthquake characterized by gradually attenuating long-tail oscillations containing a harmonic spectrum structure was observed frequently in Asama volcano. These events are called N-type earthquakes, named after the non-dumping oscillation. Forty-one N-type earthquakes occurred between November 20, 2011, and December 1, 2011. Figure 10 exhibits two broad-band seismograms of an N-type earthquake at KAW and KAE which occurred on 22 November, spectra of vertical components at six broadband stations. In Fig. 11, these hypocenters determined using the double-difference algorithm are denoted by star marks, and those of other volcanic earthquakes that occurred in the same period are done by open circles. N-type earthquakes occurred in a region limited to several hundred square meters with a centroid depth around 800 m beneath the crater bottom.

Prior to 2003, the near-field region of the summit crater did not have a permanent seismic station and almost all installed seismometers were short-period sensors. Additionally, the recording systems of seismograms before the 1960s were smoked-paper recorders. The last magmatic moderate-sized eruption in the 1950s occurred on November 10, 1958. The volcano released an intensive outburst, which was regarded as one of

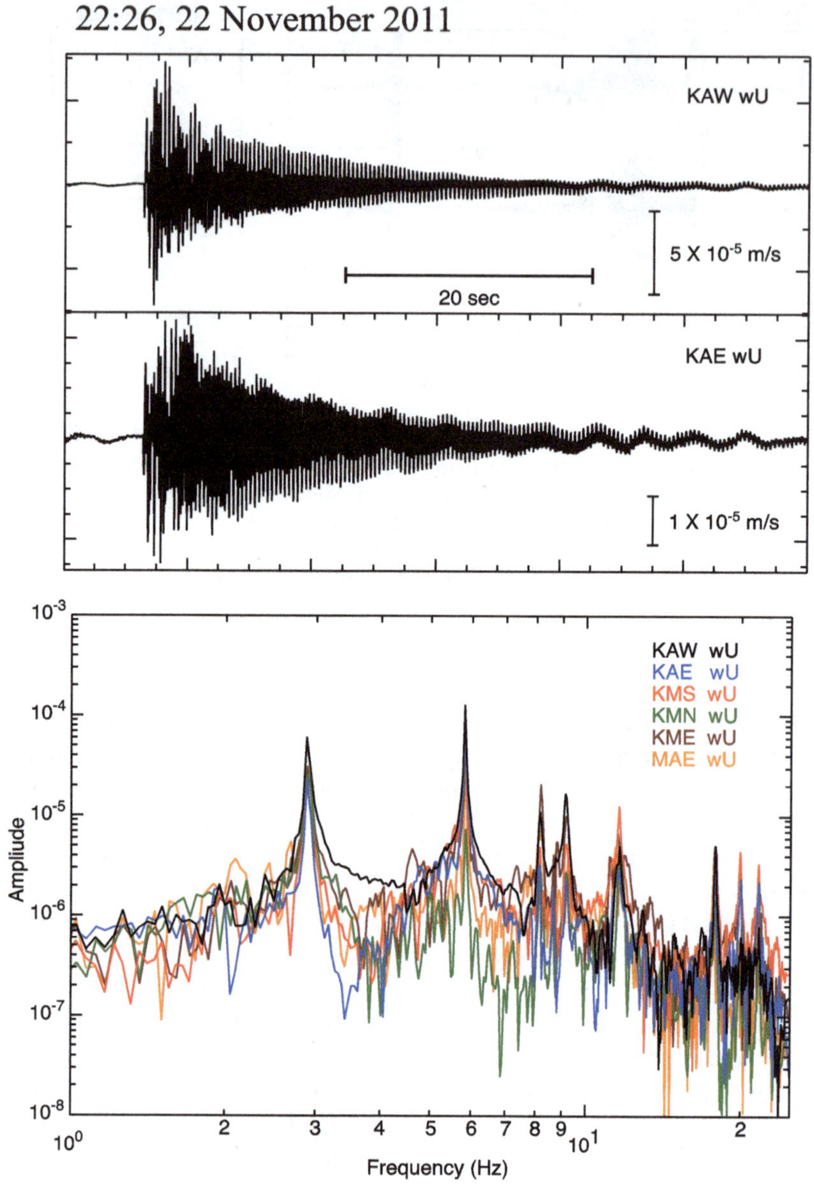

Fig. 10 Broadband seismograms of N-type earthquake. Waveforms in the upper panel are vertical component seismograms at KAW and KAE which recorded at 22:26 on November 22, 2011. Lower panel shows spectra of vertical components at six broadband stations (KAW, KAE, KMS, KMN, KNE, and MAE)

the strongest explosive eruptions in recent decades. Temporal evolution of daily counts of volcanic earthquakes and explosions in 1958 is exhibited in Fig. 12. The explosions were classified by Minakami based on his definition (Minakami 1935): Eruption A: violent explosion with strong detonation; Eruption B: explosion with strong detonation in less amount of projection than A; Eruption C: explosion with or without small detonation without air shock; Eruption D: small non-detonating eruption with projection of only volcanic gravels and ashes. Each of Eruptions A, B, and C is marked by a triangle with different color, and the counts of Eruption D are represented by bar graphs. The eruption on November 10, 1958, (Eruption A) was preceded by a high activity of volcanic earthquakes beginning in August 1958, and a repetition of minor ash eruptions (Eruption D) began in October 1958. This eruption was followed by other outbursts (Eruptions B and C) and minor eruptions (Eruption D) through the remainder of the year.

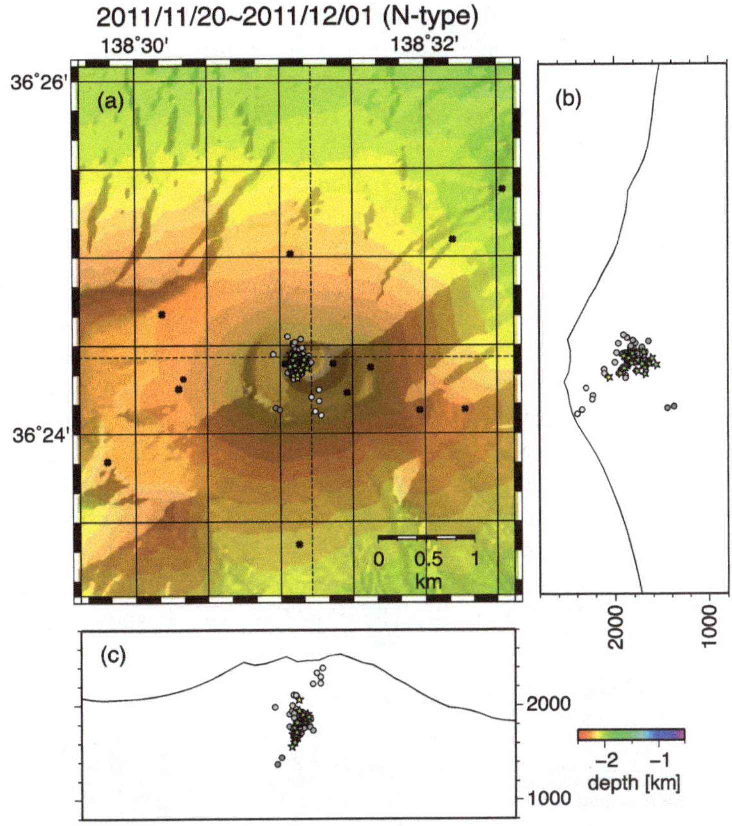

Fig. 11 Distribution of the relocated hypocenters of N-type earthquakes occurring between November 20 and December 1, 2011. Each hypocenter shown by a star is colored by its depth. Gray circles indicate other volcanic earthquakes occurring during the same period. All hypocenters are relocated using the double-difference algorithm (Waldhauser and Ellsworth 2000). (**a**) Distribution of epicenters. (**b**) North–south cross section of hypocenters. (**c**) East–west cross section of hypocenters

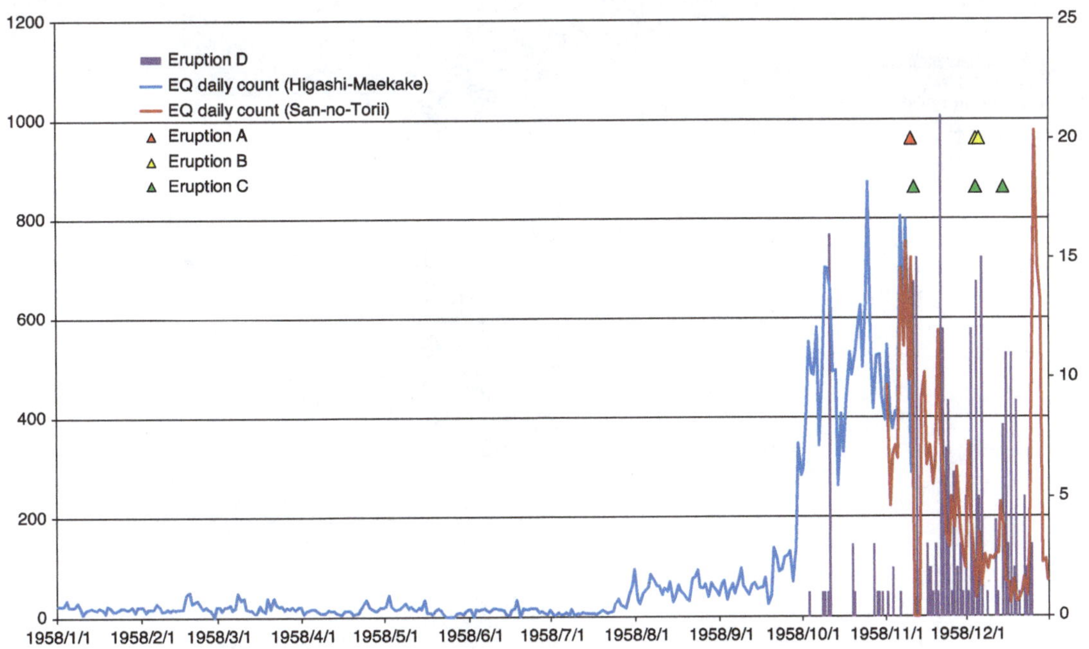

Fig. 12 Temporal evolution of daily counts of volcanic earthquakes at Higashi-Maekake and at San-no-Tori stations and explosions in 1958. The explosions were classified by Minakami based on his definition (Minakami 1935)

Unique short-period seismograms were observed in the seismic activity preceding the 1958 eruption and suspected the activity to be similar to LP tremors and LP earthquakes. Figure 13a shows a large short-period seismogram recorded at San-no-Torii station (SAN) on September 10, 1958. This short-period event exhibited a small preceding signal with a duration longer than 10 s. Figure 13b shows closeup seismograms of the LP earthquake recorded by the broadband seismometer at the summit station (KAW) and by the short-period seismometer at SAN on June 12, 2004. The feature of the lasting small preceding signal resembled the initial portion of the LP earthquake recorded by the short-period seismometer at SAN.

Figure 14a exhibits a short-period tremor composed of intermitted wave packets recorded on October 5, 1958, at the Higashi-Maekake station, which was about 1.5 km east of the summit crater. Figure 14b exhibits the LP tremor at 20:30 (JST) on June 23, 2004, recorded at KAW. The upper trace is the raw signal, while the lower trace is filtered by a high-pass filter with a cutoff frequency of 1 Hz. The lower trace in Fig. 14b showed a feature similar to that of the old smoked-paper seismogram in Fig. 14a. These unique smoked-paper seismograms suggested that LP tremors and LP earthquakes also occurred during the seismic activity preceding the 1958 eruption. In addition, an intensive activity of N-type earthquake had started on 9 November, 1 day before the strongest explosive eruption (Eruption A) in 1958 eruption activity. A lot of large long-lasting N-type earthquakes were recorded at Higashi-Maekake, whose smoked-paper record is shown in Fig. 15.

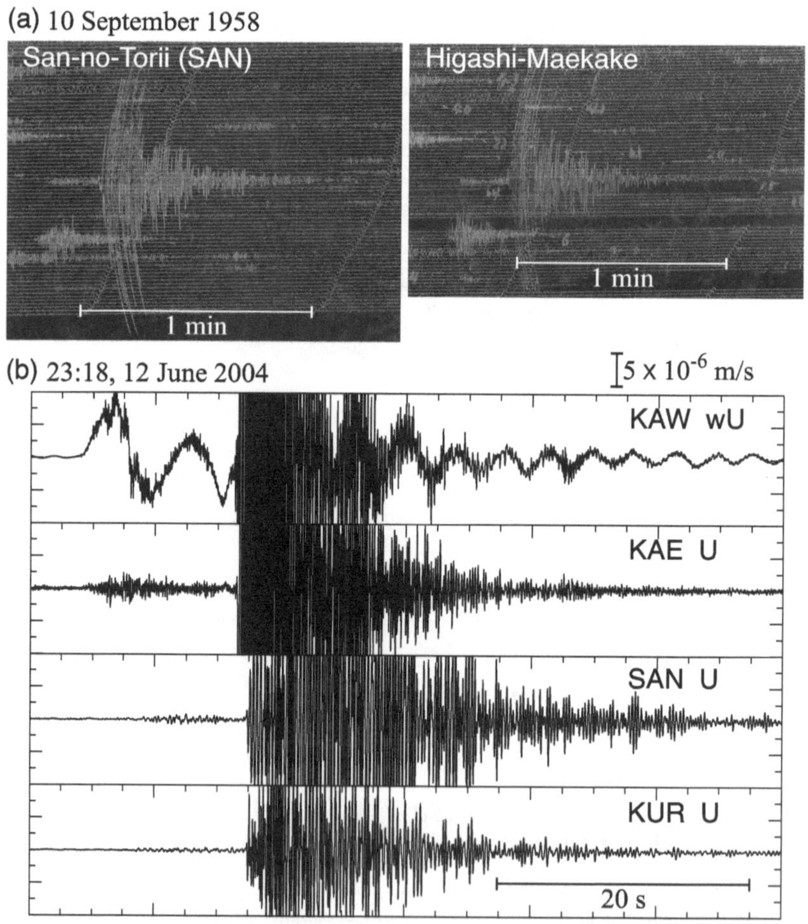

Fig. 13 (**a**) Smoked-paper records of short-period seismometers at San-no-Torii and Higashi-Maekake stations on September 10, 1958. San-no-Torii station is at the same place as the present SAN station. (**b**) Magnified view of the LP earthquake seismograms from 23:18 (JST) on June 12, 2004. Seismogram at KAW is recorded by the broadband seismometer and other records are by the short-period seismometers

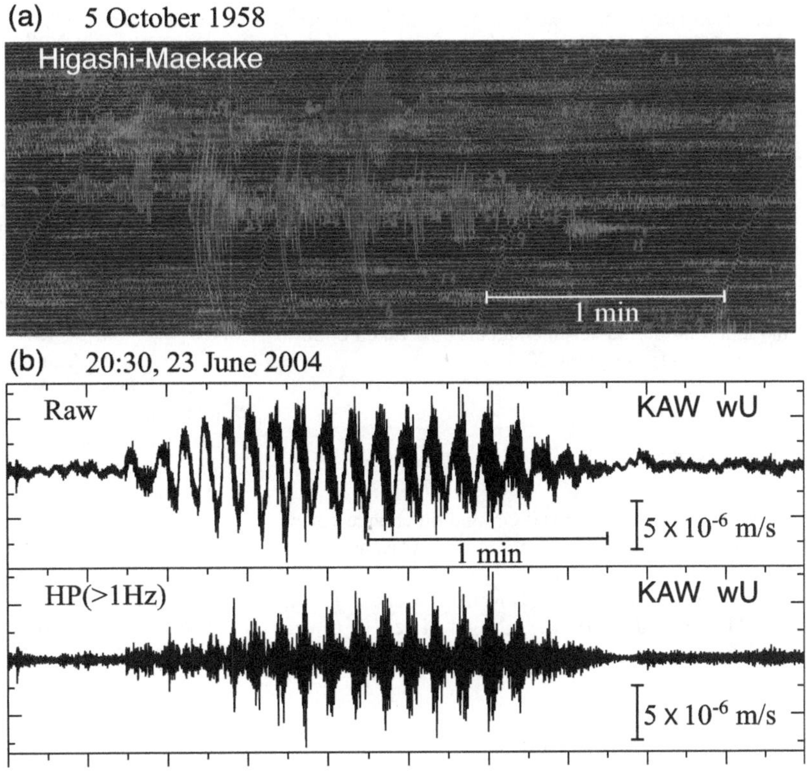

Fig. 14 (**a**) Smoked-paper record of a short-period seismometer at Higashi-Maekake station on October 3, 1958. (**b**) LP tremor occurring at 20:30 (JST) on June 23, 2004, recorded by the broadband seismometer at KAW. Upper trace is the vertical component of the raw record. Lower trace is filtered by a high-pass two-pole Butterworth IIR filter with a cutoff frequency of 1 Hz

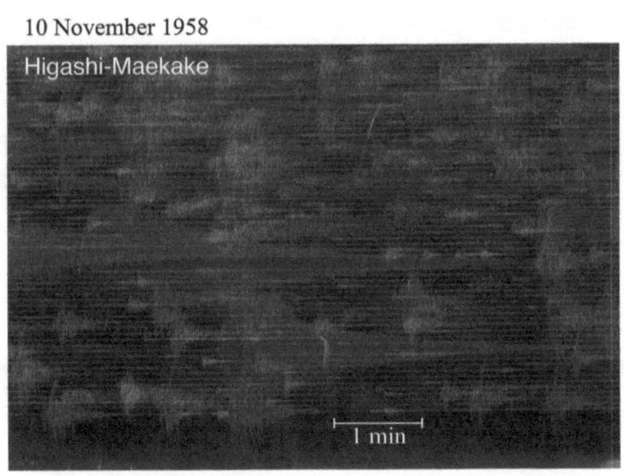

Fig. 15 Smoked-paper record of N-type earthquakes at Higashi-Maekake station on November 10, 1958

4 Summary

The Vulcanian eruption with a VEI of 2 occurred at 20:02 (JST) on September 1, 2004, following by Strombolian eruptions on 14–18 September. The estimated total amount of ejected tephra associated with the 1 September eruption was about 1.6×10^8 kg. Other five Vulcanian eruptions had been occurred by the end of 2004. After 4 years repose, three minor eruptions occurred in August 2008. The small eruption occurred on February 2, 2009, whose ejected mass was about 20% of that associated with the September 1, 2004, eruption, following with several minor eruptions as of May 2009. After more than 6 years repose, two minor eruptions happened on June 16 and 19, 2015.

The broad-band seismometers installed on the west and east rims of the summit crater recorded several kinds of unique long-period seismic signals which had not been observed for tectonic earthquakes. The VLP pulse is an impulsive waveform with a dominant period longer than 5 s accompanied by a tilt change. The three unique long-period volcanic tremors (LP tremor) occurred on June 23, 2004, exhibiting tips or sawtooth waveforms with periods longer than 5 s and durations longer than 100 s. LP and quasi-LP earthquakes characterized by the gradually attenuating motion with upward convex were intensively recorded between late-May and mid-June 2004. The unique short-period seismograms were observed in the seismic activity preceding the 1958 eruption, suggesting that LP tremors and LP earthquakes also occurred during the 1958 eruption activity.

Acknowledgments We thank the Japan Academy for permission to use several figures and contents from the paper entitled, "Recent volcanic activity at the Asama volcano and long-period seismic signals" by Takeo et al. in Proc. Jpn. Acad., Ser. B 98, 2022.

References

Churei M, Katayama H (2006) Ground tilt change associated with the 2004 eruption at Asama volcano, Japan. Bull Volc Soc Jpn 51:91–101

Maeno F, Suzuki Y, Nakada S, Koyama E, Kaneko T, Fujii T, Miyamura J, Onizawa S, Nagai M (2010) Course and ejecta of the eruption of Asama volcano on 2 February 2009. Bull Volc Soc Jpn 55:147–154

Minakami T (1935) The explosive activities of volcano Asama in 1935 (part 1). Bull Earthq Res Inst Univ Tokyo 13:629–644

Nakada S, Yoshimot M, Koyama E, Tsuji H, Urabe T (2005) Comparative study of the 2004 eruption with old eruptions at Asama volcano and the activity evaluation. Bull Volc Soc Japan 50:303–313

Takeo M, Aoki Y, Ohminato T, Yamamoto M (2006) Magma supply path beneath Mt. Asama volcano, Japan. Geophys Res Lett 33:L15310

Takeo M, Aoki Y, Koyama T (2022) Recent volcanic activity at the Asama volcano and long-period seismic signals. Proc Jpn Acad Ser B 98:416–438

Waldhauser F, Ellsworth WL (2000) A double-difference earthquake location algorithm: methods and application to the northern Hayward fault, California. Bull Seismol Soc Am 90:1353–1368

Yamamoto M, Takeo M, Ohminato T, Oikawa J, Aoki Y, Ueda H, Nakamula S, Tsuji H, Koyama E, Osada N, Urabe T (2005) A unique earthquake activity preceding the eruption at Asama volcano in 2004. Bull Volc Soc Jpn 50:393–400

Volcanic Events and Gas Emission at Asama Volcano

Minoru Takeo

1 Introduction

The source mechanisms and/or the temporal evolutions of the unique long-period seismic signals help us to understand the eruption process at Asama volcano. Here, we reveal the focal mechanism of the VLP pulse using a waveform inversion analysis and compile the temporal evolutions of multidisciplinary data including seismic, geodetic, geomagnetic, and geochemical data to make clear the internal process in the shallow regions of the conduit.

2 VLP Pulse at Asama Volcano

The VLP pulse observed at Asama volcano is characterized by the similar initial pulse shapes in the horizontal and vertical components with a much longer transient signal appearing only in the latter part of the horizontal component. As of August 2008, the summit area had four permanent broadband seismic stations (KAE, KAW, KME, and KMS). To reveal the location and the source mechanism of the VLP pulse, ten temporary broadband seismic stations were installed inside the Maekake crater. The temporal observations were carried out between September 2008 and April 2009. Maeda et al. (2011) developed a new moment-tensor inversion method to deal with horizontal seismograms contaminated by tilt motions without any assumptions regarding the tilt motions. Their idea included the effect of tilt motions into Green's functions instead of decomposing the horizontal seismograms into the translational and tilt responses.

Maeda and Takeo (2011) selected the VLP pulse at 17:18 (JST) on October 28, 2008, as a representative event and conducted inversion analysis. They applied the moment-tensor inversion method to the VLP pulses observed at Asama volcano so that the time histories of the moment-tensor could be elucidated. The locations of the VLP pulses were surveyed using a grid search algorithm in which candidates of the source location were distributed in and around the summit crater at 50-m grid intervals and elevations ranging from 1850 to 2300 m above sea level. Many previous moment-tensor inversion studies at

This chapter is a revised version of Takeo et al. (2022) with the inclusion of a reference in the References section

M. Takeo (✉)
Earthquake Research Institute, The University of Tokyo, Tokyo, Japan

Japan Meteorological Agency, Tokyo, Japan
e-mail: takeo@eri.u-tokyo.ac.jp

S. Aramaki et al. (eds.), *The Asama Volcano*, Active Volcanoes of the World,
https://doi.org/10.1007/978-981-95-5799-8_16

volcanoes have considered single forces along with the moment-tensor. However, the waveforms of the VLP pulses were well reconstructed without the single forces. Since the source parameters did not take the single forces into account, the resolution and the uniqueness of the solution were improved due to the reduction of the degree of freedom in the inversion problem.

The observed particle motions at each station (red lines) are compared with the synthetic particle motions (blue dashed lines) in Fig. 1, in which the best source location is shown by a red star. These particle motions are obtained from the seismograms which are filtered with a four-pole zero-phase Butterworth low-pass filter in the frequency range below 0.1 Hz. Even after taking the uncertainty of the source location of 50–100 m in each direction into account, the source was located to the north of the crater center at a depth of 100–300 m below the summit. Figure 2 compares the observed (red lines) and the synthetic

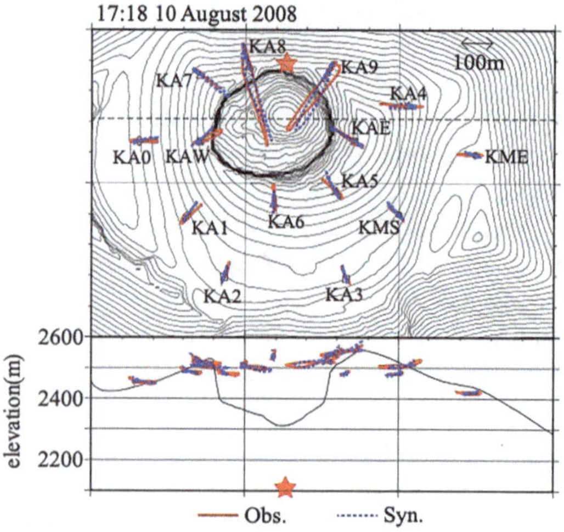

Fig. 1 Comparison of the observed particle motions (red lines) and synthetic particle motions (blue dashed lines) for the event at 17:18 (JST) on August 10, 2008 (Maeda and Takeo 2011). Particle motions are projected onto the horizontal plane (upper panel) and onto the east–west cross section (lower panel). Gray stars indicate the location of the event. Contour interval is 10 m, and the elevation of KAE is 2568 m above sea level. Guralp CMG-3 T (360 s) sensors are installed at KAW, KAE, KA2, and KA3. Guralp CMG-3 T (100 s) sensors are installed at KA0 and KA1. Broadband sensors at KMS and KME are Nanometrics Trillium (120 s) sensors, while those at the other stations are Nanometrics Trillium (40 s) sensors

(blue lines) seismograms at seven stations. The components of the moment-tensor had similar time histories (see Fig. 3), which consisted of an initial rapid inflation phase and a subsequent gradual deflation phase. This similarity allowed a unique focal mechanism for the event to be determined.

Taking the peak amplitude of each time history as the values of the corresponding moment-tensor component, the focal mechanism of the VLP pulse was estimated. Maeda and Takeo (2011) performed moment-tensor inversions for 23 VLP pulses to examine the stability of the focal mechanism. Nineteen of the VLP pulses displayed moment-tensor time histories consistent among the components. Hence, the time-invariant moment-tensor solutions were obtained for these 19 events. The eigenvectors (principal axes) of these moment-tensor solutions showed only small variations. The eigenvalues (principal values) exhibited some variations, which depended on the amplitude. The ratio of the principal values (M_{max}:M_{int}:M_{min}) was 5:3:2, where M_{max}, M_{int}, and M_{min} indicate the maximum, intermediate, and minimum principal values, respectively. Assuming a Poisson material, this is interpreted as the principal ratio for a synchronized volumetric change of a tensile crack and a cylinder with symmetry axes oriented orthogonal to each other. The strike and dip angles of the tensile crack were both nearly 70°, and the symmetry axis of the cylinder was inclined about 30° from the vertical. The time histories of the moment tensor were indicative of an initial sudden pressurization followed by a gradual depressurization at the source region. They sought to model the VLP pulse at Asama volcano by considering the inflow and outflow of gas at the source cavity, compensated by boiling of water or by exsolution of volatiles from a deeper reservoir.

Kazahaya et al. (2011) revealed a relationship between the VLP pulses and volcanic gas emissions through a combination of high temporal resolution SO_2 flux measurements, vent process monitoring using a video camera set on the summit crater rim, and broadband seismic observations at the stations. The volcanic gas observations

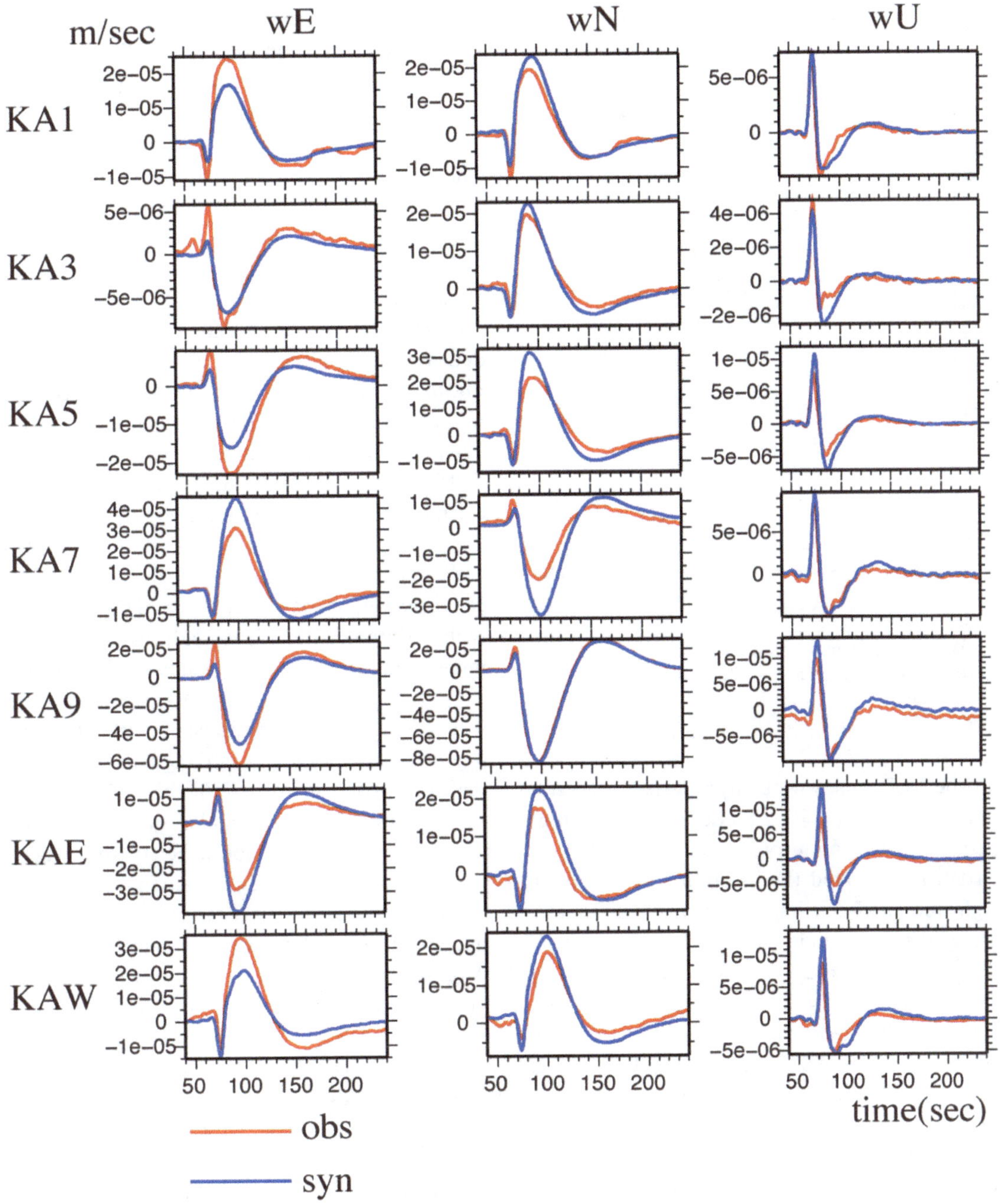

Fig. 2 Comparison of observed (red lines) and synthetic (blue lines) seismograms at seven stations (KA1, KA3, KA5, KA7, KA9, KAE, and KAW). These seismograms are filtered with a four-pole zero-phase Butterworth low-pass filter in the frequency range below 0.1 Hz

were conducted on June 2, 2009. Seven degassing bursts accompanied the VLP pulses. These bursts originated from the vent located at the crater bottom and were clearly recognized by the video camera. The seismic moments of VLP pulses and volcanic gas emissions were linearly related, strongly supporting the VLP pulse source model proposed by Maeda and Takeo (2011).

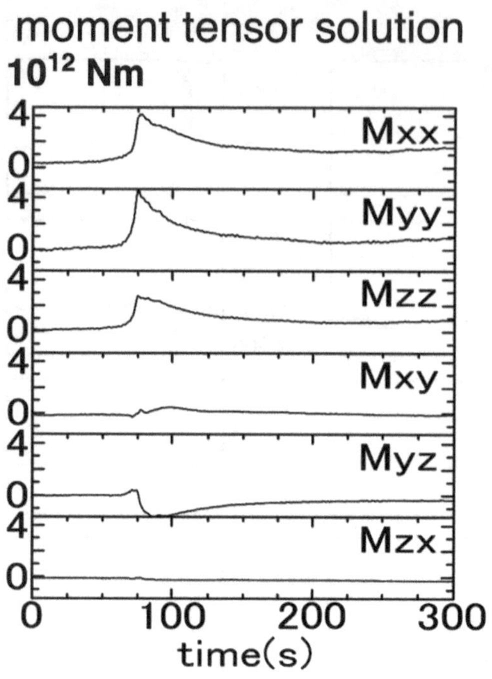

Fig. 3 Moment tensor solutions. The x-, y-, and z-axes represent the east, north, and upward directions, respectively

The VLP pulses were characterized by a transient signal with a 10–20 s duration and a high-frequency oscillation of 5–10 Hz. Maeda et al. (2019) investigated the location and size of the high-frequency oscillation sources (HF sources) using the amplitude source location method (Battaglia and Aki 2003). Most HF sources were deeper than the VLP pulse source by about 150 m. Additionally, the signals in two frequency bands of 0.00–0.1 Hz (VLP band) and of 5–10 Hz (HF band) did not show a time lag, suggesting a strong connection between the two sources, in which one immediately responds to the other. The location of the VLP pulse source was considered as the centroid of the inflation source, which was near the roof of the cavity where volcanic gas accumulated, whereas the HF oscillation seemed to be emitted more intensely from the lower and narrower portions of the cavity. This resulted in the source depth differences between the two signal bands.

3 Temporal Evolution of Seismic Activity, LP Seismic Signal Preceding 2004 Eruption

The activities of volcanic earthquakes were compared with the LP seismic signals from October 2003 to the first eruption in 2004. VLP pulse events were detected automatically using a VLP pulse detection algorithm (Kazahaya et al. 2015a). The displacement amplitudes of the detected VLP pulses were calculated by integrating the vertical component of the velocity seismograms, and the activity was evaluated by a daily cumulative amplitude.

Figure 4 shows the daily counts of volcanic earthquakes (orange line) and the daily cumulative amplitude of VLP pulses (dark red line). The daily counts of LP tremors, LP earthquakes, and N-type earthquakes are represented by bar graphs with different colors. The scale is indicated by integers in round brackets beside the left vertical axis. Before June 2004, the VLP pulse activity was synchronized with the seismicity. However, it gradually decreased toward the eruption despite the increase in seismicity. At this turning point, LP tremors and LP earthquakes occurred in clusters. VLP pulses were not detected from August 24, 2004, to the first Vulcanian eruption on September 1, 2004. By contrast, an intense seismic swarm started beneath the summit crater on August 30, 2004, and lasted until the first eruption. A reliable interpretation is that the occurrences of LP tremors and LP earthquakes indicated the blockage in the conduit (Takeo et al. 2022), resulting in the decline of the VLP pulse activity due to shielding gas emission and the increase of seismicity due to stress accumulation in and around the conduit.

N-type earthquakes mainly occurred in July–August 2004, which was before the first eruption for that year. One of the most reliable models of N-type earthquakes is the fluid-filled crack model proposed by Chouet (1986) and Chouet (2003). He showed that a fluid-filled crack generated a very slow wave propagating along the crack wall,

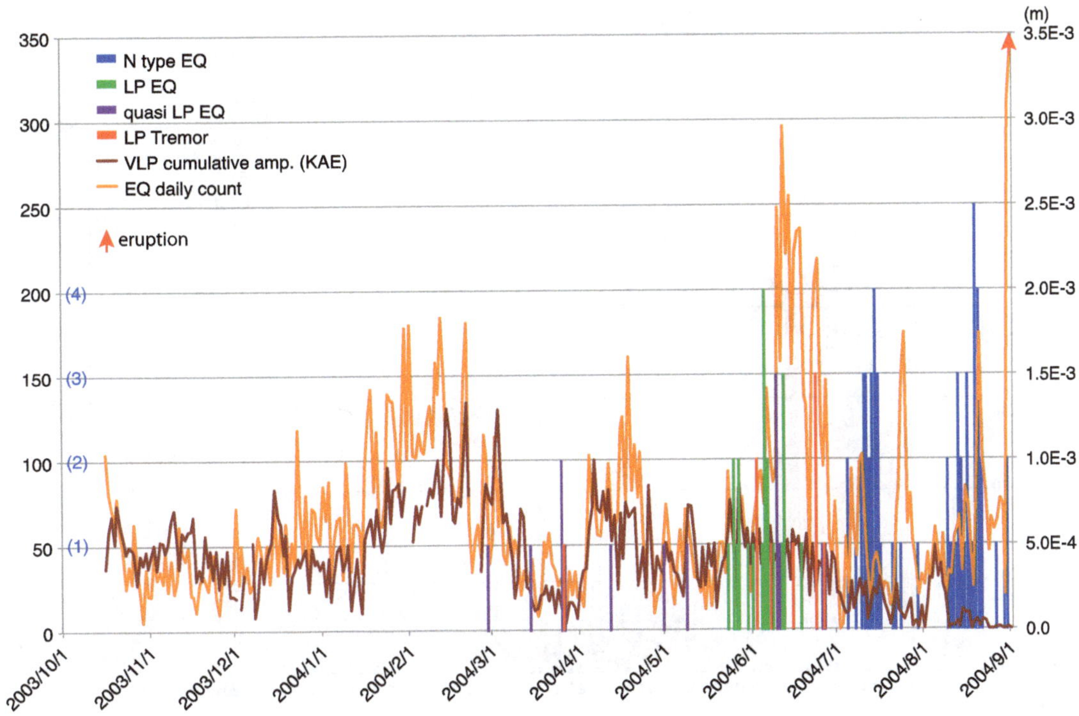

Fig. 4 Temporal evolution of the seismic activity, long-period (LP) seismic signals between October 16, 2003, and September 1, 2004. Orange and dark red graphs indicate daily counts of volcanic earthquakes and daily cumulative amplitude of the VLP pulses, respectively. Integers beside the left vertical axis indicate the scale of the daily counts of volcanic earthquakes. Numbers beside the right vertical axis show the scale of the cumulative amplitudes. Daily counts of the LP tremors, LP earthquakes, and N-type earthquakes are represented by bar graphs with different colors, where the scale is indicated by the blue integers in round brackets beside the left vertical axis

which was termed the "crack wave." The crack wave leads to more realistic estimates of the resonator size relative to a spherical resonator. Chouet's model consists of a single isolated fluid-filled crack in an infinite elastic solid body; hence, external mass transfer into or out of the crack was not taken into consideration.

The synthetic seismograms calculated by the fluid-filled model exhibited strong similarities to the observed N-type earthquakes in terms of the spectral peaks and long-lasting oscillations (e.g., Chouet et al. 1994; Nakano et al. 1998; Kumagai and Chouet, 2002; Kumagai et al. 2002; Nakano and Kumagai 2005). If the resonance of fluid-filled cracks caused N-type earthquakes, then the cracks should be isolated. This indicated that the fluid in a certain region of the conduit is confined in the cracks and does not connect with fluid in other regions. The highly frequent occurrence of N-type earthquakes in July and August 2004 suggested that the blockage in the conduit progressed during this period. Thus, the temporal evolution of all seismic data, including the seismicity, the activities of the VLP pulse, LP tremors, LP earthquakes, and N-type earthquakes, indicated that the blockage in the conduit was accelerated about 3 months before the first eruption on September 1, 2004.

4 Temporal Evolution of Seismic Activity, Ground Deformation, and Total Magnetic Force from October 2007 to January 2018

The summit broadband seismic stations were destroyed by the first Vulcanian eruption on September 1, 2004. They were repaired in

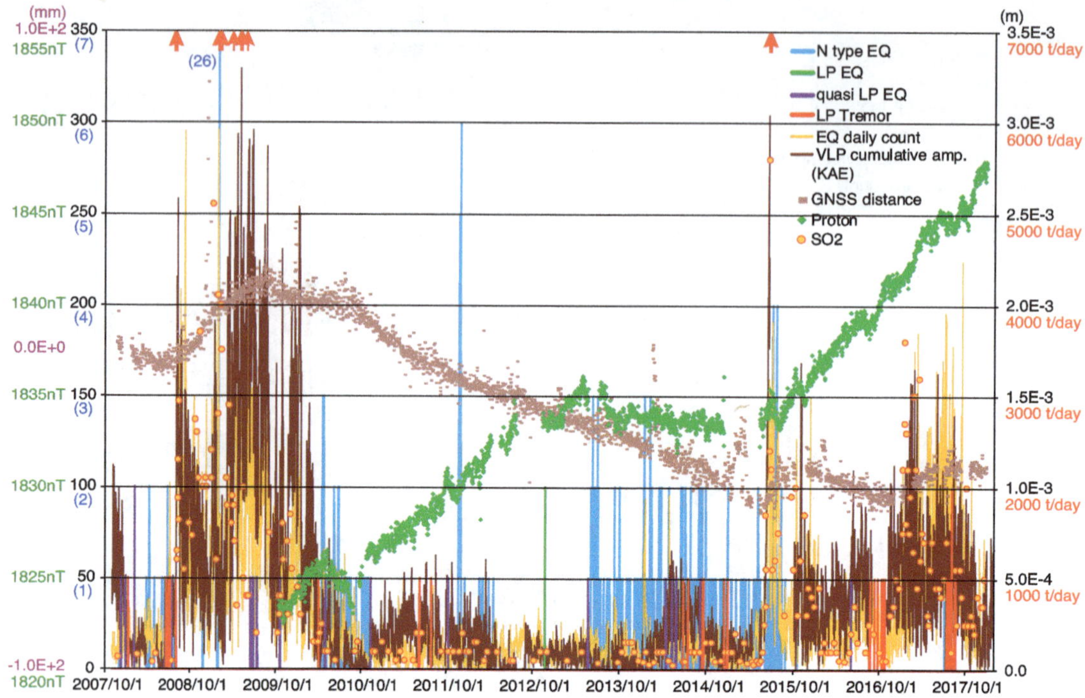

Fig. 5 Temporal evolution of the seismic activity, ground deformation, and geomagnetic total intensity anomalies between October 31, 2007, and January 31, 2018. Orange line graph, dark red line graph, and small pink dots show the daily counts of volcanic earthquakes, daily cumulative amplitude of the VLP pulses, and relative change in the GNSS baseline length for sites TASH and KVCO, respectively. Integers beside the left vertical axis indicate the scale of daily counts of volcanic earthquakes. Numbers beside the right vertical axis indicate the scale of cumulative amplitudes. Small orange circles with red circuits show the SO_2 emissions from the summit crater reported by the Japan Meteorological Agency (JMA). Green small dots represent the estimated volcanic signals of the geomagnetic total intensity anomalies at KMS. Orange numbers with unit t/day beside the right vertical axis denote the scale of SO_2 emissions. Green numbers with unit nT beside the left vertical axis represent the scale of total intensities. Daily counts of the LP tremors, LP earthquakes, and N-type earthquakes are represented by the bar graphs with different colors, whose scale is indicated by blue integers in round brackets beside the left vertical axis

October 2007. Figure 5 compiles the daily counts of volcanic earthquakes (orange line), the daily cumulative amplitudes of the VLP pulses (dark red line), and the relative changes in the GNSS baseline length for sites TASH and KVCO (small pink dots). The small orange circles with red circuits represent the SO_2 emissions from the summit crater reported by the Japan Meteorological Agency (JMA) (JMA, personal communication). The small green dots indicate the estimated volcanic signals of the geomagnetic total intensity at KMS. (The derivation of the estimation is detailed below.) The bar graphs with different colors represent the daily counts of the LP tremors, LP earthquakes, and N-type earthquakes, where the scale is indicated by integers in round brackets beside the left vertical axis. From the temporal evolution, the volcanic activity consisted of four periods: October 2007–mid-2010 (first term), mid-2010–mid-2013 (second term), mid-2013–June 2015 (third term), and June 2015–January 2018 (last term).

A magma budget of the magma plumbing system of Asama volcano was estimated based on the volcanic gas emission rates and the ground deformation data (Kazahaya et al. 2015b). During both the extension and contraction stages, the volume of degassed magma was one order of magnitude larger than the volume change of the dyke. This suggested that a significant amount of magma was supplied from the magma chamber to the dyke. Then degassed magma descended back to the magma chamber. During the contraction stages, the shrinkage volumes of the dyke were comparable to or slightly higher than the volume decreases of the magma by the volcanic gas discharge (Kazahaya et al. 2015b). This fact implied that some amount of volcanic gas is perennially supplied to the shallow region of the conduit, even in contraction stages.

The GNSS baseline switched from a contraction stage to an extension stage in the early phase of the first term. The average levels of the SO_2 emission and the VLP pulse activity were high during the extension stage, even though the existence fluctuated. An SO_2 emission greater than 2000 tons/day lasted from November 2008 to February 2009 despite the low VLP pulse activity. However, this relation flipped from April to August 2009. The VLP pulse activity represented the rapid gas emission through the northern part of the conduit as shown by the focal mechanism and its location. The SO_2 emission was an indicator of the total amount of gas emissions from the conduit. The variability of correlation between the SO_2 emission and the VLP pulse activity suggested that multi outgassing pathways existed in the shallow region of the conduit and a heterogeneous distribution of gas phases occurred within the conduit (Kazahaya et al. 2015a).

The SO_2 emission and the VLP pulse activity were maintained at a low level when the GNSS baseline was in the contracting stage through the second and third terms. However, the contraction of the GNSS baseline indicated that the extra magma intrusion beneath the western flank of Asama volcano was dormant, resulting in a basic amount of volcanic gas supply to the summit crater. The second and third terms were distinguished by the variation patterns of the geomagnetic total intensity and occurrence frequencies of N-type earthquakes. In the last term, which began with a minor eruption in June 2015, the GNSS baseline exhibited repeated extension and contraction stages with short durations. The higher levels of the SO_2 emission and the VLP pulse activity than those in the second and third terms were sustained during the last term.

The minor eruptions in August 2008 and June 2015 were preceded by short-term rapid increases in VLP pulse activity and SO_2 emissions, which lasted several days. Figure 6 compiles the daily counts of volcanic earthquakes (orange lines) and the daily cumulative amplitude of VLP pulses (dark red lines and dark red dashed lines) between April 1, 2008, and September 10, 2008 (Fig. 6a) and between February 1, 2015, and June 22, 2015 (Fig. 6b). The daily counts of VLP pulses are also shown by green lines and green dashed lines. The daily counts of LP earthquakes and N-type earthquakes are represented by bar graphs with different colors. The small orange squares represent the SO_2 emissions from the summit crater reported by JMA (JMA, personal communication). N-type earthquakes occurred frequently for several months before the eruptions. The LP earthquakes and LP tremors also occurred one and a half months before the eruptions in August 2008. Figure 7a exhibits large VLP pulses which preceded all three minor eruptions in August 2008 by about 2 min. The minor eruption in June 2015 was also preceded by a large VLP pulse, which occurred about 3 min before the eruption (Fig. 7b). Even if N-type earthquakes which frequently occurred for several months before the eruptions suggested shielding in the conduit, the rapid increment of the SO_2 emission and VLP pulse activity implied that multi outgassing pathways were not blocked completely, and certain portions of the pathways were still alive. The intensive gas emissions eventually caused these four minor eruptions.

Figure 8 compiles the daily counts of volcanic earthquakes (orange line) and the daily cumulative amplitude of VLP pulses (dark red line) with the daily counts of N-type earthquake by blue bar graphs. The daily counts of VLP pulses and SO_2 emission are also shown by a green line and small orange squares, respectively. The small eruption on February 2, 2009, was preceded by a short-term decrement of the VLP pulse activity for 9 days, and 26 N-type earthquakes occurred in a focused manner 1 day before the eruption. This intense activity of N-type earthquakes and the falling-off of the VLP pulse activity suggested that the shielding in the overall area of the conduit rapidly progressed, resulting in the largest eruption during the 2008–2009 crisis. Since a

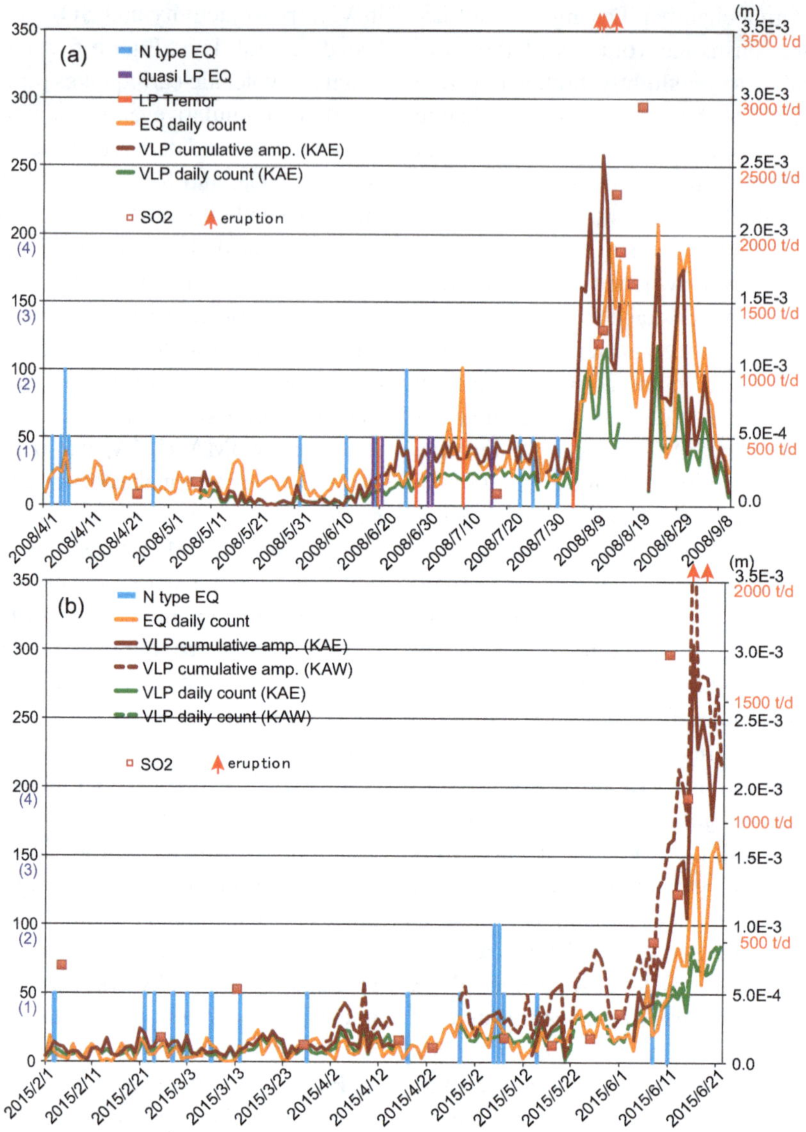

Fig. 6 (**a**) Temporal evolution of the seismic activity, long-period (LP) seismic signals between April 1, 2008, and September 10, 2008. (**b**) Temporal evolution of the seismic activity, long-period (LP) seismic signals between February 1, 2015, and June 22, 2015. Orange and dark red graphs indicate daily counts of volcanic earthquakes and daily cumulative amplitude of the VLP pulses, respectively. Integers beside the left vertical axis indicate the scale of daily counts of volcanic earthquakes. Numbers beside the right vertical axis indicate the scale of cumulative amplitudes. Small orange squares show the SO_2 emissions from the summit crater reported by JMA. Orange numbers with unit t/d beside the right vertical axis denote the scale of SO_2 emissions. Daily counts of the LP tremors, LP earthquakes, and/or N-type earthquakes are represented by the bar graphs with different colors, whose scale is indicated by blue integers in round brackets beside the left vertical axis. The daily counts of VLP pulses are also shown by green lines and green dashed lines

broadband seismometer has high sensitivity in a low-frequency range, the horizontal component of the signal is easily affected by a tilt motion (e.g., Graizer 2006; Aoyama and Oshima 2008). Based on the equation of motion for a seismometer's pendulum derived by Graizer (2006), Aoyama (2008) showed that the theoretical response agrees well with experimental results.

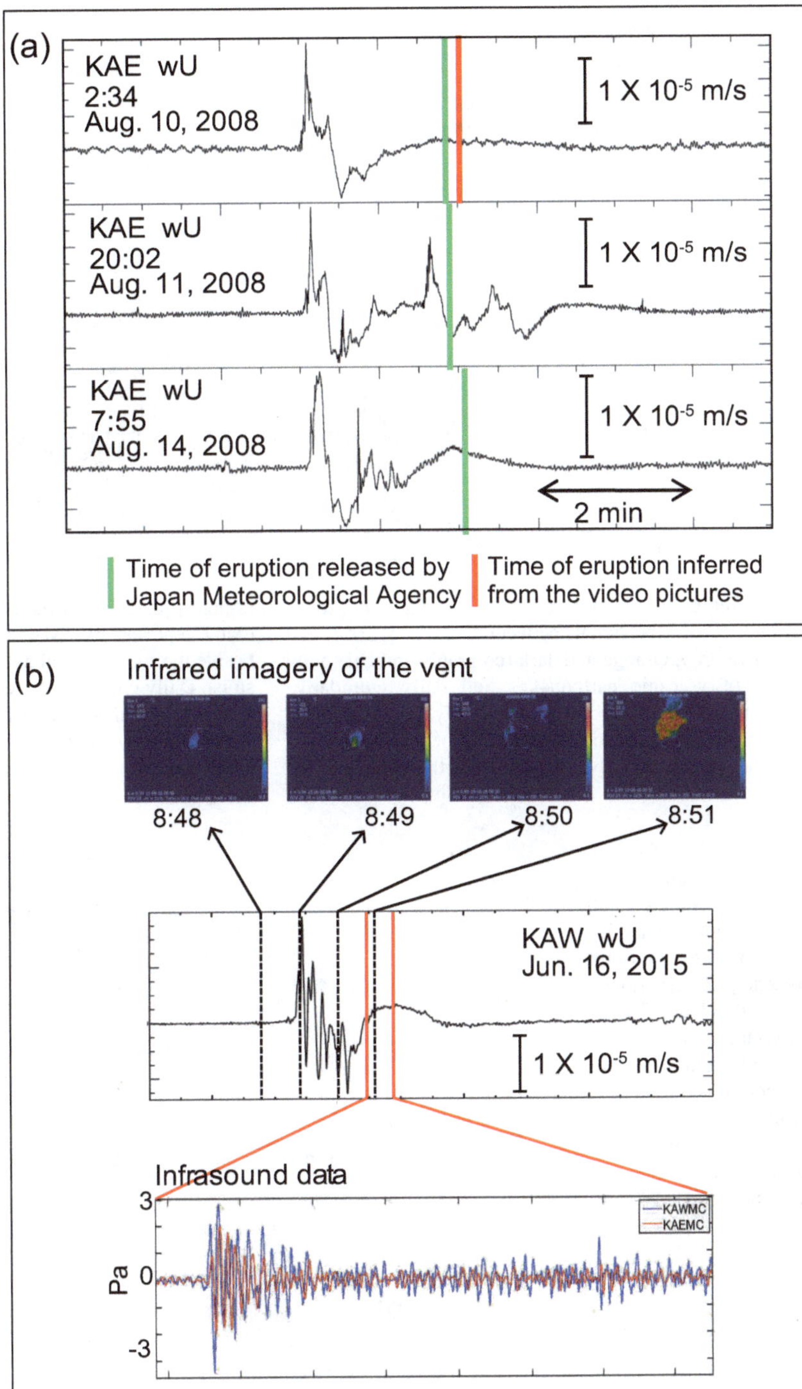

Fig. 7 (**a**) Large VLP pulses preceded three minor eruptions in August 2008. Green and red vertical lines indicate the occurrence times of the minor eruptions confirmed by video pictures or infrasound data. (**b**) A large VLP pulse preceded a minor eruption on June 16, 2015. The occurrence time of the eruption was confirmed by a series of infrared images of the vent and an infrasound data

Geno and Ripepe (2010) derived tilt motions from broadband seismometers at the Stromboli volcano using the same theoretical response. The same approach as Aoyama (2008) and Geno and Ripepe (2010) was employed to obtain tilt motions at KAE (on the east rim of the summit crater) associated with the small eruption on February 2, 2009. The tilt motions recorded at KAE are shown in Fig. 9a with the vertical component of the broadband seismogram; the peak-

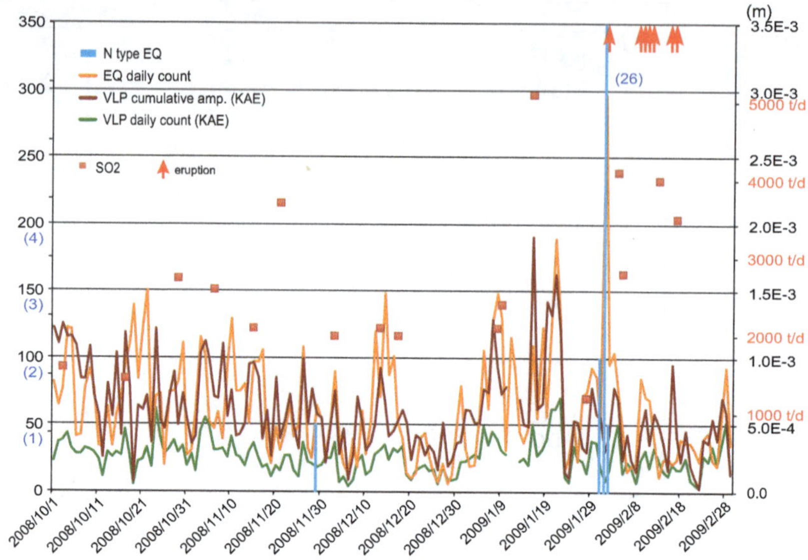

Fig. 8 Temporal evolution of the seismic activity, long-period (LP) seismic signals between October 1, 2008, and March 1, 2009. Orange and dark red graphs indicate daily counts of volcanic earthquakes and daily cumulative amplitude of the VLP pulses, respectively. Integers beside the left vertical axis indicate the scale of daily counts of volcanic earthquakes. Numbers beside the right vertical axis indicate the scale of cumulative amplitudes. Small orange squares show the SO_2 emissions from the summit crater reported by JMA. Orange numbers with unit t/d beside the right vertical axis denote the scale of SO_2 emissions. Daily counts of N-type earthquakes are represented by the blue bar graphs, whose scale is indicated by blue integers in round brackets beside the left vertical axis. The daily counts of VLP pulses are also shown by green lines and green dashed lines

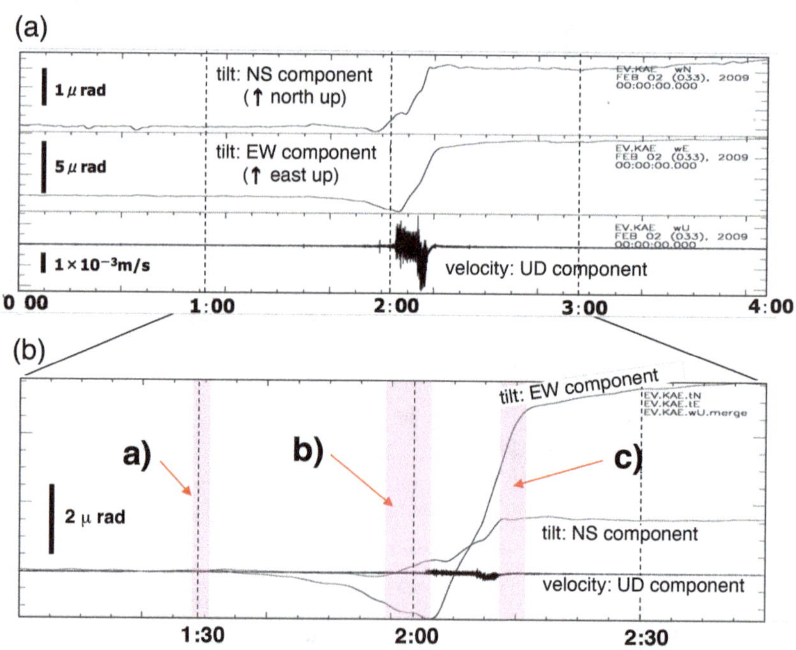

Fig. 9 Tilt motion associated with the small eruption on February 2, 2009. (**a**) The tilt motion preceding the eruption recorded at KAE. The tilt vector is directed toward the crater, so the tilt motion is dominated by the EW component. (**b**) Enlarged illustration of the tilt motion at KAE between 1:00 and 3:00 on February 2, 2009

to-peak amplitude of the tilt exceeded 6 μrad in east-west component. These tilt motions revealed a preceding inflation of the crater side, which started about 20 min before the start of the eruption at 1:51 (a) in Fig. 9b. After the eruption started, the tilt motion turned from inflation to deflation of the crater side (b) in Fig. 9b, and the rate of tilt motion decreased gradually in the final stage of the eruption (c) in Fig. 9b. Such large inflations were not recorded during the minor eruptions in 2008 and in 2015. This large inflation also pointed to the overall shielding in the conduit before the small eruption.

In a global sense, the variations of the geomagnetic field are caused by the currents in the ionosphere and magnetosphere along with the dynamo mechanism in the Earth's outer core. The variations of the geomagnetic field due to volcanic activity are much smaller. Hence, ionospheric, magnetospheric, and core effects must be removed to investigate the thermal conditions in a volcano. A common method used to eliminate non-volcanic variations from the observed total magnetic intensity at a volcano site is to simply subtract the intensity measured at a reference site close to the observation site, assuming that large-scale variations are common to both sites due to their long wavelengths (Takahashi and Fujii 2014).

Total geomagnetic intensity observations started in October 2009. Two total geomagnetic intensity sensors were installed at KMS and KUR by ERI. KMS lies on the southeast side of the summit crater, while KUR is about 3 km northeast. Therefore, it is a reasonable hypothesis that KUR and KMS have common large-scale variations of the geomagnetic field. A simple difference method was employed to estimate the volcanic component of continuous observations due to the volcanic activity in the summit area (in Fig. 5, green dots).

The geomagnetic total intensity due to the volcanic activity in the summit area exhibited an increasing secular trend over the entire period with a long dormant period from mid-2013 to mid-2015. Assuming the volcano-magnetic change represented the thermomagnetic effect, the increasing secular trend at KMS suggested an increase in the magnetization of the rocks beneath the summit crater Kamayama in response to cooling. The increment of the geomagnetic total intensity rested for 2 years from mid-2013, but began to increase just after the minor eruption on June 16, 2015. During this dormant period, the geomagnetic total intensity decreased by 10 nT until June 2015 compared with the linear secular trend from January 2010 to August 2013.

The GNSS baseline constructing stage was characterized by the small amount of SO_2 emissions and the restful VLP pulse activity. The geomagnetic total intensity increased linearly along the secular trend during the first half of the constructing stage, while the increment stopped completely in the last half of the constructing stage. This quiet interval of the total intensity variation was well synchronized with a period of the frequent occurrence of N-type earthquakes. The GNSS baseline turned to an extension from May 2015 and was followed by the rapid increment of SO_2 emissions and the activation of VLP. This resulted in an eruption on June 16, 2015. Afterward, the geomagnetic total intensity started to increase linearly along the secular trend. The gradient was similar to that before the quiet interval, and the activity of N-type earthquakes dropped dramatically.

The synchronization between the variation of the geomagnetic total intensity and the activity of N-type earthquakes supports the following hypothesis. A basic amount of volcanic gas was supplied to the conduit even in the contraction stage. Shielding in certain portions of the outgassing pathways began in mid-2013. This trapped high-temperature volcanic gas in the shielding area within the conduit. The trapped gas gradually inhibited cooling of the rocks beneath the summit crater, and N-type earthquakes occurred frequently in the shielding area. The north–south extension of the GNSS baseline at the western side of the volcano from May 2015 indicated a new dyke intrusion to the west of the summit, supplying more volcanic gas to the shallow region of the conduit. The intensive gas emission led to the minor eruption on June 16, 2015. After the eruption, the shielding portions were destroyed, and the volcanic gas passed through

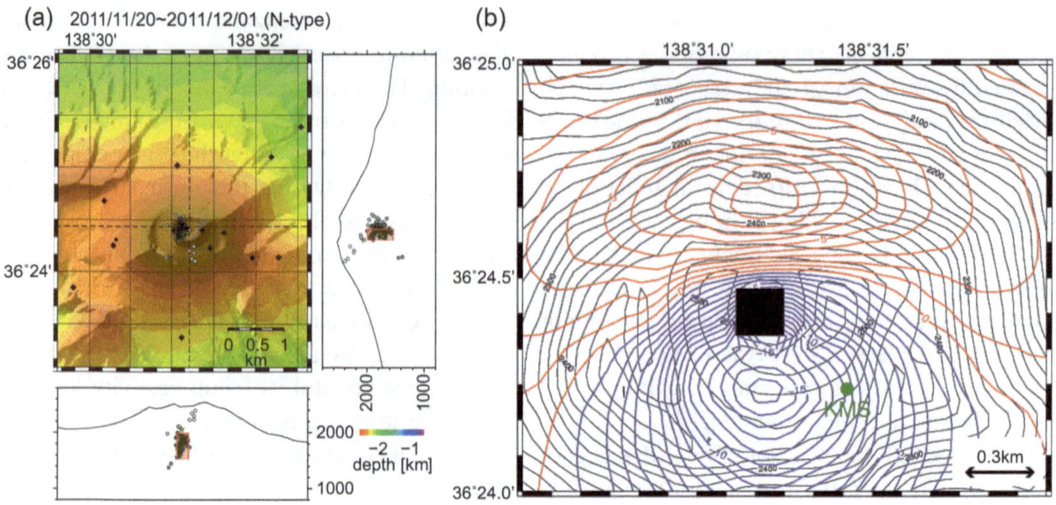

Fig. 10 (**a**) Distribution of the relocated hypocenters of N-type earthquakes occurring between November 20 and December 1, 2011, and the projection views of the cubic source on the north–south and the east–west cross sections. Pink shaded rectangles represent the projection views. Each hypocenter shown by a star is colored by its depth. Gray circles indicate other volcanic earthquakes occurring during the same period. (**b**) Spatial distribution of the geomagnetic total intensity anomaly for the cubic source model. Black rectangle represents the projection view of the cubic source. Contour interval is 1 nT

the conduit freely. Then the cooling process beneath the summit crater recovered, causing an increase in the geomagnetic total intensity along the secular trend. The dissolution of the shield also depressed the activity of N-type earthquakes.

Thermomagnetic modeling was conducted to verify this hypothesis. The concentric occurrences of N-type earthquakes in the last 2 months of 2011 provided a precise hypocenter distribution of N-type events (Fig. 10). We assumed that N-type events in the dormant period of the total intensity also occurred in the same area of the conduit and a pause in magnetization arose in this area. The change of the total intensity in the summit area was calculated by uniform demagnetization in a prism source with a width of 200 m and height of 500 m. The prism source was beneath the summit crater with a top elevation of 2000 m above sea level (Fig. 10a, pink shaded red rectangles). The magnetization intensity of the rocks in the source was assumed to be 2.2 A/m (Nakatsuka et al. 2009). A magnetic anomaly change on the surface induced by demagnetization in the prism source was exhibited by contours with intervals of 1 nT (Fig. 10b). The magnetic anomaly of −10 nT at KMS was consistent with the observed change of the geomagnetic total intensity over the entire dormant period. Although an important assumption is that the thermomagnetic effect yields the volcano magnetic change, the geothermal activity is the most intensive phenomenon in an active conduit of a volcano, including Asama volcano. Therefore, our modeling is plausible to explain the observed anomaly, and the hypothesis regarding the internal process within the conduit seems to be a feasible interpretation.

5 Summary

From the 2004 eruptions to the beginning of 2018, all eruptions occurred in the extension stages of the GNSS baseline change, indicating that the magma intrusion beneath the western flank of Asama volcano is an essential process for the volcanic activity in the last 20 years. The supplied amount of magma from the magma chamber was located about 8 km west of the summit and spanned to the dykes beneath the western flank of the volcano. The internal conditions in the shallow regions of the conduit seemed to control the eruptive style.

Acknowledgments We thank the Japan Meteorological Agency for access to their SO_2 emission data. We also thank the Japan Academy for permission to use several figures and contents from the paper entitled, "Recent volcanic activity at the Asama volcano and long-period seismic signals" by Takeo et al. in Proc. Jpn. Acad., Ser. B 98, 2022.

References

Aoyama H (2008) Simplified test on tilt response of CMG-40T seismometers. Bull Volc Soc Jpn 53:35–46

Aoyama H, Oshima H (2008) Tilt change recorded by broad-band seismometer prior to small phreatic explosion of Meakan-dake volcano, Hokkaido, Japan. Geophys Res Lett 35:L06307

Battaglia J, Aki K (2003) Location of seismic events and eruptive fissures on the piton de la Fournaise volcano using seismic amplitudes. J Geophys Res 108(B8):2364

Chouet B (1986) Dynamics of a fluid-driven crack in three dimensions by the finite difference method. J Geophys Res 91(B14):13967–13992

Chouet B (2003) Volcano seismology. Pure Appl Geophys 160:739–788

Chouet B, Page RA, Stephens CD, Lahr JC, Power JA (1994) Precursory swarms of long-period events at redoubt volcano (1989-1990), Alaska: their origin and use as a forecasting tool. J Volcanol Geotherm Res 62(1-4):95–135

Geno R, Ripepe M (2010) Inflation-deflation cycles revealed by tilt and seismic records at Stromboli volcano. Geophys Res Lett 37:L12302

Graizer V (2006) Tilts in strong ground motion. Bull Seismol Soc Am 96:2090–2102

Kazahaya R, Mori T, Takeo M, Ohminato T, Urabe T, Maeda Y (2011) Relation between single very-long-period pulses and volcanic gas emissions at Mt. Asama, Japan. Geophys Res Lett 38:L11307

Kazahaya R, Maeda Y, Mori T, Shinohara H, Takeo M (2015a) Changes to the volcanic outgassing mechanism and very-long-period seismicity from 2007 to 2011 at Mt. Asama. Jpn Earth Planet Sci Lett 418:1–10

Kazahaya R, Aoki Y, Shinohara H (2015b) Budget of shallow magma plumbing system at Asama volcano, Japan, revealed by ground deformation and volcanic gas studies. J Geophys Res Solid Earth 120:2961–2973

Kumagai H, Chouet BA (2002) Waveform inversion of oscillatory signatures in long-period events beneath volcanoes. J Geophys Res 107(B11):2301

Kumagai H, Chouet BA, Nakano M (2002) Temporal evolution of a hydrothermal system in Kusatsu-Shirane volcano, Japan, inferred from the complex frequencies of long-period events. J Geophys Res 107(B10):2236

Maeda Y, Takeo M (2011) Very-long-period pulsed at Asama volcano, Central Japan, inferred from dense seismic observations. Geophys J Int 185:265–282

Maeda Y, Takeo M, Ohminato T (2011) A waveform inversion including tilt: method and simple tests. Geophys J Int 184:907–918

Maeda Y, Takeo M, Kazahaya R (2019) Comparison of high- and low-frequency signal sources for very-long-period seismic events at Asama volcano, Japan. Geophys J Int 217:389–404

Nakano M, Kumagai H (2005) Response of a hydrothermal system to magmatic heat inferred from temporal variations in the complex frequencies of long-period events at Kusatsu-Shirane volcano, Japan. J Volcanol Geotherm Res 147(3-4):233–244

Nakano M, Kumagai H, Kumazawa M, Yamaoka K, Chouet BA (1998) The excitation and characteristic frequency of the long-period volcanic event: an approach based on an inhomogeneous autoregressive model of a linear dynamic system. J Geophys Res 103(B5):10031–10046

Nakatsuka T, Utsugi M, Okuma S, Tanaka Y, Hashimoto T (2009) Detection of aeromagnetic anomaly change associated with volcanic activity: An application of the generalized mis-tie control method. Tectonophysics, 478:3–18

Takahashi K, Fujii I (2014) Long-term thermal activity revealed by magnetic measurements at Kusatsu-Shirane volcano, Japan. J Volcanol Geotherm Res 285:180–194

Takeo M, Aoki Y, Koyama T (2022) Recent volcanic activity at the Asama volcano and long-period seismic signals. Proc Jpn Acad Ser B 98:416–438

Shallow Seismic Structure of Asama Volcano

Yosuke Aoki, Minoru Takeo, and Janire Prudencio

1 Introduction

One of the effective ways to understand how magma or hydrothermal fluid in active volcanoes is transported to the surface from depth is to understand the structure underneath the volcano's surface. Here, we investigate the shallow subsurface structure of Asama Volcano from active seismic sources and densely deployed temporal seismic network to gain more insights into the magma transport beneath Asama Volcano. Employing active sources can take advantage of known locations and origin time. See Aoki et al. (2009a, b) for experiment details.

Y. Aoki (✉)
Earthquake Research Institute, The University of Tokyo, Tokyo, Japan
e-mail: yaoki@eri.u-tokyo.ac.jp

M. Takeo
Earthquake Research Institute, The University of Tokyo, Tokyo, Japan

Japan Meteorological Agency, Tokyo, Japan

J. Prudencio
Department of Theoretical Physics and Cosmos, Universidad de Granada, Granada, Spain

Andalusian Institute of Geophysics and Earthquake Disaster Prevention, Universidad de Granada, Granada, Spain

2 Active Source Seismic Experiment

The seismic exploration was conducted in October 2006 with five dynamite sources of 250–300 kg and 464 temporary seismometers with a natural frequency of 2 Hz (Mark Products L22D; Fig. 1). Table 1 describes information on the active sources. Because seismic and geodetic observations had already indicated that a dike intrusion associated with the 2004 eruption took place to the ~5–8 km to the west of the summit (e.g., Takeo et al. 2006), we aligned seismometers striking roughly north-south and east-west, crossing the area of dike intrusion.

Each site recorded seismic waves by a sampling interval of 250 Hz. The observed waveforms indicate that seismic waves not traversing the area of dike intrusion have the maximum energy at first arrivals (Aoki et al. 2009a; Yamamoto and Sato 2010). For example, waves radiated from S1 with epicentral distances greater than 12 km (Fig. 2a) and those emitted from S2 with epicentral distances greater than 8 km (Fig. 2d) have more energy in later arrivals than the first arrival. This feature is due to the scattering of seismic waves at the area of dike intrusion where the seismic structure is inhomogeneous. Also, seismic waves traversing the summit are subject to substantial scattering. For example, waves radiated from S3 with epicentral distances greater than 5 km (Fig. 2e) and those emitted

S. Aramaki et al. (eds.), *The Asama Volcano*, Active Volcanoes of the World,
https://doi.org/10.1007/978-981-95-5799-8_17

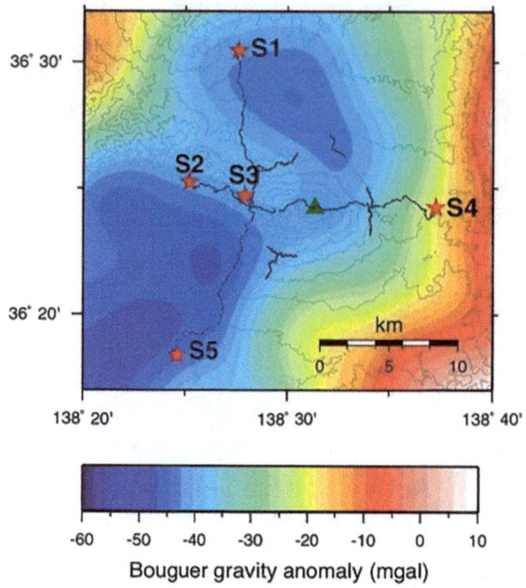

Fig. 1 Spatial distribution of active sources (red stars) and seismometers (black dots). The summit of Asama Volcano is shown by a green triangle. Bouguer gravity anomaly with a reduction density of 2670 kg/m^3 (Geological Survey of Japan 2000). Adopted and modified from Aoki et al. (2009a)

from S4 with epicentral distances greater than 10 km (Fig. 2f) have more energy in later arrivals than the first arrival because of the scattering.

3 P-wave Velocity Structure

As a first step to delineate the P-wave velocity structure, we manually picked the first arrivals of each trace. We left pickings with uncertainties of less than 0.1 s for the following analysis. This procedure resulted in 346 and 299 picks left along the north-south and east-west profiles, respectively. Then, we constructed P-wave velocity models to be consistent with first arrivals in each trace first by forward modeling (Zelt and Smith 1992) and then applying a regularized inversion of the travel times with the initial model obtained through the previous forward modeling (Zelt and Barton 1998). The forward modeling before the inversion was required because a small number of shots along each profile (three for each profile; Fig. 1b) does not allow convergence to

Table 1 Location and size of explosions

Shot	Latitude (°N)	Longitude (°E)	Altitude (m)	Depth (m)	Charge (kg)
S1	36.50761	138.46083	1210	60	300
S2	36.42072	138.42000	1730	60	300
S3	36.41211	138.46589	1847	60	250
S4	36.40431	138.62117	1171	60	300
S5	36.30626	138.40955	781	60	300

the optimum velocity structure without a reasonable initial model. Figure 3 shows the initial P-wave velocity and compares the observed and calculated travel times. It shows that the initial velocity model is already good regarding the fit between observed and estimated travel times, notwithstanding the starting model being spatially smooth enough not to produce an arbitrary result through an inversion.

Figure 4 depicts the P-wave velocity structure for both profiles, in which Fig. 4a, b compares the observed and calculated travel times and P-wave velocity structure along the north-south profile, respectively. Figure 4d, e depict those along the east-west profile, respectively. Figure 4a, d indicates that the calculated travel times fit well with the observed travel times. The root mean squares of the travel time residuals along the north-south and east-west profiles are 0.07 and 0.10 s, respectively (Aoki et al. 2009a).

Figure 4b clearly shows a high-velocity zone around S3 (distance = 0 km) at a depth of 2 km below sea level and shallower, the area of dike intrusion during the 2004 eruptions inferred from seismic and geodetic measurements (Aoki et al. 2005; Takeo et al. 2006). Figure 4e shows a high-velocity zone >3 km to the west of the summit. Combining this high-velocity body with the velocity structure along the north-south profile suggests that the high-velocity zone strikes east-west with its top deepening to the east. The north-south extent of the high-velocity zone is as narrow as approximately 5 km. We interpret that this high-velocity anomaly is due to solidified magma resulting from repeating dike intrusions because the location of the high-velocity body roughly coincides with the area of dike intrusion during the 2004 eruptions. We must note that a single dike intrusion does not form this high-

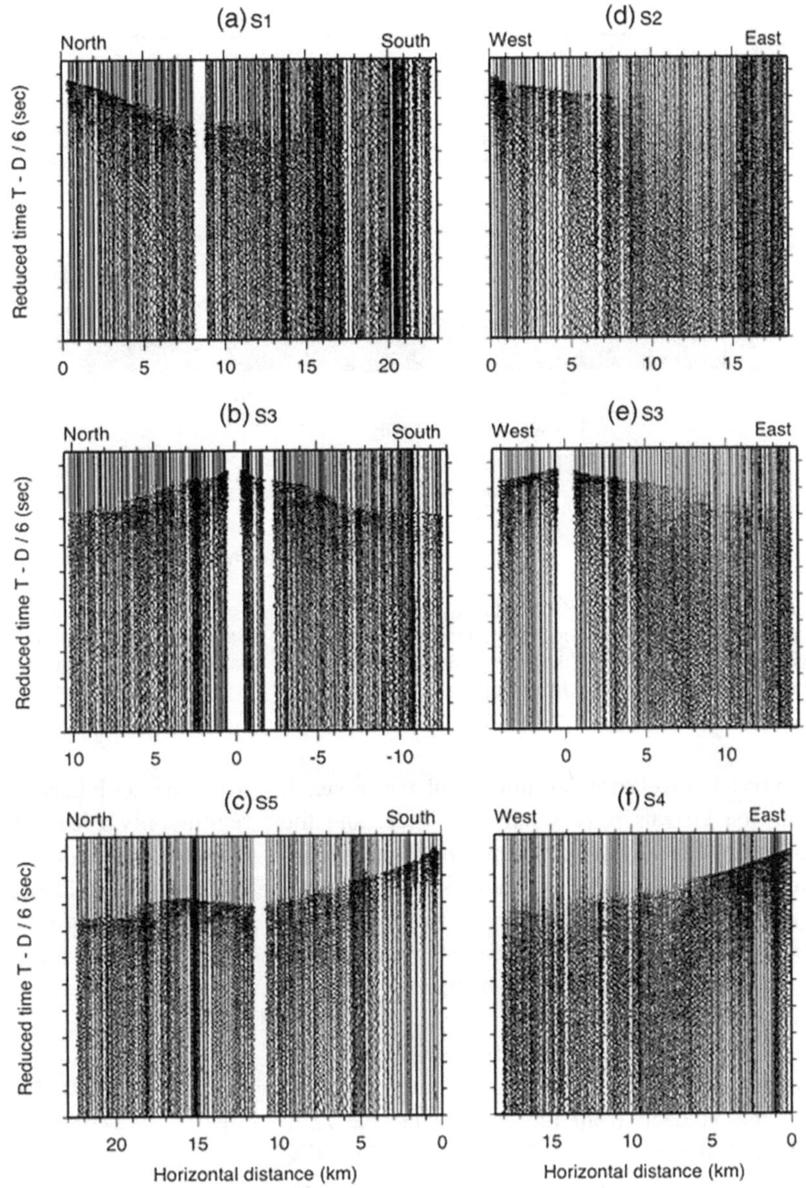

Fig. 2 Record sections of vertical seismograms along the the north–south profile ((**a**) for S1, (**b**) for S3, and (**c**) for S5, respectively) and east–west profile ((**d**) for S2, (**e**) for S3, and (**f**) for S4), respectively. Each trace is normalized by its maximum amplitude and reduced by 6 km/s. The horizontal and vertical axes represent the reduced time from the shot time and the horizontal distance from each shot point, respectively. Adopted from Aoki et al. (2009a)

velocity body. Geodetic data show that the dike intruded in 2004 is less than 1 m thick (Aoki et al. 2005; Takeo et al. 2006), too thin to be imaged in this analysis.

Figure 4c, f displays the ray paths calculated by Um and Thurber's (1987) method over the inferred P-wave velocity structure along the north-south and east-west profiles, respectively. They show that the ray coverage is good to a depth of at least 1–2 km below sea level for the north-south profile (Fig. 4c) and 1 km below sea level along the east-west profile, endorsing the robustness of the high seismic body to the west of the summit.

4 Scattering of Seismic Waves

The P-wave velocity structure, as described above (Aoki et al. 2009a, b), only explains the time of first arrivals. Although an artificial explosion represents the impulsive seismic source, the

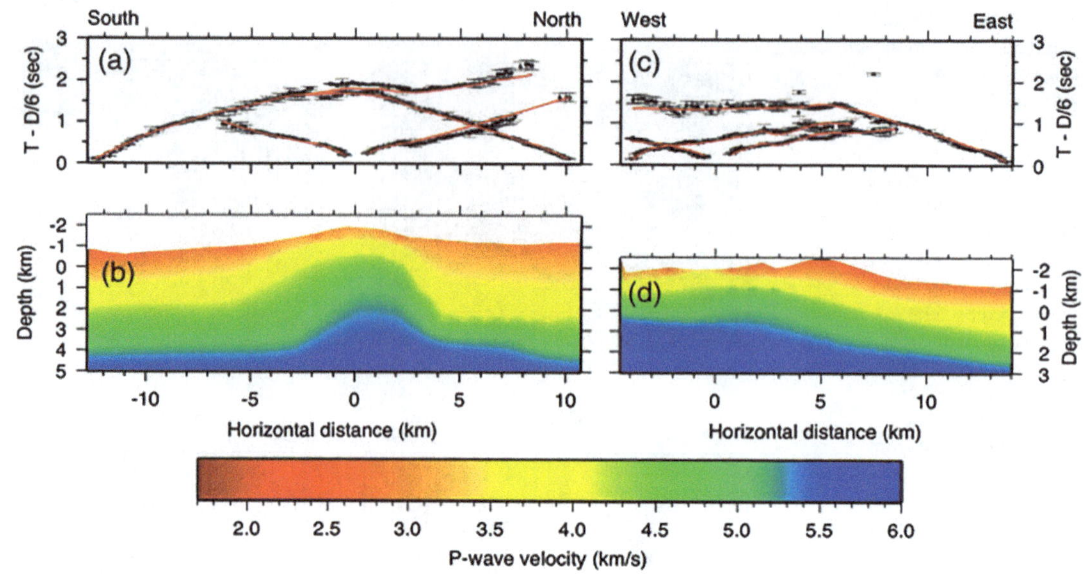

Fig. 3 (**a**) Comparison between the observed (black) and calculated (red) first arrival times along the north-south profile. Error bars denote the 1σ confidence interval. The travel times are reduced by 6 km/s. (**b**) Initial P-wave velocity structure along the north–south profile. (**c**, **d**) Same as (**a**, **b**), respectively, except that these are for the east–west profile. Adopted from Aoki et al. (2009a)

observed waveforms are not impulsive. Instead, the first arrivals have small amplitude, and the observed waveforms contain significant energy in the later arrivals and lack S-wave arrivals (Fig. 2). This observation is because the medium beneath the volcano is highly inhomogeneous, resulting in substantial scattering of seismic waves.

Yamamoto and Sato (2010) investigated the inhomogeneities of the shallow subsurface of Asama Volcano from the same data set. They modeled the seismic wavefield using the radiative transfer theory, including mode conversions between P- and S-waves through multiple anisotropic scattering. They inferred that the mean free paths of P- and S-waves at 8–16 Hz are 1.10 and 1.21 km, respectively. Because they assumed a uniform seismic velocity as Vs = 1.5 km/s and Vp/Vs = 1.73, where Vp and Vs represent P- and S-wave velocities, respectively, the corresponding mean free times of P- and S-waves are 0.41 and 0.78 s, respectively. These mean free paths are typical of volcanic subsurface; previous studies show that they range between 0.1 and 4.0 km. Also, the inferred mean free times indicate that the wavefield becomes the multiple scattering regime only a few seconds after an excitation of the wavefield. Yamamoto and Sato (2010) showed that the wavefield becomes an equilibrium state with a stationary ratio and P- and S-wave energies after ~8.8 s after excitation of the wavefield at a seismic site 1500 m from the epicenter.

5 Attenuation of Seismic Waves

Seismic waves attenuate not only by scattering but also by intrinsic dissipation. The former and latter are called scattering and intrinsic attenuations, respectively. Inhomogeneity and rheology control the scattering and intrinsic attenuations, respectively. Prudencio et al. (2017) separated the contribution between scattering and intrinsic attenuations from the observed seismogram of the active source experiment.

In contrast to Yamamoto and Sato (2010), who considered conversions between P- and S-waves, Prudencio et al. (2017) assumed the diffusion of seismic energy for simplicity. Their assumption

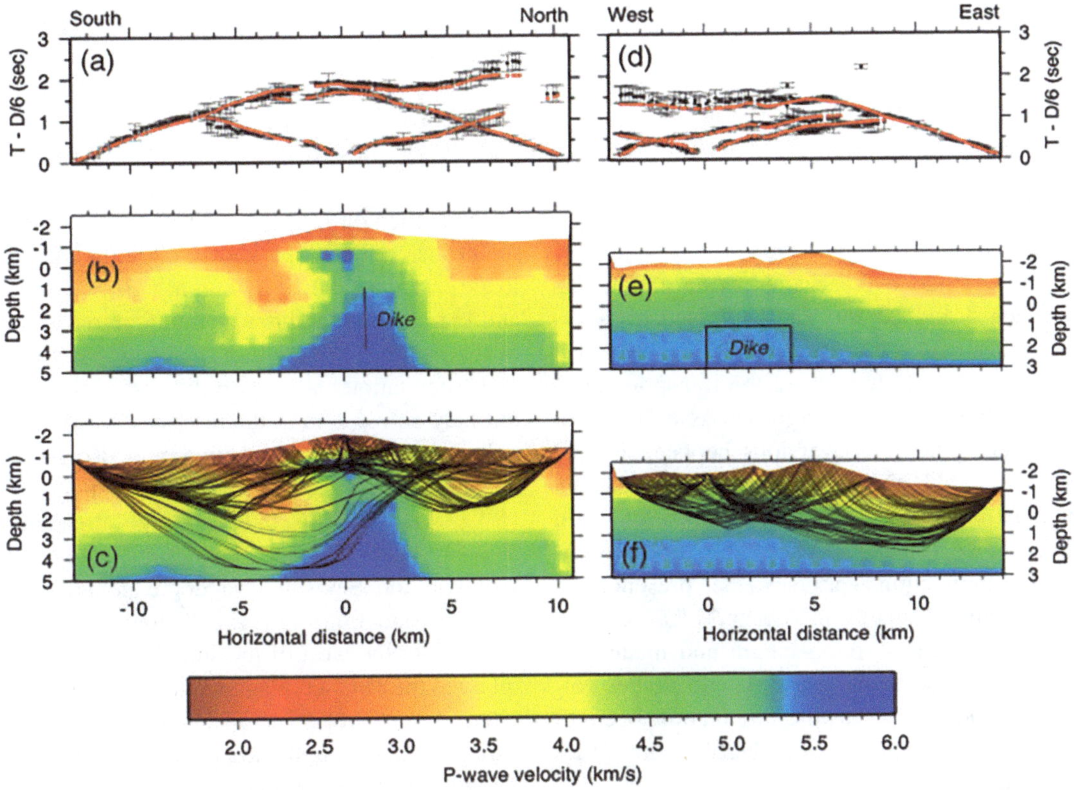

Fig. 4 (**a**) Comparison between the observed (black) and calculated (red) first arrivals along the north–south profile. Error bars denote the 1σ confidence interval. (**b**) P-wave velocity structure along the north–south profile. (**c**) Ray paths along the north–south profile calculated by the method of Um and Thurber (1987). (**d–f**) Same as (**a–c**), respectively, except that these are along the east–west profile. The approximate position of the intruded dike during the 2004 eruptions (Aoki et al. 2005; Takeo et al. 2006) is projected in (**b**) and (**e**). Adopted from Aoki et al. (2009a)

corresponds to the regime in which P- and S-wave energies are equipartitioned. The diffusion model describes that the seismic energy envelope E as a function of the distance between the source and the seismometer r and time t is given by

$$E = \frac{E_0}{(4\pi dt)^{-\frac{p}{2}}} \exp\left(-\frac{r^2}{4dt} - bt \right) \tag{1}$$

where E_0, d, and b represent the source energy, diffusivity, and lapse time from the origin time, respectively. p denotes the geometrical spreading terms; $p = 2$ and 3 correspond to the surface and body waves, respectively. This study employs $p = 3$ because the observed wavefield mainly comprises body waves. Intrinsic and scattering Quality factors, Q_i and Q_s, are given by

$$Q_i = \frac{2\pi f}{b} \tag{2}$$

and

$$Q_s = \frac{2\pi fpd}{v^2}, \tag{3}$$

respectively, where f and v represent frequency and the average S-wave velocity. This study assumes a constant velocity of $v = 1.5$ km/s, the average S-wave velocity at the top 2 km derived by Aoki et al. (2009a).

The total quality factor Q is given by $Q^{-1} = Q_i^{-1} + Q_i^{-1}$. The associated absorption length l_i and the transport mean free path l_{tr} are given by

$$l_i = \frac{3v}{b} \quad (4)$$

and

$$l_{tr} = \frac{3d}{v}, \quad (5)$$

respectively.

Table 2 lists average attenuation parameters between 4 and 20 Hz. The estimated transport mean free path is consistent with Yamamoto and Sato (2010), who evaluated the transport mean free path with a more rigorous assumption of considering mode conversions between P- and S-waves. Table 2 indicates that the scattering attenuation dominates the total attenuation, as in other active volcanoes. The dominance of scattering attenuation endorses the presence of strong inhomogeneity in Asama Volcano. Also, the estimated absorption length and mean free path are similar to those estimated in other volcanoes, indicating that the attenuation strength of Asama Volcano is comparable with other volcanoes.

Spatial variation of intrinsic and scattering attenuations was estimated by a back-projection method by Del Pezzo et al. (2016). While the obtained attenuation parameters from a single seismogram are most sensitive to those at the source and receiver, they approximated the weighting function as a two-dimensional Gaussian distribution with foci at the source and receiver for simplicity. Indeed, they justified this approximation through numerical experiments. See Del Pezzo et al. (2016) and Prudencio et al. (2017) for methodological details of the spatial mapping of the attenuation parameters.

Figures 5 and 6 depict spatial variations of perturbations from the average values (Table 2) of intrinsic attenuation, Q_i^{-1}, and scattering attenuation, Q_s^{-1}, respectively, at 4, 6, 8, 12, 16, and 20 Hz. These figures show that intrinsic and scattering attenuation perturbations do not depend much on frequency. The spatial variation of intrinsic attenuation shows high attenuations to the east and south of the summit, depicted as I1 and I3 in Fig. 5a, and low attenuation to the west of the summit, depicted as I2 in Fig. 5a. The I1 high attenuation anomaly to the east of the summit may correspond to the unconsolidated material ejected from Hotokeiwa and Ko-Asama volcanoes, which were active between 24 and 15 ka (Aramaki 1963). The I3 high attenuation anomaly to the south of the summit may correspond to the Sekison lava dome generated in 24 ka (Aramaki 1963). The I2 low attenuation anomaly to the west of the summit might correspond to the remnant magma intrusion during the activity of Kurofu volcano until 24 ka. The intruded magma might have cooled slowly to create dense and fractureless rocks. These rocks are imaged as a high-velocity (Aoki et al. 2009a) or high-resistivity (Aizawa et al. 2008) body.

The spatial variation of scattering attenuation shows high attenuation to the east (Sc1) and west (Sc2; Fig. 6). The Sc1 high attenuation body corresponds to the high intrinsic attenuation body to the east of the summit (I1; Fig. 5). Combining high intrinsic and scattering attenuations to the east of the summit confirms that these high attenuations are due to the unconsolidated material ejected from Hotokeiwa and Ko-Asama volcanoes, which were active between 24 and 15 ka (Aramaki 1963). The Sc2 high attenuation anom-

Table 2 The estimated average Q_i^{-1}, Q_s^{-1}, Q_t^{-1}, l_i, and l_{tr} between 4 and 20 Hz

Frequency (Hz)	Q_i^{-1}	Q_s^{-1}	Q_t^{-1}	l_i (m)	l_{tr} (m)
4	0.012 ± 0.005	0.12 ± 0.02	0.11 ± 0.02	5	718
6	0.010 ± 0.004	0.06 ± 0.01	0.07 ± 0.01	4	836
8	0.008 ± 0.003	0.04 ± 0.01	0.05 ± 0.04	4	806
12	0.006 ± 0.002	0.03 ± 0.05	0.04 ± 0.05	4	836
16	0.005 ± 0.002	0.02 ± 0.03	0.03 ± 0.02	4	896
20	0.004 ± 0.002	0.015 ± 0.003	0.019 ± 0.004	4	968

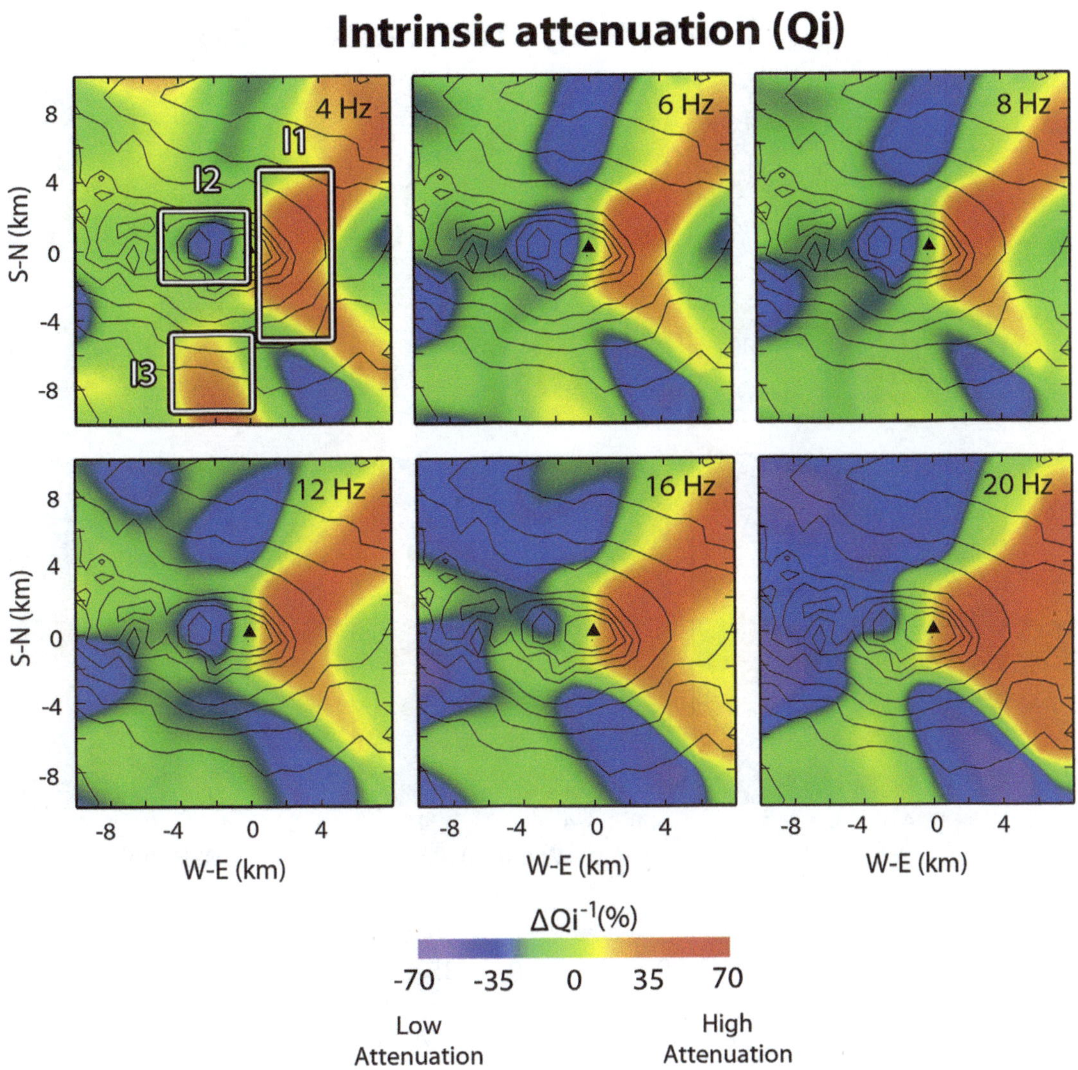

Fig. 5 Spatial variations of the perturbation from the average values of intrinsic attenuation, Q_i^{-1}, at (**a**) 4, (**b**) 6, (**c**) 8, (**d**) 12, (**e**) 16, and (**f**) 20 Hz. Black triangles in each panel denote the summit of Asama volcano. I1, I2, and I3 in (**a**) represent high (I1 and I3) and low (I2) attenuation regions which the main text describes in details. Adopted from Prudencio et al. (2017)

aly corresponds to the low intrinsic attenuation to the west of the summit (I2; Fig. 5). We interpret this contrasting attenuation property as due to the solidified magma during the activity of Kurofu volcano until 24 ka. The intruded magma has low intrinsic attenuation because of its slow solidification but high scattering attenuation because of its inhomogeneity.

6 Mt Asama's Shallow Seismic Structure

Active and passive seismic surveys found high-velocity bodies in many active volcanoes. As previous studies did, we interpret the high-velocity body as solidified magma due to previous magma intrusions. Here, we would like to note that the

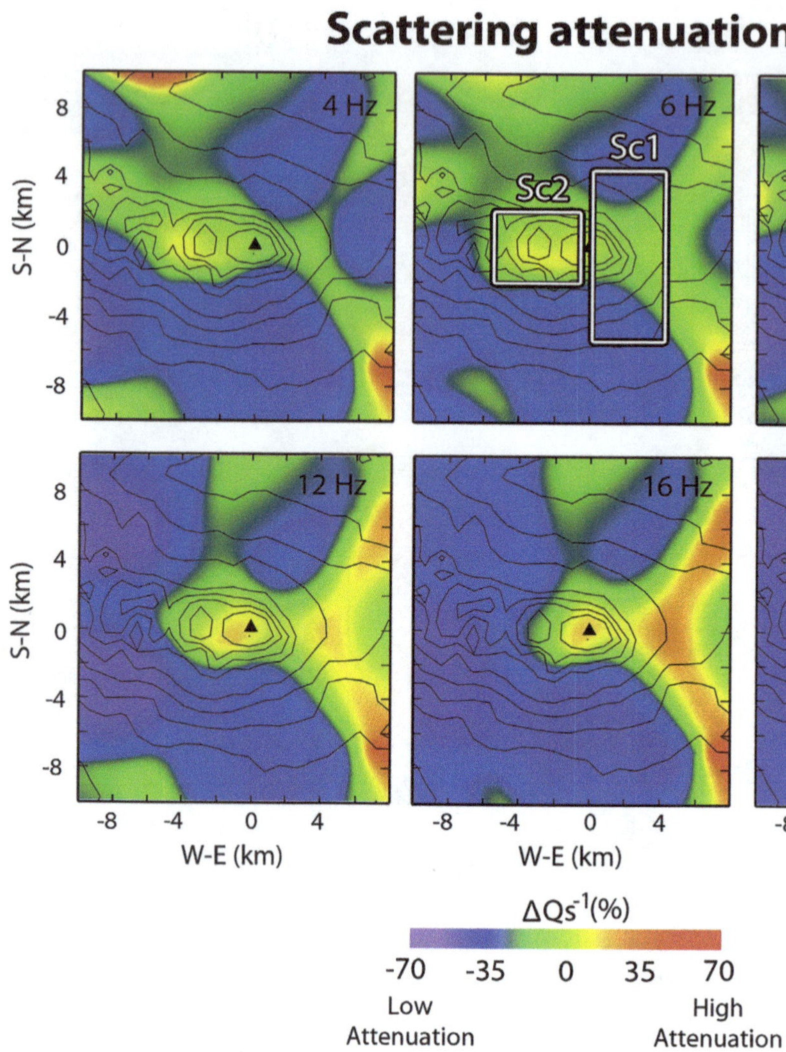

Fig. 6 Same as Fig. 5, but for the scattering attenuation, Q_s^{-1} . Sc1 and Sc2 in (b) denote high attenuations regions which the main text describes in details. Adopted from Prudencio et al. (2017)

high-velocity zone we found is consistent with the regional geology (Aramaki 1963) that Quaternary volcanoes distribute roughly east-west, the easternmost and youngest being Asama Volcano. Our result is also compatible with insights gained from other geophysical observations, i.e., precise earthquake locations, ground deformation, gravity anomaly, and resistivity structure.

Figure 1b indicates that the Bouguer gravity anomalies in the area have east-west trending local maxima to the west of the summit. This saddle-shaped gravity highly suggests that the high-velocity body corresponds to a high-density body, consistent with our hypothesis that the solidification of repeatedly intruded magma forms the high-velocity body. When magma is cooled slowly, it will become fractureless, thus dense, rock with high seismic velocity.

Electromagnetic surveys indicate that the high-velocity body to the west of the summit is less conductive than the surroundings (Aizawa et al. 2008; Usui et al. 2017). This resistivity structure is consistent with our hypothesis because slowly solidified magma has low permeability, resulting in high resistivity. Aizawa et al. (2008) attributed this high resistivity, extending up to about 1 km above sea level, to solidified magma associated with eruptions of Kurofu Volcano, which was active until 24,000 years ago (Aramaki 1963).

Our hypothesis that the high-velocity body is due to solidified magma implies that some magma should fail to make the surface during eruptions. In fact, geodetic data, field observations, and Airborne Synthetic Aperture Radar observations all suggest that in the 2004 eruptions, only a small fraction of intruded magma was ejected (Nakada et al. 2005; Oki et al. 2005). We speculate that in Asama Volcano, much of the intruded magma did not make the surface but was arrested and solidified in previous eruptions as well as recent ones.

7 Summary

We conducted an active seismic experiment with five dynamites and more than 400 temporary seismometers to delineate the seismic structure of the shallow subsurface of Asama Volcano. The analysis of first arrival times indicates a high P-wave velocity body to the west of the summit. This high-velocity body coincides with the area of the dike intrusion during 2004, 2008, and 2009 eruptions delineated by earthquake relocations and modeling of geodetic data. We interpret that this high-velocity, low intrinsic attenuation, and low scattering attenuation are due to the solidification of repeatedly intruded magma. Our interpretation is endorsed by the low conductivity in the area inferred from electromagnetic surveys.

References

Aizawa K, Ogawa Y, Hashimoto T et al (2008) Shallow resistivity structure of Asama volcano and its implications for magma ascent process in the 2004 eruption. J Volcanol Geotherm Res 173:165–177. https://doi.org/10.1016/j.jvolgeores.2008.016

Aoki Y, Watanabe H, Koyama E et al (2005) Ground deformation associated with the 2004–2005 unrest of Asama volcano, Japan. Bull Volcanol Soc Jpn 50:575–584. https://doi.org/10.18940/kazan.50.6_575

Aoki Y, Takeo M, Aoyama H et al (2009a) P-wave velocity structure beneath Asama volcano, Japan, inferred from active source seismic experiment. J Volcanol Geotherm Res 187:272–277. https://doi.org/10.1016/j.jvolgeores.009.09.004

Aoki Y, Takeo M, Morita Y, Tsutsui T (2009b) Active-source seismic experiment confirms the magma pathway of mount Asama, Japan. EOS Am Geophys Union 90:482–483. https://doi.org/10.1029/2009EO500002

Aramaki S (1963) Geology of Asama volcano. J Fac Sci Univ Tokyo Sec II 14:229–443

Del Pezzo E, Ibáñez J, Prudencio J et al (2016) Absorption and scattering 2-D volcano images from numerically calculated space-weighting functions. Geophys J Int 206:742–756. https://doi.org/10.1093/gji/ggw171

Geological Survey of Japan (2000) Gravity of Japan. Geological Survey of Japan (CD-ROM)

Nakada S, Yoshimoto M, Koyama E et al (2005) Comparative study of the 2004 eruption with old eruptions at Asama volcano and activity evaluation. Bull Volcanol Soc Jpn 50:303–313. https://doi.org/10.18940/kazan.50.5_303

Oki S, Murakami M, Watanabe N et al (2005) Topographic change of the summit crater of the Asama volcano during 2004 eruption derived from repeated airborne synthetic aperture radar (SAR) measurements. Bull Volcanol Soc Jpn 50:401–410. https://doi.org/10.18940/kazan.50.5_401

Prudencio J, Aoki Y, Takeo M et al (2017) Separation of scattering and intrinsic attenuation at Asama volcano (Japan): evidence of high volcanic structural contrasts. J Volcanol Geotherm Res 333:96–103. https://doi.org/10.1016/j.jvolgeores.2017.01.014

Takeo M, Aoki Y, Ohminato T, Yamamoto M (2006) Magma supply path beneath Mt. Asama volcano, Japan. Geophys Res Lett 33:L15310. https://doi.org/10.1029/2006GL026247

Um J, Thurber C (1987) A fast algorithm for two-point seismic ray tracing. Bull Seismol Soc Am 77:972–986. https://doi.org/10.1785/BSSA0770030972

Usui Y, Ogawa Y, Aizawa K et al (2017) Three-dimensional resistivity structure of Asama volcano revealed by data-space magnetotelluric inversion using unstructured tetrahedral elements. Geophys J Int 208:1359–1372. https://doi.org/10.1093/gji/ggw459

Yamamoto M, Sato H (2010) Multiple scattering and mode conversion revealed by an active seismic experiment at Asama volcano. Jpn J Geophys Res 115:B07304. https://doi.org/10.1029/2009JB007109

Zelt CA, Barton PJ (1998) Three-dimensional seismic refraction tomography: a comparison of two methods applied to data from the Faeroe Basin. J Geophys Res 103:7187–7210. https://doi.org/10.1029/97JB03536

Zelt CA, Smith RB (1992) Seismic travel time inversion for 2-D crustal velocity structure. Geophys J Int 108:16–34. https://doi.org/10.1111/j.1365-246X.1992.tb00836.x

Upper-Crustal Seismic Structure of Asama Volcano

Yosuke Aoki, Yutaka Nagaoka, Kiwamu Nishida, and Minoru Takeo

1 Introduction

In arc volcanoes, magma that ascends from depth is typically stored at a specific depth where buoyancy forces are negligible. The location where magma is stored is referred to as a magma reservoir. A precise understanding of the location and dimensions of these magma reservoirs within active volcanoes is crucial for comprehending and anticipating eruption behavior, as the state of the magma at its point of residence significantly influences the eruption style. However, they remain poorly understood, largely due to the limitations of conventional local earthquake tomography regarding imaging capability. Obtaining the optimal distribution of earthquakes for local earthquake tomography in volcanic regions is uncommon. Furthermore, it is challenging to identify reliable initial arrivals of S-waves, which are more susceptible to the presence of a magma chamber than P-waves. A robust low-velocity region with a radius of approximately 5 km is observed west of the summit. The technique of deriving Rayleigh wave phase velocities at a frequency of 0.1–0.2 Hz from seismic ambient noise records is emerging (e.g., Campillo and Roux 2015). The S-wave velocity structure derived from the phase velocities indicates the presence of a low-velocity body at depths of 5 km or greater. When our results are considered alongside those of seismic exploration and geodetic data, it becomes evident that the inferred low-velocity region represents a magma reservoir.

Asama Volcano has been the subject of rigorous monitoring due to its historically active status. The shallow magma plumbing system is well-defined. During the 2004 eruption, magma-filled cracks (dikes) intruded to a distance of approximately 5–7 km to the west of the summit, preceding the summit eruption (Takeo et al. 2006). An active source seismic experiment was conducted to image the area of the dike intrusion, which exhibited a high P-wave velocity anomaly (Aoki et al. 2009). The high-P wave velocity anomaly was interpreted as the remnant of repeating intrusions, indicating the persistence of the inferred magma pathway.

While the active source seismic experiment can resolve shallow seismic structures, it cannot resolve structures deeper than 3–4 km below sea level (Aoki et al. 2009). Conversely, a regional seismic tomography study visualizes broader-scale structures extending from the lower crust to

Y. Aoki (✉) · K. Nishida
Earthquake Research Institute, The University of Tokyo, Tokyo, Japan
e-mail: yaoki@eri.u-tokyo.ac.jp

Y. Nagaoka
Meteorological College, Japan Meteorological Agency, Chiba, Japan

M. Takeo
Earthquake Research Institute, The University of Tokyo, Tokyo, Japan

Japan Meteorological Agency, Tokyo, Japan

S. Aramaki et al. (eds.), *The Asama Volcano*, Active Volcanoes of the World,
https://doi.org/10.1007/978-981-95-5799-8_18

the upper mantle (e.g., Nakajima and Hasegawa 2010; Nakajima 2014). Nevertheless, the method cannot image the upper crust sufficiently to identify the locations of crustal magma chambers. Nagaoka et al. (2012) addressed this limitation by imaging the seismic structure of the upper crust beneath Asama Volcano using seismic interferometry with seismic ambient noise. For further technical details on seismic interferometry, see Campillo and Roux (2015). In this chapter, we describe how Nagaoka et al. (2012) imaged the seismic structure of the upper crust beneath Asama Volcano.

2 Data Analysis

The vertical seismograms utilized in this study were recorded between July 2005 and July 2007 at 89 stations. Of these, 18 employed broadband sensors, while the remaining 71 utilized short-period sensors with a natural frequency of 1 Hz. Figure 1 illustrates the spatial distribution of the seismic sites. The improved station coverage afforded by this period was utilized by temporally deploying 6 broadband and 11 short-period sensors. The entire region was initially divided into three subregions (Fig. 1) based on the prior knowledge of the spatial variation of seismic velocities from a previous study (Nishida et al. 2008). This earlier study exhibited a clear velocity contrast, with slower velocities to the northwest and faster velocities to the southeast. Subsequently, cross-correlations were conducted for all possible pairs within each subregion, employing appropriate preprocessing techniques. For further details, please refer to the paper by Nagaoka et al. (2012).

Figure 2a–c illustrate the cross-correlations as a function of interstation distance for each region. The figures demonstrate the propagation of Rayleigh wave trains across all regions, albeit at varying speeds. The primary wave packet traverses at approximately 3 km/s in Region 2, a considerably faster rate than in other regions, where it moves at approximately 2 km/s. This discrepancy is consistent with the findings of previous studies (e.g., Nishida et al. 2008). The wave packets travel a shorter distance in Region 3 (Fig. 2c) than in Region 2 (Fig. 2b) due to the

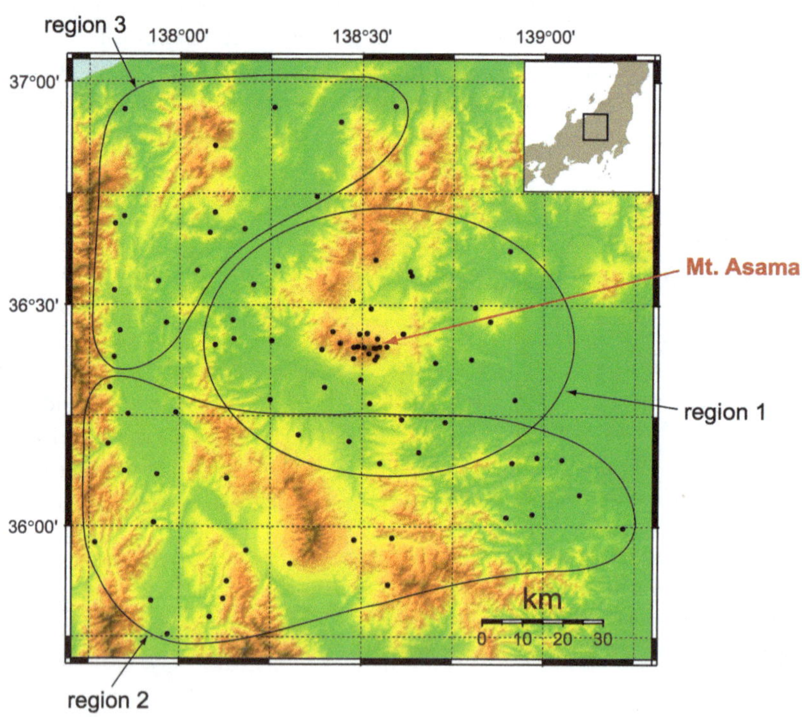

Fig. 1 Spatial distribution of seismic sites used in this study. Three regions for phase velocity measurements are circled. The location of Mt. Asama is also shown. The rectangle in the inset shows the the area of this study

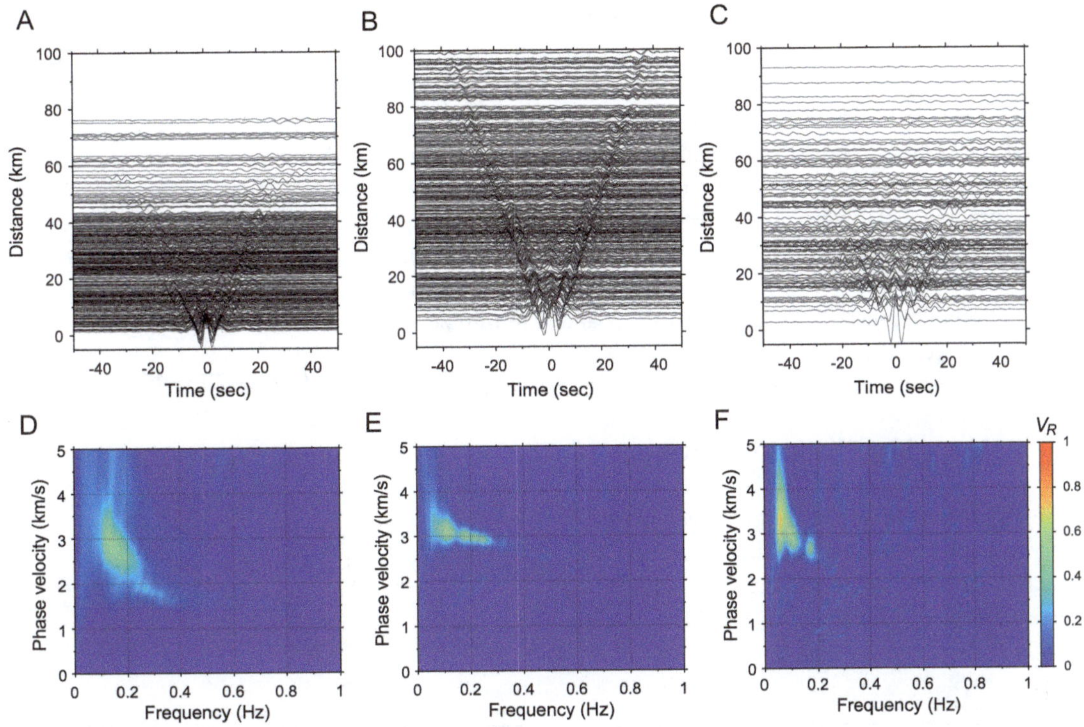

Fig. 2 An example of cross-correlations and variance reductions (V_R) in phase velocity measurements as a function of frequency and phase velocity. Top panels show an example of cross-correlations with a passband between 0.1 and 0.4 Hz in Regions (**a**) 1, (**b**) 2, and (**c**) 3, respectively. The bottom panels depict variance reductions of phase velocity measurements as a function of frequency and phase velocity in Regions (**a**) 1, (**b**) 2, and (**c**) 3, respectively

presence of greater intrinsic attenuation and scattering by thicker sediment, as evidenced by the presence of low-velocity bodies (Nishida et al. 2008). A visual inspection of Fig. 2 also demonstrates that the causal and acausal components of the cross-correlations are comparable, thereby supporting the quality of the obtained cross-correlations.

In each region, reference dispersion curves were measured on the assumption that (1) the seismic velocity within the region varies only with depth and (2) the ambient noise is azimuthally isotropic and uncorrelated. In that situation, the real part of the cross-spectrum Φ_{ij}^{syn} at given frequency ω is a function of the power spectrum $A(c,\omega)$, interstation distance r, and phase velocity c as (e.g., Aki 1957; Sánchez-Sesma and Campillo 2006)

$$\Phi^{syn}(c,\omega) = A(c,\omega) J_0\left(\frac{\omega r}{c}\right) \tag{1}$$

where A and c, the phase velocity, are unknowns. ij represents the pair of stations i and j. J_0 represents the zeroth-order Bessel function of the first kind. The reference dispersion curve for each region is defined to maximize the variance reduction, V_R, as (Fig. 2d–f)

$$V_R(c,\omega) = 1 - \frac{\sum_{i,j}\left(\Phi_{ij}^{obs}(c,\omega) - \Phi_{ij}^{syn}(c,\omega)\right)^2}{\sum_{i,j}\left(\Phi_{ij}^{obs}(c,\omega)\right)^2} \tag{2}$$

where the superscripts *obs* and *syn* represent the observed and calculated cross-spectrums, respectively. To estimate $A(c,\omega)$, the least-squares method is employed to maximize V_R, given c and ω. The obtained reference dispersion curves indicate that the phase velocity in Region 2 is faster than that in the other regions, consistent with the findings of a previous study (Nishida et al. 2008) and visual inspection of cross-correlations. The high density of seismometers in Region 1 permits the measurement of phase velocity up to 0.4 Hz, which is higher than that of the other regions. The lower density of seismometers in Regions 2 and 3 restricts the measurement of reliable phase velocities up to 0.3 and 0.25 Hz, respectively.

3 Mapping Phase Velocities

Once the reference phase velocity was derived, the travel time anomaly was measured as a perturbation from the reference phase velocity for all available station pairs (Fig. 3a–c). Phase velocity anomalies were measured for three frequency bands (0.1–0.2, 0.15–0.3, and 0.2–0.4 Hz) in Regions 1 and 2 and one frequency band (0.1–0.2 Hz) in Region 3, respectively.

We determined the phase shift between the observed and the synthetic waveforms derived from the reference dispersion curve in the frequency domain. The optimum phase velocity perturbation γ for each given frequency band minimizes

$$S(B,\gamma) = \sum_{\omega} \left(\Phi^{obs}(r,\omega) - B(\gamma)\Phi^{syn}(r,\omega,\gamma) \right)^2 \tag{3}$$

where

$$\Phi^{syn}(r,\omega,\gamma) = A(\omega) J_0\left(\frac{\omega r}{(1+\gamma)c_0(\omega)} \right) \tag{4}$$

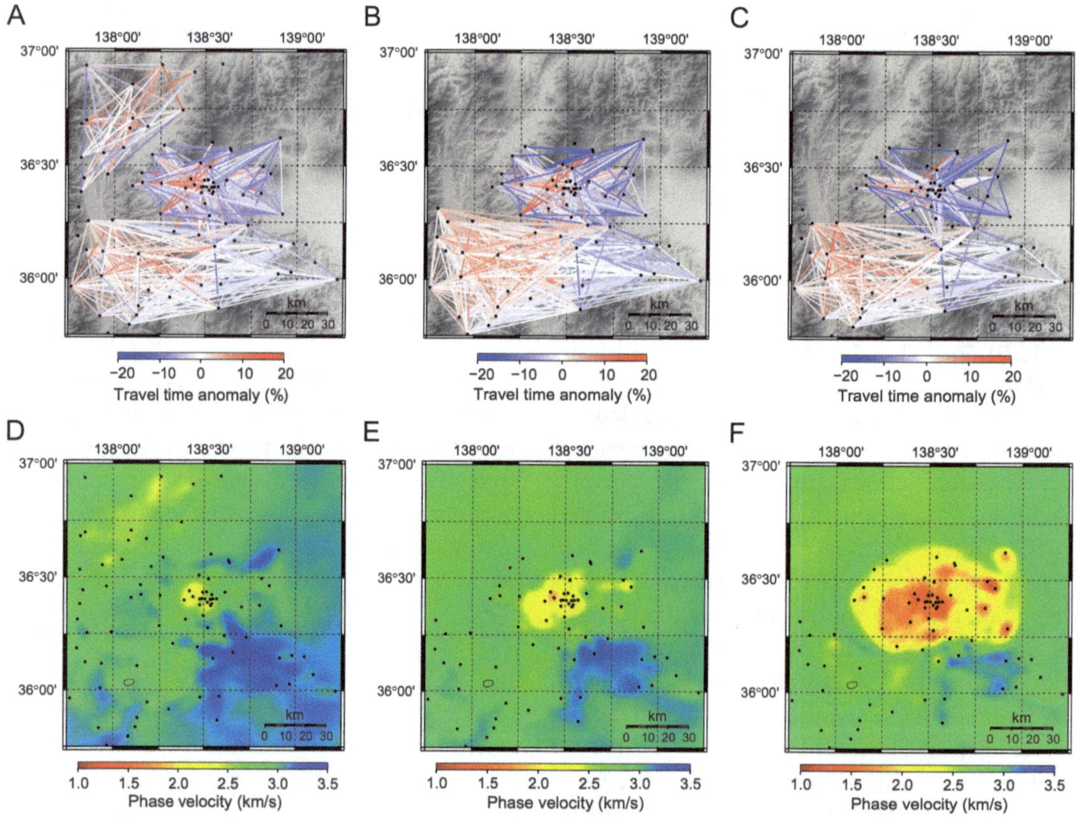

Fig. 3 Perturbations in travel times with respect to the reference dispersions and the obtained phase velocities. (**a**) and (**d**) represent the travel time perturbations from those expected from the reference dispersions and the obtained phase velocities at 0.1–0.2 Hz. (**b**) and (**e**) are the same as (**a**) and (**c**) but at 0.15–0.3 Hz. (**c**) and (**f**) are the same as (**a**) and (**c**) but at 0.2–0.4 Hz

c_0 represents the reference phase velocity at frequency ω, and B represents the amplitude correction factor to be estimated. We assume small velocity perturbations γ. The introduction of B signifies that the amplitude information of cross-correlations is disregarded. For further insight into the constraints imposed on the velocity perturbation γ, see Nagaoka et al. (2012).

Figure 3a illustrates that the travel time at a frequency range of 0.1–0.2 Hz is relatively slow to the west of the volcano, exhibiting a maximum deviation of 10%. This observation suggests the presence of a low-velocity body to the west of the summit. In light of these observations, we employed a nonlinear inversion technique to derive the spatial distribution of phase velocities with a spatial resolution of 0.03 degrees, or approximately 3 km. For further technical details, refer to the paper by Nagaoka et al. (2012).

Figure 3d illustrates that the Rayleigh wave phase velocity at a frequency of 0.1–0.2 Hz is approximately 10 km slower than the average value to the west of the summit, exhibiting a deviation of up to 20%. It is necessary to determine whether the local low phase velocity at 0.1–0.2 Hz is a genuine phenomenon or an artifact. Nagaoka et al. (2012) demonstrated that this velocity anomaly is a genuine phenomenon through resolution tests.

4 Three-Dimensional S-Wave Velocity Structure

The regionalized dispersion curves were inverted for local one-dimensional velocity models at each point using a method developed by Montagner (1986). Figure 4 illustrates the sensitivity kernels of the initial model at 36°24′ N 138°27′ E, where a low phase velocity anomaly was identified (Fig. 3). The sensitivity kernels are presented at 0.15, 0.225, and 0.3 Hz, assuming that the P-wave velocity and density are scaled as a function of the S-wave velocity. The derived sensitivity kernel (Fig. 4) indicates that lower frequencies are more sensitive to structures at depth. Figure 5 illustrates the three-dimensional S-wave velocities derived from the local one-dimensional S-wave velocity collection, which suggests the presence of a low S-wave velocity body at depths between 5 and 10 km. The low-velocity region to the west of the summit is no longer discernible at higher frequencies (Fig. 3e, f). Figure 5, however, depicts high S-wave velocity anomalies at depths shallower than 3 km near the repeating diking region, as inferred from active source seismic exploration (Aoki et al. 2009). The S-wave velocity structure is consistent with the P-wave velocity derived from an active source seismic experiment (Aoki et al. 2009, Fig. 4). Furthermore,

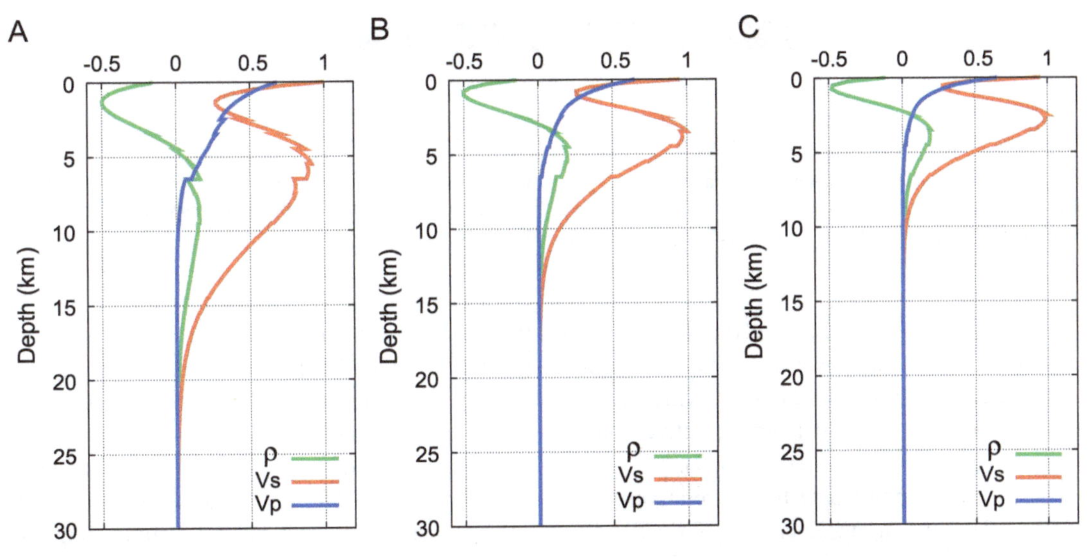

Fig. 4 Sensitivity kernels of density (green), S-wave velocity (red) and P-wave velocity (blue) at (**a**) 0.15 Hz, (**b**) 0.225 Hz, and (**c**) 0.3 Hz

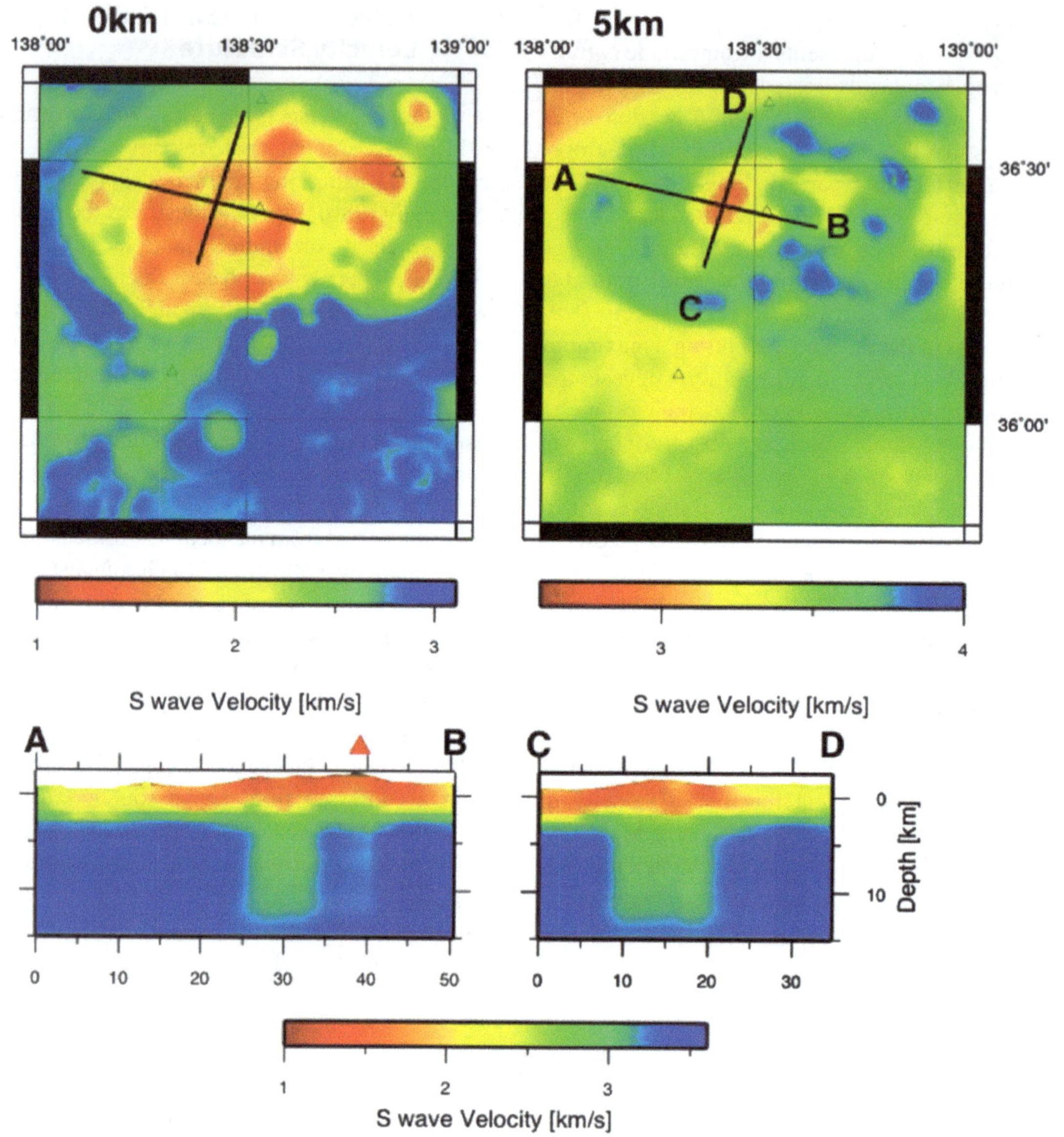

Fig. 5 S-wave velocity around Asama Volcano. The upper panels represent the plan view of the S-wave velocity at 0 and 5 km below sea level, respectively. The lower panels represent the vertical view of the S-wave velocity along the A–B and C–D profiles, respectively, shown in the upper panel

the inferred S-wave velocity at the depth of sea level demonstrates a slower phase velocity to the east of the summit than to the west, which is also consistent with the result from an active source seismic experiment (Aoki et al. 2009).

What factors contribute to the relatively low S-wave velocity ranging from 5 to 10 km? In this study, we propose that the observed characteristics indicate a magma reservoir supported by the available evidence. Tilt motions were observed during the February 2009 eruption, primarily from the horizontal record of broadband seismometers, which were utilized to capitalize on the capability of broadband seismometers to record tilt motions. In a previous study, Nagaoka et al. (2012) posited that the observed tilt motion was partially attributable to dike deflation occurring immediately above the low S-wave velocity anomaly identified in the present investigation. A more detailed discussion of this topic can be found in chapter "Ground Deformation at Asama Volcano Measured by Geodetic Data".

The observation of a dike participating in an eruption indicates that the low S-wave velocity anomaly represents a magma reservoir beneath the dike deflated during the 2009 eruption. Due to technical limitations, conventional seismic imaging studies have not imaged the crustal magma reservoir (Aoki et al. 2009) or only as a broad low-velocity zone (Nakajima and Hasegawa 2010; Nakajima 2014). In contrast, our method can delineate the size of a crustal magma chamber. Furthermore, the results demonstrate that ambient noise cross-correlations can image an anomaly with a radius of approximately 5 km at a depth of 5 km or greater, provided that optimal station coverage is attained.

5 Discussion

While many volcanoes possess a magma chamber directly beneath their summit, a magma chamber horizontally offset from the volcano's summit could be a ubiquitous feature. Indeed, tiltmeters at Kirishima volcano, located in southwest Japan, indicated a tilt toward a point situated approximately 8 km west of the summit in response to an eruption on January 26, 2011. This point aligns with the location of a deflation source derived from GPS observations. These observations demonstrate the ubiquity of a horizontally offset magma reservoir and indicate that the magma pathway of an active volcano is primarily subject to structural controls, resulting in a sinuous path that ultimately reaches the surface. The deployment of dense seismic instrumentation, as observed in the case of Asama volcano, will facilitate the detection of analogous magma reservoirs in other volcanoes, thereby enhancing our understanding of the magma transport processes occurring beneath active volcanoes.

References

Aki K (1957) Space and time spectra of stationary stochastic waves, with special reference to microtremors. Bull Earthq Res Inst Univ Tokyo 35:415–456. https://doi.org/10.15083/0000033938

Aoki Y, Takeo M, Aoyama H et al (2009) P-wave velocity structure beneath Asama volcano, Japan, inferred from active source seismic experiment. J Volcanol Geotherm Res 187:272–277. https://doi.org/10.1016/j.jvolgeores.009.09.004

Campillo M, Roux P (2015) Crust and lithospheric structure–seismic imaging and monitoring with ambient noise correlations. Treatise Geophys 1:391–417. https://doi.org/10.1016/B978-0-444-53802-4.00024-5

Montagner JP (1986) Three-dimensional structure of the Indian Ocean inferred from long period surface waves. Geophys Res Lett 13:315–318. https://doi.org/10.1029/10.1029/GL013i004p00315

Nagaoka Y, Nishida K, Aoki Y et al (2012) Seismic imaging of magma chamber beneath an active volcano. Earth Planet Sci Lett 333-334:1–8. https://doi.org/10.1016/j.epsl.2012.03.034

Nakajima J (2014) Seismic attenuation beneath Kanto, Japan: evidence for high attenuation in the serpentinized subducting mantle. Earth Planet Space 66:12. https://doi.org/10.1186/1880-5981-66-12

Nakajima J, Hasegawa A (2010) Cause of M~7 intraslab earthquakes beneath the Tokyo metropolitan area, Japan: possible evidence for a vertical tear at the easternmost portion of the Philippine Sea slab. J Geophys Res 115:B04301. https://doi.org/10.1029/2009JB006863

Nishida K, Kawakatsu H, Obara K (2008) Three-dimensional crustal S-wave velocity structure in Japan using microseismic data recorded by hi-net tiltmeters. J Geophys Res 113:B10302. https://doi.org/10.1029/2007JB005395

Sánchez-Sesma FJ, Campillo M (2006) Retrieval of the green's function from cross correlation: the canonical elastic problem. Bull Seismol Soc Am 96:1182–1191. https://doi.org/10.1785/0120050181

Takeo M, Aoki Y, Ohminato T, Yamamoto M (2006) Magma supply path beneath Mt. Asama volcano, Japan. Geophys Res Lett 33:L15310. https://doi.org/10.1029/2006GL026247

Leveling Observation of Asama Volcano

Masayuki Murase

1 Introduction

This chapter will introduce a vertical deformation detected by a repetitive survey of precise leveling in the Asama volcano. The GNSS and InSAR are useful techniques to detect volcanic deformation accurately. However, these techniques have only been around for the last 30 years or so.

A volcano generally alternates between an active period and a quiescent period. An eruption may occur after a quiescence period of over 100 years. Therefore, given the cycle of an active volcano eruption, an observation scheme should be considered for possible implementation in the medium to long term.

A leveling survey is a simple method to measure the relative height with high accuracy between two benchmarks. It has been conducted in Japan since the end of the nineteenth century. We can discuss a vertical deformation in Japan over 100 years using leveling data.

In and around the Asama volcano, a leveling survey started in 1902 and continues today. We use these data to consider vertical deformation and temporal volume change of pressure sources over 100 years.

2 Leveling Survey and Its Errors

A leveling survey is a method to measure the relative height between two benchmarks along a leveling route at a particular time. A vertical deformation is calculated as the difference in the relative height between two benchmarks from data measured at different times.

A leveling survey is a low temporal resolution measurement, because it takes time to conduct it on foot (2–4 km/day). On the other hand, the accuracy of vertical deformation obtained by a leveling survey is higher than that of GNSS and InSAR. We believe that a leveling survey is suitable for detecting deformations in the preparatory process of eruption, because such deformations may be small, slowly progressing, and localized to a very small region around the volcano.

Supplementary Information The online version contains supplementary material available at https://doi.org/10.1007/978-981-95-5799-8_19.

M. Murase (✉)
Department of Earth and Environmental Sciences, College of Humanities and Sciences, Nihon University, Tokyo, Japan
e-mail: murase.masayuki@nihon-u.ac.jp

S. Aramaki et al. (eds.), *The Asama Volcano*, Active Volcanoes of the World,
https://doi.org/10.1007/978-981-95-5799-8_19

Here, we consider an error in the leveling. It is difficult to calculate the error of a leveling survey, and we estimate a standard deviation using closures of round-trip measurements (e.g., Yamamoto 2022). Since an error of a leveling survey increases with the square root of the measurement distance, the equation of the standard deviation e_i (mm) on the distance from the benchmark (BM) 0 (reference point) to BM i as

$$e_i = \sigma_0 \sqrt{\sum_{j=1}^{i} S_j}$$

with

$$\sigma_0 = \sqrt{\frac{1}{4} \sum_{j=1}^{n} \left(\frac{\mu_j^2}{S_j} \right) \frac{1}{n}}$$

where σ_0 (mm/km$^{1/2}$) is a coefficient of a standard deviation, S_j (km) is a measurement distance between BM $j - 1$ and BM j, μ_j is closures of round-trip measurements between point $j - 1$ and point j, n is number of benchmarks (j = 1, 2,..., n). There are other ways, omitted details here, to calculate σ_0 using loop closures of the leveling survey.

Using the nationwide precise leveling survey conducted by the Geospatial Information Authority of Japan (GSI), the Geodetical Society of Japan (1974) calculated the values of σ_0 to be 1.77, 1.50, 1.85, and 1.44 mm/km$^{1/2}$ from 1883 to 1913, from 1923 to 1945, from 1946 to 1962, and from 1962 to 1968, respectively. Murase et al. (2023) calculated σ_0 to be 0.33, 0.34, and 0.53 mm/km$^{1/2}$ using the precise leveling data around the Asama volcano (Kurumazaka route) in 2019, 2021, and 2022, respectively. Although errors sometimes become larger due to deterioration in measurement conditions due to bad weather, they generally decrease over time due to technological advances.

The expected maximum error was estimated using the above equation, assuming the case where the value of σ_0 was 1.85 from 1946 to 1962, and the distance was 27 km on the longest leveling route in our analysis. A maximum error of only 9.6 mm was estimated, and this value was considered sufficiently small not to affect the results significantly. Therefore, we concluded that the results were accurate enough to detect a greater than one-centimeter deformation.

3 Leveling Route and Vertical Deformation

As shown in Fig. 1, six leveling routes exist in and around the Asama volcano. The survey periods of the precise leveling used in our analysis during 1902–2022 are shown in Fig. 2. The location, observation history, and detected deformation of each leveling route are as follows.

3.1 Route 18

In the southern region of the Asama volcano (Route 18), the Military Land Survey conducted a precise leveling survey to prepare the national large-scale map in 1902; thereafter, its successor, GSI, has repeated the surveys ten times. The results of the first-order leveling surveys, including the relative height in Route 18, were downloaded from the GSI website (https://vldb.gsi.go.jp/sokuchi/level/KENSOKUSYUROKU/).

The temporal changes in the vertical deformation along Route 18 exhibit a significant uplift from 1902 to 1952 and a moderate subsidence from 1952 to 1992 (Fig. 3a, b). The deformation showed a short-term uplift from 1992 to 2001, returning to a moderate subsidence from 2001 to 2010.

BM 549 exhibits the largest deformation of approximately 6 cm, and BM 546 shows only a temporal change of less than 1 cm. The deformations along Route 18 strongly correlate with distance from the mountain summit. The temporal change in the vertical deformation also strongly correlates with the number of eruption days per year (Fig. 3c). Therefore, these deformations were due to the volcanic activity in the Asama volcano.

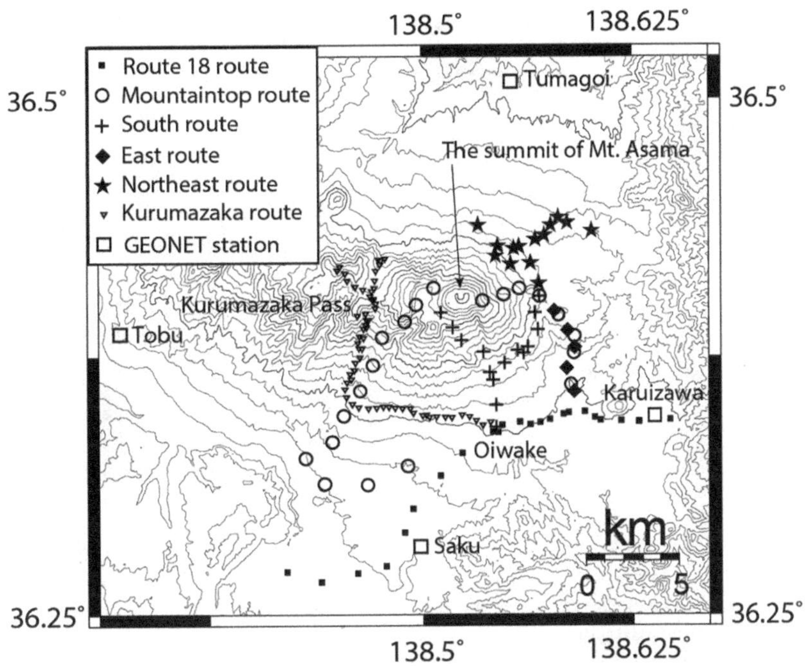

Fig. 1 Location of Mount Asama volcano showing benchmarks for the precise leveling and GEONET stations. The symbols with square, open circle, plus sign, diamond, star, and inverted triangle denote the locations of the benchmarks in the Route 18, the Mountaintop route, the South route, the East route, the Northeast route, and the Kurumazaka route, respectively. The open squares indicate the location of the GEONET stations

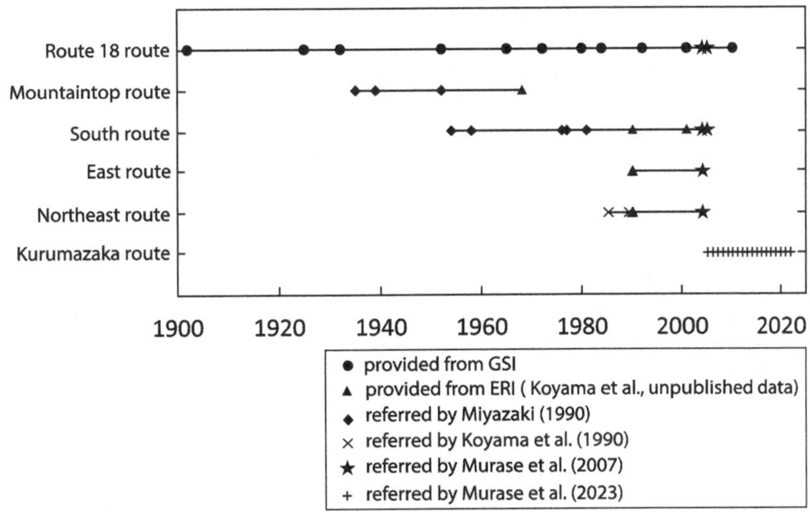

Fig. 2 The survey periods of the precise leveling from 1902 to 2022. The symbols with circle, triangle, diamond, cross mark, star, and plus sign denote the time when the leveling was performed

3.2 Mountaintop Route

To determine a detailed spatial distribution of the vertical deformation in Asama volcano, Earthquake Research Institute (ERI), University of Tokyo, established a leveling route along a trail up the mountain (Mountaintop route) in 1934; the leveling surveys were conducted in 1935, 1939, and 1950/1953 (Takahashi 1936; Miyazaki 1990).

From 1935 to 1939, uplift around the southwest part of the mountain was detected (Fig. 4a). Then, a significant uplift around the summit and subsidence around the southwest part of the mountain were detected from 1939 to 1950/53 (Fig. 4b). They are valuable near-field data observed at the risk of eruption.

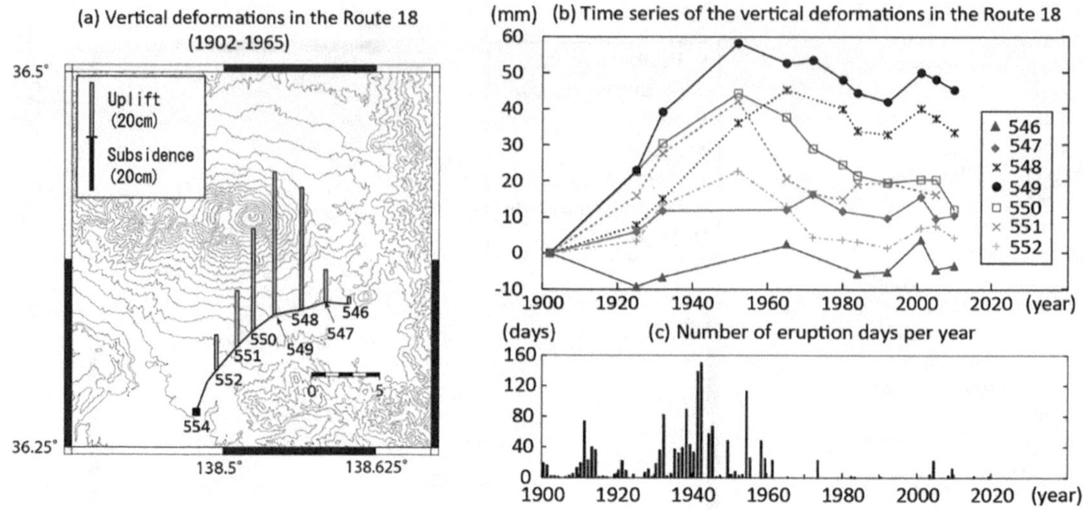

Fig. 3 Vertical deformations on Route 18. (**a**) The spatial distribution of the vertical deformations during 1902–1965 on Route 18. The black square denotes the reference point of deformations. (**b**) Time series of vertical deformation along Route 18 from 1902 to 2010. Relative deformation was measured with respect to the reference point (benchmark 554). (**c**) The number of eruption days per year from 1900 to 2022

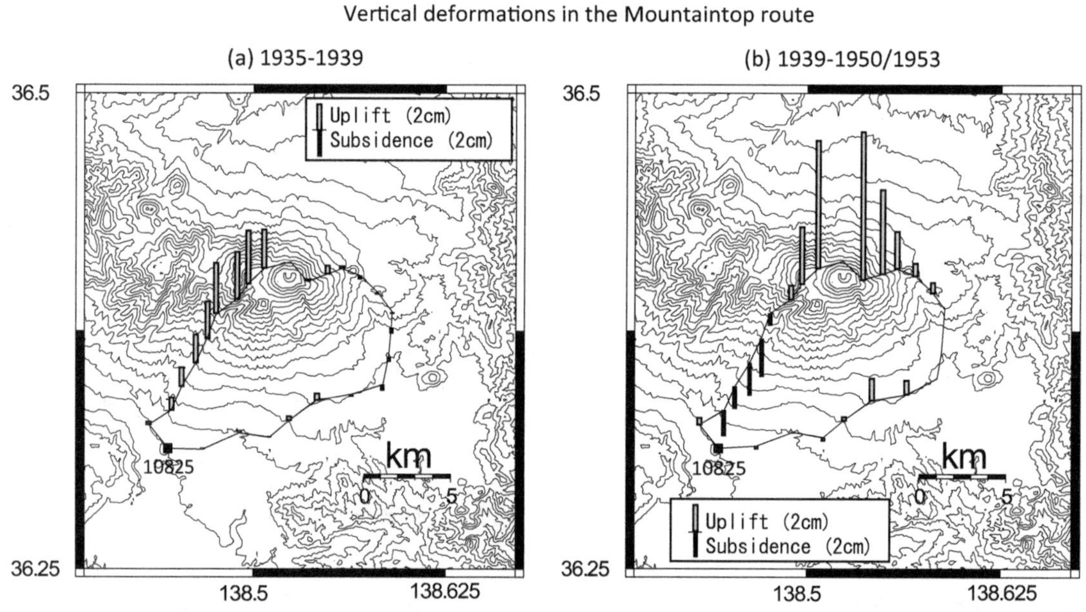

Fig. 4 Vertical deformations in the Mountaintop route. (**a**) The spatial distribution of the vertical deformations during 1935–1939. Black square (benchmark 10825) denotes the reference point of deformations. (**b**) The spatial distribution of the vertical deformations during 1939–1950/1953

3.3 South Route

On the route over the southern slope of the mountain (South route), deformations were detected in 1954, 1958, 1976, 1977, and 1981 (Miyazaki et al. 1978; Miyazaki 1990).

Uplift and subsidence were detected at the southwest part of the mountain from 1954 to

1976 and from 1976 to 1981 (Figs. 5a, b). In BM10823, located at the southeast part of the volcano, a moderate subsidence was continued from 1954 to 2010 (Fig. 5b). It is worth noting that different temporal deformation changes were detected in the volcano's southeast and southwest parts.

3.4 Northeast Route and East Route

The ERI established a new observation route for leveling and electro-optical distance measurement along the northeast of the mountain (Northeast route) in 1977; the leveling surveys were conducted twice a year during the period from 1977 to 1990 (Koyama and Takeda 1998).

The temporal change of the Northeast route from 1977 to 1990 has been described by Koyama et al. (1990). The temporal changes in the vertical deformation exhibit a slight uplift during the period from 1977 to 1982 and a subsidence during the period from 1983 to 1987. And then, the subsidence became small and went back to slight uplift from 1988 to 1990. From 1977 to 1990, deformations were 1 cm or less than 1 cm on the Northeast route.

The numerical data on the Northeast route from 1977 to 1990 were not published, and we only use deformation from 1985 to 1989 in our analysis (Fig. 6a). After the 2004 eruption event, Nagoya University and ERI surveyed around the northeast toward the south of Asama volcano (Route 18, South, East, and Northeast routes). They detected the deformations from 1990 to 2004 (Fig. 6b).

3.5 Kurumazaka Route

During the 2004 eruption, a dike intrusion was estimated west of the summit based on the deformation detected by the GPS measurements (Aoki et al. 2005; Murakami 2005). The deformation due to the dike intrusion was too small to be detected. Nagoya University and ERI established a leveling route along the Kurumazaka Pass (Kurumazaka route) in 2005 and conducted the survey of this route; Nihon University and the

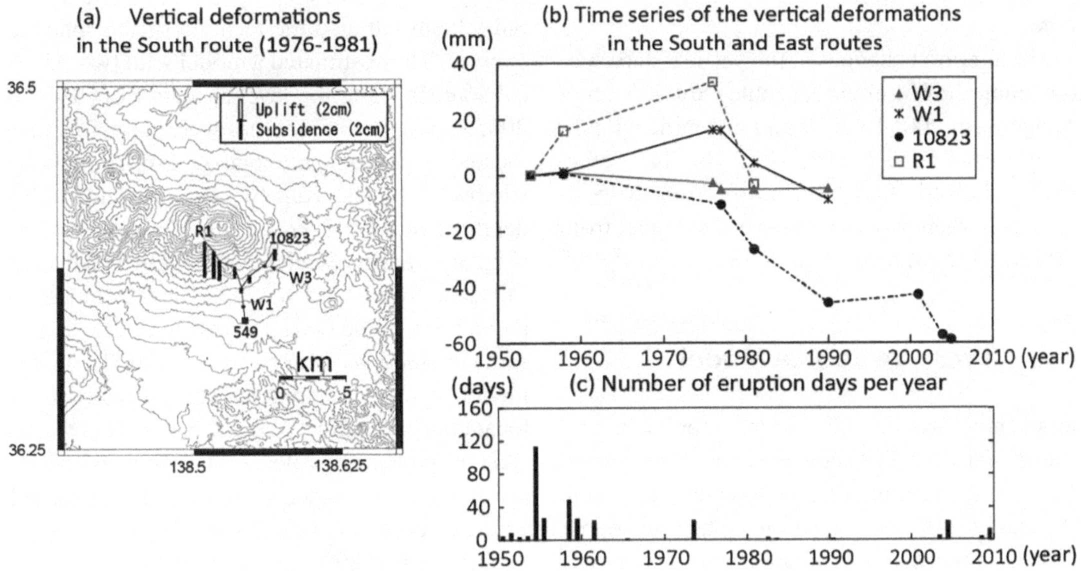

Fig. 5 Vertical deformations in the South route. (**a**) The spatial distribution of the vertical deformations during 1976–1981 in the South route. The black square denotes the reference point of deformations. (**b**) Time series of vertical deformation along the South route from 1954 to 2005. Relative deformation was measured with respect to the reference point (benchmark 549). (**c**) The number of eruption days per year from 1950 to 2010

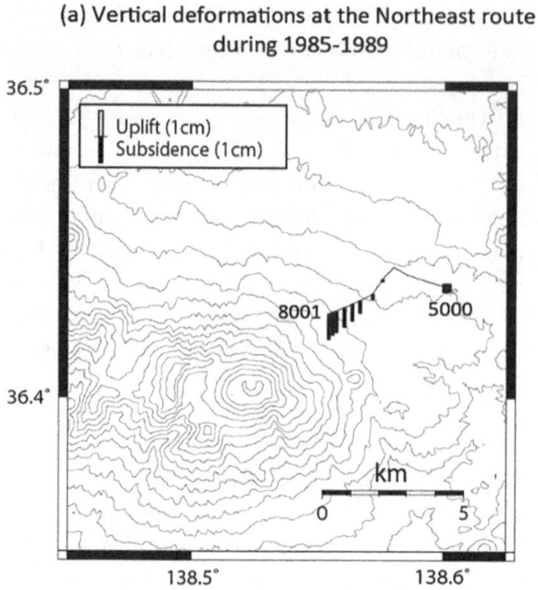

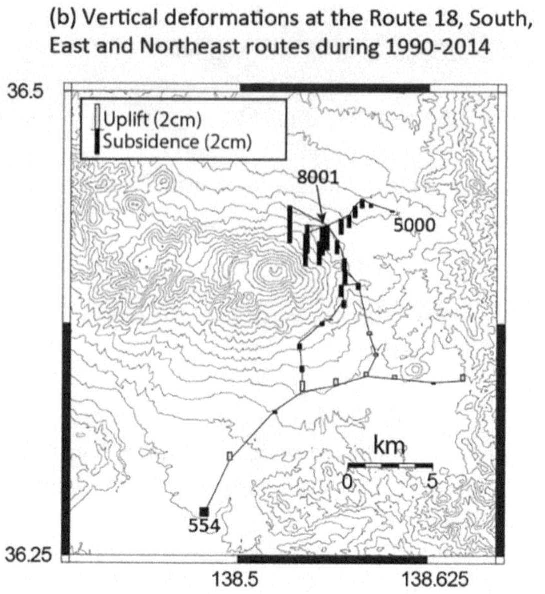

Fig. 6 (**a**) Vertical deformations in the Northeast route during 1985–1989. Black square (benchmark 5000) denotes the reference point of deformations. (**b**) Vertical deformations in the Route 18, South, East, and Northeast routes during 1990–2014. Black square (benchmark 554) denotes the reference point of deformations

Japan Meteorological Agency have succeeded in this survey and carried out it almost once a year until now (2022) (Table S1 lists the relative heights of benchmarks on the Kurumazaka route).

The temporal changes in the vertical deformation along the Kurumazaka route exhibit a small change from 2005 to 2009 and a significant subsidence from 2009 to 2015 (Fig. 7b). The deformation showed short-term uplift from 2015 to 2017 and then went back to subsidence from 2017 to 2022 (Murase et al. 2023).

4 Pressure Source Model

Some pressure source models for spherical source and dike intrusion have been proposed. Based on the leveling data in Mountaintop route, Miyazaki (1990) estimated one spherical source located at the depth of about 5 km (below sea level) beneath the Kurofu volcano (at 2 km to the west of the summit) during the period from 1935 to 1939 and the other one located at the depth of about 0 km beneath the Maekake volcano (summit) during the period from 1939 to 1950/1953 (Fig. 8a). Since Miyazaki (1990) calculated only the depth of the source at the fixed location for each period, Murase et al. (2007) calculated not only depth but also the location of the spherical sources. They estimated a model with two spherical sources using the leveling data from 1902 to 2005. The spherical sources were estimated to be located at a depth of about 6.5 km, about 4 km southwest of the Kurofu volcano, and with a depth of about 1.5 km at the Maekake volcano (Fig. 8a). In this research, we freshly added leveling data from 2005 to 2022, and estimate the parameters of the two spherical sources using the same method as Murase et al. (2007). In this model, one of the sources was estimated to be located at a depth of about 7.5 km, about 2 km west of the Kurofu volcano; the other source was estimated to be located close to the Maekake volcano at a depth of about 2 km (Fig. 8a).

During the 2004 eruption event, GPS measurements estimated a dike intrusion west of the summit (Aoki et al. 2005; Murakami 2005). The leveling data from 1990 to 2004 include the deformation due to this dike intrusion. However,

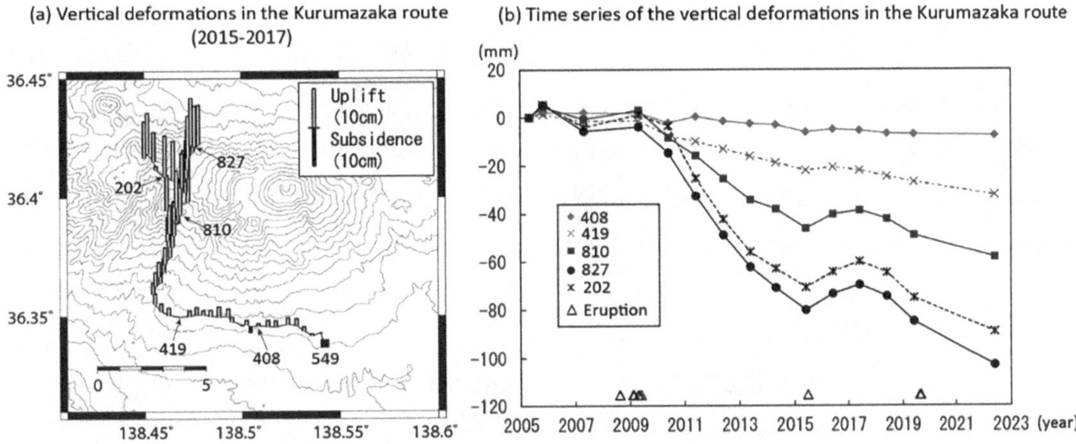

Fig. 7 Vertical deformations in the Kurumazaka route. (**a**) The spatial distribution of the vertical deformations during 2015–2017 in the Kurumazaka route. The black square denotes the reference point of deformations. (**b**) Time series of vertical deformation along the Kurumazaka route from 2005 to 2022. Relative deformation was measured with respect to the reference point (benchmark 549). Triangle symbols denote the day when the eruption occurred

benchmarks were placed only in the regions at a distance exceeding 10 km from the dike. We checked the influence of the leveling data due to dike intrusion using optimal parameters of the dike estimated by Aoki et al. (2005). The vertical deformation of less than 1 mm was expected, and we concluded that the deformation due to dike intrusion can be negligibly small.

Oikawa et al. (2006) discussed the hypocenter distribution of volcanic earthquakes during the dormant period from 1996 to 1999. The earthquake depths beneath the Maekake volcano are shallower than 0 km (Fig. 8a). The region around the magma source is subject to low seismicity due to its high temperature. Therefore, we suggest a magma source beneath the Maekake volcano.

Based on electromagnetic measurements, the low resistivity region in the Kurofu volcano was estimated to be located at a depth of approximately 10 km (Ogawa and Hashimoto 2006). Our estimated source is located in the upper section of the low-resistivity region. This region indicates the existence of fluids; generally, it is considered to be indicative of the existence of magma in the deeper regions of an active volcano. Based on these results, we propose the existence of a magma reservoir beneath the Kurofu volcano.

5 Temporal Volume Change of Pressure Sources Estimated Using Precise Leveling Data

Murase et al. (2007) estimated the temporal volume change of two pressure sources using precise leveling data, assuming that two stable spherical sources existed simultaneously from 1902 to 2005. In our analysis, the temporal volume change of two pressure sources was estimated by Murase et al. (2007) method from 1902 to 2022. The vertical deformations at the Northeast route from 1980 to 1989, Route 18 from 2000 to 2010, and the Kurumazaka route from 2005 to 2022 were new additions to the data.

The temporal evolution of the spherical source beneath the Kurofu volcano can be explained as follows (Fig. 8b). From 1902 to 1943, a large inflation of 7.8×10^7 m^3 (0.5–4.2×10^6 m^3/year) is estimated. After 1943, a deflation of -3.8×10^7 m3 (-2.8 to -0.03×10^6 m^3/year) continued until 1964, and then two small inflation events occurred around 1976 and 2005. This inflation changed to deflation in 2005, which continued until 2022.

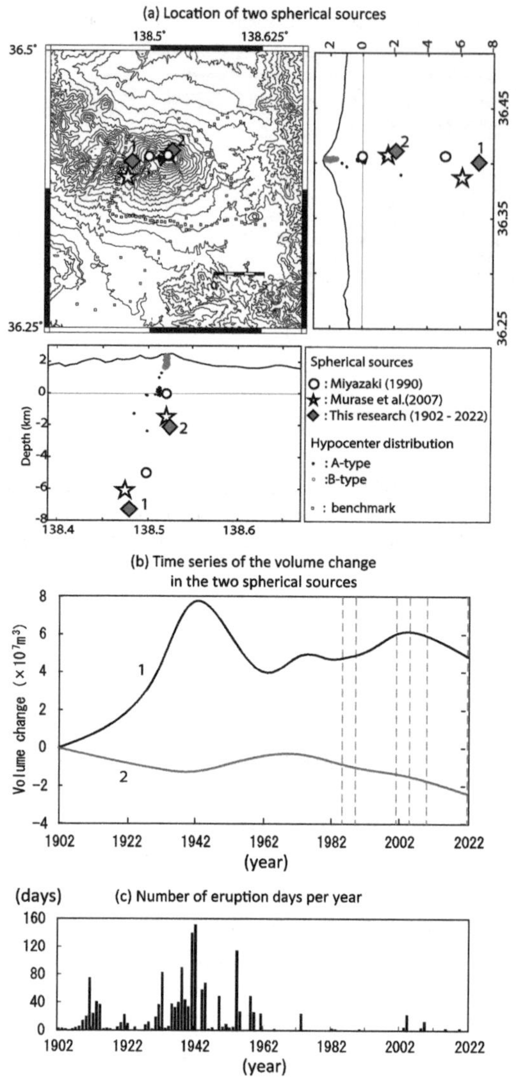

Fig. 8 The estimated two spherical sources. (**a**) Location map with a cross-sectional view of the estimated sources, including the hypocenters of volcanic earthquakes (Oikawa et al. 2006). The black and white dots denote the hypocenters of A- and B-type volcanic earthquakes, respectively. The circles and stars denote the locations of spherical sources estimated by Miyazaki (1990) and Murase et al. (2007). The diamond marks of 1 and 2 denote the location of spherical sources beneath Kurofu and Maekake volcanoes, respectively. (**b**) Volume change in the two spherical sources. The black and gray lines denote the volume change in the spherical sources 1 and 2, respectively. The dashed lines denote the observation time of the newly added data. (**c**) The number of eruption days per year from 1902 to 2022

The number of eruption days increases from 1930 to 1942 and decreases from 1943 to 1961 (Fig. 8c). The temporal change of the number of eruption days corresponds with the temporal volume change in the source beneath the Kurofu volcano. After 1962, the Asama volcano appeared more dormant than in the first half of the twentieth century. The small inflations were estimated before the medium and small-scale eruptions occurred in 1973 and 2004.

The volume change in the source beneath the Maekake volcano is about five times smaller than that in Kurofu volcano. Not only the Kurofu volcano but also the Maekake volcano appear to have changed the volume change rate due to the eruptions in 1973 and 2004. Since the eruption frequencies are high when compared with the observed frequencies around the Maekake volcano, we consider it challenging to detect relations between eruption and volume change in the first half of the twentieth century.

The temporal change in pressure source beneath the Kurofu volcano strongly correlates with the eruption frequency. The inflation is high during the active period from 1930 to 1942. The leveling data do not suggest a significant volume change during the dormant period of the latter half of the twentieth century. We proposed that the magma supply increased from 1930 to 1940, which resulted in the active period.

6 Summary

The six types of leveling routes have been established around the Asama volcano. Route 18 had the longest history among the leveling routes. The temporal changes in the vertical deformation along Route 18 exhibit a significant uplift from 1902 to 1952 and a moderate subsidence from 1952 to 1992. The deformation showed a short-term uplift from 1992 to 2001, returning to a moderate subsidence from 2001 to 2010.

Based on the leveling data, the model with two spherical sources was estimated, and the temporal volume change of these two pressure sources was estimated from 1902 to 2022.

From 1902 to 1943, a large inflation is estimated in the spherical source beneath the Kurofu volcano. After 1943, a deflation continued until 1964, and then two small inflation events occurred around 1976 and 2005. This inflation changed to deflation in 2005, which continued until 2022. The volume change in the source beneath the Maekake volcano is about five times smaller than that in the Kurofu volcano. Not only the Kurofu volcano but also the Maekake volcano appear to have changed the volume change rate due to the eruptions in 1973 and 2004.

The temporal change in pressure source beneath the Kurofu volcano strongly correlates with the eruption frequency. We proposed that the magma supply increased from 1930 to 1940, which resulted in the active period.

Acknowledgments We thank Dr. Yosuke Aoki for his careful reading of our manuscript and his numerous constructive suggestions that helped to improve this paper. We gratefully acknowledge the Geospatial Information Authority of Japan for providing us with the leveling data on Route 18. We are also grateful to the Japan Meteorological Agency for providing us with the number of eruption days. We would like to thank Dr. Fumiaki Kimata, Dr. Hitoshi Y. Mori, and the staff of the Japan Meteorological Agency for their support of the leveling survey on the Kurumazaka route. Some figures were prepared using generic mapping tools (Wessel and Smith 1995).

References

Aoki Y, Watanabe T, Koyama E, Oikawa J, Morita Y (2005) Ground deformation associated with the 2004–2005 unrest of Asama volcano, Japan. Bull Volcanol Soc Jpn 50(6):575–584

Geodetical Society of Japan (1974) Overview of geodesy, 511

Koyama E, Takeda T (1998) Improving method of Leveling surveys at the Asama volcano. Tech Res Rep Earthq Res Inst Univ Tokyo 4:1–6

Koyama E, Takeda T, Gyoda N, Tsuji H, Ida Y (1990) Crustal deformation surveys on the Northeastern flank of Asama volcano. Progr Abstr Volcanol Soc Jpn 2:74

Miyazaki T (1990) Recent crustal movements of active volcanoes in Japan as revealed by leveling survey. Bull Earthq Res Inst Univ Tokyo 65(3):669–807

Miyazaki T, Gyoda N, Koyama E, Osada N (1978) Leveling surveys at Asama volcano. Bull Earthq Res Inst Univ Tokyo 53:487–495

Murakami M (2005) Magma plumbing system of Asama volcano inferred from continuous measurements of GPS. Bull Volcanol Soc Jpn 50(5):347–361

Murase M, Endo Y, Sano T, Iijima S, Otsuka Y, Teguri Y, Kudo N (2023) Deformations Detected by Precise Levelling Surveys in the Asama volcano (2005–2022), Japan Geoscience Union Meeting 2023 abstract

Murase M, Ono K, Ito T, Miyajima M, Mori H, Aoyama H, Oshima H, Yoshida Y, Terada A, Koyama E, Takeda T, Watanabe H, Kimata F, Fujii N (2007) Time-dependent model for volume changes in pressure sources at Asama volcano, Central Japan due to vertical deformations detected by precise leveling during 1902-2005. J Volcanol Geotherm Res 164:54–75

Ogawa Y, Hashimoto T (2006) Asama volcano EM field experiment group Magnetotelluric observation across the dyke intrusion of Asama volcano. Japan Geoscience Union Meeting abstract, V101-019

Oikawa J, Ida Y, Tsuji H (2006) Precise hypocenter determination for volcanic earthquakes observed at Asama volcano, Japan. Bull Volcanol Soc Jpn 51(2):117–124

Takahashi R (1936) Precise leveling at the Asama volcano. Bull Earthq Res Inst Univ Tokyo 14:18–25

Wessel P, Smith WHF (1995) New version of the Generic Mapping Tools released. Eos, Transactions of the American Geophysical Union 76, 329

Yamamoto K (2022) Introduction to the precise Leveling survey: toward the detection of minute vertical ground deformation associated with volcanic activity. Bull Volcanol Soc Jpn 67(2):221–232

Ground Deformation at Asama Volcano Measured by Geodetic Data

Yosuke Aoki, Minoru Takeo, Takao Ohminato, Yutaka Nagaoka, and Xiaowen Wang

1 Introduction

Volcanic unrest often accompanies pressurization, transport, or both of magmatic or hydrothermal fluids, resulting in Earth's deformation measured at the surface. While volcano deformation processes have multiple time scales, the shortest of which is less than a second, this chapter deals with deformation of time scales longer than a few minutes, which can be detected by geodetic measurements, such as Global Navigation Satellite System (GNSS), Synthetic Aperture Radar (SAR), tiltmeters, and broadband seismometers.

This chapter overviews insights gained from observations with such geodetic data. Leveling surveys have also been conducted in the Asama volcano (Murase et al. 2007) to measure longer-term deformation, but this chapter does not deal with leveling surveys. Chapter "Leveling Observation of Asama Volcano" discusses ground deformation with a time scale longer than a few years measured by leveling surveys.

Y. Aoki (✉) · T. Ohminato
Earthquake Research Institute, The University of Tokyo, Tokyo, Japan
e-mail: yaoki@eri.u-tokyo.ac.jp

M. Takeo
Earthquake Research Institute, The University of Tokyo, Tokyo, Japan

Japan Meteorological Agency, Tokyo, Japan

Y. Nagaoka
Meteorological College, Japan Meteorological Agency, Chiba, Japan

X. Wang
Department of Surveying and Geo-Informatics, Faculty of Geosciences and Environmental Engineering, Southwest Jiaotong University, Chengdu, China

2 Inflation and Deflation of Asama Volcano Measured by GNSS

2.1 Ground Deformation Before the 2004 Eruption

The Asama volcano has been monitored by over ten Global Navigation Satellite System (GNSS) sites, as illustrated in Fig. 1. The data recorded by each GNSS site is updated every 30 seconds, with the site position calculated daily (every 24 h). Figure 2 illustrates that the baseline length between two GNSS sites traversing the Asama volcano fluctuates over time, exhibiting a correlation with seismic activity. The number of volcano-tectonic earthquakes increased by a factor of two on two occasions: from October 2000 to April 2001 and from May to August 2002, preceding the 2004 eruption. The occurrence of both seismic unrest and the extension of GPS baselines indicates a potential correlation between the volcano-tectonic earthquakes and magma intru-

S. Aramaki et al. (eds.), *The Asama Volcano*, Active Volcanoes of the World,
https://doi.org/10.1007/978-981-95-5799-8_20

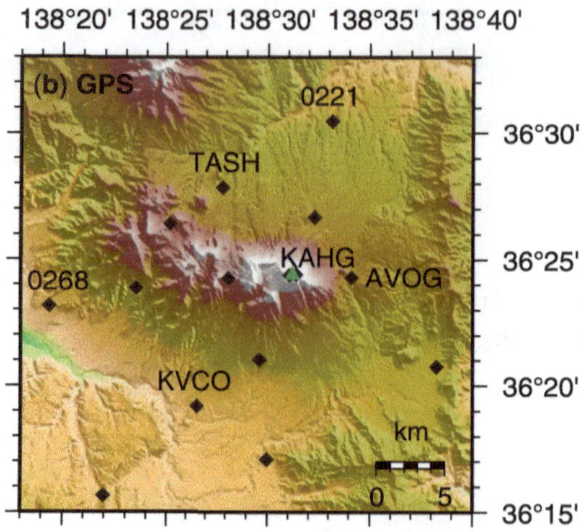

Fig. 1 Spatial distribution of GNSS sites around Asama Volcano. Adopted from Aoki et al. (2013)

sion beneath the western flank of the Asama volcano.

The GPS baseline contraction was observed three times before the 2004 eruption. The first instance occurred before March 2000, the second between July 2001 and February 2002, and the third between March 2003 and April 2004. The observed contraction is of a magnitude that cannot be attributed to tectonic motion of non-volcanic origin. It is therefore postulated that the contraction is due to the downward migration of intruded magma to the west of the Asama volcano or the viscoelastic relaxation of the host rock in response to the magma intrusion during the unrest. Seismicity was relatively low during the initial contraction period but relatively high during the subsequent two periods (Fig. 2a). The maximum temperature of the crater exceeded 200 °C since autumn 2002, leading to a minor eruption in February 2003, indicating that the shallow portion of the vent was activated at that time. These observations suggest significant differences between the initial and subsequent contractions.

2.2 Ground Deformation Associated with the 2004 Eruption

Aoki et al. (2005, 2013) employed a modeling approach to investigate the ground deformation field between June 2004 and March 2005, encompassing the 2004 eruption. This involved assuming the intrusion of a dike with a uniform opening into an elastic, homogeneous, and isotropic medium. Neither a spherical source nor a fault dislocation model can adequately explain the observed ground deformation pattern. Instead, the observations are better explained by a dike intrusion to the western flank of Mt. Asama (Fig. 3). The estimated volume of the intrusion was 6.8×10^6 m^3, which is approximately three times larger than the volume of the ejected magma during the eruption (2×10^6 m^3: Nakada et al. 2005). Although the observed deformation is not sufficiently large to constrain all of the model parameters with precision, the horizontal location and the total volume of the intruded dike are relatively well constrained, with uncertainties of ±2 km and ±30%, respectively. The eastern portion of the modeled dike aligns with the western end of the relocated hypocenter distribution, as documented by Takeo et al. (2006; Chapter 15). The depth at the top of the dike, for which a standard deviation of 1.3 km has been estimated, coincides with the depth range of volcano-tectonic earthquakes. The distribution of dike-induced seismicity reflects the distribution of ambient stresses near failure. Thus, the region near the dike tip is characterized by a concentration of earthquakes.

The GPS network indicated a sudden increase in the baseline length between 0221 and MIYO (Fig. 1b) between 21 and 22 July 2004. This suggests a nearly vertical magma intrusion to the western flank of Mt. Asama at that time (Aoki et al. 2005, 2013; Murakami 2005; Takeo et al.

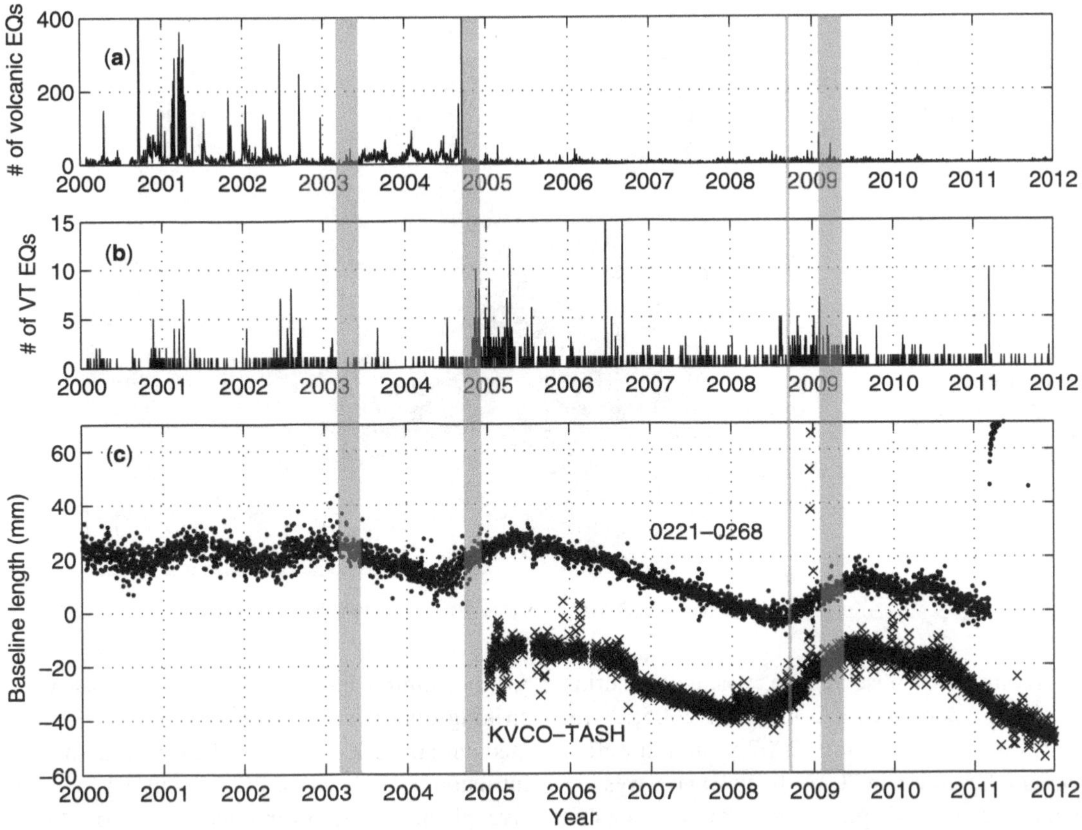

Fig. 2 Temporal evolution of seismic activity and ground deformation between 2000 and 2011. Gray shades denote the periods of unrest and eruptions. (**a**) Daily counts of volcanic earthquakes with unclear P- and S- arrivals. (**b**) Daily counts of volcano-tectonic earthquakes with clear P- and S-wave arrivals. (**c**) Relative change of the baseline length between 0221 and 0268 (dots) and KVCO and TASH (crosses), respectively. The baseline length between 0221 and 0268 was significantly extended due to the 2011 Tohoku-oki earthquake. In contrast, the KVCO-TASH baseline was not significantly extended because the 2011 Tohoku-oki earthquake significantly expanded the area around the volcano in the east-west direction but not in the north-south direction as much as the east-west direction. Outliners of the KVCO-TASH baseline length in late 2008 are likely due to instrumental problems. Adopted from Aoki et al. (2013)

2006). Furthermore, the volcanic glow became discernible during the final 10 days of July 2004, with the maximum temperature at the crater's base exceeding 500 °C following the abrupt extension of the baseline length. These observations suggest that the elevated temperature at the shallow part of the vent resulted from the magma intrusion.

2.3 Ground Deformation Associated with the 2008–2009 Eruption

The 2008–2009 eruption also resulted in surface deformation, monitored by continuous GPS sites. Before the summer of 2008, the area surrounding the volcano exhibited compression, a phenome-

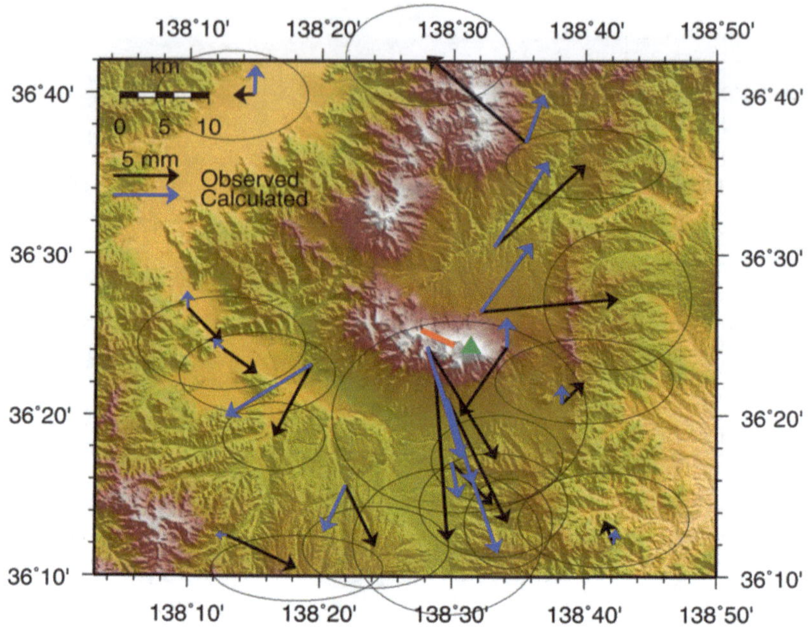

Fig. 3 Comparison between the observed displacements between June 2004 and March 2005, spanning the 2004 eruption, and the calculated displacements from a dike to the west of the summit (red line). Ellipses represent 1σ uncertainties of the observed displacements. The green triangle indicates the summit of Asama Volcano. Adopted from Aoki et al. (2013)

non commonly observed in volcanoes during periods of dormancy (Aoki et al. 2005, 2013; Murakami 2005). In July 2007, a month before the initial eruption in 2008, the volcano's western flank began to display north-south extension (Fig. 2). Subsequently, the volcano exhibited quasi-linear expansion until June 2009. This extension was attributed to the intrusion of a dike to the west of the summit, as observed in 2004. Furthermore, the immediate vicinity of the summit also exhibited expansion. KAHG (Fig. 1), situated on the eastern rim of the summit, exhibited an east-southeastward displacement of approximately 25 mm in the latter part of 2008. However, the lack of observations precludes the identification of the onset time (Fig. 4a). Subsequently, KAHG exhibited a northwest displacement of approximately 10 mm during the early months of 2009. The next site closest to the summit, AVOG (Fig. 1), located 4 km from the summit, did not exhibit significant deformation (Fig. 4b), suggesting that the source of deformation at the summit is likely shallow.

The displacement field during the period of ongoing summit expansion and expansion to the west (June–December 2008) is best explained by an east-west-striking dike with a top at 0.7 km below sea level and a shallow source responsible for the summit inflation (Fig. 5). The east-west-striking dike is situated close to the one associated with the 2004 unrest, indicating that the dike intrusion to the west of the summit was a pervasive phenomenon during the unrest of Asama volcano. This observation also suggests the existence of a magma pathway to the west of the summit. The estimated volume of intrusion is 1.64×10^6 m^3, representing approximately 25% of the volume observed during the 2004 unrest.

The lack of multiple GPS sites on the summit precludes the possibility of constraining the shape of the source responsible for the summit deformation, even if the source is assumed to be located directly beneath the summit. The expansion of a spherical source can be excluded as a potential mechanism for the observed deformation, as the requisite pressure change well above the rock fracture toughness is inconsistent with the observed horizontal displacements. These displacements, which range from approximately 25 mm at a distance of 250 m from the summit to less than 5 mm at a distance of 4 km, are not aligned with the predicted expansion of a spherical source. A similar line of reasoning excludes the possibility of conduit pressurization, regardless of whether the conduit is open or closed at its ends. Traction on the conduit with a radius of

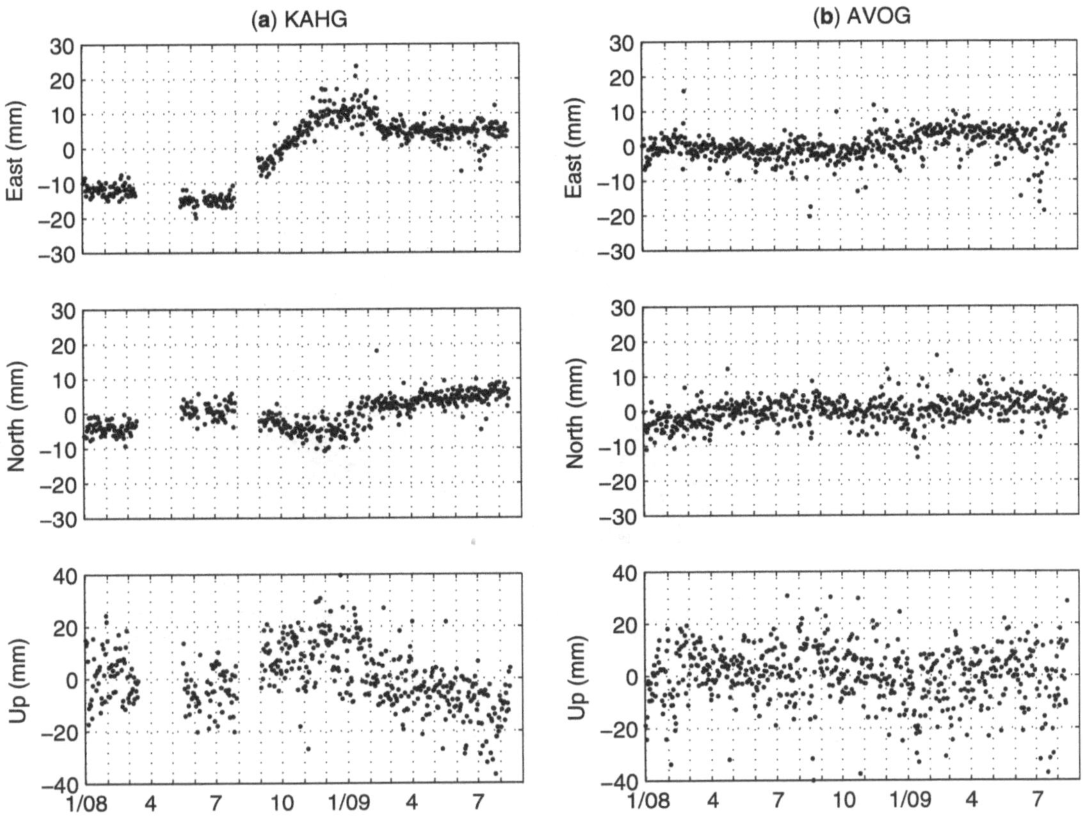

Fig. 4 An example of time series of (**a**) KAHG and (**b**) AVOG with respect to a GPS site far enough from Asama Volcano. Figure 1 shows the location of these sites. Adopted from Aoki et al. (2013)

approximately 50 meters may be a feasible mechanism, consistent with the observed displacements and the fracture toughness of rocks. If we assume a conduit with a radius of 50 m extending from the summit and 1 km below sea level, where the vertically elongated seismicity marks its bottom (Takeo et al. 2006), the required traction on the conduit wall is approximately 15 MPa, which is reasonable given that a confining pressure at a depth of 1.5 km below the surface, for instance, is approximately 30 MPa with the average rock density of 2000 kg/m^3.

2.4 Ground Deformation After the 2008–2009 Eruption

Assessing more recent ground deformation in the Asama volcano is more complicated than before because the postseismic deformation of the 2011 Tohoku-oki earthquake (Mw = 9.0) contaminates the observed deformation field. The postseismic deformation is primarily ENE-WSW extension with secondary NNW-SSE contraction. Munekane (2021) removed the postseismic deformation and deformation associated with tectonic strain accumulation by removing the long-wavelength component from the observed deformation field and then extracted the volcano deformation.

Munekane (2021) then modeled the extracted deformation field by the temporal evolution of the pressurization and depressurization, or volume increase and decrease, of a dike to the west of the summit at 1 km below sea level and an oblate spheroid below the dike at a depth of 12 km. The aspect ratio of the oblate spheroid Munekane (2021) modeled is so large that it is approximated as a sill. See Munekane (2021) for technical details on constraining the source geometry and location.

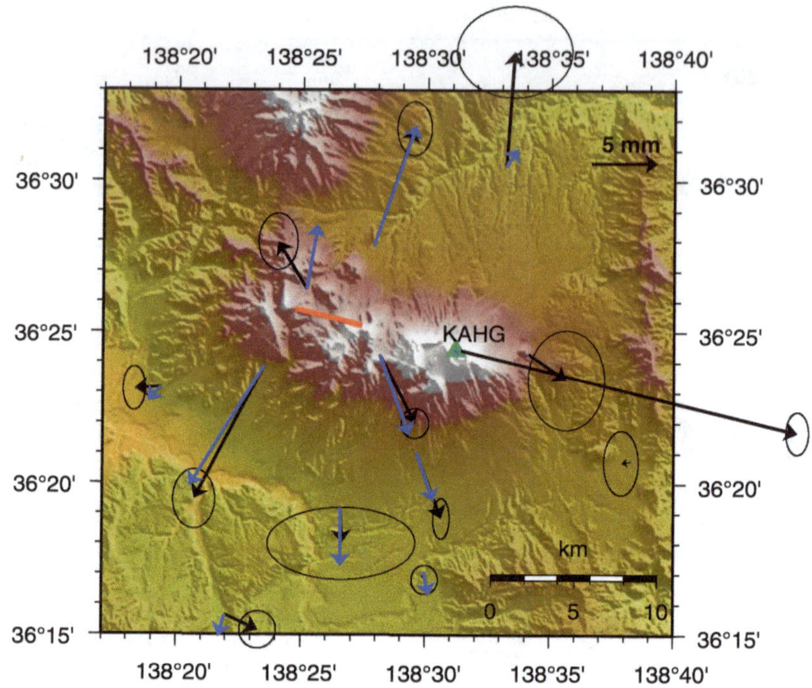

Fig. 5 Comparison between the observed displacements between June and December 2008 (black) and the calculated displacements with the modeled dike (red) that fits the observed displacements, excluding those at KAHG obviously affected by the shallow source beneath the summit. Adopted from Aoki et al. (2013)

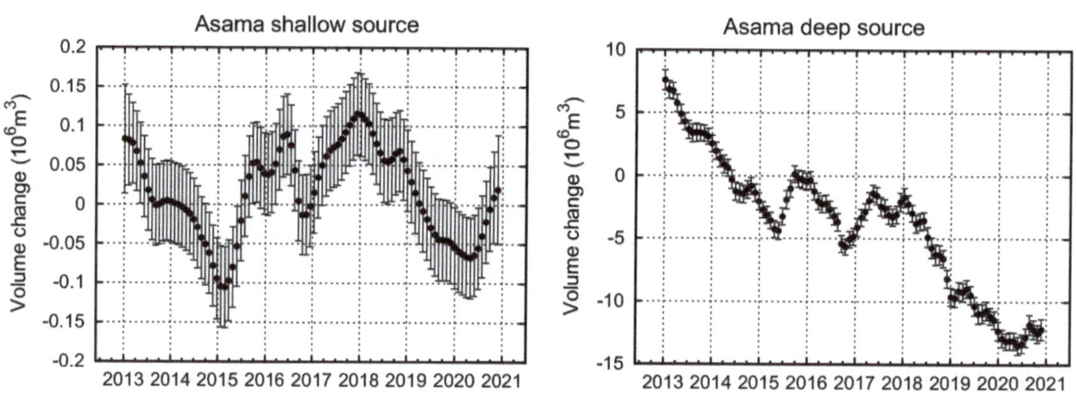

Fig. 6 Temporal evolution of the volume change of (left) the dike at 1 km below sea level and (right) the spheroidal pressure source at 12 km. Modified from Munekane (2021)

The temporal evolution of the pressurization of the dike and oblate spheroid between 2013 and 2021 exhibits a monotonous volume decrease of the deep oblate spheroid and oscillatory volume changes of the shallow dike (Fig. 6). While the Asama volcano experienced minor eruptions in June 2015 and August 2019 and increased seismic activity in June 2020 and March 2021, the temporal evolution of the dike volume does not exhibit an apparent correlation with these activities.

Munekane (2021) interpreted the monotonous volume decrease of the deep source as the long-term decrease in magma supply from depth, leading to a long-term decline in volcanic activity of the Asama volcano. Notwithstanding the monotonous reduction in the volume of the deep source beneath the Asama volcano, the decreasing rate is not steady with time (Fig. 6). Munekane (2021) found that the short-term fluctuation of the volume change anti-correlates with the rate of volume increase of a magma reservoir of

Kusatsu-Shirane volcano, ~25 km to the north of Asama volcano, at ~4 km below sea level. In other words, a rapid volume decrease of the deep source of Asama corresponds to a rapid volume increase of the magma reservoir of Kusatsu-Shirane volcano. This finding might reflect the connectivity of the magma pathway at Asama and Kusatsu-Shirane volcanoes at the upper or mid-crustal level. However, independent information, such as insights from petrology, is required to confirm this conjecture.

3 Tilt Change by the 2009 Eruption

Tilt motions were observed during the February 2009 eruption, primarily derived from the horizontal component of broadband seismic records (Nagaoka et al. 2012). We took advantage of the fact that the horizontal component of broadband seismic records contains both translational and rotational (tilt) motions (e.g., Rodgers 1968; Maeda et al. 2011) to extract the tilt motion from these records. Subsequently, the co-eruptive tilt records at the summit, which recorded up to 7 microradians of tilt changes indicative of mass loss at the summit, were excluded. In other locations, the tilt is observed to be oriented towards the west of the summit, reaching up to 1.1 microradians (Fig. 7), which suggests a source located to the west. We employed a nonlinear optimization to model the observed deformation field as a deflating dike embedded in a homogeneous, elastic, and isotropic halfspace (Okada 1985). The optimal location for the dike is at its summit, approximately 1 km below sea level, with a northwest-southeast strike (Fig. 7), near the site of previous dike intrusions during periods of unrest. This observation suggests that the intruded dike is somehow connected to the summit, thereby allowing the intruded dike to be ejected by an eruption.

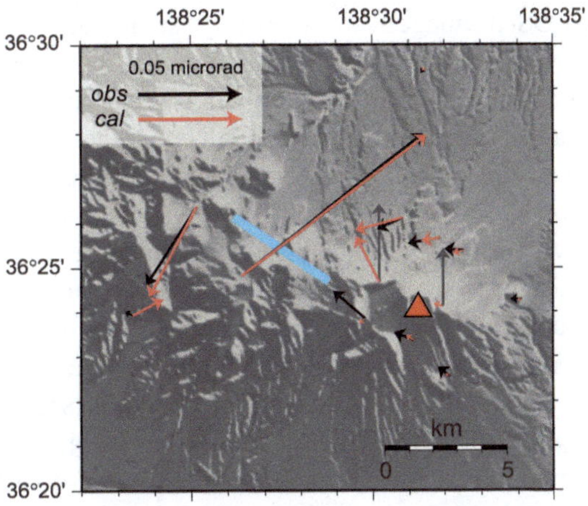

Fig. 7 Tilt motion associated with the eruption on 2 February 2009. The blue line denotes the modeled location of the intruded dike. Black and pink vectors represent the observed and calculated tilt directions. The red triangle denotes the location of the summit of Asama Volcano. Adopted from Nagaoka et al. (2012)

4 Potential Volcano Instability Measured by SAR

Given the insignificant or gradual nature of ground deformation at Asama volcano, as observed through SAR Interferometry (InSAR), no studies have investigated the potential ground deformation associated with inflation and deflation. Nevertheless, Wang et al. (2019) identified surface deformation of the volcano that may be related to surface instability based on analysis of SAR images.

We collected 140 SAR images, comprising 120 C-band Sentinel-1 images and 20 L-band ALOS-2 images. The objective was to detect surface deformation between 2014 and 2018. The Sentinel-1 images were acquired from ascending and descending paths, whereas the ALOS-2 images were acquired solely from the descending path. Consequently, the Sentinel-1 images differentiate the vertical and (quasi-)east components of displacements. The temporal interval of the Sentinel-1 images is 12 or 24 days, whereas the

ALOS-2 images are unevenly sampled, with intervals ranging from 14 to 224 days. The brief sampling interval of Sentinel-1 SAR images enables the maintenance of high coherence, even when the radar wave's wavelength is relatively short (~56 mm). The ALOS-2 images, which have a longer wavelength (~236 mm), are more resistant to temporal decorrelation. The analysis is constrained to a limited area of a few kilometers from the summit, given that the coherence is sufficient only within this region due to the sparse vegetation. Even ALOS-2 images are subject to temporal decorrelation in areas with dense vegetation.

We generated SAR interferograms from ALOS-2 images with a temporal separation of less than 400 days and a perpendicular baseline smaller than 400 m. This procedure reduces the number of interferograms and the processing time. Also, we formed two groups of Sentinel-1 interferograms: one group consists of pairs with a temporal separation of less than 60 days and a perpendicular baseline of less than 200 m; the other group consists of pairs with a temporal separation of less than 400 days and a perpendicular baseline of less than 100 m. This procedure is because the Asama volcano is covered with snow most of the time in winter, leading to significant temporal decorrelation for some winter interferograms. We then added inter-annual summer interferograms for the Sentinel-1 data to avoid unconnected SAR images. We used the Digital Elevation Map of a resolution of 10 m provided by the Geospatial Information Authority of Japan.

Subsequently, persistent and distributed scatterers were identified based on the temporal stability of the amplitude of each pixel in the SAR images. Once the scattereres have been identified, a time series is obtained for each persistent and distributed scatterer, with an appropriate correction for atmospheric disturbance. For the sake of brevity, the technical details are omitted here; readers are referred to Wang et al. (2019) and the references therein for further information. Figure 8a shows the topography with a resolution of 10 m used in this study, in which the black dotted line indicates the extent of our measurement points. Figure 8b–d denotes the mean line-of-sight velocities from the ALOS-2, ascending Sentinel-1, and descending Sentinel-1 images, respectively, with respect to the reference point shown in Fig. 8a. The standard deviations of the line-of-sight velocities are mostly less than 3 mm/year.

The line-of-sight measurements from the three datasets display asymmetric patterns, with deformation occurring predominantly at the volcano's eastern flank (Fig. 8b–d). The observed deformation can be roughly divided into two regions: the northeast and southeast flanks, which encompass approximately 3.68 km^2 and 2.14 km^2, respectively. The northeast flank displays line-of-sight extension across all three datasets, with maximum extension rates of approximately 14, 12, and 10 mm/year observed in the ALOS-2, ascending Sentinel-1, and descending Sentinel-1 images. In contrast, the southeast flank exhibits a fan-shaped structure with apparent line-of-sight shortening and extension along descending and ascending paths. The maximum mean velocities across the three datasets range between 9 and 12 mm/year.

Figure 9 illustrates the line-of-sight velocities along the P_1P_1' and P_2P_2' profiles, which have a width of 60 m (Fig. 8a) and encompass the northeast and southeast flanks, respectively. Gray lines represent the topography. The displacements along P_1P_1' indicate that regions near the summit exhibit considerably higher line-of-sight extension rates, with minimal velocities observed at distances approximately 1.2 km or further from the summit. The displacements along P_2P_2' display a more dispersed pattern than those along P_1P_1', with the highest velocities occurring at the central portion of the slope. The deformation rate along P_2P_2' from the ascending path is considerably higher than that from the descending paths, indicating a notable eastward movement.

Figure 10 illustrates the temporal evolution of LOS displacements at P_{NEF} and P_{SEF}, with Figs. 10b–d denoting the locations. The time series displays a linear trend throughout the observation period, indicating that the controlling mechanisms for deformation remain stationary. No significant deformation anomalies were iden-

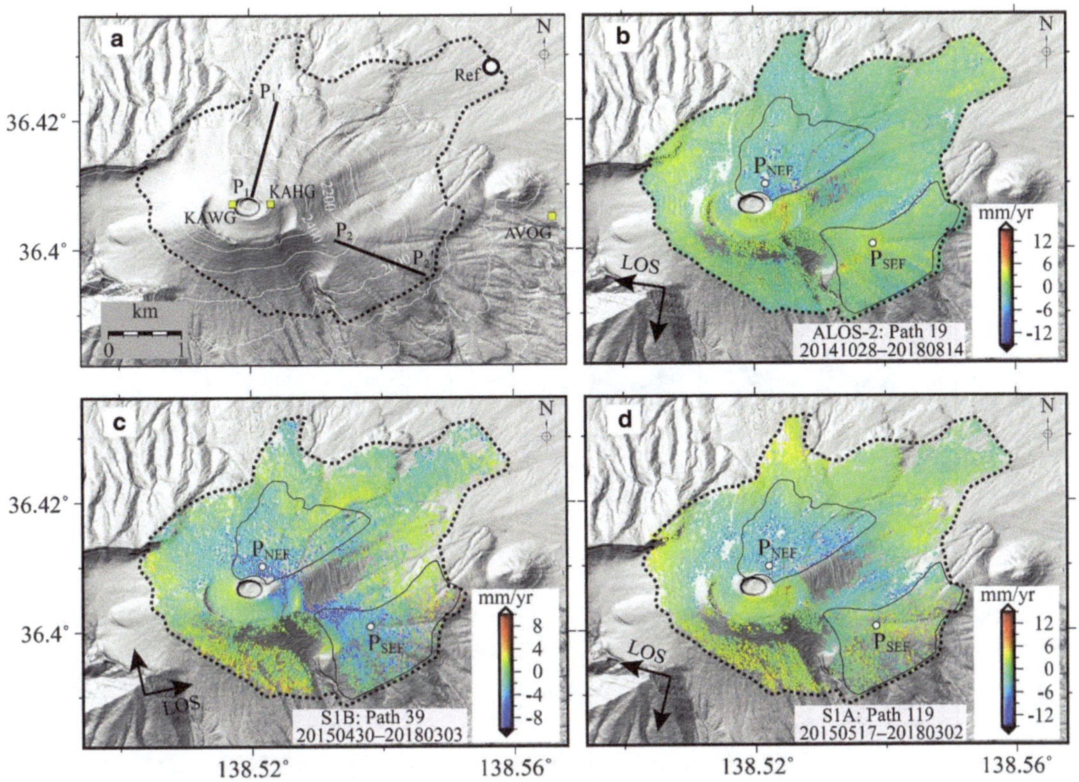

Fig. 8 (**a**) Topography around Asama Volcano. Two solid lines indicate profiles along which Fig. 9 shows line-of-sight displacements. The mean line-of-sight displacements from (**b**) ALOS-2, (**c**) ascending Sentinel-1, and (**d**) descending Sentinel-1 images. Black dotted lines denote the area with sparse vegetation so that Sentinel-1 images can extract surface displacements. P_{NEF} and P_{SEF} represent the points at which Fig. 10 shows the temporal evolution of line-of-sight displacements. Adopted from Wang et al. (2019)

tified in association with the minor eruption in June 2015, which suggests that the mass transfer during this eruption was limited.

Additionally, the quasi-vertical and quasi-east–west velocities were calculated by decomposing the line-of-sight velocities from the three SAR datasets. InSAR cannot account for the north-south component due to its inherent insensitivity to line-of-sight (LOS) measurements. Most points within the northeast flank (65%) exhibit subsidence, while most within the southeast flank (59%) exhibit sliding deformation. At PNEF, the observed subsidence is 6.0 ± 0.2 mm/year, accompanied by a minor east-west displacement. At PSEF, the apparent eastward velocity is 6.7 ± 0.08 mm/year, with 2.1 ± 0.05 mm/year of subsidence. We can thus propose that the deformation observed at the northeast flank is predominantly due to subsidence. In contrast, the deformation observed at the southeast flank is presumably in the direction of the slope.

To validate our InSAR observations, we compared GNSS measurements obtained from the KAHG GNSS site at the crater's eastern rim. The 3D displacements from KAHG at the volcano's east side were projected into the radar LOS direction for comparison with InSAR measurements. The comparisons (Fig. 11) indicate that the displacement time series derived from ascending Sentinel-1 and descending ALOS-2 data are consistent with the GNSS measurements.

The InSAR observations do not indicate the presence of a discernible transition between inflation and deflation. The temporal evolution of displacements from InSAR demonstrates a linear trend at the volcano's northeast and southeast

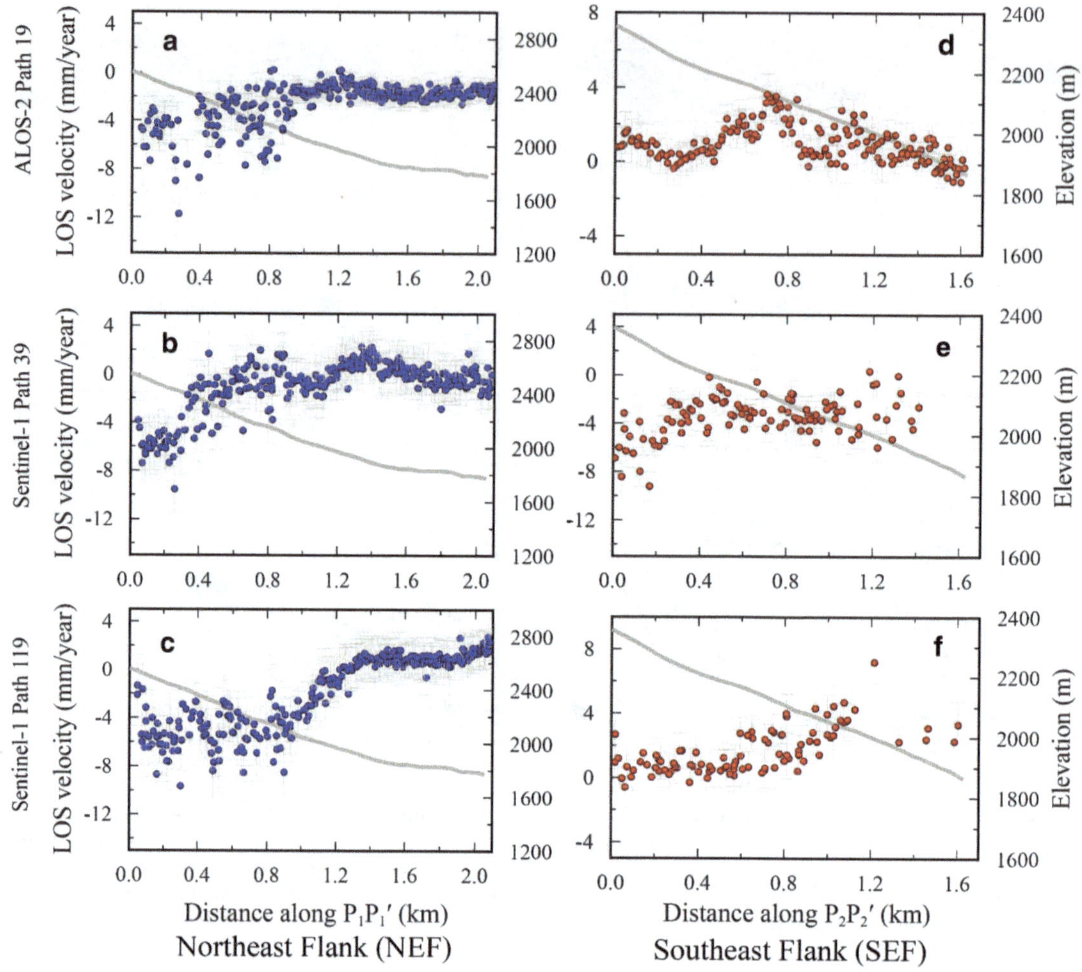

Fig. 9 Line-of-sight velocities along the two profiles P_1-P_1' and P_2-P_2' (Fig. 8) encompassing the northeastern and southeastern flanks. Line-of-sight velocities along the P_1-P_1' profile from (**a**) ALOS-2, (**b**) ascending Sentinel-1, and (**c**) descending Sentinel-1 images. (**d**)–(**f**) Same as (**a**)–(**c**) but for the P_2-P_2'profile. The positive and negative velocities represent the line-of-sight shortening and extension, respectively. Error bars represent the standard deviation. Gray lines represent the topography along the profiles. Adopted from Wang et al. (2019)

flanks. One potential explanation for the persistent temporal patterns observed in the deformation is the prevalence of shallow processes, such as thermal contraction or flank instability, over deeper magma pressurization or depressurization. Furthermore, the ground-based measurements were predominantly situated near the summit, whereas our InSAR measurements predominantly covered the summit region and the upper flanks. The deformation induced by the inflation or deflation of magma sources beneath the Asama volcano may be spatially extensive and insensitive to our application of InSAR.

The northeast flank of Asama volcano displays a sustained pattern of subsidence throughout the observation period. Recent studies have demonstrated that the constant subsidence observed at active volcanoes can be attributed to several mechanisms, including magma withdrawal, consolidation of erupted deposits, viscoelastic relaxation of the host rock, thermal contraction of lava deposits or intruded magma

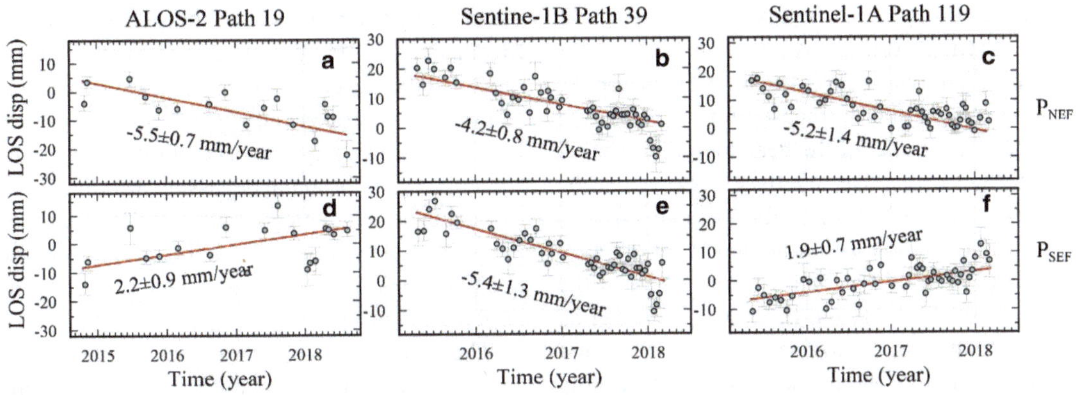

Fig. 10 Temporal evolution of line-of-sight displacements at P_{NEF} and P_{SEF} (Fig. 8). (**a**)–(**c**) Time series of line-of-sight changes at P_{NEF} from ALOS-2, ascending Sentinel-1, and descending Sentinel-1 images. (**d**–**f**) same as (**a**–**c**) but at P_{SEF}. Error bars represent the standard deviation. Adopted from Wang et al. (2019)

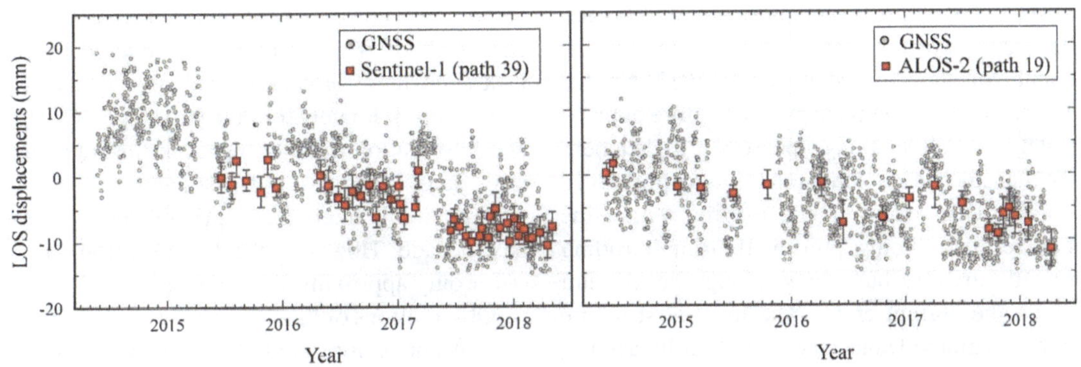

Fig. 11 Comparison of displacement time series between InSAR measurements (red squares) and GPS measurements (gray circles) at KAHG. The three-dimensional GPS displacements are projected onto the line-of-sight direction of (**a**) ascending Sentinel-1 and (**b**) descending ALOS-2 measurements. Error bars indicate the estimated standard deviation of InSAR measurements. Adopted from Wang et al. (2019)

body, and the pore pressure drop associated with hydrothermal fluid circulations. We initially excluded magma withdrawal and viscoelastic relaxation as these two mechanisms typically induce broadly symmetric deformation around the magma source, which is inconsistent with the observed deformation. In addition, these two mechanisms result in deformation on a spatial scale that is broader than that observed in the present study.

It is hypothesized that the observed subsidence may be attributed to thermal contraction of the lava flow. The geological map indicates that a lava flow resulting from the 1783 eruption covers the northeast flank. Yasui and Koyaguchi (2004) estimated the thickness of the lava deposits on the northeast flank to range from approximately 20 m at the flank to approximately 90 m near the summit. The observed deformation is found to correlate with the thickness of the lava deposit. The duration of thermoelastic contraction is contingent upon the volume and depth of the source, as well as the thermal diffusivity of the surrounding crust. Recent studies have demonstrated that surface subsidence requires more than 160 years to reduce its rate of change to less than 1 mm/year. This is contingent upon assuming a heat source with a volume of 5×10^6 m^3 at a depth of approximately 100 m, a temperature of 1000 K, and a thermal diffusivity of 1×10^{-5} m²/s. Assuming an average thickness of 50 m for the lava deposits at the northeast summit of Asama volcano, the volume of the heat source is comparable to the case above. We thus propose that the thermoelastic contraction of the 1783 lava flow is an insufficient explanation for the observed subsidence of approximately 10 mm per year at the northeast flank of Asama. Furthermore, it can be posited that meteorological precipitation and groundwater flow would accelerate the cooling process, necessitating additional mechanisms to account for the observed subsidence at the northeast flank.

An additional potential cause of the observed subsidence is a reduction in pore pressure associated with hydrothermal activity beneath the Asama volcano. Aoki et al. (2009, 2013) posited that the ancient magma intrusions before approximately 24,000 years ago possess a higher density than the recent intrusions due to prolonged cooling. This results in the magma within the deep reservoir being prevented from intruding upward directly, but rather along the conduit toward the current crater. The internal structure of the Asama volcano gives rise to a hydrothermal fluid flow from depth to the volcano's eastern flank, where Usui et al. (2017) reported the presence of a porous layer with elevated water content at a depth of approximately 1 km above sea level. Assuming that the surface subsidence is entirely due to the loss of pore volume in the porous layer, the relationship between the vertical displacement u_z and porosity loss is given by (Todesco et al. 2014)

$$u_z = \frac{z(\varepsilon_0 - \varepsilon)}{1-\varepsilon} \tag{1}$$

where z is the thickness of the porous rock layer, ε_0 and ε denote the porosity before and after the porosity loss. Equation (1) indicates that a 1% loss of porosity in a 50 m rock layer results in subsidence of 213–495 mm when the initial porosity is 0.3 and 0.5, respectively. Thus, we propose that the compaction of the porous layer east of the Asama volcano may be identified as a potential cause for the observed subsidence at the northeast flank. Morita et al. (2016) observed a significant release of CO_2 at the eastern portion of the Asama volcano summit. They proposed that permeable zones connected to the deep hydrothermal fluid layer are the source of this substantial degassing. Modifications to the rate of magmatic gas influx at the base of the hydrothermal system or alterations to the permeability of the porous layer may precipitate a decline in pore pressure.

The observation indicates that the surface of the southeast flank exhibits a predominantly southeastern movement in the slope direction. Given an average slope angle of approximately 20° for the southeast flank, it can be postulated that the observed displacements result from slope instability. The geological map indicates that the east flank of Asama volcano is covered by old lava deposits of Hotokeiwa volcano, which was active approximately 11,000 years ago. The linear deformation trend observed at the southeastern flank (Fig. 10d–f) suggests that the deformation is likely a constant process over the past decade. However, the InSAR observations cover only approximately 4 years.

Observations of flank instability at large basaltic volcanoes indicated that a weak substrate beneath the sliding surface should accommodate persistent motion. Furthermore, several factors, including the magnitude of the magma supply rate, the thickness of the substrate, and the edifice's gravitational spreading, influence the rate of flank motion. However, the rock types at Hotokeiwa are predominantly dacite to rhyolite (Yasui 2018). Furthermore, there is a lack of knowledge regarding the internal structure and deposit thicknesses of Hotokeiwa volcano. In light of the considerations above, it is not feasible to infer the properties of the sliding substrate based solely on our InSAR measurements. Nevertheless, in light of the minimal magmatic activity observed at Asama without significant surface deformation, we propose that gravitational loading on the southeast flank influences the slope motion. The spreading of small horizontal velocity (less than 10 mm per year) also indicates that the depth of the sliding surface is considerable.

The likelihood of persistent, deep-seated flank motion at active volcanoes evolving into large-volume landslides is reduced due to the relaxation of elastic stress before failure. This argument is supported by the absence of scarps or benches on the surface of the southeast flank at Asama volcano and the lack of a substantial slope failure record in the region. Nevertheless, the possibility of a sudden trigger to failure through the interaction of volcanic activity and slope instability cannot be discounted.

InSAR measurements revealed surface deformation that was not discernible through GPS or tiltmeter observations. These findings suggest the possibility of flank sliding along the northeastern and southeastern flanks of the Asama volcano. While this observation may be associated with the volcano's flank instability, the likelihood of a catastrophic sector collapse appears remote, given the slow (<10 mm/year) movement and the absence of evidence for previous sector collapses.

5 Summary

Geodetic observations indicate that the Asama volcano has undergone fluctuations in elevation over the past few decades. The volcano's inflation and deflation are synchronous with volcanic activity's occurrence. The inflationary phase coincides with increased seismic activity and often increased degassing from the summit, while the dormant phase of volcanic activity coincides with the deflationary phase. A dike intrusion to the west of the summit at approximately 1 km below sea level provides a plausible explanation for the observed inflation of the volcano in association with the 2004 and 2008–2009 eruptions. This observation leads us to hypothesize that the magma pathway from the west of the summit is well established, with unrest typically initiated by a dike intrusion to the west. This hypothesis is supported by the co-eruptive tilt motions observed during the 2 February 2009 eruption, which were attributed to a dike deflation to the west.

The mechanics of the deflation of the volcano during its dormancy remain poorly understood. The observed deflation can be attributed to either magma migration from the intruded dike or the viscoelastic relaxation of the host rock around the intruded dike. To verify the mechanics, assessing whether the viscoelastic relaxation can generate the observed deflation is necessary. If the viscoelastic relaxation is deemed implausible, the observed deflation may be attributed to the depressurization of a magma reservoir at the mid-crustal level (~12 km).

References

Aoki Y, Watanabe H, Koyama E et al (2005) Ground deformation associated with the 2004—2005 unrest of Asama volcano, Japan. Bull Volcanol Soc Jpn 50:575–584. https://doi.org/10.18940/kazan.50.6_575

Aoki Y, Takeo M, Aoyama H et al (2009) P-wave velocity structure beneath Asama volcano, Japan, inferred from active source seismic experiment. J Volcanol Geotherm Res 178:273–278. https://doi.org/10.1016/j.jvolgeores.2009.09.004

Aoki Y, Takeo M, Ohminato T et al (2013) Structural controls on magma pathways beneath Asama volcano, Japan. Geol Soc London Spec Publ 380:67–84. https://doi.org/10.1144/SP380.6

Maeda Y, Takeo M, Ohminato T (2011) A waveform inversion including tilt: method and simple tests. Geophys J Int 184:907–918. https://doi.org/10.1111/j.1365-246X.2010.04892.x

Morita M, Mori T, Kazahaya R, Tsuji H (2016) Diffuse carbon dioxide emissions from hidden subsurface structures at Asama volcano. Japan Bull Volcanol 78:17. https://doi.org/10.1007/s00445-016-1008-5

Munekane H (2021) Modeling long-term volcanic deformation at Kusatsu-Shirane and Asama volcanoes, Japan, using the GNSS coordinate time series. Earth Planet Space 73:192. https://doi.org/10.1186/s40623-021-01512-2

Murakami M (2005) Magma plumbing system of the Asama volcano inferred from continuous measurements of GPS. Bull Volcanol Soc Jpn 50:347–361. https://doi.org/10.18940/kazan.50.6_347

Murase M, Ono K, Ito T et al (2007) Time-dependent model for volume changes in pressure sources at Asama volcano, Central Japan due to vertical deformations detected by precise leveling during 1992–2005. J Volcanol Geotherm Res 164:54–75. https://doi.org/10.1016/j.jvolgeores.2007.04.001

Nagaoka Y, Nishida K, Aoki Y et al (2012) Seismic imaging of magma chamber beneath an active volcano. Earth Planet Sci Lett 333:1–8. https://doi.org/10.1016/j.epsl.2012.03.034

Nakada S, Yoshimoto M, Koyama E et al (2005) Comparative study of the 2004 eruption with old eruptions of Asama volcano and the activity evaluation. Bull Volcanol Soc Jpn 50:575–584. https://doi.org/10.18940/kazan.50.5_303

Okada Y (1985) Surface deformation due to shear and tensile faults in a half-space. Bull Seismol Soc Am 75:1135–1154. https://doi.org/10.1785/BSSA0750041135

Rodgers PW (1968) The response of the horizontal pendulum seismometer to Rayleigh and love waves, tilt, and free oscillations of the earth. Bull Seismol Soc Am 58:1385–1406. https://doi.org/10.1785/BSSA0580051385

Takeo M, Aoki Y, Ohminato T, Yamamoto M (2006) Magma supply path beneath Mt. Asama volcano, Japan. Geophys Res Lett 33:L15310. https://doi.org/10.1029/2006GL026247

Todesco M, Costa A, Comastri A et al (2014) Vertical ground displacement at Campi Flegrei (Italy) in the fifth century: rapid subsidence driven by pore pressure drop. Geophys Res Lett 41:1471–1478. https://doi.org/10.1002/2013GL059083

Usui Y, Ogawa Y, Aizawa K et al (2017) Three-dimensional resistivity structure of Asama volcano revealed by data-space magnetotelluric inversion using unstructured tetrahedral elements. Geophys J Int 208:1359–1372. https://doi.org/10.1093/gji/ggw459

Wang X, Aoki Y, Chen J (2019) Surface deformation of Asama volcano, Japan, detected by time series InSAR combining persistent and distributed scatterer. Earth Planet Space 71:121. https://doi.org/10.1186/s40623-019-1104-9

Yasui M (2018) Textures of the eruptive products of Asama-Maekake volcano from the 12th century: indicator of eruptive processes. Nihon University, Tokyo, pp 37–50

Yasui M, Koyaguchi T (2004) Sequence and eruptive style of the 1783 eruption of Asama volcano, Central Japan: a case study of an andesitic explosive eruption generating fountain-fed lava flow, pumice fall, scoria flow, and forming a cone. Bull Volcanol 66:243–262. https://doi.org/10.1007/s00445-003-0308-8

Electrical Resistivity of Asama Volcano

Takao Koyama, Yoshiya Usui, and Koki Aizawa

1 Introduction

Asama volcano is one of the most active ones in Japan and has been frequently active since ancient times, sometimes causing many casualties nearby. Its proximity to Tokyo has also impacted the city through ash dispersal, and its volcanic activities have been the focus of public attention (e.g., Aramaki 1956, 1957; Hayakawa and Nakajima 1998). It was observed using measurement apparatus such as seismometers since the early twenty century (Omori 1911). Electromagnetic measurements have also been done for monitoring of volcanic eruptions. The disturbance of the magnetic field due to the eruption was measured (Minakami 1935a, b), while magnetic surveys were also conducted (Minakami 1938). In addition, the electromagnetic disturbance during eruptions by using an electric wire, which is an antenna, has also been done, because the residents in the vicinity of Asama volcano have claimed that the radio has caught some noises during the eruption (Minakami 1935b). Nagaoka and Ikebe (1937a, b) have reported that the sudden variation of the electromagnetic field during the eruption may occur due to the change of the magnetization distribution caused by the intrusion of the magmatic materials with high temperatures. However, the distribution and the change of the electrical conductivity structure were not focused on.

Kawasaki (1965) measured electrical resistivity in the northern or northeastern foot of Asama volcano, using the direct current method with the L10 model ground resistivity meter by Yokogawa Electric Works Ltd. It is covered by the Agatsuma and Kotaki pyroclastic flow, which is underneath. It shows the resistive property of about 10 kΩm at a top 100 m beneath the surface. Kawasaki (1965) also found that the Kotaki pyroclastic flow is less resistive than the Agatsuma pyroclastic flow, and its boundary is much less resistive than the bulk resistivity of both layers. They conclude that it becomes an aquifer to be shown as less resistive.

Honkura and Koyama (1978) conducted telluric measurements as well as a geomagnetic survey to check the potential of this kind of measurement to detect the change of the electrical conductivity by using a leakage current from the train as a source current.

In 1982, the magnetotelluric measurements in VLF and ELF bands have been conducted under the project of the multidisciplinary observation of Asama volcano (Yatsugatake Geomagnetic Observatory, Earthquake Research Institute, University of Tokyo 1982). The survey was done

T. Koyama (✉) · Y. Usui
Earthquake Research Institute, The University of Tokyo, Tokyo, Japan
e-mail: tkoyama@eri.u-tokyo.ac.jp

K. Aizawa
Faculty of Science, Institute of Seismology and Volcanology, Kyushu University, Fukuoka, Japan

S. Aramaki et al. (eds.), *The Asama Volcano*, Active Volcanoes of the World,
https://doi.org/10.1007/978-981-95-5799-8_21

on two lines: one is 8 km long in a north-south direction in the eastern foot of Asama volcano, and the other is 3 km long in an east-west direction on the volcano's east flank. The data of apparent resistivity show that, at a couple of kilometers depth, the Koasama lava dome is resistive as hundreds to one thousand Ωm, and the north area of Asama volcano is less resistive than one hundred Ωm. It also shows that the apparent resistivity drastically decreases close to the summit of Asama volcano.

In the early twenty-first century, a comprehensive magnetotelluric survey has been carried out, and its details are shown in the next section.

2 Broadband Magnetotelluric Survey in the Vicinity of Asama Volcano.

Asama volcano became unrest in 2004 and 2005, and eruptions occurred in September 2004. Some emitted volcanic ashes reached even Tokyo, over 100 km from the volcano. A geodetic deformation has also been observed, and the dike intrusion in the west of the volcano was inferred (Aoki et al. 2005). Takeo et al. (2006) analyzed seismic data of more than 500 events and revealed the locations of hypocenters. It indicated the characteristic magma pathway, which curves westward at a depth of 1 km beneath the sea level. The subsurface structure, however, was still unknown.

After the unrest of Asama volcano, a broadband magnetotelluric (MT) survey has been conducted in 2005 and 2006 under the seventh national project for the volcano eruption prediction in order to reveal the subsurface electrical conductivity structure (Aizawa et al. 2008). The electromagnetic measurements have been conducted at 74 sites in a 15 km by 15 km area, in which 39 sites required the instrument MTU5-A of Phoenix Geophysics to acquire the high-frequency band of 10,000–0.3 Hz (acoustic magnetotelluric: AMT), 22 sites for low-band data of 300–0.0005 Hz by using MTU5 (MT). The other 13 sites have conducted only telluric measurements for low-band data. The dipole length of the telluric measurements was about 20 m at every site. One-hour to one-night measurements have been done at each AMT site, and 2–5 days measurements have been conducted at each MT site (Fig. 1).

Aizawa et al. (2008) has performed two-dimensional modeling of the subsurface electrical conductivity in four profiles: three in North-South lines and another in East-West line across the current main summit of Asama volcano, Kamayama, by using off-diagonal elements of the MT impedance tensor, that is, TE and TM modes, accounting for topographic effect and correction of static shift effects which is a kind of erroneous site effect (Ogawa and Uchida 1996). The results show the conductivity model down to 3 km beneath the ground surface. It indicates two-layer background: the top layer, about 1 km below the ground surface, is resistive with a resistivity of 50–5000 Ωm, and the bottom layer is as conductive as 1.6–30 Ωm. It also detects a characteristic resistive body, exceeding 100 Ωm, even beneath 1 km below the surface west of the summit.

Usui et al. (2017) conducted the three-dimensional modeling, accounting for precise three-dimensional topographic effects and galvanic shift correction using data sets acquired by Aizawa et al. (2008). They have used diagonal components of MT impedance tensor and geomagnetic transfer functions, as well as off-diagonal components of MT impedance used in Aizawa et al. (2008). As Aizawa et al. (2008) roughly checked the two-dimensionality, the overall results are consistent with each other. Figure 1 shows the horizontal cross-sections at six elevations. It can be seen that a resistive layer is overlaid over the whole surface. A highly conductive layer of more than 1 S/m is widely spread beneath the top resistive layer (C1 and C2 in Fig. 1). It is worth noting that the resistivity of the eastern side is lower than that of the western side. In this conductive layer, however, a circular resistive body was found west of the caldera at 0.5–1.5 km above sea level (R2 in Fig. 1), which is also detected in the two-dimensional model by Aizawa et al. (2008). In addition, the three-dimensional model reveals another resistive body to the west of the aforementioned resistive body (R1 in Fig. 1). However, it is unknown if they are connected due to the data's low sensitivity.

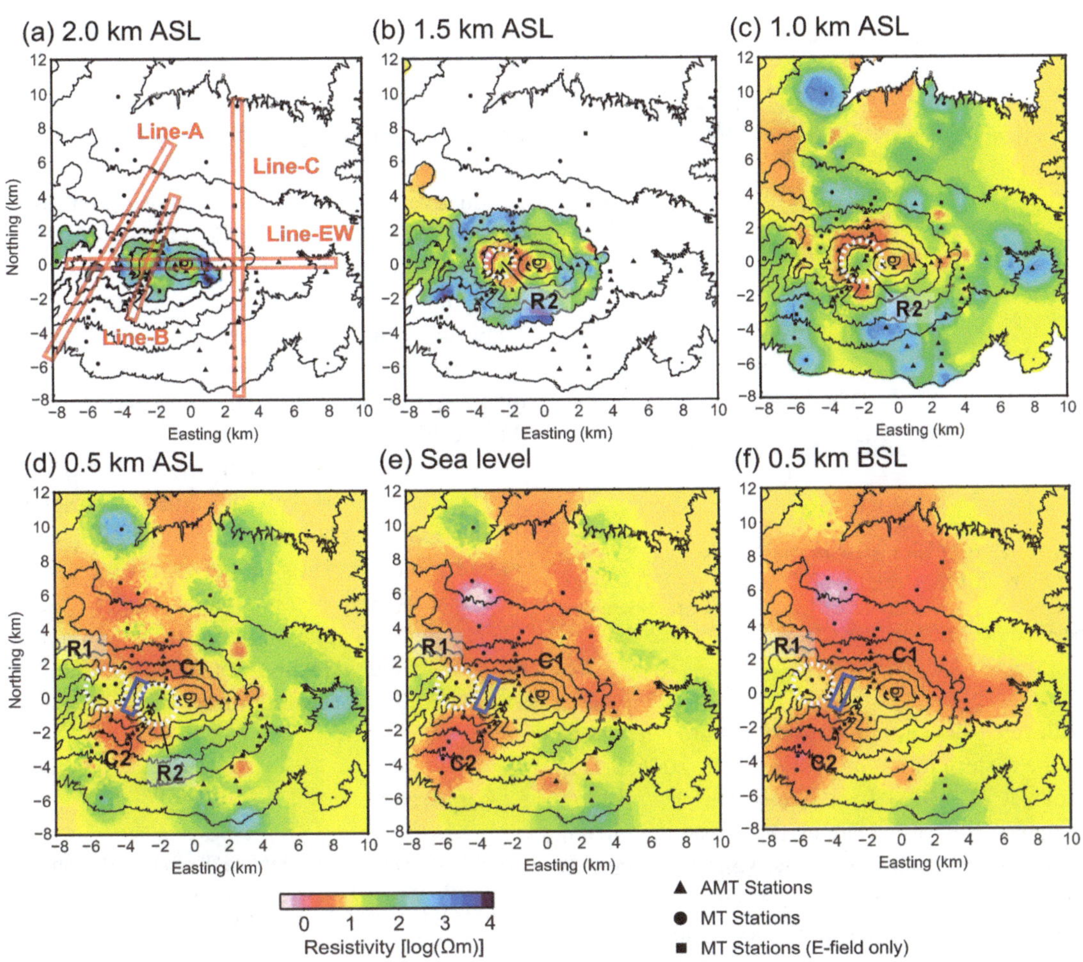

Fig. 1 Horizontal cross-sections of the subsurface resistivity model by three-dimensional inversion. (Fig. 8 of Usui et al. (2017))

3 Subsurface Structure of Asama Volcano Inferred from Electrical Resistivity

The electrical conductivity is well correlated with fluid and altered minerals. The top resistive layer at the whole subsurface corresponds to the unsaturated zone, which cannot keep fluid or precipitation in it and is highly porous, which is consistent with the result of the direct current method (Kawasaki 1965). 0.5–1.0 km beneath the ground surface, a conductive zone is distributed especially in the north, east, and southwest, which is also consistent with the result of the previous survey done in 1982 (Yatsugatake Geomagnetic Observatory, Earthquake Research Institute, University of Tokyo 1982). CO_2 mapping in Asama volcano revealed the high CO_2 flux at the current main summit, indicating that the active hydrothermal system exists underneath (Morita et al. 2016). Thus, this widespread conductive zone is fluid-rich in a hydrothermally altered clay layer, also commonly found in other volcanoes.

However, a resistive zone exists on the western side of the main crater at the same elevation or depth. There are old calderas in the west of the current caldera, Eboshi volcano complex, Kurofu volcano, and Hotokeiwa volcano (Aramaki 1956, 1957), and there may be intrusive and solidified magma beneath it; they may show resistive properties. Indeed, comparison with the locations of hypocenters of earthquake shows that earthquakes do not happen inside the resistive body but at the edge of it (Fig. 2). Also, the magma

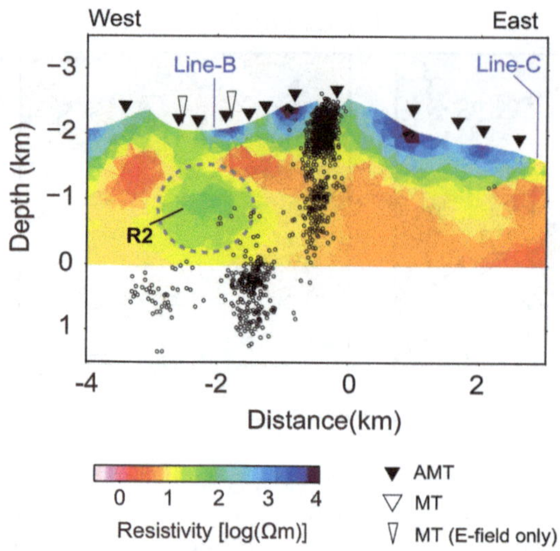

Fig. 2 Vertical profile of the resistivity model of the Line-EW in Fig. 1 (Fig. 9 of Usui et al. (2017)). Small open circles are locations of hypocenters of earthquakes between Jan. 2004 and Jan. 2015 (Takeo et al. 2022)

pathway is not straightforward upward but curved to the eastern direction (Takeo et al. 2006, 2022; Aoki et al. 2013) It is likely that the original magma reservoir exists at the west of volcano (Nagaoka et al. 2012) but the solidified massive blocks impede the ascending magma. It was also detected as a high seismic velocity anomaly by the active seismic survey after the MT survey (Aoki et al. 2009). The existence of these massive bodies may influence the distribution of the hydrothermal system and its evolution, and thus lead to uneven distribution of the conductive layer, that is, a wide conductive region exists in the north, east, and south of the current main crater.

4 Summary

Over 100 years, electromagnetic measurements have been done at and in the vicinity of Asama volcano to reveal the volcanic eruptions and mitigate the hazard risk for the residents.

In the twenty-first century, comprehensive MT surveys and three-dimensional modeling revealed the electrical conductivity structures beneath the volcano at almost 3 km depth. Basically, it consists of two layers in the background: a top layer is resistive, which is an unsaturated and porous layer, and a conductive layer is underneath it, which is a fluid-rich clay layer. The conductive layer, however, shows uneven distribution, and the western side of the volcano is relatively resistive. The hypocenter locations show that there is no seismicity in the resistive body, and it indicates that this resistive body may be previously intrusive and solidified massive magma; it impedes the magma ascendance from the western magma reservoir and changes the direction of the magma pathway.

The electrical resistivity mainly corresponds to the existence of fluid and hydrothermally altered minerals. Thus, it can unveil the subsurface distribution of the hydrothermal system. Therefore, the magnetotelluric survey is a powerful tool to elucidate the volcano's mechanisms and dynamics and its eruptions.

References

Aizawa K, Ogawa Y, Hashimoto T et al (2008) Shallow resistivity structure of Asama volcano and its implications for magma ascent process in the 2004 eruption. J Volcanol Geotherm Res 173:165–177

Aoki Y, Watanabe H, Koyama E et al (2005) Ground deformation associated with the 2004–2005 unrest of Asama Volcano, Japan. Bull Volcanol Soc Jpn 50:575–584

Aoki Y, Takeo M, Aoyama H et al (2009) P-wave velocity structure beneath Asama volcano, Japan, inferred from active source seismic experiment. J Volcanol Geotherm Res 187:272–277

Aoki Y, Takeo M, Ohminato T et al (2013) Magma pathway and its structural controls of Asama volcano, Japan. Geol Soc Lond Spec Publ 380:67–84

Aramaki S (1956) The 1783 activity of Asama volcano part 1. Jpn J Geol Geogr 27:189–229

Aramaki S (1957) The 1783 activity of Asama volcano part 2. Jpn J Geol Geogr 28:11–33

Hayakawa Y, Nakajima H (1998) Volcanic eruptions and hazards of Asama written in historical records. Bull Volcanol Soc Jpn Ser 2:213–221

Honkura Y, Koyama S (1978) Magnetic and electric investigations of the Asama volcano. Bull Earthq Res Inst Univ Tokyo 53:523–532

Kawasaki I (1965) Seasonal variation of the results surveyed by means of the electrical prospecting method (I). J Jpn Soc Eng Geol 6(2):115–120

Minakami T (1935a) The explosive activities of volcano Asama in 1935 (part 1). Bull Earthq Res Inst Univ Tokyo 13(3):629–644

Minakami T (1935b) The explosive activities of volcano Asama in 1935 (part 2). Bull Earthq Res Inst Univ Tokyo 13:790–800

Minakami T (1938) Magnetic surveys of volcano Asama. Bull Earthq Res Inst Univ Tokyo 16:100–116

Morita M, Mori T, Kazahaya R, Tsuji H (2016) Diffuse carbon dioxide emissions from hidden subsurface structures at Asama volcano, Japan. Bull Volcanol 78(3):17. https://doi.org/10.1007/s00445-016-1008-5

Nagaoka H, Ikebe T (1937a) Sudden magnetic variations observed on the volcano Asamayama. Proc Imperial Acad 13(3):62–64

Nagaoka H, Ikebe T (1937b) Magnetic variation during an explosion of Asamayama and its mechanism. Proc Imperial Acad 13(7):251–256

Nagaoka Y, Nishida K, Aoki Y et al (2012) Seismic imaging of magma chamber beneath an active volcano, earth planet. Sci Lett 333:1–8

Ogawa Y, Uchida T (1996) A two-dimensional magnetotelluric inversion assuming Gaussian static shift. Geophys J Int 126:69–76

Omori F (1911) The eruptions of Asama-yama. Rep Imp Earthq Invest Comm 73:29–101

Takeo M, Aoki Y, Ohminato T, Yamamoto M (2006) Magma supply path beneath Mt. Asama volcano, Japan. Geophys Res Lett 33. https://doi.org/10.1029/2006GL026247

Takeo M, Aoki Y, Koyama T (2022) Recent volcanic activity at the Asama volcano and long-period seismic signals. Proc Jpn Acad Ser B 98:416–438

Usui Y, Ogawa Y, Aizawa K et al (2017) Three-dimensional resistivity structure of Asama volcano revealed by data-space magnetotelluric inversion using unstructured tetrahedral elements. Geophys J Int 208:1359–1372

Yatsugatake Geomagnetic Observatory, Earthquake Research Institute, University of Tokyo (1982) Measurements of geomagnetic total intensity and earth resistivity on Asama volcano. Rep Coord Committee Predict Volcanic Erupt 24:12–15

Gravity Change of Asama Volcano

Takahito Kazama

1 Introduction

Gravimetry is one of the most effective methods for monitoring spatiotemporal mass variation in volcanoes, because the gravity value $g(t)$ at time t and point (x_0, y_0, z_0) can be written as the spatial integral of the mass distribution $dm(x, y, z, t)$:

$$g(t) = \int \frac{G\, dm(x,y,z,t)}{r^2} \cdot \frac{z_0 - z}{r} \qquad (1)$$

$$r = \sqrt{(x - x_0)^2 + (y - y_0)^2 + (z - z_0)^2}, \qquad (2)$$

where G denotes the gravitational constant. Superconducting and absolute gravimeters have often been used to monitor temporal changes in gravity at gravity stations. Regional and long-term gravity changes can also be monitored by campaign gravity measurements by using spring-type portable relative gravimeters. Typical observation errors of superconducting, absolute, and spring-type portable relative gravimeters are approximately 0.1, 2, and 10 μGal, respectively ($1\ \mu\text{Gal} = 1 \times 10^{-8}\ \text{m/s}^2$) (Hinderer et al. 2015; Niebauer 2015).

T. Kazama (✉)
Graduate School of Science, Kyoto University, Kyoto, Japan
e-mail: takujin@kugi.kyoto-u.ac.jp

Volcanic gravity changes are mainly caused by the following three mechanisms (Hagiwara 1978; Okubo 1992). First, the gravity change is negatively correlated with the vertical displacement at a gravity measurement point. Assuming a homogeneous semi-infinite medium, the proportional coefficient becomes equal to $(\beta + 2\pi\rho G)$, where β and ρ indicate the free-air vertical gravity gradient (< 0) and the medium density, respectively. The coefficient is approximately -0.1967 μGal/mm, using typical values of -0.3086 μGal/mm and 2670 kg/m^3 for β and ρ, respectively. Second, gravity changes also occur owing to the spatiotemporal variation in the medium-density distribution, which can be caused by inflation and/or deflation of pressure sources under volcanoes. The gravity change due to this effect can be calculated by the spatial integral of the medium-density change, numerically or analytically. Third, the gravity value can change with density variation and redistribution of the volcanic fluid. For example, the inflation of a pressure source contributes to a decrease in gravity, because the higher density medium is replaced by the lower density volcanic fluid at the surface of the source.

At Asama volcano, absolute gravity values have been repeatedly measured since the 2000s. In particular, absolute gravity data were continuously collected at Asama Volcano Observatory (AVO; star in Fig. 1a) from September to November 2004, to monitor the magma mass

S. Aramaki et al. (eds.), *The Asama Volcano*, Active Volcanoes of the World,
https://doi.org/10.1007/978-981-95-5799-8_22

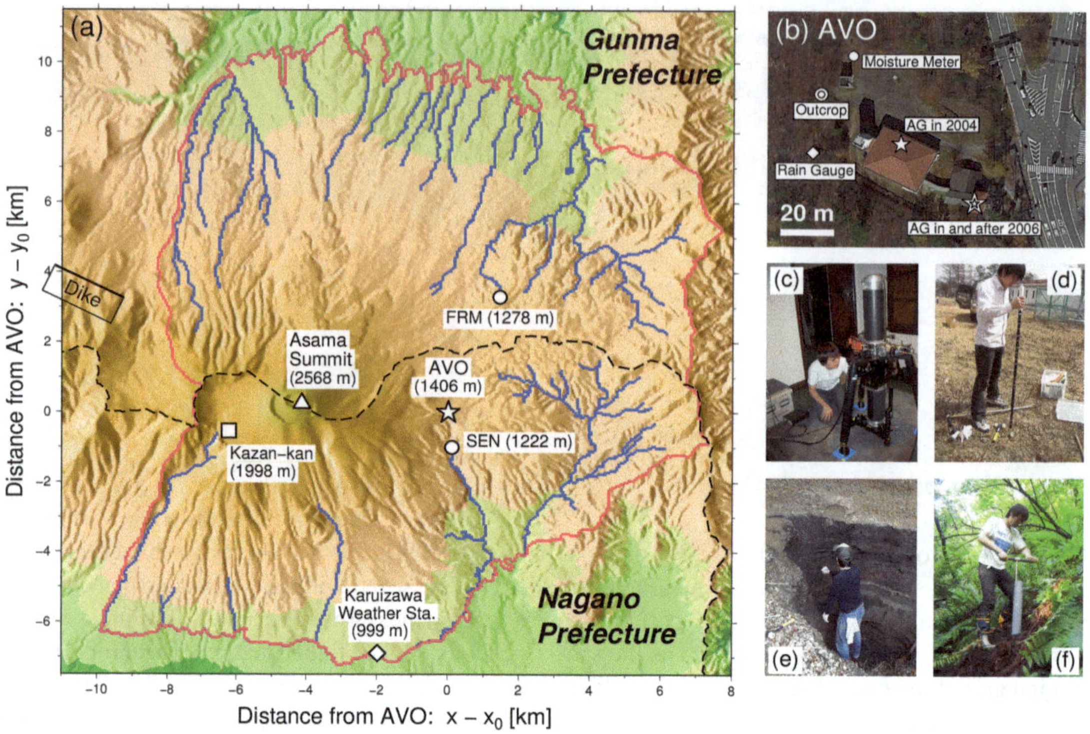

Fig. 1 Topographic map and gravity/hydrological observations at Asama volcano. (**a**) Topography of Asama volcano, drawn using a digital elevation model provided by Geospatial Information Authority of Japan Geographical Survey Institute (1997). The area enclosed by the red line indicates the computational region of the hydrological model in Kazama et al. (2015). Blue lines denote the positions of rivers and springs, and the black dashed line denotes the boundary between Gunma and Nagano Prefectures. The triangle, the star, two circles, the square and the diamond indicate the locations of the Asama's summit, Asama Volcano Observatory (AVO), water height stations (FRM and SEN), Kazan-kan and Karuizawa Weather Station, respectively. The open rectangle shows the horizontal position of an inflating dike during the 2004 eruptive event, determined from continuous GNSS observations (Aoki et al. 2005). (**b**) Satellite image of AVO, obtained at Google Earth Pro. Single and double stars indicate the locations of gravity points where absolute gravity was measured in 2004 and in/after 2006, respectively. The single circle, double circle and diamond denote the positions of a moisture meter, an outcrop and a precipitation gauge, respectively. (**c**) Photograph of the FG5 absolute gravimeter on the basement rock in the AVO's old building (the double star in panel (**b**)), taken in June 2006. (**d**) Photograph of the installation of the profile moisture meter outside the AVO buildings (the single circle in panel (**b**)), taken in April 2006. (**e**) Photograph of the sampling of soil cores at the outcrop outside the AVO buildings (the double circle in panel (**b**)), taken in 2007. (**f**) Photograph of the boring of a well to measure water height at SEN (one of the circles in panel (**a**)), taken in August 2005

redistribution associated with the eruptive activity of Asama volcano in 2004. Okubo (2005) demonstrated a 5-μGal absolute gravity change owing to the ascent of magma mass in the volcanic conduit, using the absolute gravity data collected in September 2004 (Okubo 2005). However, during October–November 2004, the volcanic gravity signal was obscured by 20-μGal hydrological gravity disturbance, associated with several heavy rainfall events and the consequent increase in underground water mass. In fact, the infinite flat plate of a water mass with unit thickness causes a gravity change of $2\pi\rho_w G = 0.0419$ μGal/mm, where ρ_w is water density. Therefore, the gravity change due to the water mass increase of 50 mm thickness is calculated to be 2.095 μGal, which exceeds the typical observation error of absolute gravimeters (2 μGal).

Kazama et al. (2015) thus calculated hydrological gravity change at AVO numerically, by modeling the four-dimensional water mass distribution in Asama volcano based on physical equations of water flow in the soil. They showed that the calculated hydrological gravity change agreed within approximately 2.6 μGal with the absolute gravity data observed during the noneruptive periods of 2006 and 2007. After subtracting the calculated hydrological gravity change from the absolute gravity values observed during the eruptive period of September–November 2004, they also found a volcanic gravity signal of approximately 5 μGal amplitude, which is consistent with the results of the previous study (Okubo 2005) and other volcanic observations.

This chapter reviews the hydrological and volcanic gravity changes observed at Asama volcano during the 2004 eruption, according to Kazama et al. (2015). This chapter also discusses the possible usage of continuous gravimetry for real-time monitoring of the magma mass distribution in Asama volcano during future eruptions.

2 Gravity and Hydrological Observations

Absolute gravity data were collected at AVO (star in Fig. 1a), 4 km east of the Asama's summit (triangle in Fig. 1a), using FG5 absolute gravimeters (Micro-g LaCoste Inc.; Fig. 1c) of Earthquake Research Institute, the University of Tokyo. The gravimeters were installed in the storage room of the AVO's new building (single star in Fig. 1b) during the eruptive period in 2004. The gravity data were collected continuously from September 9 to November 15 to monitor volcanic gravity changes. Gravity data were also acquired on the basement rock in the AVO's old building (double star in Fig. 1b) during the noneruptive periods of 2006 and 2007 to quantify hydrological gravity changes.

One set of gravity data was composed of 50–100 times of a drop gravity value which was measured by the free fall of a corner cube reflector every 10 sec, and the set gravity values were continuously collected every 30–120 min. The frequencies of drops-per-set and sets-per-day were increased during the gravity measurements in 2004, to clarify volcanic gravity changes with a high temporal resolution.

Hydrological data have also been collected at and around AVO (Fig. 1b) since 2005, to observe water mass variations directly and to utilize these data for hydrological modeling. The water height of the unconfined groundwater was measured continuously at FRM and SEN (Fig. 1f), where unconfined groundwater discharges into the rivers (Fig. 1a), using the Mini-Diver water height gauges (Van Essen Instruments B.V.). A PR2 profile moisture meter (Delta-T Devices Ltd.; Fig. 1d) was installed on the ground of AVO in 2006, to measure the soil moisture content at a depth of 0–1 m from the ground surface at AVO. The porosity and permeability values were measured through soil tests to the soil cores sampled at the outcrop of AVO (Fig. 1e) in 2007, and these parameters were determined to be 0.50 m^3/m^3 and 2.0×10^{-6} m/s, respectively.

Meteorological data obtained at and around the volcano were utilized as boundary conditions in hydrological modeling. Precipitation data were collected by Nagano Prefectural Government at the Mine-no-Chaya precipitation gauge, located in the AVO's compound (diamond in Fig. 1b), and time series of precipitation were put in the ground surface of the numerical hydrological model. Time series of air temperature, relative humidity, wind speed, and daylight hours, collected at Karuizawa Weather Station of Japan Meteorological Agency (diamond in Fig. 1a), were used to estimate the time variation in evapotranspiration on the ground of AVO.

3 Physical Modeling of Hydrological Gravity Disturbance

In the field of hydrology, the uppermost ground is composed of an unsaturated layer (or soil-water layer), a saturated layer (or an unconfined groundwater aquifer) and an impermeable layer (Fig. 2). In Fig. 2, $h_s = h_s(x, y)$, $h = h(x, y, t)$, and $h_b = h_b(x, y)$ denote the altitudes of the ground surface, groundwater table, and top surface of the impermeable layer, respectively. Water precipitates onto the ground and vertically infiltrates the unsaturated layer. After the water reaches the water table, it flows horizontally and discharges into rivers as spring water. These water flows in the unsaturated and saturated layers can be expressed by nonlinear diffusion equations of volumetric water content ($\theta = \theta(x, y, z, t)$) and groundwater height ($h(x, y, t)$), respectively (Jury and Horton 2004; Haitjema 1995). Therefore, four-dimensional water distributions (i.e., $\theta(x, y, z, t)$ and $h(x, y, t)$) can be obtained by numerically solving these equations under adequate boundary/initial conditions and realistic soil parameters.

Once the four-dimensional water distributions are estimated, hydrological gravity change can be calculated by the spatial integral of Eq. (1) for the water mass distribution. If the integral is performed for the unsaturated and saturated layers, the value of $dm(x, y, z, t)$ in Eq. (1) can be written as

$$dm = \begin{cases} \rho_w \, n \, dx \, dy \, dz & (\text{for } h_b(x,y) \le z < h(x,y,t)) \\ \rho_w \, \theta(x,y,z,t) \, dx \, dy \, dz & (\text{for } h(x,y,t) \le z < h_s(x,y)), \end{cases} \quad (3)$$

where n is the porosity of the saturated layer.

Kazama et al. (2015) applied the above hydrological modeling to Asama volcano and confirmed that the modeled spatiotemporal water distributions agreed well with the observed distributions at the volcano. $g_{w1}(t)$ and $g_{w2}(t)$ (red and blue lines) in Fig. 3a indicate the hydrological gravity changes in 2004, calculated from the spatial integral of the water mass in the unsaturated and saturated layers, respectively. In the integral calculation, the coordinate value of AVO was set for the gravimeter position (x_0, y_0, z_0) (see Eqs. (1)–(2)). $g_{w1}(t)$ increased rapidly during rainfall events because the water mass supplied to the unsaturated layer affected the downward gravitational force (i.e., gravity increase) on a gravimeter installed on the ground. $g_{w1}(t)$ decreased gradually after the rainfall, because the water mass moved away from the vicinity of the gravimeter according to evapotranspiration and soil-water infiltration. Consequently, $g_{w1}(t)$ changed by approximately 25 μGal from August 27 (day of year = 240) to November 25 (day of year = 330) in 2004. By contrast, the temporal variation of $g_{w2}(t)$ was less than 0.2 μGal, because a significant change in water height was not expected in the unconfined aquifer, whose water level was modeled to be located 120 m below the ground surface of AVO.

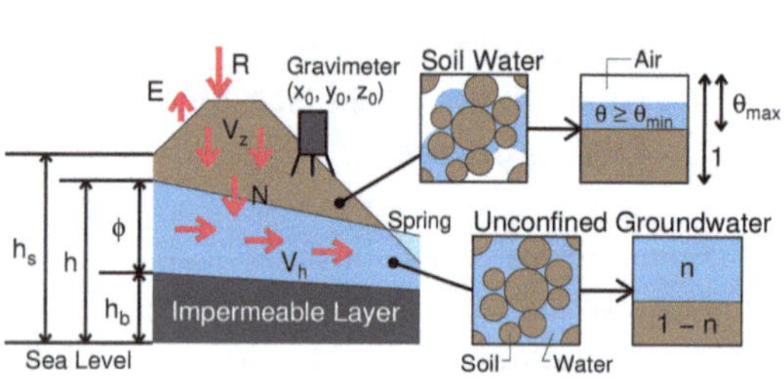

Fig. 2 Schematic structure of underground water in the uppermost ground

The red lines in Fig. 3b–d show the sum of the modeled hydrological gravity changes, $g_w(t) = g_{w1}(t) + g_{w2}(t)$, and the gray circles indicate the set gravity values observed with absolute gravimeters installed at AVO, $g_{obs}(t)$. $g_w(t)$ was quantitatively consistent with $g_{obs}(t)$ for the gravity increase during heavy rainfall events (black triangles in Fig. 3). However, the $g_w(t)$'s value relative to $g_{obs}(t)$ gradually increased after rainfall events, and the rate of difference between $g_{obs}(t)$ and $g_w(t)$ was approximately -0.20 μGal/day for every year in common. In fact, Kazama et al. (2015) observed the excess discharge of unconfined groundwater at SEN (1 km south of AVO; Fig. 1) immediately after heavy rainfall, due to the existence of a local shallow impermeable layer exposed below the surface soil around SEN. Because the hydrological model does not consider the existence of the local impermeable layer around SEN, more water mass recharges after heavy rainfall events in the hydrological model than in the actual Asama volcano.

If the additional hydrological effect related to the excess water discharge at SEN is expressed as $\Delta g_w(t) = -0.20 \cdot (t - t_0)$ (green line in Fig. 3a), where t_0 is the time of the heavy rainfall event (triangles in Fig. 3), then the sum

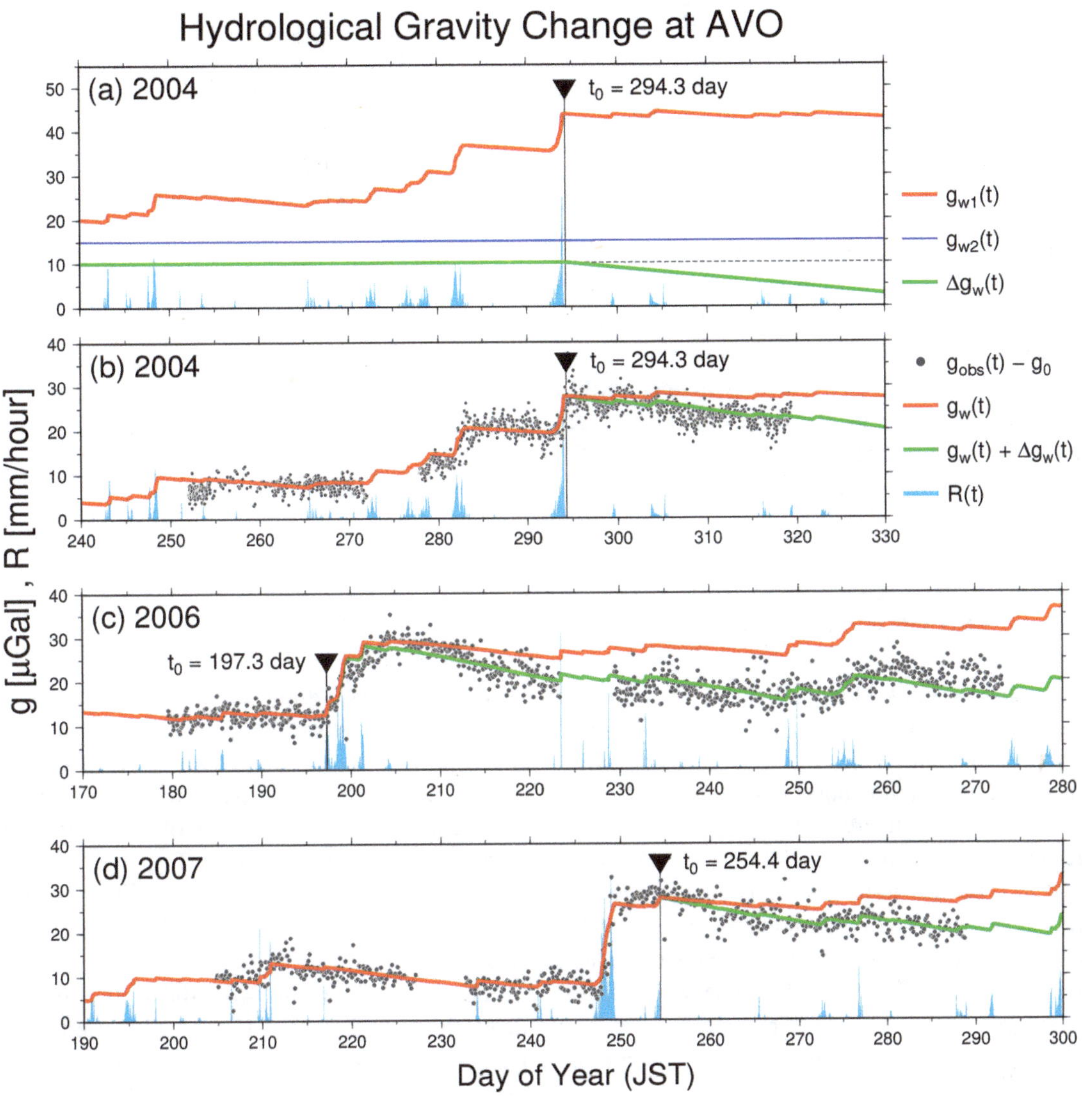

Fig. 3 Hydrological gravity changes at AVO. Panel (**a**) shows the hydrological gravity change calculated using the physical hydrological model of Kazama et al. (2015). $g_{w1}(t)$ and $g_{w2}(t)$ (red and blue lines) indicate the gravity values expected from the water mass in the unsaturated and saturated layers, respectively. $\Delta g_w(t)$ (green line) indicates the additional effect of the excess water discharge at the SEN's spring. Note that in panel (**a**), intercepts of $g_{w1}(t)$, $g_{w2}(t)$ and $\Delta g_w(t)$ are set to be 20, 15 and 10 μGal, respectively, so as to show all of the time series in this panel. Panels (**b**)–(**d**) compare the calculated hydrological gravity changes with the absolute gravity data observed in 2004, 2006 and 2007, respectively. In each panel of (**b**)–(**d**), gray circles indicate $g_{obs}(t) - g_0$, the set gravity data observed by FG5 absolute gravimeters at AVO. An arbitrary value g_0 is subtracted from the observed absolute gravity data $g_{obs}(t)$, so as to show all of the observed data in each panel. The value of g_0 for each year is set to be 979527895 μGal for 2004, 979528015 μGal for 2006, and 979528020 μGal for 2007. The red line indicates $g_w(t) = g_{w1}(t) + g_{w2}(t)$, i.e. the sum of the hydrological gravity change due to the water mass in the unsaturated and saturated layers. The green line indicates $g_w(t) + \Delta g_w(t)$, i.e. all of the hydrological effects including the water mass in the ground and the excess water discharge. In all panels, cyan bars show the hourly precipitation observed at AVO, and the inverted triangle denotes the time when the excess water discharge is expected to start

of the hydrological gravity change ($g_w(t)+\Delta g_w(t)$; green lines in Fig. 3b–d) is consistent with the observed gravity change ($g_{obs}(t)$; gray circles in Fig. 3b–d). The root-mean-square values of $g_{obs}(t)-(g_w(t)+\Delta g_w(t))$ during volcanically quiet periods were calculated to be 2.66 and 2.49 μGal for 2006 and 2007, respectively. Kazama et al. (2015) thus concluded that their hydrological model, with the additional effect of excess water discharge, reproduced the hydrological gravity disturbances observed at Asama volcano within approximately 2.6 μGal, which is nearly equal to the typical observation error of the FG5 absolute gravimeters (Niebauer 2015).

4 Volcanic Gravity Change During the 2004 Eruption

Kazama et al. (2015) corrected for the effects of hydrological gravity disturbances in the absolute gravity value observed during the eruptive period in 2004, by subtracting $(g_w(t)+\Delta g_w(t))$ from $g_{obs}(t)$. The green circles in Fig. 4a show the corrected gravity value, $g_{obs}(t)-(g_w(t)+\Delta g_w(t))$, and the red line indicates the daily-average corrected gravity, which is denoted by $\Delta g(t)$. $\Delta g(t)$ changes within 5 μGal and is negatively correlated with the eruptive activity of Asama volcano. For example, $\Delta g(t)$ decreased by approximately 2 μGal during a continuous eruption from September 14 to 18. $\Delta g(t)$ also decreased by approximately 3 μGal during three very small eruptions from October 16 to 20. These results suggest that the time variation in $\Delta g(t)$ is related to some volcanic mass variation(s) in Asama volcano.

Volcanic gravity signals have often been explained by pressure changes in magma chambers such as dikes and spherical pressure sources (Hagiwara 1978; Okubo 1992). In the case of Asama volcano, a quasi-vertical dike on the western side of the Asama summit (open rectangle in Fig. 1a) was inferred from GNSS observations around the volcano (Aoki et al. 2005, 2013). However, only -0.12 μGal change in gravity is expected from the dike's inflation during the absolute gravity measurement in 2004, because AVO was located in the strike direction of the dike (Fig. 1a). The observed gravity change ($\Delta g(t)$; red line in Fig. 4a) cannot be fully explained by other possible causes such as the mass redistributions of volcanic gas and ash.

Kazama et al. (2015) therefore explained the observed gravity change ($\Delta g(t)$) by the distribution of magma mass in the open conduit of Asama volcano (Okubo 2005). By performing the spatial integral of Eq. (1) for the magma mass in the conduit, its gravity change ($g_c(t)$) can be expressed as

$$\Delta g(t) \simeq g_c(t) = G\sigma\left[\frac{1}{\sqrt{d^2+(h_m(t)-z_0)^2}}-\frac{1}{d}\right], \tag{4}$$

where σ, $h_m(t)$, and d indicate the linear density, magma head altitude, and horizontal distance between the conduit and AVO, respectively (Fig. 4c). The $g_c(t)$'s variation explains the negative correlation between the observed gravity ($\Delta g(t)$) and volcanic activity (Fig. 4a), because $g_c(t)$ decreases with the ascent of the magma head located above the gravimeter altitude (Fig. 4d).

The red lines in Fig. 4b indicate the magma head altitude ($h_m(t)$), which is converted from the observed gravity residual ($\Delta g(t) \simeq g_c(t)$) using Eq. (4). Although $h_m(t)$ has two solutions to a $g_c(t)$'s value (dashed line in Fig. 4d), one unique solution can be chosen according to the continuity and boundary conditions of magma head variation. The solution of $h_m(t) < z_0$ should be selected for the time series before September 16, because Geospatial Information Authority of Japan found a lava dome in the Asama's crater on September 16 for the first time during the eruptive activity in 2004 (Oki et al. 2005). The solution of $h_m(t) \geq z_0$ should be chosen for the time series on and after September 16, because the volcanic glows were intermittently observed by Japan Meteorological Agency from September 16 to December 10.

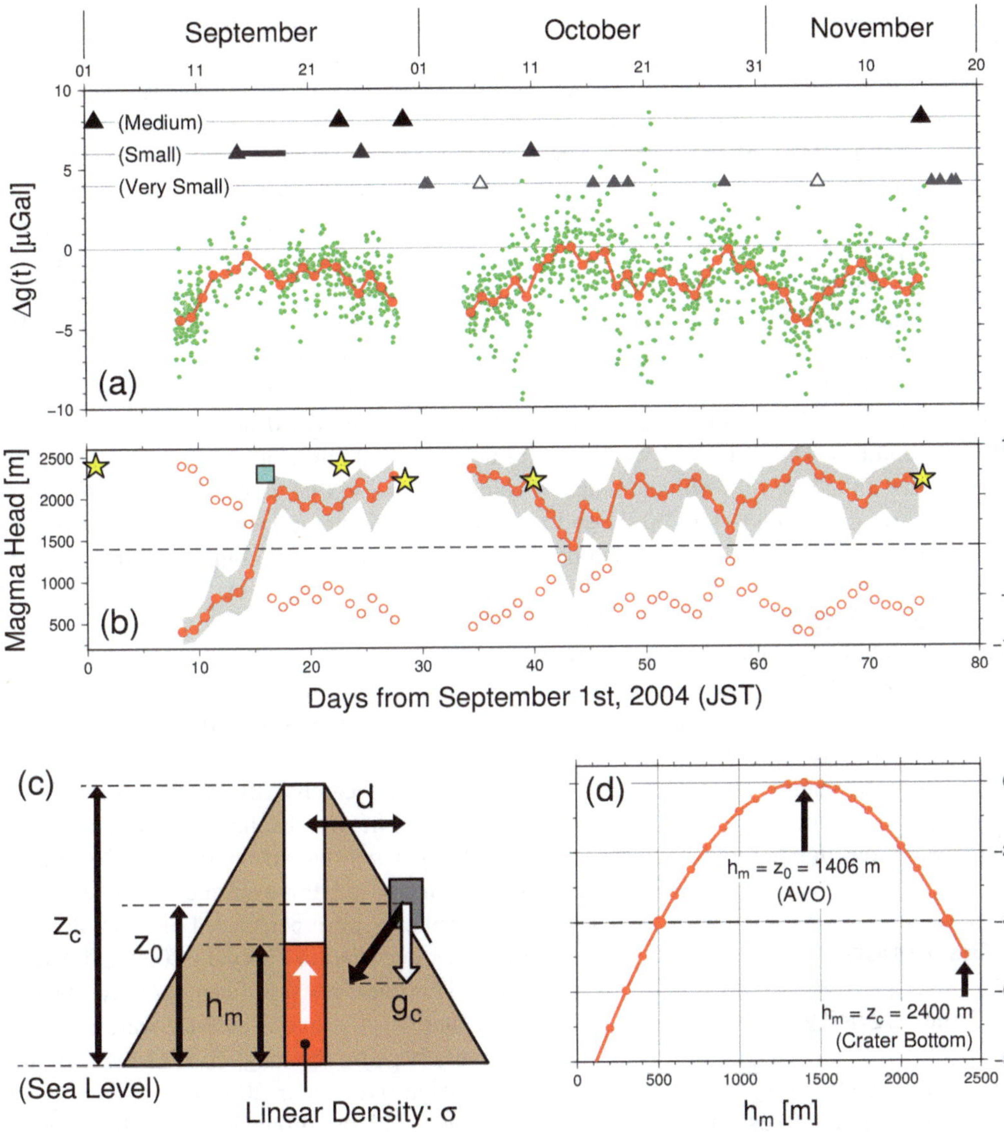

Fig. 4 Residual gravity change due to the time variation in magma head altitude, associated with the 2004 eruption of Asama volcano. (**a**) Green circles and red lines show the raw and daily average of the gravity residual ($g_{obs}(t)-g_w(t)-\Delta g_w(t)$ and $\Delta g(t)$), respectively. An arbitrary constant value was subtracted from these gravity values, so as to orient the $\Delta g(t)$'s maximum to 0 μGal. Black and gray triangles indicate the times of eruptions observed at Asama volcano, and the size of each triangle is consistent with the eruption magnitude. The gray thick line indicates the duration of continuous small eruptions observed at Asama volcano. White triangles indicate the times of other volcanic phenomena observed at Asama volcano; small tilt changes were observed 2.5 km north-northeast of the Asama summit on October 6, and low volcanic sounds were heard 4 km north-northeast of the Asama summit on November 5. All data on eruptions and other volcanic activities are from Japan Meteorological Agency (2013). (**b**) Red lines indicate the magma head altitude in the Asama conduit ($h_m(t)$), to which the residual gravity ($\Delta g(t)$) was converted using Eq. (4). Although $h_m(t)$ has two solutions for one $\Delta g(t)$ value (see panel (**d**)), one solution (solid circles) was chosen and the other solution (open circles) was rejected based on several volcanic observation results. Gray areas indicate the error ranges of the magma head, calculated from standard deviation values of the daily residual gravity. The square and stars represent the times/locations of a lava dome discovered in the Asama crater (Oki et al. 2005) and single-force earthquakes detected by seismometers around Asama volcano (Ohminato et al. 2006), respectively. (**c**) Schematic structure of the Asama conduit. (**d**) Gravity change of g_c to the magma head altitude h_m (Eq. (4)), calculated using the constants of $d = 4000$ m, $z_0 = 1406$ m and $\sigma = 1.0\times10^8$ kg/m

The magma head was located below the gravimeter altitude before September 15 and passed the gravimeter's altitude on or around September 15. The rising speed of magma head reached a maximum of approximately 450 m/day between September 15 and 17, which does not conflict with the typical ascent rates observed for other volcanoes (Rutherford et al. 2000). After September 15, which corresponds to the major eruptive period during the 2004's volcanic activity, the magma head was located close to the Asama's crater (~ 2400 m) and fluctuated between the altitudes of AVO and the crater with a period of 10–15 days. The estimated variation in the magma head altitude agrees with the times/altitudes of the lava dome (Oki et al. 2005) and single-force earthquakes (Ohminato et al. 2006) (cyan square and yellow stars in Fig. 4b, respectively). The magma head variation was also found to be consistent with the frequency of the B-type earthquakes (Takeo et al. 2006) and emission rate of SO_2 gas (Ohwada et al. 2013), according to Kazama et al. (2015). In this respect, the estimated variation in the magma head altitude (the red lines in Fig. 4b) can comprehensively explain the temporal variations in other volcanic data.

5 Discussion

5.1 Linear Density

When converting $g_c(t)$ to $h_m(t)$ according to Eq. (4), Kazama et al. (2015) used the linear density value of $\sigma = 1.0 \times 10^8$ kg/m for Asama volcano such that the variation range of $g_c(t)$ corresponded to the observed gravity change ($\simeq$ 5 μGal). The linear density can be expressed as the product of a few constant values:

$$\sigma = \pi r_c^2 \rho_m n_c , \tag{5}$$

where r_c, ρ_m, and n_c denote the radius of the volcanic conduit, magma density, and porosity in the conduit, respectively. The value of $\sigma = 1.0 \times 10^8$ kg/m can be explained by typical values of $r_c = 155$ m, $\rho_m = 2600$ kg/m^3, and $n_c = 0.5$.

However, these parameters (r_c, ρ_m, n_c) cannot be determined independently from the gravity data, because gravity is an integral value of the mass distribution (Eq. (1)). Therefore, other volcanic observation results should be considered to determine these parameters and the resultant linear density value. In the case of Asama volcano, the density structure below the crater was estimated from cosmic-ray muon radiography in Tanaka et al. (2007). According to their results, the radius of the volcanic conduit is approximately 100–150 m; this was consistent with the typical value described above. Note that muon radiography data may not be collected in typical volcanoes, but in that case, seismic and/or infrasound data may be utilized for the determination of some parameters contained in Eq. (5) (Nishimura 2009; Ishii et al. 2023).

5.2 Gravimeter's Position

Earthquake Research Institute, the University of Tokyo has collected absolute gravity data at AVO repeatedly, when the volcanic activity of Asama volcano increased. For example, absolute gravity measurements were performed after small eruptions that occurred in February 2009 and August 2019. However, no significant gravity changes associated with volcanic activity were detected, probably because the magnitude of the activities was smaller than that in 2004.

As $g(t)$ is proportional to r^{-3} as shown in Eq. (1), the position of the gravimeters should be closer to that of the mass variation sources to obtain gravity changes with higher amplitude. According to Eq. (4), a gravity change of approximately 80 μGal is expected if a gravimeter is installed at Kazan-kan, a mountain hut located along the western climbing route of Asama volcano ($d = 2200$ m, $z_0 = 2000$ m; square in Fig. 1a) (Kazama 2019).

In addition, the magma head altitude could be monitored more accurately, if two gravimeters are simultaneously installed at two independent stations at different altitudes. Assume that the altitude of station B (z_B) is higher than that of station A (z_A). When the altitude of the ascending magma head is $h_m(t) < z_A < z_B$, gravity increases at both stations. However, when $h_m(t)$ exceeds z_A, the gravity change at A decreases, whereas gravity continues to increase at B. The gravity value also starts to decrease at B after $h_m(t)$ exceeds z_B; in particular, it can be understood that the magma head is approaching the volcanic summit if gravity decreases simultaneously at the two stations.

5.3 Real-Time Monitoring of Magma Head Altitude

The magma head altitude in the conduit of Asama volcano can be monitored from continuous gravity observations, and information on the magma head altitude may be used for the real-time prediction of eruptions at the summit of the volcano. As shown in Fig. 4d, gravity is expected to increase when the magma head altitude ($h_m(t)$) approaches the gravimeter altitude (z_0), whereas gravity is expected to decrease when $h_m(t)$ deviates from z_0. Although $h_m(t)$ has two solutions for one gravity value (Fig. 4d), only one solution of $h_m(t)$ must be chosen to explain other volcanic data and surface phenomena observed at the crater.

To accurately determine the magma head altitude in real time, hydrological gravity disturbances should be adequately subtracted from the observed gravity data at the same time. One of the easiest ways of hydrological disturbance correction is to model hydrological gravity change by assuming a linear gravity response to precipitation (Harnisch and Harnisch 2006; Imanishi et al. 2006). However, water flows in unsaturated layer and unconfined aquifer are governed by nonlinear equations (Jury and Horton 2004; Haitjema 1995); hence, the gravity response to precipitation is nonlinear. Instead of linear hydrological models, nonlinear hydrological models, such as those developed by Kazama et al. (2015), should be applied to accurately correct the hydrological gravity disturbances observed at Asama volcano. Real-time correction can be achieved by calculating the hydrological effect ($g_w(t)$) along with gravity observations and then subtracting $g_w(t)$ from the observed gravity value ($g_{obs}(t)$).

Note that the additional gravity decrease ($\Delta g_w(t)$; green line in Fig. 3a) relating to the excess water discharge at SEN must also be subtracted from $g_{obs}(t)$, when gravity observation is conducted at AVO during the rainy season. According to Kazama et al. (2015), the excess water discharge occurs just after a heavy rainfall event every year, and the discharging rate doubles from 1700 m^3/day to 3400 m^3/day. Therefore, the water discharge needs to be continuously monitored at SEN, to determine whether to correct for the effect of $\Delta g_w(t)$ in gravity value observed at AVO ($g_{obs}(t)$).

6 Summary

The continuous gravity change of 5 μGal in amplitude was obtained from the absolute gravity data collected at Asama Volcano Observatory (1406 m in altitude; 4 km east of the Asama's summit) during the eruptive period in 2004, after the hydrological gravity disturbance was corrected from the raw gravity data using a physical hydrology model. The gravity value decreased during the eruptions, because the ascending magma in the upper part of the volcanic conduit contributed to the upward attraction force to the gravimeter installed at the observatory. The magma head altitude, which was converted from the observed gravity change, varied in the range of 400–2400 m and approached to the Asama's crater (~ 2400 m in altitude) during the eruptions in 2004. The time variation of the magma head altitude agreed comprehensively with those of other independent volcanic observations such as single-force earthquakes and volcanic gas emission rate. These results suggest that the magma head altitude in the volcanic conduit can be monitored in real time by collecting gravity changes continuously, as long as the hydrological gravity disturbance is corrected from the raw gravity data simultaneously.

References

Aoki Y, Watanabe H, Koyama E, Oikawa J, Morita Y (2005) Ground deformation associated with the 2004–2005 unrest of Asama volcano, Japan. Jpn. Bull. volcanol. Soc. Jpn. 50(6):575–584, https://doi.org/10.18940/kazan.50.6_575, in Japanese with English abstract

Aoki Y, Takeo M, Ohminato T, Nagaoka Y, Nishida K (2013) Magma pathway and its structural controls of Asama Volcano, Japan. Geol. Soc. Lond. Special Publ. 380(1):67–84. https://doi.org/10.1144/SP380.6

Geographical Survey Institute (1997) Digital map 50m grid (elevation), vol 2. Japan Map Center, Tokyo, Japan, CD-ROM

Hagiwara Y (1978) The Mogi model as a possible cause of the crustal uplift in the eastern part of Izu peninsula and the related gravity change. Bull. Earthquake Res. Inst. Univ. Tokyo 52:301–309. https://doi.org/10.15083/0000033210, in Japanese with English abstract

Haitjema HM (1995) Analytic element modeling of groundwater flow. Elsevier, Amsterdam

Harnisch G, Harnisch M (2006) Hydrological influences in long gravimetric data series. J. Geodyn. 41(1):276–287. https://doi.org/10.1016/j.jog.2005.08.018

Hinderer J, Crossley D, Warburton R (2015) 3.04—superconducting gravimetry. In: Schubert G (ed) Treatise on geophysics, 2nd edn. Elsevier, Oxford, pp 59–115. https://doi.org/10.1016/B978-0-444-53802-4.00062-2

Imanishi Y, Kokubo K, Tatehata H (2006) Effect of underground water on gravity observation at Matsushiro, Japan. J. Geodyn. 41(1):221–226. https://doi.org/10.1016/j.jog.2005.08.031

Ishii K, Yokoo A, Ohkura T, Kazama T (2023) Temporal variation in the depth of the magma surface at Aso volcano in 2014–2015. Bull. Volcanol. 85(1):1–21. https://doi.org/10.1007/s00445-022-01616-x

Japan Meteorological Agency (2013) National catalogue of the active volcanoes in Japan (the fourth edition, English version). https://www.data.jma.go.jp/svd/vois/data/tokyo/STOCK/souran_eng/menu.htm

Jury WA, Horton R (2004) Soil physics, 6th edn. John Wiley & Sons, London

Kazama T (2019) Gravity observations at volcanic areas in Japan: current status and hydrological disturbance problems. Bull. Volcanol. Soc. Jpn. 64(3):189–212. https://doi.org/10.18940/kazan.64.3_189, in Japanese

Kazama T, Okubo S, Sugano T, Matsumoto S, Sun W, Tanaka Y, Koyama E (2015) Absolute gravity change associated with magma mass movement in the conduit of Asama volcano (central Japan), revealed by physical modeling of hydrological gravity disturbances. J. Geophys. Res. Solid Earth 120(2):1263–1287. https://doi.org/10.1002/2014JB011563

Niebauer T (2015) 3.03—gravimetric methods—absolute and relative gravity meter: instruments concepts and implementation. In: Schubert G (ed) Treatise on geophysics, 2nd edn. Elsevier, Oxford, pp 37–57. https://doi.org/10.1016/B978-0-444-53802-4.00057-9

Nishimura T (2009) Ground deformation caused by magma ascent in an open conduit. J. Volcanol. Geothermal Res. 187(3):178–192. https://doi.org/10.1016/j.jvolgeores.2009.09.001

Ohminato T, Takeo M, Kumagai H, Yamashina T, Oikawa J, Koyama E, Tsuji H, Urabe T (2006) Vulcanian eruptions with dominant single force components observed during the Asama 2004 volcanic activity in Japan. Earth Planets Space 58(5):583–593. https://doi.org/10.1186/BF03351955

Ohwada M, Kazahaya K, Mori T, Kazahaya R, Hirabayashi J, Miyashita M, Onizawa S, Mori T (2013) Sulfur dioxide emissions related to volcanic activity at Asama volcano. Jpn. Bull. Volcanol. 75(12):1–14. https://doi.org/10.1007/s00445-013-0775-5

Oki S, Murakami M, Watanabe N, Urabe B, Miyawaki M (2005) Topographic change of the summit crater of the Asama volcano during 2004 eruption derived from repeated airborne synthetic aperture radar (SAR) measurements. Bull. Volcanol. Soc. Jpn. 50(5):401–410. https://doi.org/10.18940/kazan.50.5_401, in Japanese with English abstract

Okubo S (1992) Gravity and potential changes due to shear and tensile faults in a half-space. J. Geophys. Res. Solid Earth 97(B5):7137–7144. https://doi.org/10.1029/92JB00178

Okubo S (2005) Gravity changes associated with volcanism: observation, theory, and analysis. Bull. Volcanol. Soc. Jpn. 50(Special):S49–S58. https://doi.org/10.18940/kazan.50.Special_S49, in Japanese with English abstract

Rutherford MJ, Gardner JE, Sigurdsson H (2000) Rates of magma ascent. In: Encyclopedia of volcanoes, pp 207–217

Takeo M, Aoki Y, Ohminato T, Yamamoto M (2006) Magma supply path beneath Mt. Asama volcano, Japan. Geophys. Res. Lett. 33(15):L15310. https://doi.org/10.1029/2006GL026247

Tanaka HKM, Nakano T, Takahashi S, Yoshida J, Takeo M, Oikawa J, Ohminato T, Aoki Y, Koyama E, Tsuji H, Niwa K (2007) High resolution imaging in the inhomogeneous crust with cosmic-ray muon radiography: The density structure below the volcanic crater floor of Mt. Asama, Japan. Earth Planetary Sci. Lett. 263(1):104–113. https://doi.org/10.1016/j.epsl.2007.09.001

Muography of Asama Volcano

Hiroyuki K. M. Tanaka

1 Introduction

While similar in principle to medical radiographic applications, the muon acts as a probe for muography instead of the X-ray. When muons travel through matter with heterogeneous density distribution, a proportion of the muons stop and decay (depending on the density of the matter which is traversed) which results in a shadow image reflecting the targeted interior distribution. Muography differs from radiography in two important ways. Firstly, the source of high-energy muons is natural, abundant, and available worldwide. In the Earth's atmosphere, primary cosmic rays and atmospheric nuclei collide, and after a series of subsequent interactions, cosmic ray muons are eventually produced by the decay process of mesons. Muons arrive only from the upper hemisphere, and their vertical flux is 70 m^{-2} s^{-1} (Zyla et al. 2020). Due to the relativistic effect, these high-energy muons can travel to the Earth's surface and also beyond in shallow underground/underwater locations (a few km/~10 km respectively). Secondly, the muon is suitable as a probe for subterranean and/or underwater imaging. Despite sharing many characteristics with electrons, being a charged lepton and being free from strong interaction and sensitive to electromagnetic force, muons are 200 times heavier than electrons. Therefore, unlike the electron, the strongly penetrative characteristics of muons make it possible to image larger and/or more dense objects (for example, a few kilometers of solid rock) with muography.

The muon's energy and intensity at sea level are also strongly influenced by its arrival angle. These angular-dependent energy spectra are not determined only by galactic cosmic rays (GCRs). High-energy muons tend to be azimuthally isotropic; however, since the meson's (from which muons are generated) mean free path varies in the atmosphere, the average energy of cosmic-ray muons tends to be determined by the angle of its parent meson. Mesons that are more vertically aligned are more likely to lose more energy in comparison to horizontally aligned parent mesons since they collide with atmospheric nuclei before they decay into muons. On the other hand, horizontally aligned mesons tend to decay before they collide with atmospheric nuclei; therefore, they maintain and transfer more energy to the muons they generate. Consequently, the muon energy spectrum varies for different arriving zenith angles. As a result, horizontal muon flux is lower than vertical flux, but the average muon energy is higher. This high-energy characteristic of horizontal muons means that these particles are more likely to penetrate into denser and larger objects; this makes it possible to image, for example, volcanoes within

H. K. M. Tanaka (✉)
Earthquake Research Institute, The University of Tokyo, Tokyo, Japan
e-mail: ht@eri.u-tokyo.ac.jp

S. Aramaki et al. (eds.), *The Asama Volcano*, Active Volcanoes of the World,
https://doi.org/10.1007/978-981-95-5799-8_23

reasonable timeframes ranging from hours (Tanaka 2016) to a few months (Olah et al. 2023), depending on the size and average density of the target.

Underground measurements were utilized in the first stages of cosmic ray studies. Since there were a lot of carefully mapped-out mine galleries available to study, these were the first sites used for research of this kind. In 1955, the Australian scientist E.P. George was the first to propose using cosmic-ray-derived muons for surveying geological structures (George 1955). With Geiger counters, he made rudimentary measurements of the thickness of the rock overburden above a hydroelectric plant tunnel in the Australian Snowy Mountains by comparing the reduction in the muon flux after passing through the rock with the value of the open-sky flux. However, this setup was not capable of resolving structural information within the overlying rock. Luis Alvarez modified the design of George's muography apparatus with the goal of revealing the internal structures of inaccessible, massive objects. During the late 1960s, Alvarez's group applied muography in the Pyramid of Khafre (Egypt) to search within for unknown mortuaries (Alvarez et al. 1970). Vertical trajectories of muons traveling through the pyramid were counted with gaseous detectors. The researchers were hoping to discover new chambers (the Belzoni chamber was the only known chamber) inside the Pyramid of Khafre, because Khafre's father, Cheops, had more complicated internal structures (three chambers) inside his pyramid; researchers had surmised that more hidden chambers might be located inside the Pyramid of Khafre. Alvarez's group discovered no unknown chambers inside the pyramid; however, this trailblazing experiment showed the potential of muography for imaging man-made and natural objects.

High-energy (in the TeV range) muons can penetrate through more than a kilometer of rock; therefore, it was recognized that muography was also applicable to mountains. By counting muons and determining the trajectories of muons emerging from the other side of the mountain (at angles almost parallel to the ground surface), the internal structure of the mountain could be accurately imaged with muography. In 1995, Nagamine et al. (1995) reported on a new muography breakthrough. His group pointed out that near-horizontal cosmic ray muons have the potential to be utilized for imaging the interior of a gigantic geophysical structure, such as the top of a volcano. By tracking muon trajectories from the detector to the exiting point of the volcano, these muons act as probes to define the density distribution within the volcano itself. However, when collecting muons from this angle, the amount of matter the muons have to traverse is much larger than the pyramid. Only higher-energy muons, which are also rare, can travel through a sufficient thickness of matter for practical muography of this kind.

The first successful experimental evidence obtained with muography was from an experiment conducted on the summit of Asama volcano, Japan, in 2007. For the first time, the density distribution of the volcano's internal structure was revealed with muographic imagery (Tanaka et al. 2007a). After this first experiment, several volcanoes worldwide have been muographically studied. The unique strength of muography is that a density profile of a volcano's interior structure can be characterized. Density is an essential property to elucidate in order to evaluate various volcanic phenomena, for example: the dynamics of magma within the conduit, volcanic cone growth, and landslide/fracturing analysis. Other features that are ideal for muographically derived density distribution include: composition and state of a given target, indirect studies of common driving forces of geodynamics (such as gravity and buoyancy), and time-dependent variations in volcanic structures, along with identification of volcanic materials such as magma (Tanaka et al. 2014; Olah et al. 2023), water (Tanaka et al. 2011), ash (Tanaka 2020; Oláh et al. 2021), and the hollow pathways (Tanaka et al. 2007a) within volcanos that these materials settle into or are ejected from.

Prerequisite to determining the muon flux after passing through the mountain, the muon energy spectrum (Zyla et al. 2020) must be combined and evaluated with the muon energy-range relationship (Groom et al. 2002). The densimetric rock thickness along the muon path (X) is derived from the average density (ρ) and the exterior shape of the target volume; this can be defined by ρx, where x is the geometrical rock thickness along the muon path. Furthermore, by utilizing the range-energy relationship and the value of X, the minimum muon energy (E_c) can be determined. It is also necessary to know the muon flux after passing through the mountain, which can be derived by integrating the muon energy spectrum with the energy over the range between E_c and infinity. Then, the average density along the muon path can be inversely derived from the transmitted muon flux. Since it is rare to detect muons with sufficient energy to travel more than 2–3 km through rock, only shallow regions of volcanoes are targeted. However, there are many volcanoes and other geological features that are ideal for muographic studies. Since the summits of most volcanoes have vents and/or active craters that are both of scientific interest and also present safety concerns to the local community, these are often excellent targets for muography imaging. Thus far, lava dome formation (Kusagaya and Tanaka 2015) and collapse (Tanaka 2016), magma ascent and descent (Tanaka et al. 2014), vulcanian explosion mechanism (Tanaka et al. 2009a), and magma convection (Tanaka et al. 2009b; Shinohara and Tanaka 2012) have been investigated with muography. Muography is also well suited to the study of volcanic eruptions which have similar mechanisms but often vastly different eruption styles.

Muographic images range from a static image taken after several months of exposure to sequential images (with much shorter exposure times) which can help to clarify more dynamic processes. Examples of volcanic processes that can be captured with muographic images are: the explosion of solidified magma deposits on a crater floor (Tanaka et al. 2009a), ascent and descent of magma underneath a crater floor (Tanaka et al. 2014), enlargement of the vent size and crack generation by recurrent fumarolic explosions (Jourde et al. 2016), plug formation (Olah et al. 2023), crater floor collapse (Lo Presti et al. 2020), and tephra deposition and erosion (Tanaka 2020; Oláh et al. 2021). Recently, machine-learning-based eruption forecasts (Nomura et al. 2020; Oláh and Tanaka 2022) and joint measurements combining muography with other techniques are increasing, for instance in conjunction with gravity (Nishiyama et al. 2014; Gibert et al. 2022), electric resistivity (Gibert et al. 2022), and satellite-based interferometric synthetic aperture radar (InSAR) (Olah et al. 2023). Joint measurements improve the accuracy of measurements of the magma rise evolution process in the conduit of a volcano. This is essential information to get as early as possible (Olah et al. 2023), as it will buy critically important time for those who are responsible for local alarm and evacuation protocols.

2 Principle of Muography

2.1 Source

2.1.1 Muon Spectrum

The following section will describe in more detail the process by which muons are naturally produced in the Earth's atmosphere. As previously mentioned, high-energy muons are produced in the interaction of the primary cosmic rays and the nuclei in the Earth's atmosphere. These muons are called atmospheric or cosmic ray muons. Extensive studies of the differential atmospheric muon spectrum at various zenith angles have been conducted by various authors. By taking into account the fact that primary particles are isotopically present within the outer sections of the Earth's atmosphere and both pions and kaons (Zyla et al. 2020) retain the directions of their producer after the following processes:

$$\text{Charged pions}\,\pi^{\pm} \rightarrow \mu^{\pm} + \nu_{\mu}\left(\overline{\nu}_{\mu}\right)(\sim 100\%), \quad (1)$$

$$\text{Charged Kaons}\,K^{\pm} \rightarrow \mu^{\pm} + \nu_{\mu}\left(\overline{\nu}_{\mu}\right)\ (\sim 63.6\%), \quad (2)$$

$$\text{Neutral Kaons}\,K_{L} \rightarrow \pi^{\pm} + e^{\pm} + \nu_{e}\left(\overline{\nu}_{e}\right)\ (\sim 40.6\%), \quad (3)$$

$$K_{L} \rightarrow \pi^{\pm} + \mu^{\pm} + \nu_{\mu}\left(\overline{\nu}_{\mu}\right)\ (\sim 27.0\%), \quad (4)$$

$$\text{Neutral Kaons}\,K_{S} \rightarrow \pi^{+} + \pi^{-}\ (\sim 69.2\%), \quad (5)$$

where the numbers in parentheses indicate branching ratios. Validations and improvements of these authors' analytical expressions have been made by comparing with the spectrum of muons that can be taken from experiments made at sea level (Allkofer et al. 1975; Jokisch et al. 1979; Achard et al. 2004; Haino et al. 2004), and the current expression matches the observed muon flux. The muon spectrum is estimated by assuming a pion and kaon generation spectra value of $1.8 \times 10^4 E^{-2.7}$ (nucleons m^{-2} s^{-1} sr^{-1} GeV^{-1}) as follows (Zyla et al. 2020):

$$dN / dEd'$$

$$\approx 0.14 E^{-2.7} [1 / \left(1 + 1.1 E \cos\theta / 115\,\text{GeV}\right)$$

$$+0.054 / \left(1 + 1.1 E \cos\theta / 850\,\text{GeV}\right)]\text{cm}^{-2}\,\text{s}^{-1}\,\text{sr}^{-1}\,\text{GeV}^{-1} \quad (6)$$

where the first term and the second term, respectively, indicate the contribution of pions and charged kaons. As previously described, the average muon energy decreases in vertical muons because there is a stronger likelihood that the mesons will collide with atmospheric nuclei before having a chance to decay into muons while at larger angles (from zenith). There is also a tendency for low-energy muons to decay before reaching the surface, and/or pions will tend to decay before they further interact.

The decay rate of a muon at rest is $\tau = 2.2$ μs. However, since a high-energy muon travels at almost the speed of light in vacuum (c), this decay constant extends due to the relativistic effect. Consequently, the muon's survival probability (W) and its decay lengths (L) are expressed by the muon's path length (L_{PATH}) as follows:

$$W = \exp\left(-L_{\text{PATH}} / L\right), \quad (7)$$

$$L = c\beta\gamma\tau = 3 \times 10^8 \left(\text{ms}^{-1}\right) \times E / m \times 2.2 \times 10^{-6}, \quad (8)$$

where E is the muon's kinetic energy and m is the muon mass. As shown in Eq. (8), it is only possible for relativistic muons to arrive at sea level.

By the time muons have traveled to sea level from the atmosphere, they lose energy ~2 GeV on average. Thus, the muon energy spectrum varies at different atmospheric depths, especially in the energy regions lower than 2 GeV, and sub-GeV muons originated in the upper atmosphere cannot reach sea level. For this reason, depth of the atmosphere has a much stronger influence on lower energy muon flux. Based on the results of the CAPRICE 94 muon measurements, the low-energy (<2 GeV) muon flux increases by a factor of 5 if the atmospheric depth is reduced by one half of the value at sea level (500 hPa) (Engel et al. 2001). In contrast, muon flux with higher energies (3–20 GeV) increases only by a factor of 1.5 in the same scenario (Engel et al. 2001).

2.1.2 Seasonal and Atmospheric Effects in Muon Flux

One of the factors which can modify the muon count rate is seasonal atmospheric temperature variations since these also affect barometric variations that are connected to tropospheric pressure. Generally, within a day of temperature variations, there are negative correlations in the muon count. Barometric pressure generally only varies by around 1% at regions closer to the equator; thus, these muon count differences tend to be more noticeable in regions located far from the equator at which temperature changes are more strongly tied to barometric pressure changes. For example, for muons with energies above 0.4 GeV (corresponding to the muon range of 200 g cm^{-2} in rock), the daily flux variation effected by temperature variations measured at 44°21′N, 76 m asl was ±2.5% (Saftoiu et al. 2012).

The region of the atmosphere in which seasonal temperature variation occurs has different effects depending on the energy levels of the muon flux. Low-energy muons are most influenced by tropospheric temperature variations, whereas high-energy muons are most influenced by stratospheric temperature variations. Mesons are more copiously produced in the stratosphere, and temperature changes the MFP (mean free path) of the meson-nucleon interactions. As a consequence, high-energy muons have a positive stratospheric temperature correlation. A study on this subject was conducted with the Daya-Bay detectors, which were located at three different atmospheric depths (ranging 250–860 hg cm^{-2}) and recorded the seasonal variations in the muon flux at each depth (Daya Bay Collaboration 2018). There was a tendency that the seasonal variations were more enhanced for deeper detectors (detectors placed further from the Earth's surface): 0.5% at a depth of 250 hg cm^{-2} and 1% at a depth of 860 hg cm^{-2}. At the point of larger depths (3800 hg cm^{-2}), stronger variations (~3%) were observed (Borexino Collaboration 2019). After collecting data, time-dependent muography measurements must be adjusted for the (A) high-energy stratospheric effect and the (B) low-energy tropospheric effect to verify the detection accuracy. Since there is only vertical muon data regarding (A) so far, the aforementioned experiments with the underground detectors should be repeated for horizontal muons in order to further improve time-sequential volcano muography in the future.

2.1.3 Geomagnetic and Interplanetary Magnetic Field Effects in the Muon Flux

While cosmic-ray muons are available worldwide, there are some differences in the average flux. Two factors, (A) positively charged primary GCRs and (B) the geomagnetic field which is directed toward the geomagnetic North Pole, contribute to differences in the Eastern trajectories of the incoming cosmic rays which are suppressed (East-West effect). Cut-off values range from less than 10^9 V near the geomagnetic poles to about 1.6×10^{10} V for vertical particles near the equator (Smart and Shea 1994). Consequently, cosmic-ray muons from the East generally have a weaker flux than those from the West. This East-West geomagnetic effect is more pronounced at higher altitudes. On the other hand, at sea level in middle latitude regions, only primary GCRs having lower energies are more affected by the geomagnetic field. However, since these GCRs don't tend to produce muons, this effect is negligible. For this reason, there has been a lot of research done to try to observe the East-West effect in sea-level muon flux (Hansen et al. 2005). Recent measurements at Fermi National Accelerator Laboratory, Illinois, USA, have shown the East-West asymmetry of cosmic-ray muons to be ~1% (Petrova 2019). No significant zenith-angular dependence was present in the asymmetry up to ~80° from zenith.

One essential step for evaluating time-sequential muographic images is to verify the effect of daily and seasonal variations on the muon flux. Both spatiotemporal variations in the interplanetary magnetic field (IMF) and the atmospheric temperature are responsible for these variations. Various solar events can influence the anisotropies of the IMF: for example, 11-year and 22-year solar cycles and large-scale solar bursts of plasma, which may also affect the magnetic field, such as coronal mass ejections (CME). While these solar cycles tend to have much stronger effects on the muon flux as a whole, CME effects are usually only apparent for sub-GeV muons ($E > 100$ MeV) for much shorter periods (within a few days) which modulate the muon flux by~10% (de Mendonca et al. 2016).

As previously mentioned, this geomagnetic effect strongly depends on the latitude where the cosmic-ray muons are measured. When it was first measured 70 years ago, the vertical flux of muons with momentum around 0.33 GeV/c at latitude 60 degree was 1.8 times higher with respect to the flux at the equator (Conversi 1950). More recently, this latitudinal geomagnetic effect has been measured between the middle latitude region (the cutoff values of 2.7×10^9 V) and the region near the equator (the cutoff values of 1.4×10^{10} V), respectively, indicating variations by 20% and 10% in the flux of muons with energies of ~0.3 GeV/c and ~1 GeV/c; no significant variations could be seen for muons above 3 GeV/c (Allkofer et al. 1975). To summarize, within the mid-latitude region, this latitudinal effect is virtually irrelevant for muography. However, in regions near the equator, it can influence muons below ~5 GeV at sea level.

2.2 Muon Flux Attenuation Within Matter

2.2.1 Muon Energy Loss

Muons often collide with electrons during the ionization process, and after each collision, they lose a very small fraction of their energy. Since a muon's energy loss via ionization is uniquely determined by the distances traversed by muons,

muons of a given energy would have an almost entirely unique range. However, the radiative process also has a very strong influence on muon flux when muons penetrate a dense material (such as rock or water) since these materials present significantly higher possibilities to collide with nuclei in comparison to the matter in the atmosphere. Muon fluctuations are attributed to these radiative processes: bremsstrahlung, direct pair production, and photonuclear interactions which account for a large but random fraction of energy loss. Radiative processes are responsible for virtually all energy loss in high-energy muons (>~1 TeV). For example, the radiative process dominates above 708 GeV in SiO_2. In conclusion, the type and rate of muon energy loss will depend on the materials it traverses. In less dense materials with lower concentrations of electrons and nuclei, ionization is the prevalent cause of energy loss. In contrast, muons tend to lose more energy via radiative processes when traversing more dense materials with larger nuclei.

For the calculation of the analytical range, the energy loss relationship is expressed by:

$$-\mathrm{d}E/\mathrm{d}X = a + bE, \tag{9}$$

where the first term and the second term, respectively, refer to the contribution of the ionization process and the radiative processes. The parameters a and b in Eq. (9) depend on the material the muons traverse, and are given in the continuous slowing down approximation (CSDA) range presented by Groom et al. (2002). For example, the 1-GeV muon's range is 550 g cm^{-2} in SiO_2; however, 1-GeV muons can penetrate only 470 g cm^{-2} of water since it is more electronically dense (Groom et al. 2002). On the other hand, while 10-TeV muons can penetrate 6900 hg cm^{-2} in SiO^2, they can penetrate 7800 hg cm^{-2} in water (Groom et al. 2002).

2.2.2 Muon Flux After Passing Matter

Range is a function of incident muon energy; therefore, it can be incorporated in the open-sky muon energy spectrum (Eq. (6)). If both the muon path length and the average density along the muon path are known, the densimetric thickness (x) can be calculated by multiplying them, and thus the minimum energy (E_c) of muons that can penetrate through a material with this thickness can be determined using Eq. (1). By integrating the open-sky spectrum $I(E, \theta)$ from E_c to infinity, we obtain the integrated muon intensity $N(E_c, \theta)$, which equates to the number of muons with sufficient energy to traverse the entire volume of the target region of interest:

$$N(E_c\theta) = \int_{Ec}^{\infty} I(E,\theta)\,dE, \tag{10}$$

where $I(E,\theta)$ is the zenith-angle-dependent open-sky muon energy spectrum. The integrated muon intensities calculated with Eq. (2) are shown for various near-horizontal angles in Fig. 1.

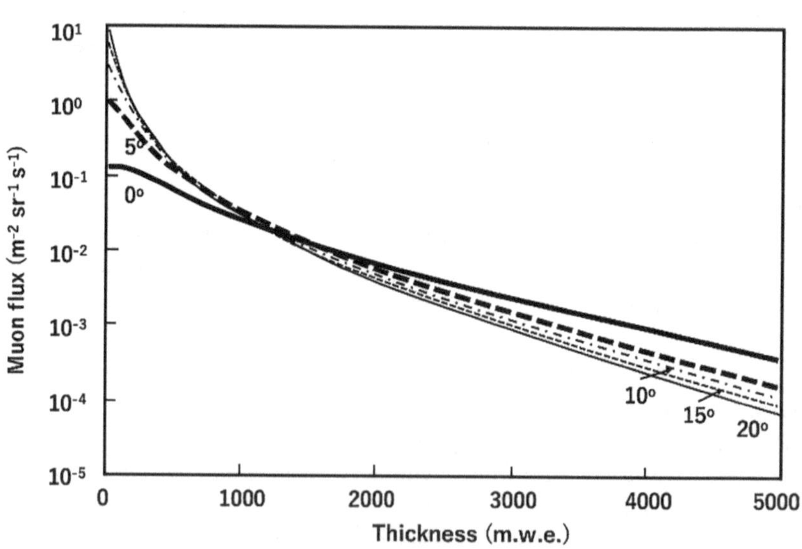

Fig. 1 Muon flux after passing through rock. The muon flux is shown as a function of the rock thickness in units of meter water equivalent (m.w.e.) for various elevation angles

2.3 Limitations of Muography and Potential Targets

2.3.1 Background Noise in Muography

The Coulomb force between moving muons and the nuclei in the medium they are traveling through leads to a large number of small deflections in the muon trajectories. As a result, their angular distribution tends to broaden as they propagate through matter. However, after penetrating matter thicker than 500 hgcm^{-2}, the angular spread is suppressed to ~12 mrad (also termed "half width at half maximum" or HWHM); this value is virtually independent from the total thickness of the material they have traversed owing to the steep nature of the muon spectrum (Taira and Tanaka 2010). Due to this angular spread, positioning resolution can be somewhat compromised at the target; for example, there could be a positioning error of up to 12 m at a distance of 1 km from the detector. Nevertheless, given an appropriately chosen target and muography detector, the high-energy muon's strong penetrating nature (coupled with its low divergence) enables efficient and sufficiently accurate muographic imaging of a distant target; thus, this technique is suitable for surveying volcanoes with several types of long-range applications.

Background events negatively affect the accuracy of muographic images, and these events have different origins: extrinsic background (outside the target), intrinsic background (inside the target), and the background associated with the observation system itself (being the combination of detectors, radiation shields, and other related equipment). Primary GCRs produce cascades of protons, neutrons, mesons, and other electromagnetic particles along with muons, and extrinsic background is originated in these other secondary particles. While most of these particles experience decay, convert to other particles, or are shielded by the atmosphere, some survive and are present at sea level. Compared to the number of muons above 1 GeV surviving at sea level, survival rates of corresponding particles (in terms of energy level) are: 4 orders of magnitude lower for pions, 3 orders of magnitude lower for electrons, and 2 orders of magnitude lower for protons (Zyla et al. 2020).

In situations where muon flux travels through high-density materials (e.g., a kilometer-scale section of a volcano), muon survival rate strongly decreases and the background event fraction increases. Due to being heavier than muons, protons can travel longer distances within matter (~10 cm in Pb) and therefore have a higher likelihood of creating false hits in detector muon counts. Electromagnetic components such as electrons and gammas are easier to distinguish (the electron's radiation length, for example, is less than 6 mm in Pb) and can be rejected by verifying a linear muon trajectory with multiple layers of detectors and radiation shields. In addition to extrinsic background, intrinsic background contributes to tracking errors and is caused by muons scattered inside the observation system. When radiation shields are increased to reduce extrinsic background rates, intrinsic background noise tends to increase which (if unchecked) tends to blur resultant images.

2.3.2 Duration of Muographic Measurements

Muographic images are generated in an angular space formed by elevation and azimuth angles (θ, φ). These angles of an incident particle can be computed as follows by connecting two points in a space $S(x, y, z)$ with a straight line:

$$\varphi = \tan^{-1}\left[\left(x_1 - x_2\right)/D\right], \tag{11}$$

$$\theta = \tan^{-1}\left[\left(y_1 - y_2\right)/D\right], \tag{12}$$

where L is the distance between the upstream detectors (pointed toward the target object) and downstream detectors (pointed away from the target object). The subscripts of x and y indicate the first and second points, respectively. The positioning resolution of the points (Δx, Δy) therefore gives the detector's angular resolution ($\Delta\theta$, $\Delta\varphi$).

The maximum (or highest) definition of a muographic image, R, is defined by $\Phi\Theta\Delta\varphi^{-1}\Delta\theta^{-1}$ pixels, where Φ and Θ are respectively the horizontal and vertical viewing angles of the detector.

Bin sizes (pixel size) of the image can be optimized to acquire sufficient statistics for muon counts recorded in each bin. Since the time required for data acquisition is anti-proportional to the square of the distance between the target and the detector, this pixel size and the distance between the target and the detector dictate the measurement time required for attaining a given level of statistics for counts of muons arriving from a given section of the target volume. Therefore, for effective muographic measurements, it is ideal to get the detector as close to the target as possible. However, in most cases, accessibility and infrastructure availability (e.g., electricity) at geological sites limit the options for the muon detector placement (Tanaka 2016).

2.3.3 Potential Volcanological Targets

Muography has some of the following limitations. (a) If the target has a strong density contrast, this contrast will show up on a muographic image. However, because muography only resolves the average density distribution along individual muon paths, the user must end up making assumptions or interpretations about the more localized structure along these muon paths; alternatively, the user could use more than one detector to resolve the three-dimensional density structure (Tanaka et al. 2010). (b) Muography is limited to near-surface depths, and results are only obtained for the volumes located at elevations higher than the detector. Therefore, local topography has a large impact on these measurements. For example, the detector usually must be placed on a slope pointing toward a topographically prominent feature of interest. Otherwise, the detector needs to be installed underneath the target of interest by utilizing tunnels, boreholes, etc. Also, (c) potential target choices are constrained by the fact that muography is usually only applicable to targets within a specific range: measuring several meters to a few km, and (d) the quality of the resultant muographic image depends on the heterogeneity of the geological environment, the density contrast of the structure being surveyed, and the period of data collection. Time-dependent variations of the subterranean densimetric heterogeneity, such as dynamics of geofluids, can be captured by time-sequential muographic observations, but its time resolution depends on the total size of the detector.

3 Muography Detector

The most essential elements of muography detector design are to efficiently track incoming cosmic-ray muons as well as to keep costs low and keep power consumption at a minimum. Large solar panels are often used for powering the detector in outdoor environments; however, in rainy conditions (for example, the rainy season in Japan), even with a safety factor of 10, the electric power supply from the solar panel becomes unstable. In the following section, these factors and how they affect the effectiveness of the three main types of muon detectors, (a) nuclear emulsion detectors, (b) scintillation detectors, and (c) gaseous detectors, will be discussed.

3.1 Nuclear Emulsion Detectors.

This type of detector uses nuclear emulsion which is composed of AgBr micro-crystals that record a particle trajectory when a charged particle passes through the detector. These paths can be analyzed when the emulsion is later developed. The higher the grain density (approximately equivalent to the level of the ionization) in the emulsion, the better the quality of the measurement. The quality of the scanning microscopes used to read the tomographic image of grains and recognize the particle tracks recorded in the emulsion plate after the development is also vital. The main advantage of nuclear emulsion detectors is that they don't require electricity to operate, and thus are ideal in situations with limited or no access to commercial electricity. However, all muon events are recorded inside the nuclear emulsions without timing information. Consequently, decisions about whether or not to use this detector revolve around the following questions: (a) is time information required for the muographic image? and (b) how expensive/available is an electricity source?

3.2 Scintillation Detectors

Muon events are recorded as a flash of light generated by the incident charged particles within scintillation detectors; these events are subsequently detected by photo-multiplier tubes (PMTs) attached to the end of the plastic scintillator bar to later be analyzed as muon counts. The excellent light insulation of scintillators allows the vertex point to be uniquely determined. Particle trajectories of muons can be tracked by connecting two or more vertex points determined by the geometrical address of the scintillator bars and their corresponding timings. This is achieved with a reconstruction algorithm which contains parameters coming from the geometry of the detector (the positioning resolution, the size of the active area, and the distance between the scintillator bars). If a readout from a detector with a large number of channels is required, then silicon multipliers or multi-anode PMTs (MAPMTs) are used as cost-effective photodetectors, typically by using a large matrix of independent photocathodes (Taira and Tanaka 2010; Lesparre et al. 2012; Lo Presti et al. 2020). Scintillation detectors are real-time detectors, meaning that they yield the timing of individual muon events unlike nuclear emulsions, which only measure the integrated muon events.

3.3 Gaseous Detectors

Similar to scintillation detectors, gaseous detectors can also be utilized as real-time detectors based on a similar tracking technique. Recent developments in the multi-wire proportional chamber (MWPC) have shown gaseous detectors' great potential for application to volcano muography (Oláh et al. 2021). In the MWPC, a signal (denoting a particle count) is shared between wires. Therefore, a single particle hit should always generate two signals from two connected wires. At higher incident angles, more than two anode wires generate signals. If there is a particularly high energy deposit, more wires can output the signal threshold which creates a cluster of output signals. Thus, the track reconstruction algorithm is not exactly the same as scintillation detectors. Another difference between scintillator and gaseous detectors is that the gaseous detector requires a constant gas flow (typically ranging from 0.1 to 1.0 litters/h) over the entire period of the measurement (Caffau et al. 1997; Barnafoldi et al. 2012). The benefits of these detectors are that they are relatively cheap and very power efficient with an excellent spatial resolution (up to a few mm) which make this a practical detector for large-scale muographic measurements (Olah et al. 2023).

4 Asama Volcano Muography

It is usually not feasible for muography to image the deepest regions of a volcano such as a magma chamber. Direct evidence of eminent eruptions from the deepest regions of the volcano is therefore usually not possible with muography. However, in most volcanoes, shallower regions are ideal targets. At this level of depth, many volcanic phenomena can be studied such as: identifying the most likely type of eruption to occur in the future, lava dome growth cycles, and/or dormant/active region evolutions. A series of muography images can, for instance, indicate growth of a plug within pathways of magma by visualizing the shape and size of a low-density region under a deposit of solidified magma.

4.1 2004 Asama Volcano Eruption

Silicic volcanoes tend to be relatively large in general; therefore, usually only the upper regions can be surveyed with muography. Volcanoes featuring vulcanian explosions are, on the other hand, particularly well-suited to muography since they are characterized by repeated explosions and consequently the tephra is dispersed over a wide area. Usually, a vulcanian explosive eruption ends with a viscous lava flow. Tanaka et al. (2007a) derived the near-surface density distribution of a volcano of this type (Asama volcano), and imaged the lava deposited inside the crater and a magma pathway in the shallow region of this volcano in order to reveal the eruption style of this particular volcano.

Volcanic activities of Asama volcano have increased slightly since around 2000, including seismic activity in the shallow area just below the summit, increased volcanic plume activity, and an increase in crater floor temperature. There was also a very small eruption from February to April 2003 that caused ash to fall in a localized region near the crater. Moreover, an explosive eruption, the first explosive eruption since April 1983, occurred at 20:02 on September 1, 2004, at Asama volcano.

In the 2007 muography measurement campaign, as shown in Fig. 2, the lava deposited inside the crater due to the 2004 Asama volcano eruption was imaged, and the same measurement also imaged a low-density region underneath the crater floor. This low-density region was interpreted as a magma pathway that was plugged by lava deposits on the crater floor. Shortly after the 2004 eruption, the lava deposited on the crater floor would have cooled and solidified. Unsolidified magma in the pathway that remains within the volcano after the eruption would drain, presumably would withdraw further underground and leave an empty, hollowed-out pathway. Muographic images of the low-density region of Asama volcano after the 2004 eruption revealed this theorized kind of structure with a vacant pathway in the anticipated region below the crater floor. If a large-scale eruption occurs, the magma body will then ascend. It was therefore predicted that if magma ascended and gas was released from the depressurized magma in this pathway, high gas pressure would cause the plug to explode. Removal of the plug would release fragments of the old magma deposit as well as the newly ascended magma rapidly from the crater in the form of volcanic ash and debris.

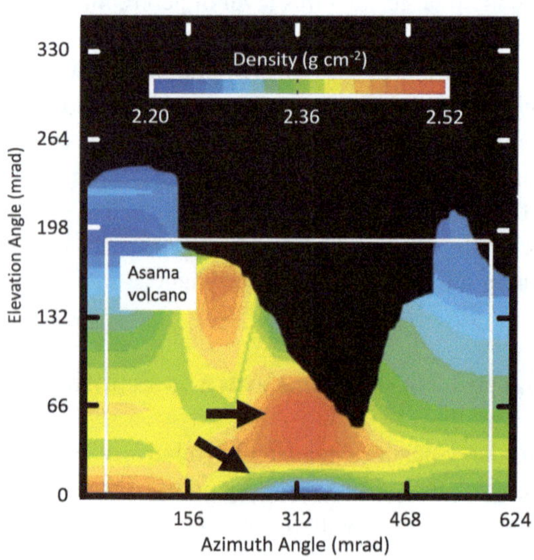

Fig. 2 Muographic image of the summit region of Asama volcano. The white box marks the region shown in the time series in Fig. 3. The upper arrow indicates dense solidified magma deposited on the crater floor. The lower arrow indicates the lower-density region found underneath this magma deposit

4.2 2009 Asama Volcano Eruption

Since July 2008, volcanic activity has continued to increase at Asama volcano, but in January 2009 (during the 2009 muography measurement campaign), volcanic earthquakes and volcanic gas emissions also began. Eruption warning levels were elevated. From February 1, 2009, the inclinometer at the foot of the mountain showed signs of a slight change, and in the morning of the same day, the number of volcanic earthquakes increased. At around 01:51 on the following day, a small eruption occurred at the summit. There has been no increase in volcanic earthquakes since that time and the gas discharge rate has gradually decreased.

One of the goals of muography is to achieve accurate estimates of eruption sequences, and the measurements taken before and after the 2009 Asama volcano eruption were the first step (Tanaka et al. 2009a). Combining the data obtained from the 2009 measurement campaign with the 2007 measurement campaign proved that muographic data can, in principle, be used to predict an eruption sequence. Figure 3 shows a comparison of the muographic image acquired between 6 January and 1 February (upper) with

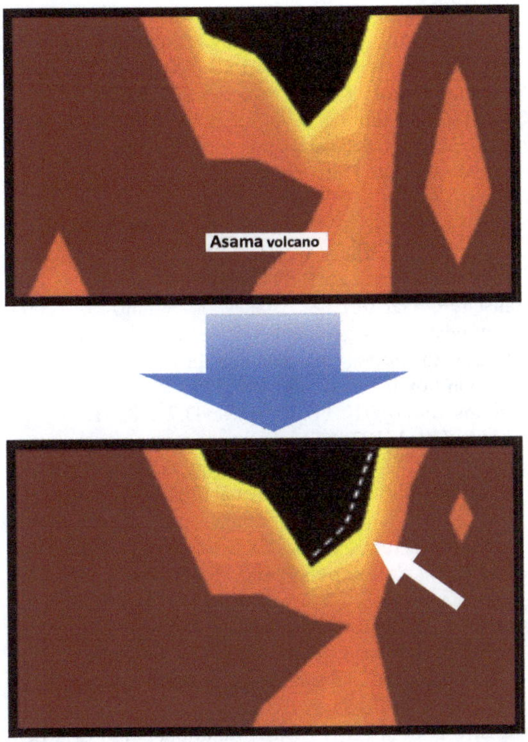

Fig. 3 Time-sequential muographic images before and after the 2009 Asama volcano eruption. The dashed lines drawn in the bottom image show the shape of the crater floor observed before the eruption

that between 2 February (the day of the eruption) and 5 March (lower). According to this image, the mass change occurred only within the crater with no significant difference visible below the crater floor. A dotted line is added to indicate the shape of the crater before the eruption to make the difference before and after the eruption more apparent. Figure 2 also indicates how the northern section of the magma deposit disappeared after the eruption; however, underneath the crater floor, there is no sign of increased density.

Since the average density would be expected to increase if magma had ascended through this passage, we can conclude that magma probably did not flow up this pathway in the 2009 Asama volcano eruption. Instead, it is likely that high-pressure vapor simply blasted through the old magma deposit which had been acting as a "plug" of the pathway. Furthermore, a petrological study of the 2009 eruption ejecta verified that the chemical composition of the magma matched the magma deposit on the crater, adding more evidence to confirm this muographic image. Based on this 2009 muography measurement result, the 2009 Asama volcano eruption could then be evaluated as a small eruption. By conducting several more muographic measurements of volcano data worldwide, a database could be created to further improve the accuracy and speed of predictions like this.

5 Summary

As previously discussed, several density distributions have been muographically acquired since 2007 from volcanoes worldwide including Usu, Japan (Kusagaya and Tanaka 2015), Showa-Shinzan lava dome, Japan (Tanaka et al. 2007b), Asama volcano, Japan (Tanaka et al. 2009b), Izu peninsula, Japan (Tanaka 2015), Omuroyama, Japan (Miyamoto et al. 2022), Unzen, Japan (Tanaka 2016), Sakurajima, Japan (Olah et al. 2023), Satsuma-Iwojima, Japan (Tanaka et al. 2009a), La Soufriere, France (Gibert et al. 2022), Puy de Dome, France (Ambrosino et al. 2015), Etna, Italy (Lo Presti et al. 2020), Vesuvius, Italy (Macedonio et al. 2022), and Stromboli, Italy (Tioukov et al. 2019). Also, plug formation and disappearance associated with activation and deactivation of eruption activities have been muographically observed in other volcanoes which will be applicable to Asama volcano when volcanic activity increases in the future.

A standard method to appraise the magnitude of explosive volcanic eruptions worldwide was established in 1982 with the Volcanic Explosivity Index (VEI) (Newhall and Self 1982). VEI numbers are assigned after eruptions and are based on the mass of erupted material (tephra). It would be profitable to us if we could predict the VEI of a particular volcano prior to an eruption. Besides VEI, the following three questions are some of the most important ones to answer: (a) "When does the eruption begin?", (b) "where does the eruption occur?", and (c) "how long it will last?". Recent progress in technologies that monitor volcanic tremors and land deformation has come far in helping to answer the question (a). However, answering (b) and (c) is more challenging. As more muography data accumulate from several volcanos, we get closer to answering questions (b) and (c) and also closer to the realization of the ultimate dream to predict the VEI of particular volcanoes prior to an eruption.

References

Achard P et al (2004) Measurement of the atmospheric muon spectrum from 20 to 3000 GeV. Phys Lett B 598:15

Allkofer OC, Clausen K, Dau WD (1975) The low-momentum muon spectrum near the equator. Lett Nuovo Cim 12:107–110

Alvarez LW et al (1970) Search for hidden chambers in the pyramid. Science 167:832–739

Ambrosino F, Anastasio A, Bross A, Béné S, Boivin P, Bonechi L et al (2015) Joint measurement of the atmospheric muon flux through the Puy de Dôme volcano with plastic scintillators and resistive plate chambers detectors. J Geophys Res Solid Earth 120(11):7290–7307. https://doi.org/10.1002/2015jb011969

Barnafoldi GG et al (2012) Portable cosmic muon telescope for environmental applications. Nucl Instrum Methods 689:60

Borexino Collaboration (2019) Modulations of the cosmic muon signal in ten years of Borexino data. J Cosmol Astropart Phys 2019:46. https://doi.org/10.1088/1475-7516/2019/02/046

Caffau E, Coren EF, Giannini G (1997) Underground cosmic-ray measurement for morphological reconstruction of the "Grotta Gigante" natural cave. Nucl Instrum Methods A 385:480

Conversi M (1950) Experiments on cosmic-ray mesons and protons at several altitudes and latitudes. Phys Rev 79:749. https://doi.org/10.1103/PhysRev.79.749

Daya Bay Collaboration (2018) Seasonal variation of the underground cosmic muon flux observed at Daya bay. J Cosmol Astropart Phys 2018:1. https://doi.org/10.1088/1475-7516/2018/01/001

de Mendonca RRS, Braga CR, Echer E, Dal Lago A, Rockenbach M, Such NJ, Munakata K (2016) Deriving the solar activity cycle modulation on cosmic ray intensity observed by Nagoya muon detector from October 1970 until December 2012. Proc Int Astronom Union 12(S328):130–133

Engel R, Gaisser TK, Stanev T (2001) The flux of atmospheric muons. Proc ICRC 2001:1029–1032

George EP (1955) Cosmic rays measure overburden of tunnel. Commonw Eng 1955:455–457

Gibert D, de Bremond J, Carlus B, Deroussi S, Ianigro J-C, Jessop DE et al (2022) Observation of the dynamics of hydrothermal activity in La Soufrière of Guadeloupe volcano with joint muography, gravimetry, electrical resistivity tomography, seismic and temperature monitoring. Geophys Monograph Ser 270:55–73

Groom DE, Mokhov NV, Striganov SI (2002) Muon stopping power and range tables 10 MeV–100 TeV. Atomic Data Nucl Data Tables 76:183–356. https://doi.org/10.1006/adnd.2001.0861

Haino S et al (2004) Measurements of primary and atmospheric cosmic-ray spectra with the BESS-TeV spectrometer. Phys Lett B 594(1-2):35

Hansen P, Gaisser TK, Stanev T, Sciutto SJ (2005) Influence of the geomagnetic field and of the uncertainties in the primary spectrum on the development of the muon flux in the atmosphere. Phys Rev D 71:083012. https://doi.org/10.1103/PhysRevD.71.083012

Jokisch H et al (1979) Cosmic-ray muon spectrum up to 1 TeV at 75° zenith angle. Phys Rev D 19(5):1368

Jourde K, Gibert D, Marteau J et al (2016) Muon dynamic radiography of density changes induced by hydrothermal activity at the La Soufrière of Guadeloupe volcano. Sci Rep 6:33406. https://doi.org/10.1038/srep33406

Kusagaya T, Tanaka HKM (2015) Muographic imaging with a multi-layered telescope and its application to the study of the subsurface structure of a volcano. Proc Jpn Acad Ser B 91:501–510

Lesparre N et al (2012) Design and operation of a field telescope for cosmic ray geophysical tomography. Geosci Instrum Method Data Syst 1:33–42

Lo Presti D, Riggi F, Ferlito C et al (2020) Muographic monitoring of the volcano-tectonic evolution of Mount Etna. Sci Rep 10:11351. https://doi.org/10.1038/s41598-020-68435-y

Macedonio G, Saracino G, Ambrosino F, Baccani G, Bonechi L, Bross A et al (2022) Muography of the volcanic structure of the summit of Vesuvius. Geophys Monogr Ser 270:123–136. https://doi.org/10.1002/9781119722748.ch9

Miyamoto S et al (2022) A muographic study of a scoria cone from 11 directions using nuclear emulsion cloud chambers. Geosci Instrum Method Data Syst 11:127–147. https://doi.org/10.5194/gi-11-127-2022

Nagamine K et al (1995) Method of probing innerstructure of geophysical substance with the horizontal cosmic-ray muons and possible application to volcanic eruption prediction. Nucl Instr Methods A 356:585–595

Newhall CG, Self S (1982) The volcanic explosivity index (VEI): an estimate of explosive magnitude for historical volcanism. J Geophys Res 87:1231–1238. https://doi.org/10.1029/JC087iC02p01231

Nishiyama R, Tanaka Y, Okubo S, Oshima H, Tanaka HKM, Maekawa T (2014) Integrated processing of muon radiography and gravity anomaly data toward the realization of high-resolution 3-D density structural analysis of volcanoes: case study of Showa-Shinzan lava dome, Usu, Japan. J Geophys Res 119:699–710

Nomura Y, Nemoto M, Hayashi N, Hanaoka S, Murata M, Yoshikawa T et al (2020) Pilot study of eruption forecasting with muography using convolutional neural network. Sci Rep 10:5272. https://doi.org/10.1038/s41598-020-62342-y

Oláh L, Tanaka HKM (2022) Machine learning with muographic images as input: an application to volcano eruption forecasting. Geophys Monogr Ser 270:43–54. https://doi.org/10.1002/9781119722748.ch4

Oláh L, Tanaka HKM, Hamar G (2021) Muographic monitoring of hydrogeomorphic changes induced by post-eruptive lahars and erosion of Sakurajima volcano. Sci Rep 11:17729. https://doi.org/10.1038/s41598-021-96947-8

Olah L et al (2023) Muon imaging of volcanic conduit explains link between eruption frequency and ground deformation. Geophys Res Lett 50:e2022GL101170

Petrova O (2019) Particle and cosmology, 16th Baksan School on Astroparticle Physics (2019) East-west asymmetry effect in atmospheric muon flux in the Far Detector of NOvA. http://www.inr.ac.ru/~school/lectures/Petrova.pdf

Saftoiu A, Bercuci A, Brancus M, Duma M, Haungs A, Mitrica B, Petcu M, Rebel H, Sima O, Toma G (2012) Measurements of the cosmic muon flux with the WILLI detector as a source of information about solar events. Rom J Phys 56:664–672

Shinohara H, Tanaka HKM (2012) Conduit magma convection of a rhyolitic magma: constraints from cosmic-ray muon radiography of Iwodake, Satsuma-Iwojima volcano. Jpn Earth Planet Sci Lett 349:87–97

Smart DF, Shea MA (1994) Geomagnetic cutoffs: a review for space dosimetry applications. Adv Space Res 14(10):787–796

Taira H, Tanaka HKM (2010) A potential space- and power-effective muon sensor module for imaging a volcano. Earth Planets Space 62(2):179

Tanaka HKM (2015) Muographic mapping of the subsurface density structures in Miura, Boso and Izu peninsulas. Japan Sci Rep 5:8305. https://doi.org/10.1038/srep08305

Tanaka HKM (2016) Instant snapshot of the internal structure of Unzen lava dome, Japan with airborne muography. Sci Rep 6:39741

Tanaka HKM (2020) Development of the muographic tephra deposit monitoring system. Sci Rep 10:14820. https://doi.org/10.1038/s41598-020-71902-1

Tanaka HKM et al (2007a) High resolution imaging in the inhomogeneous crust with cosmic-ray muon radiography: the density structure below the volcanic crater floor of Mt. Asama, Japan. Earth Planet Sci Lett 263:104–113

Tanaka HKM et al (2007b) Imaging the conduit size of the dome with cosmic-ray muons: the structure beneath Showa-Shinzan lava dome, Japan. Geophys Res Lett 34:053007

Tanaka HKM et al (2009a) Detecting a mass change inside a volcano by cosmic-ray muon radiography (muography): first results from measurements at Asama volcano, Japan. Geophys Res Lett 36:L17302

Tanaka HKM, Uchida T, Tanaka M, Shinohara H, Taira H (2009b) Cosmic-ray muon imaging of magma in a conduit: degassing process of Satsuma-Iwojima volcano. Jpn Geophys Res Lett 36:L01304

Tanaka HKM et al (2010) Three dimensional CAT scan of a volcano with cosmic ray muon radiography. J Geophys Res 115:B12332. https://doi.org/10.1029/2010JB007677

Tanaka HKM et al (2011) Cosmic muon imaging of hidden seismic fault zones: rainwater permeation into the mechanical fractured zones in Itoigawa-Shizuoka tectonic line. Jpn Earth Planet Sci Lett 306:156–162

Tanaka HKM et al (2014) Radiographic visualization of magma dynamics in an erupting volcano. Nat Commun 5(3381):1–9

Tioukov V, Alexandrov A, Bozza C et al (2019) First muography of Stromboli volcano. Sci Rep 9:6695. https://doi.org/10.1038/s41598-019-43131-8

Zyla PA et al (2020) The review of particle physics. Prog Theor Exp Phys 2020:083C01

Degassing Activity of Asama Volcano

Hiroshi Shinohara

1 Introduction

Asama volcano is a persistently degassing volcano with frequent eruptions. Although the degassing is intensive, the degassing vent and fumaroles, located inside the summit crater, are not accessible, and geochemical studies in the early stages are focused mainly on the low-temperature fumaroles and the groundwater on the flank (Noguchi 1935). The modern geochemical studies started with an application of COSPEC in the 1970s (Moffat et al. 1972), followed by various studies in the 2000s, such as the frequent monitoring of the SO_2 emission rates (Ohwada et al. 2013), the Multi-GAS measurement of the gas composition (Shinohara et al. 2015), the soil-CO_2 flux surveys (Morita et al. 2016), and quantitative modelling of the degassing process with the aid of geophysical data (Kazahaya et al. 2011, 2015a, b).

2 SO_2 Emission Rates

Measurements of the SO_2 emission rates by COSPEC at Asama volcano started in 1972 (Moffat et al. 1972), and then are repeated to monitor the volcanic activity (Okita and Shimozuru 1975; Ohta et al. 1978, 1982; Kyushu University, University of Tokyo 1984, 1986). Positive correlations of the SO_2 emission rates with the seismic and eruptive activities are recognized by the early studies; however, their quantitative evaluation is limited because of the infrequent measurements. Japan Meteorological Agency started the SO_2 emission rate measurements on a regular basis in 2002 and publishes the results on the website (https://www.data.jma.go.jp/vois/data/tokyo/306_Asamayama/306_So2emission.htm). These frequent measurements enabled detailed comparisons of the SO_2 emission rates and other activities.

The SO_2 emission rates of Asama volcano are characterized by their contrasting shift between the large rates during the active periods and the small rates during the inactive (quiet) periods. Ohwada et al. (2013) reported the variation of the SO_2 emission rates along with the variation of the other volcanic activities from 2004 to 2012. An active period started in Sept. 2004 with repeated Vulcanian eruptions and lava extrusion, and the large SO_2 emission rates with a peak of 53 kg/s (4600 t/day) are observed during the early stage. During the active period, a reddish glow was observed at the summit even after the eruptive activities, implying the magma at the near surface during the active period. The average SO_2 emission rate during the active period in 2004–2006 is 20 kg/s (1700 t/day). During the inactive period

H. Shinohara (✉)
Geological Survey of Japan, National Institute of Advanced Industrial Science and Technology (AIST), Tsukuba, Ibaraki, Japan
e-mail: shinohara-h@aist.go.jp

S. Aramaki et al. (eds.), *The Asama Volcano*, Active Volcanoes of the World,
https://doi.org/10.1007/978-981-95-5799-8_24

from Aug. 2006 to Jul. 2008, the SO_2 emission rates were stable at small values of 1–5 kg/s (86–170 t/day) with an average of 1.5 kg/s (130 t/day). Similar variation was observed in 2008–2012.

The SO_2 emission rates varied in parallel with the number of B-type earthquakes (Ohwada et al. 2013). The B-type earthquakes are characterized by a spindle-shaped waveform lacking distinct P and S phase signatures and are commonly low in frequency. Hypocenters of the B-type earthquakes were vertically distributed beneath the summit crater, likely along a magma conduit (Aoki et al. 2013). The close correlation of SO_2 emission rates with the number of B-type earthquakes supports the idea that B-type earthquakes are caused by flows of gas and/or magma along the shallow conduit.

The GPS observation revealed repetitive inflation and deflation of the volcanic edifice, which shows a good correlation with the seismicity and the plume height (Murakami 2005). The deformation source was estimated as a dike-shaped intrusion at a few km depth and at 6 km to the west of the summit (Aoki et al. 2013). Variation of the baseline length between the two GPS observation sites at the west flank is a good proxy of the dike volume change and dike inflation occurs in the early stages of the active period (Fig. 1). Beginning of the inflation coincides with the increase of the SO_2 emission rates, and the deflation starts when the SO_2 emission rates start to decrease, but before reaching the stable small values.

The long-term averaged SO_2 emission rate from a volcano can be an indicator of the long-term magmatic activity of the volcano, and its sum in the region or the whole earth can give a quantitative constraint to the regional or the global volatile budget in the crust-atmosphere system. By compiling a 32-year-long dataset from 1975 to 2006, Mori et al. (2013) estimated the long-term SO_2 emission rates from Japanese volcanoes, including Asama volcano. Without the regular monitoring started in 2002, the SO_2 emission rate measurement tends to be performed only during active periods that biased the emission rates to larger values. Therefore, a simple average of the measured values will not give an accurate estimate of the long-term average. In order to reduce the uncertainty, Mori et al. (2013) estimated the long-term variation of the emission rate, including the inactive period, by interpolating and extrapolating the emission rate data to fit with the variation of other observations such as plume height and size, reported by Japan Meteorological Agency (2005). The estimated long-term average SO_2 emission rate is 360 t/day for Asama volcano, that is, about 6% of the total emission rate from all the subaerial volcanoes in Japan (Mori et al. 2013) and 0.6% of the global emission rate (Carn et al. 2017).

3 Volcanic Gas Composition

Direct sampling of the volcanic gases is not realistic at Asama volcano, as the gases are discharged from a degassing vent and fumaroles located within the summit crater surrounded by the steam cliff (Fig. 2), and the volcanic gas compositions are estimated by various techniques to measure volcanic plume (diluted volcanic gas) composition. The first example of the plume measurements was reported by Noguchi and Kamiya (1963), who installed an alkaline solution on the summit crater rim to measure concentrations of S and Cl absorbed from the volcanic plume. They observed the S/Cl ratio increase prior to the eruption on November 10, 1958, from less than 10 in 1957 to 30 in September 1958, and then the ratio decreased to around five in 1960. The S/Cl ratio increase was associated with about 30 times increase in the amount of absorbed sulfur species, implying an increased sulfur flux from the crater.

Shinohara et al. (2015) performed the Multi-GAS measurements of the plume composition to estimate the volcanic gas composition by repeating the on-site measurements and the daily measurements by an automatic system installed on the summit crater rim. The estimated ratios of CO_2/SO_2, H_2O/SO_2, H_2/SO_2, and SO_2/H_2S of the volcanic gases showed large scatters (Fig. 3), in particular, in the results of the automatic measurements. The large scatters are attributed to

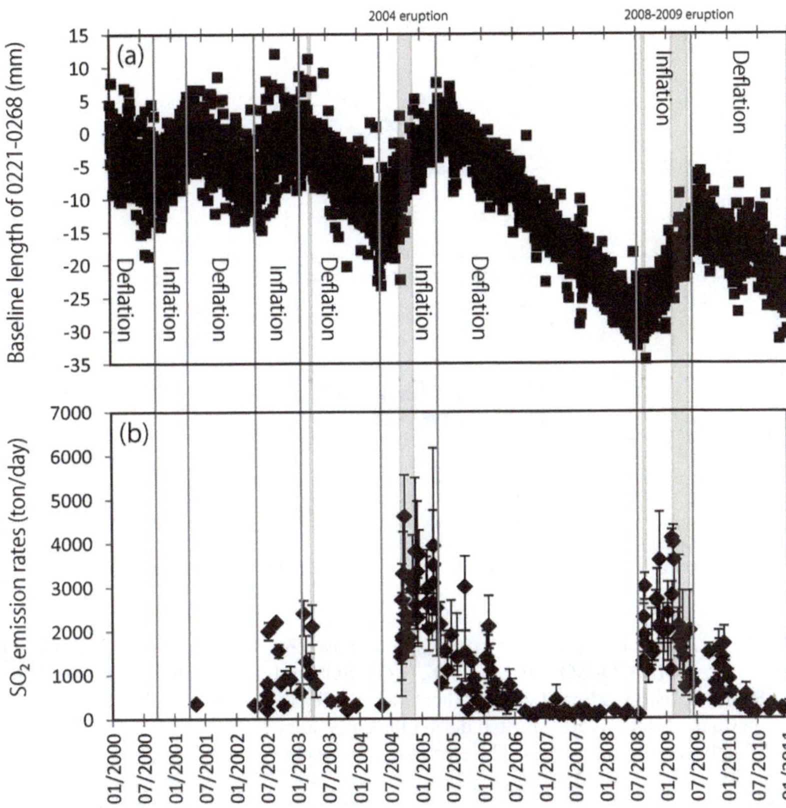

Fig. 1 Comparison between ground deformation and SO_2 emission rates from 2000 to 2011. Gray-filled areas represent periods of eruptions. (**a**) Baseline length change between the GPS observation sites 950,221 and 950,268 from Aoki et al. (2013). The vertical lines indicate the transitions between inflation and deflation. (**b**) SO_2 emission rates from Ohwada et al. (2013). Reproduced from Kazahaya et al. (2015a) with permission

Fig. 2 Photograph of the summit crater taken from the western rim. Red, blue, and green circles indicate example locations of the main degassing vent, medium-temperature fumarolic areas, and low-temperature fumarolic areas, respectively. Reproduced from Shinohara et al. (2015) with permission

variable contributions of low-temperature fumarolic gases emitted near the measurement site (Fig. 1), and the ratios estimated for high-concentration plumes are judged as more reliable. The reliable dataset of the CO_2/SO_2 and H_2O/SO_2 ratios did not show clear variation with the variation of the SO_2 emission rate as well as the eruptions occurred during the monitoring period (Fig. 3). At some volcanoes, the CO_2/SO_2 varied with eruptive activities and the variation is interpreted to be caused by variation in a degassing pressure and by a volatile differentiation in the magma reservoir, such as CO_2-rich bubble accumulation (Aiuppa et al. 2009). The apparently stable CO_2/SO_2 ratios at Asama volcano imply that the eruptions were not associated with the degassing pressure change nor the volatile differentiation process at this volcano. Based on

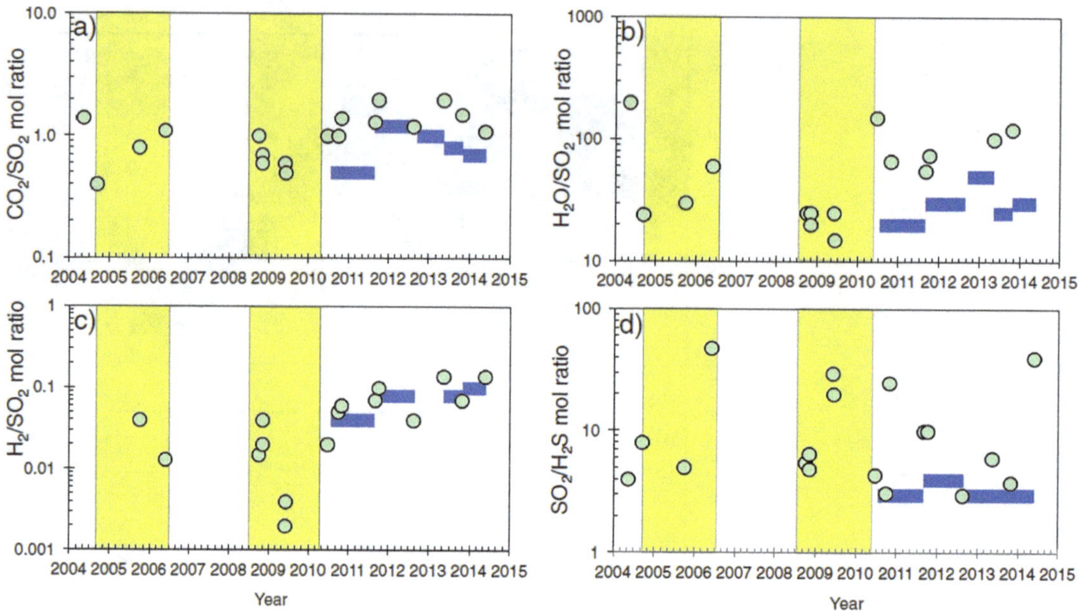

Fig. 3 Variation of volcanic gas composition shown by molar ratios of (**a**) CO_2/SO_2, (**b**) H_2O/SO_2, (**c**) H_2/SO_2, and (**d**) SO_2/H_2S. The shadowed areas indicate active degassing periods. Eruptions are repeated during the early stage of the active degassing periods. The circles are the results of the on-site measurements, and the thick bars show the estimated representative compositions of each period obtained by the automatic measurement. Reproduced from Shinohara et al. (2015) with permission

the estimated ratios of H_2O/SO_2, SO_2/H_2S, and H_2/SO_2, the apparent equilibrium temperature is calculated for the following reaction;

$$SO_2 + 3H_2 = H_2S + 2H_2O.$$

The calculated temperature ranges from 480° to 690 °C and does not show any correlation with the variation of the SO_2 emission rates, implying that the volcanic gases are always derived from a high-temperature source, such as a shallow magma degassing, even during the inactive period.

The S/Cl and Cl/F ratios of the volcanic gases are estimated by FT-IR (Mori and Notsu 2005), Alkali-Filter (Shinohara et al. 2015), and the ash leachate analyses (Nogami et al. 2008; Shinohara et al. 2015; Fig. 4). The S/Cl and Cl/F ratios obtained by the different methods globally agree with each other, suggesting the reliability of the methods. The ratios obtained by the leachate analyses often show a wide scatter, but the ranges commonly overlap with the ratios obtained by other methods. The wide range ratios were obtained by the leachate analyses even for a single eruption, which could be caused by variable conditions of the ash-plume interaction. During the continuous ash discharge on Sept. 16–18, 2004, contrasting Cl/F ratios were obtained by the FT-IR as 33 and by the ash leachate analyses as less than 1, whereas both methods give similar ratios of 5–10 for the ashes discharged on other days in the same eruption period (Figs. 4c, d). These contrasting ratios are attributed to selective fluorine fixation on the ash by an intense ash-plume interaction (Mori and Notsu 2005). The low and scattered Cl/F ratios obtained in 2009 by the ash leachate analyses are also likely the results of the selective fluorine fixation.

Except for the scattered values obtained by the ash leachate analyses, the S/Cl and Cl/F ratios are relatively constant at 3–10 and 5–20, respectively, without significant correlation with the SO_2 flux variation nor the eruptions (Fig. 4). In contrast, Noguchi and Kamiya (1963) observed the S/Cl ratio increase from less than 10 to 30 prior to the eruption in 1958. With the prejudice,

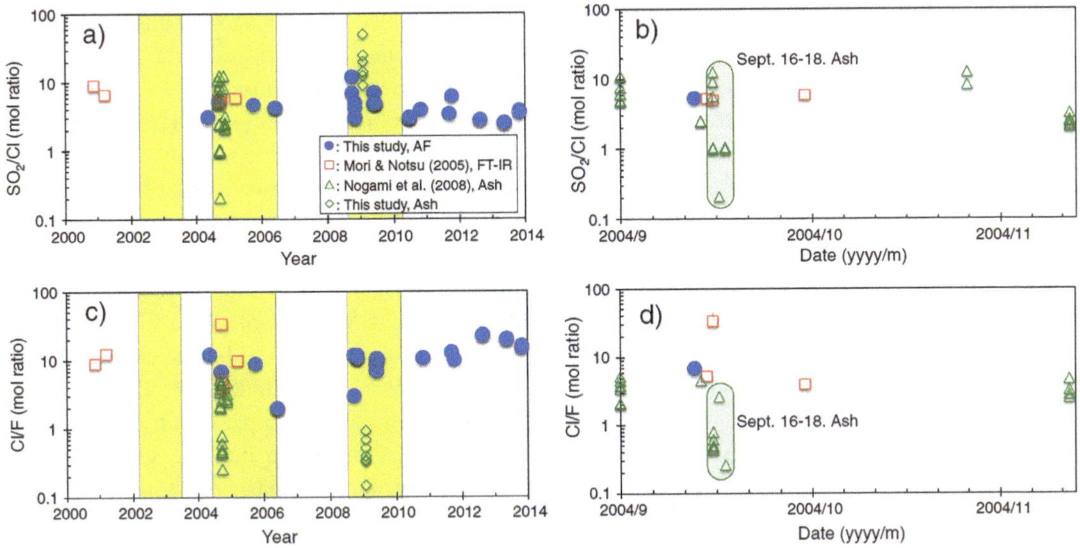

Fig. 4 Variation in SO_2/Cl and Cl/F molar ratios. (**a**) and (**b**) show the long- and short-term variation of the SO_2/Cl ratio, and (**c**) and (**d**) show the long- and short-term variation of the Cl/F ratio, respectively. The circles are the results of Shinohara et al. (2015) obtained by the alkaline-filter techniques, the squares are FT-IR data (Mori and Notsu 2005), the triangles and the diamonds are the ash leachate data by Nogami et al. (2008) and by Shinohara et al. (2015), respectively. The shadowed areas in (**a**) and (**c**) indicate active degassing periods. Range of the leachate data of September 16–18 ashes is framed on (**c**) and (**d**). Reproduced from Shinohara et al. (2015) with permission

we can see about twice decrease in the S/Cl ratios from the early stage of the eruption in 2008 to the quiet periods after 2010, suggesting that the S/Cl increase can be a common precursory change. The smaller change associated with the 2008 eruptions could be due to about an order of magnitude smaller eruption sizes than the size of the 1958 eruption. The average volcanic gas composition of Asama volcano is estimated as $CO_2/SO_2 = 0.8$, $H_2O/SO_2 = 30$, $H_2/SO_2 = 0.06$, $SO_2/H_2S = 6$, $SO_2/Cl = 5$, and $Cl/F = 10$. This composition is similar to other persistently degassing volcanoes in Japan (Shinohara 2013), suggesting a similar degassing condition of magmas with similar volatile contents in the magmas.

4 Diffuse Volatile Emissions

Geochemical surveys of waters from cold springs, rivers, and wells were performed during an early period of the volcano studies (Ossaka et al. 1957, and references therein). There is no hot spring discharge at the surface, but some hot waters were found in the wells on the southeast flank of the volcano (Ossaka et al. 1957; Takahashi and Hosoya 1978). Clear contribution of volcanic volatiles is not recognized in these waters with an exception of the sulfate-rich acid water discharge at Jigokudani, 2 km southwest of the summit (Ossaka et al. 1957). Takahashi and Hosoya (1978) performed repeated water sampling from four springs (including Jigokudani) and four boreholes. Changes in concentrations of dissolved components are found at two boreholes before and after the 1973 eruption, which was interpreted as the result of underground water flow changes due to crustal deformation associated with the volcanic activity.

Soil CO_2 degassing is another important outlet of volcanic volatiles. Morita et al. (2016) performed the diffuse carbon dioxide (CO_2) flux and soil temperature surveys by the accumulation chamber method at 211 sites around the summit of Asama volcano, and found high CO_2 flux areas on the eastern flank of the central cone Kamayama and the eastern rim of Maekake crater, the outer cone. The distribution of the high CO_2 fluxes agrees with the distribution of high anomalies of soil temperature, suggesting a common source for the CO_2 and the heat, likely a shallow hydrothermal system. Aizawa et al. (2008) modeled the electrical resistivity structure of the volcano based on the magnetotelluric surveys and estimated a low resistive zone (about 1 Ωm) at several hundred meters depth on the eastern side of the summit, which was interpreted as a hydrothermal system with a well-developed fracture network. Localization of the diffuse CO_2 anomalies probably reflects permeable zones controlled by the local topography and the hidden fractures, such as the crater rims of the 1108 Plinian eruption. The permeable zones are connected to the hydrothermal fluid layer with low resistivity through an upper sealed layer made of clay minerals.

Morita et al. (2016) conducted the four surveys of the soil CO_2 flux measurement from Oct. 2012 to Oct. 2014, when the SO_2 emission rates were low (ranging from 50 to 300 t/day with an average of 170 t/day; Ohwada et al. 2013; Japan Meteorological Agency 2014). By compiling the results of the three years, the total diffuse CO_2 flux from the summit area of 1.1 km^2 is estimated as 12.6 t/day. The CO_2/SO_2 molar ratio of the volcanic plume is estimated as 0.8 (Shinohara et al. 2015), and the CO_2 flux by the plume is calculated as 28–170 t/day with an average of 94 t/day. The diffuse flux accounts for 12% (6.7–37%) of the total (diffuse and plume) CO_2 flux from the summit area, indicating the significant contribution of the diffuse degassing during the quiet period.

5 Degassing Mechanism and Budget

Variation of the SO_2 emission rates show a good correlation with the number of B-type earthquakes, in particular of the low-frequency ones (Kazahaya et al. 2015b, Fig. 1). Detailed examination of the high temporal resolution data of the SO_2 emission rates obtained by the SO_2 imaging system revealed two types of the gas emissions at Asama volcano: sporadic gas bursts from the vent and continuous gas emissions from fumaroles and the vent in the crater (Kazahaya et al. 2011). The gas burst occurs immediately after the VLP (very long period) seismic events. The amount of SO_2 emitted by each gas burst event, quantified separately from the continuous gas emissions, linearly correlates with the seismic moment of the associated VLP event. The VLP events of Asama volcano are caused by the expansion of a tensile crack and a cylinder (Maeda and Takeo 2011), and the sudden expansion is likely caused by a rapid gas injection into the seismic source followed by a gas release resulting in the gas burst. The VLP sources are located at depths of 100–150 m north of the crater center (Maeda and Takeo 2011; Fig. 5), which corresponds to 4 MPa under a lithostatic pressure gradient. At such low pressures, most volatile components in magmas are already exsolved into the gas phase. The rapid gas injection also indicates a movement of a gas phase already accumulated at the top of a magma column rather than exsolution of volatiles from the magma at the VLP source.

Based on the linear relationship between the SO_2 amount and the VLP moment, Kazahaya et al. (2015b) estimated the amounts of SO_2 discharged by each gas burst from the seismic record (Kazahaya et al. 2015b; Fig. 6). The SO_2 amount discharged by a single VLP event is relatively constant within the range from 0.5 to 2.0 ton/event (Fig. 6c), suggesting a constant seismic source volume change. In contrast, the number of the VLP events and the total SO_2 emission rates by the gas bursts (eruptive gas emission) show a large variation (Figs. 6b, d). The SO_2 emission rates by the gas bursts are large when the total SO_2 emission rates are large, but the ratio of the gas burst/total emissions does not correlate with the total emission rates (Fig. 6e). The large ratio was observed before the 2008 eruption and also before the increase in the total emission rate. In contrast, the large ratios were not observed before the 2009 eruptions, but the ratio increased during the sharp decrease in the total emission rate during the early stage of the active period. Variation of the gas burst/total emission rate ratio might be

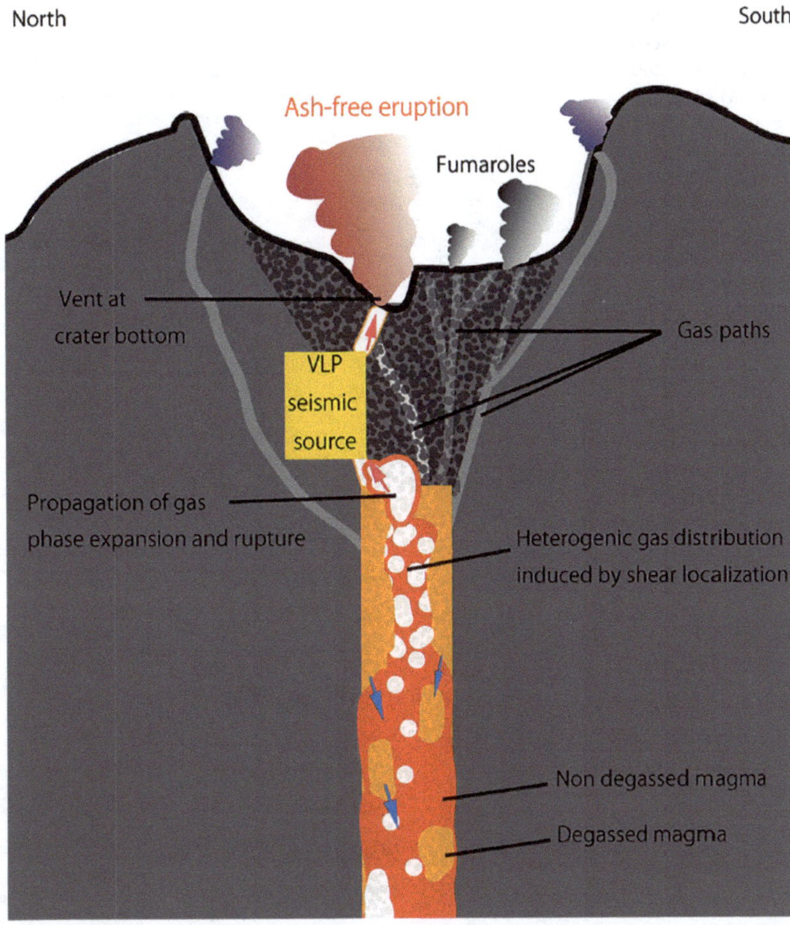

Fig. 5 Schematic diagram of the shallow conduit system at Mt. Asama. The conduit magma convection causes the intensive degassing. The VLP seismic source is located north of the crater center (Maeda and Takeo 2011). The shear localization induced by movement of the volatile-rich and outgassed magmas occurs to produce a heterogenic distribution of the gas phases within the conduit, in particular, at the top of the convecting magma column. Sudden outgassing induces rapid pressurization at the VLP seismic source to excite the VLP seismic event. Reproduced from Kazahaya et al. (2015b) with permission

related to variation of heterogeneity of the gas phase distribution at the top of the magma column, which is likely controlled by shear localization in the magma flow in the conduit. The mechanism of the ratio variation is not clear yet, but the variation of the gas burst/total emission ratios could be a key phenomenon to elucidate the shallow magma-gas processes.

The intense SO_2 emissions need a high rate of magma degassing. Ohwada et al. (2013) calculated the volume of the degassed magma (V_{magma}) with the following equation:

$$V_{magma} = \frac{M_{SO_2}}{\Delta C_S} \times \frac{32}{64} \div \rho_{magma},$$

where M_{SO2} is the weight of the discharged SO_2, ΔC_S is the difference of sulfur concentrations in the magmas before and after degassing, 32/64 is the molecular weight ratio of sulfur to SO_2, and ρ_{magma} is the density of the magma. The calculation was made with assumptions of complete degassing of an andesite melt whose S content is 2400 ppm (Ohta et al. 2007), magma density of 2500 kg/m^3, and no H_2S degassing. The last assumption results in 14% underestimates, because the average SO_2/H_2S ratio of Asama volcanic gas is 6 (Shinohara et al. 2015).

During the active periods of 2004–2006 and 2008–2010, the average SO_2 emission rates are estimated as 20 kg/s (1700 t/day) and 19 kg/s (1600 t/day), and the total emitted SO_2 is calculated to be 1.1 and 0.9 Mt., which require degassed magma volumes of 1.9×10^8 and 1.5×10^8 m^3, respectively (Ohwada et al. 2013). These volumes are more than two orders of magnitude larger than the volume of the erupted magma in 2004 (1×10^6 m^3), implying that most

Fig. 6 Variations in degassing and seismic activities from 2007 to 2011. (**a**) Monthly average total gas emission rates. (**b**) Monthly time series of the number of VLP seismic events. (**c**) Monthly time series of the displacement amplitudes of the VLP seismic events averaged over each month, and monthly average amounts of SO_2 gas emitted by a single VLP seismic event. (**d**) Monthly average eruptive gas emission rates calculated using the VLP seismic data. (**e**) Monthly time series of eruptive/total gas ratio. Vertical arrows indicate the eruptions. Broken lines signify no data. Reproduced from Kazahaya et al. (2015b) with permission

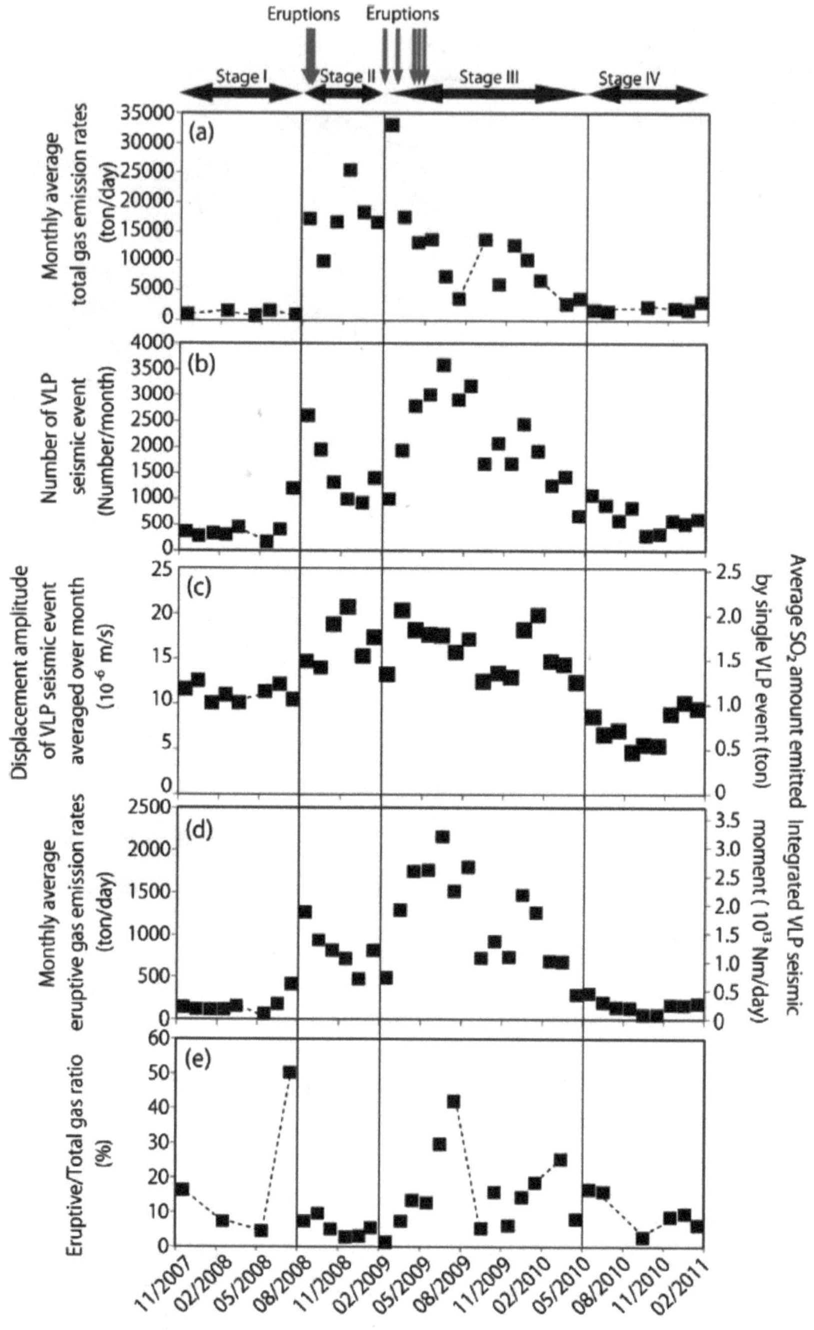

of the volcanic gases are derived from a non-erupted magma (excess degassing). The large degassed magma volume implies that the volcanic gases are supplied from a magma chamber at depth. Even during the inactive periods of 2006–2008 and 2010–2012, when 0.1 and 0.2 Mt. of SO_2 emitted with the average SO_2 emission rates of 1.5 kg/s (130 t/d) and 2.5 kg/s (220 t/day), the degassed magma volume amounts to 1.6×10^7 and 3.4×10^7 m^3, respectively (Ohwada et al. 2013).

Two models have been proposed for the mechanism of persistent degassing that requires degassing of a large magma chamber at depth: the permeable gas flow model (Edmonds et al. 2003) and the conduit magma convection model (Kazahaya et al. 1994; Shinohara 2008). Ohwada et al. (2013) examined the two models, adapting reasonable parameter sets such as conduit permeability of $>10^{-10}$ (m^2) and a magma conduit radius of 50 m (Aoki et al. 2013), and found that both models can cause the observed SO_2 emissions. Considering the constant volcanic composition during the active and inactive periods (Fig. 3), however, Ohwada et al. (2013) concluded that the intensive degassing at Asama volcano needs to be driven by the conduit magma convection. According to the magma convection model, the magma stored in a deep magma chamber ascends through a magma conduit, degasses at the top of the conduit at low pressure, and the degassed magma descends through the same conduit. The averaged SO_2 emission rate of 20 and 1.5 kg/s during the active period of 2004–2006 and the inactive period of 2006–2008 requires the magma convection rate of 2.6×10^5 and 2.0×10^4 (m^3/day), respectively.

Based on the results of various geophysical observations, Aoki et al. (2013) modeled a magma plumbing system of Asama volcano as follows: a magma supplied to the upper crust is stored in a mid-crustal magma chamber at 5–10 km depth, 8 km west from the summit, ascends to a depth of about 1 km below sea level forming an east-west-trending dike, then moves horizontally east for several kilometers to beneath the summit before ascending vertically to the summit crater (Fig. 7a). The intensive volcanic gas emission implies that the conduit magma convection occurs in the magma plumbing system. Kazahaya et al. (2015a) calculated the magma budget in the magma plumbing system based on the volcanic gas emission rates and ground deformation data. Total amount of the SO_2 emitted during 1975–2011 is 5.4 Mt., and total volume of the degassed magma is 9×10^8 m^3. Such a large amount of magma cannot be stored in the dike but is likely derived from the mid-crustal magma chamber, implying that the dike is just a pass-point of the magma counterflows connecting the deep magma chamber and top of the conduit.

The inflation and deflation of the dike at the west occurs along with the changes in the seismicity, the SO_2 emission rates, and the eruptive activities (Fig. 1). Volume change of the dike during the inflation from July 2004 to March 2005 is $+6.8 \times 10^6$ m^3 and that during the deflation from May 2005 to June 2008 is -1.4×10^7 m^3 (Kazahaya et al. 2015a). Such volume changes at the pressure source are commonly considered as the result of a magma input to or a magma discharge from the pressure source. Volumes of the magma degassed during the inflation and deflation periods are 1.0×10^8 and 0.9×10^8 m^3, respectively, implying that volumes of the magma passed through the dike are an order of magnitude larger than the magma input/output volume inferred from the inflation or the deflation. Therefore, the volume change at the dike does not indicate a result of a simple magma input to or output from the dike, but should be considered as a result of the dynamic balance between the counterflows through the dike, and change in the balance may cause the pressure change at the dike resulting in inflation/deflation (Kazahaya et al. 2015a; Fig. 7b).

Asama volcano has repeated frequent magmatic eruptions, but the erupted magma volumes are far less than the degassed magma volume. The total degassed magma volume in 1975–2011 (9×10^8 m^3) is two to three orders of magnitude larger than the volume of the ejected magma of 1×10^6 m^3 in the same periods (Aoki et al. 2013). Large eruptions are also repeated at Asama volcano, such as those in 1108 (VEI = 5 with the erupted magma volume of 6×10^8 m^8) and in 1783 (VEI = 4 with the magma of 5×10^8 m^8). The degassed magma volume in the recent 36 years is even larger than the erupted magma volumes of such large eruptions. If the volcano continues the persistent degassing with a constant SO_2 emission rate in the historical age, the total magma volume degassed during a thousand years (1012–2011) amounts to 2.6×10^{10} m^3, which is one order of magnitude larger than the erupted magma volume in the historical ages. The conduit magma convection is the dominant magmatic process of the volcano rather than the eruptive activities.

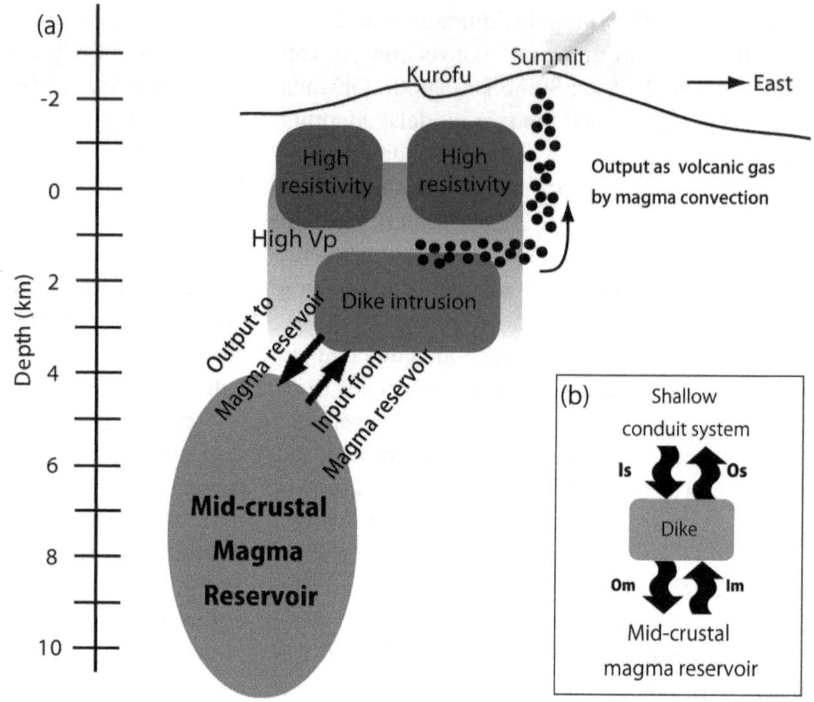

Fig. 7 Schematic diagrams of magma transport and volumetric balance of Asama Volcano. (**a**) Magma plumbing system of Asama Volcano (Aoki et al. 2013). (**b**) Volumetric balance at the dike. The dike volume is controlled by the balance among the input from the mid-crustal magma reservoir (Im), the output to the mid-crustal magma reservoir (Om), the input from the shallow conduit system (Is), and the output to the shallow conduit system (Os). Reproduced from Kazahaya et al. (2015a) with permission

6 Summary

Volcanic activities of Asama volcano are characterized by repetitive transition between the active degassing periods with large SO_2 emission rates (>10 kg/s) and the inactive periods with small but continuous SO_2 emissions (<2 kg/s SO_2). Change in the various activities, such as seismicity, deformation, and eruptions, coincides with the variation of the SO_2 emission rates. In contrast, the volcanic gas compositions do not show significant changes with time, suggesting a stable degassing condition throughout the active and inactive periods. The diffused CO_2 emissions from the summit area amount to about 16% of the amount by the total degassing. The intense persistent degassing requires one to two orders of magnitude larger amount of magmas to be degassed than the amounts of the erupted magmas and is likely caused a degassing of a large mid-crustal magma chamber by the conduit magma convection.

References

Aiuppa A, Federico C, Giudice G, Giuffrida G, Guida R, Gurrieri S, Liuzzo M, Moretti R, Papale P (2009) The 2007 eruption of Stromboli volcano: insights from real-time measurement of the volcanic gas plume CO_2/SO_2 ratio. J Volcanol Geotherm Res 182:221–230

Aizawa K, Ogawa Y, Hashimoto T, Koyama T, Kanda W, Yamaya Y, Mishina M, Kagiyama T (2008) Shallow resistivity structure of Asama volcano and its implications for magma ascent process in the 2004 eruption. J Volcanol Geotherm Res 173:165–177. https://doi.org/10.1016/j.jvolgeores.2008.01.016

Aoki Y, Takeo M, Ohminato T, Nagaoka Y, Nishida K (2013) Magma pathway and its structural controls of Asama Volcano, Japan. Geol Soc London Spec Publ 380. https://doi.org/10.1144/SP380.6

Carn SA, Fioletov VE, McLinden CA, Li C, Krotkov NA (2017) A decade of global volcanic SO_2 emissions measured from space. Sci Rep 7. https://doi.org/10.1038/srep44095

Edmonds M, Oppenheimer C, Pyle DM, Herd RA, Thompson G (2003) SO_2 emissions from Soufrière Hills Volcano and their relationship to conduit permeability, hydrothermal interaction and degassing regime. J Volcanol Geotherm Res 124(1–2):23–43. https://doi.org/10.1016/s0377-0273(03)00041-6

Japan Meteorological Agency (2005) National catalogue of the active volcanoes in Japan, 3rd edn. Japan Meteorological Agency, Tokyo, Japan. (in Japanese)

Japan Meteorological Agency (2014) Volcanic activity of AsamaVolcano in 2014. (in Japanese). http://www.data.jma.go.jp/svd/vois/data/tokyo/STOCK/monthly_v-act_doc/tokyo/2014y/306_14y.pdf

Kazahaya K, Shinohara H, Saito G (1994) Excessive degassing of Izu-Oshima volcano: magma convection in a conduit. Bull Volcanol 56(3):207–216. https://doi.org/10.1007/bf00279605

Kazahaya R, Mori T, Takeo M, Ohminato T, Urabe T, Maeda Y (2011) Relation between single very-long-period pulses and volcanic gas emissions at Mt. Asama, Japan. Geophys Res Lett 38:L11307. https://doi.org/10.1029/2011gl047555

Kazahaya R, Aoki Y, Shinohara H (2015a) Budget of shallow magma plumbing system at Asama volcano, Japan revealed by ground deformation and volcanic gas studies. J Geophys Res Solid Earth 120:2961–2973

Kazahaya R, Maeda Y, Mori T, Shinohara H, Takeo M (2015b) Changes to the volcanic outgassing mechanism and very-long-period seismicity from 2007 to 2011 at Mt. Asama, Japan. Earth Planet Sci Lett 418:1–10. https://doi.org/10.1016/j.epsl.2015.02.034

Kyushu University, University of Tokyo (1984) Variation of emission rates of sulfur dioxide associated with the 1982–1983 eruption at Asama volcano. Rep of Coordinating Comm for Prediction of Volcanic Erupt 30:111–113. (in Japanese)

Kyushu University, University of Tokyo (1986) Emission rates of sulfur dioxide from Asama volcano during the period from 1984 to 1985. Rep of Coordinating Comm for Prediction of Volcanic Erupt 37:1–2. (in Japanese)

Maeda Y, Takeo M (2011) Very-long-period pulses at Asama volcano, Central Japan, inferred from dense seismic observations. Geophys J Int 185:265–282. https://doi.org/10.1111/j.1365-246X.2011.04938.x

Moffat AJ, Nakahara T, Akitomo T, Langal L (1972) Mt. Asama volcano SO_2. Barringer Research Technical Paper 062

Mori T, Notsu K (2005) Remote FT-IR measurements of volcanic gas chemistry in the plume of Asama volcano. Bull Volcanol Soc Jpn 50:567–574. (in Japanese with English abstract)

Mori T, Shinohara H, Kazahaya K, Hirabayashi J, Matsushima T, Mori T, Ohwada M, Odai M, Iino H, Miyashita M (2013) Time-averaged SO_2 fluxes of subduction-zone volcanoes: an example of 32 years exhaustive survey for Japanese volcanoes. J Geophys Res Atmos 118(5):8662–8674. https://doi.org/10.1002/jgrd.50591

Morita M, Mori T, Kazahaya R, Tsuji H (2016) Diffuse carbon dioxide emissions from hidden subsurface structures at Asama volcano. Japan Bull Volcanol 78:17. https://doi.org/10.1007/s00445-016-1008-5

Murakami M (2005) Magma plumbing system of the Asama Volcano inferred from continuous measurements of GPS. Bull Volcanol Soc Jpn 50(5):347–361. (in Japanese with English abstract)

Nogami K, Onizawa S, Hirabayashi J (2008) Geochemical observation on the 2004 eruption of Asama volcano through analysis of water-soluble components on the ash. Bull Volcanol Soc Jpn 53:67–77. (in Japanese with English abstract)

Noguchi K (1935) Geochemical studies of Japanese volcanoes (part. 1) researches on the gas emission and water discharges from the summit and its surrounding of Asama volcano (part. 1). J Chem Soc Japan 56(12):1495–1510. (in Japanese)

Noguchi K, Kamiya H (1963) Prediction of volcanic eruption by measuring the chemical composition and amounts of gases. Bull Volcanol 26:367–378

Ohta K, Kagiyama T, Matsuo N (1978) Remote sensing measurements of sulfur dioxide emissions from the Volcano Asama. Bull Earthq Res Inst, University of Tokyo 53(2):533–542. (in Japanese with English abstract)

Ohta K, Kagiyama T, Matsuo N, Hirahara S, Hirokawa M (1982) Remote sensing measurements of emission rates of sulfur-dioxide from the Asama Volcano Rep Joint geophysical and geochemical observations of the Asama volcano in 1981:63–72,8 (in Japanese)

Ohta Y, Miyaghi I, Yamaguchi T, Yamaguchi Y (2007) Search for primary magma: water content of melt inclusions. Monthly Chikyu 29(1):16–21,8. (in Japanese)

Ohwada M, Kazahaya K, Mori T, Kazahaya R, Hirabayashi J, Miyashita M, Onizawa S, Mori T (2013) Sulfur dioxide emissions related to volcanic activity at Asama Volcano, Japan. Bull Volcanol 75:775. https://doi.org/10.1007/s00445-013-0775-5

Okita T, Shimozuru D (1975) Remote sensing measurements of mass flow of sulfur dioxide gas from volcanoes. Bull Volcanol Soc Jpn 19(3):151–157. (in Japanese with English abstract)

Ossaka J, Murai I, Minami E, Tsuya H (1957) Studies on the springs and underground waters in volcano Asama and its vicinity (1) chemical compositions of the springs and underground waters in the Zyabori river valley and its neighborhood. Bull Earthqu Res Inst, University of Tokyo 35:395–414. (in Japanese with English abstract)

Shinohara H (2008) Excess degassing from volcanoes and its role on eruptive and intrusive activity. Rev Geophys 46:RG4005. https://doi.org/10.1029/2007RG000244

Shinohara H (2013) Volatile flux from subduction zone volcanoes: insights from a detailed evaluation of the fluxes from volcanoes in Japan. J Volcanol Geotherm Res 268:46–63. https://doi.org/10.1016/j.volgeores.2013.10.007

Shinohara H, Ohminato T, Takeo M, Tsuji H, Kazahaya R (2015) Monitoring of volcanic gas composition at Asama volcano, Japan, during 2004–2014. J Volcanol Geotherm Res 303:199–208. https://doi.org/10.1016/j.jvolgeores.2015.07.022

Takahashi H, Hosoya Y (1978) Variations in chemical species in the groundwater around the Asama volcano. Relation of the explosive activity in 1973. Bull Earthqu Res Inst, University of Tokyo 53(2):551–568. (in Japanese with English abstract)

Sulfur Concentration in Magmas From Asama Volcano

Yoshiaki Yamaguchi, Andreas Auer, and Isoji Miyagi

1 Introduction

Among the magmatic volatiles H_2O, Cl, F, S, and CO_2, sulfur plays a special role for its diverse behavior during magmatic differentiation. It has a unique ability to record chemical and physical conditions during the evolution of magmatic systems up to the moment of a volcanic eruption when it is released into the atmosphere (Métrich and Mandeville 2010). In silicate melts, sulfur occurs either as sulfide (S^{2-}) or sulfate (S^{6+}) ion, which is controlled by melt composition and oxygen fugacity (Jugo et al. 2005). Oxidized, water-rich arc magmas can transport large amounts of dissolved sulfate into the crust where they can directly precipitate sulfate minerals (Luhr 2008), initiate Cu–porphyry formation (Hattori and Keith 2001), or contribute to the emissions of active eruptions (Krueger 1983).

Measuring sulfur dioxide as well as H_2S emissions from active volcanoes is among the most important volcanic monitoring efforts and is either done using ground-based techniques (Stoiber and Jepsen 1973; Galle et al. 2003; Mori and Burton 2006) or satellite-based techniques (Carn et al. 2016). The Japan Meteorological Agency monitors (among others) the amount of sulfur dioxide emitted from Mt. Asama (Chap. "Volcanic Events and Gas Emission at Asama Volcano"), Mt. Aso Nakadake, Miyakejima, Sakurajima as significant shifts in sulfur emissions can also herald an impending eruption (Daag et al. 1996).

Large scale injections of sulfur into the atmosphere can also have direct climate forcing effects (Sigurdsson 1982, 1990; Metrich et al. 1991; Thordarson and Self 2003). Therefore, accurate estimation of magmatic sulfur content is also required for historic and prehistoric eruptions. A *petrographic method* can be employed to estimate magmatic sulfur concentrations prior to eruption (Devine et al. 1984) based on the composition of melt inclusions (MI). Melt inclusions are isolated fractions of melt that became entrapped within phenocrysts at various levels and depths within the magmatic system. However, for arc volcanoes, the estimated sulfur concentrations based on the actual emissions (e.g., by satellite observation) and those from melt inclusions usually diverge by an order of magnitude, known as the excess sulfur problem (Sharma et al. 2004). For a long time, Asama Volcano was considered part of this problem,

Y. Yamaguchi
Department of Geology, Shinshu University, Matsumoto, Nagano, Japan

A. Auer (✉)
Department of Geoscience, Shimane University, Matsue, Shimane, Japan
e-mail: auer@riko.shimane-u.ac.jp

I. Miyagi
Geological Survey of Japan, National Institute of Advanced Industrial Science and Technology (AIST), Tsukuba, Ibaraki, Japan

S. Aramaki et al. (eds.), *The Asama Volcano*, Active Volcanoes of the World,
https://doi.org/10.1007/978-981-95-5799-8_25

with very low sulfur concentrations reported from melt inclusion studies (Kohno et al. 1993). Magmas erupted at Mt. Asama are of evolved silica-rich composition (Aramaki 1963), and petrographic studies have focused on melt inclusions within the phenocryst assemblage that was erupted with these magmas. The existence of "cryptic" basaltic compositions had early on been recognized by Anderson (1982), but subsequently these basaltic compositions were not considered during petrographic sulfur estimates from Asama Volcano. Here we review the existing melt inclusion data from Asama Volcano (Kohno et al. 1993) and demonstrate that the excess sulfur problem at Asama volcano can be resolved when including (rare) olivine-hosted mafic melt inclusions into consideration (Yamaguchi et al. 2023).

2 Geological Background

Mt. Asama is a long-living composite stratovolcano that has been amalgamated during several stages of activity, which are Kurofu (100–20 ka), Hotokeiwa (20–10 ka), and Maekake (10 ka–present) (Takahashi et al. 2007, 2008a, b). The Maekake period is further subdivided into three active periods of which the first lasted from 9200 to 8600 years ago, the second from 6300 to 4900 years ago, and the third from the fourth century to the present.

Magmas erupted during the earliest stages (Gippa Lava Group of the Kurofu period) are characterized by low silica content (53–58 wt.%) and show dominant fractionation of olivine + pyroxene. In contrast, the whole-rock SiO_2 content of the current Maekake stage ranges from 58 to 63 wt.% and is dominated by fractionation of pyroxene and plagioclase.

These plagioclase phenocrysts have sieve-textured rims, and the core composition is bimodal. Reverse compositional zoning is observed in orthopyroxene, and abundant additional disequilibrium features have been described (Aramaki and Takahashi 1992; Yasui and Koyaguchi 2004; Miura et al. 2007; Yasui 2016). In addition, matrix glass of the Maekake Group exhibits a wide range of compositions from 64 to 73 wt% SiO_2, showing a linear change on the Harker diagram (Yasui 2016). Based on these observations, it is believed that the andesite of the Maekake Group was produced by the mixing of mafic and felsic end-member magma (Aramaki and Takahashi 1992; Yasui and Koyaguchi 2004; Miura et al. 2007; Yasui 2016). The magma of relatively large-scale eruptions during the third active period of the Maekake Group is relatively rich in MgO. This period is also characterized by a clear increase in the erupted volume compared to previous active periods (Takahashi et al. 2007). The whole-rock chemical composition of the Maekake Group is distributed on the straight line connecting one type of felsic end-member and multiple mafic end-members in the magnesium Harker diagram (Fig. 1).

The most mafic end member compositions at Asama volcano had been described early on by Anderson (1982). However, because this study relied solely on olivine-hosted melt inclusions, these values were subsequently not discussed with respect to the proposed magma mixing results nor the primary magma volatile content at Asama Volcano. This is mainly attributed to the extremely low proportion of olivine in these samples as well as the complicated and demanding sample preparation. However, more recent studies at other volcanoes have also shown that andesitic to rhyolitic magmas can contain trace amounts of olivine phenocrysts, thought to have originated from more primitive basaltic magma that feeds these volcanoes, for example in products from the Kussharo caldera KP1 ejecta (Miyagi et al. 2012) and within the Aso 4 ejecta (Miyagi et al. 2023).

Melt inclusions give a broader perspective on magma evolution at Asama as well as melt sulfur concentrations in the least evolved products.

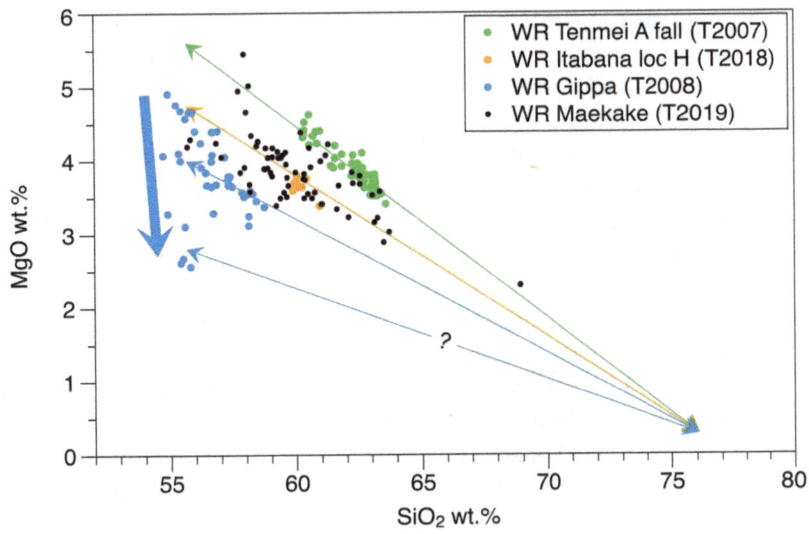

Fig. 1 Harker diagram showing whole rock composition during major eruptive stages of Asama volcano

3 Melt Inclusion Studies and Thermodynamic Modeling

Most effusive and explosive volcanic ejecta have lost the majority of volatile components, including sulfur. Phenocryst-hosted melt inclusions are an appropriate alternative study target (Wallace and Edmonds 2011) but require that the ejecta have been cooled rapidly after the eruption (Miyagi et al. 2023). We therefore choose to analyze the melt inclusions in pumice fall deposits rather than lava. To facilitate comparison with volcanic gas, the concentration of sulfur is reported as SO_2 weight. We selected the Itabana brown pumice that erupted about 20 ka during the Kurofu period (Nakamura et al. 1997) and is composed of 70.3% of matrix, 23.8% of plagioclase, 5.9% of Cpx and Opx (all by weight%), and an extremely small amount of olivine (Yamaguchi et al. 2023). In addition, we selected the Tenmei pumice from the Maekake period that erupted in 1783. In Tenmei pumice, the matrix accounts for 43.5% by volume, and the phenocrysts include 39.2% volume of plagioclase, 8.4% volume of Opx, 7.1% by volume of Cpx, 1.9% by volume of magnetite, and no or very small amount of olivine (0.028% by volume or less) (Aramaki 1957).

Because the proportion of olivine was very low, a large amount of the pumice sample was crushed and subjected to dynamic separation as previously described (Yamaguchi et al. 2023). Electron microprobe analysis (EPMA) conditions were as described in (Yamaguchi et al. 2023), and thermodynamic modeling was done using the MELTS software (Gualda et al. 2012) on the OL4MI2 composition of Yamaguchi et al. (2023) using the same method as Miyagi et al. (2023).

4 Sulfur Concentration in Magmas from Asama Volcano

Olivine-hosted melt inclusions from the Tenmei and Itabana pumice contain the most mafic compositions reported from Asama volcano in accordance with observations by Anderson (1982), suggesting that Asama's upper crustal storage system is fed by cryptic injections of basalt. The values also overlap with whole-rock compositions erupted during the earliest stages of Asama's activity (Gippa group). Most melt inclusion datapoints have been corrected for post-entrapment crystallization (PEC), and the compositional trends caused by these overgrowth corrections for melt inclusions are almost identical to the compositional distributions of whole-rock chemistry and melt (Fig. 2).

The observed similarity supports the idea that the compositional change trends in the whole-rock chemistry of the Gippa group were mainly caused by the removal of olivine (Takahashi et al. 2008b). The MgO composition of melt inclusions in Cpx, Opx, and Pl was distributed on a straight line connecting the most MgO-poor composition of the whole-rock chemistry of the Gippa group and the common felsic end member (Fig. 2). This observed mixing trend suggests that mafic end members, similar to the Gippa Lava group, have admixed not only in the Maekake stage (Tenmei) (Aramaki and Takahashi 1992; Yasui and Koyaguchi 2004; Miura et al. 2007; Yasui 2016), but also in the Hotokeiwa stage (Itabana) to a successively more extensive large felsic storage in the upper crust beneath Asama. The fact that melt compositions tend to have less MgO than whole-rock compositions (Fig. 2) may be due to mixing magma with incomplete removal of MgO-rich minerals (olivine, pyroxene) and felsic endmembers.

In that case, for both the Itabana and Tenmei eruptions, the composition of the mafic end-member melt during magma mixing can be estimated to be approximately 3% MgO (by weight) and approximately 55% SiO_2 (by weight). The sulfur concentration in Matrix glass is always low regardless of the degree of melt differentiation (Fig. 3). Importantly, the sulfur concentration of Melt inclusions in phenocrysts is high in olivine host phenocrysts and always low in all other host phenocrysts. Those included in plagioclase tend to have even lower sulfur concentrations.

The sulfur concentration of melt inclusions in olivine is roughly distributed on a straight line away from the origin in the diagram with the horizontal axis K_2O. Such a distribution suggests that either both elements were concentrated or diluted in the melt by crystallizing or dissolving phases that do not contain sulfur and K_2O or that only the sulfur concentration changed with little change in K_2O. In addition, such a distribution can also suggest that only the sulfur concentration in the melt increases or decreases in response to the crystallization or dissolution of a phase that does not contain K_2O and has an extremely high sulfur concentration. Since the maximum sulfur concentration of the melt inclusions decreases as

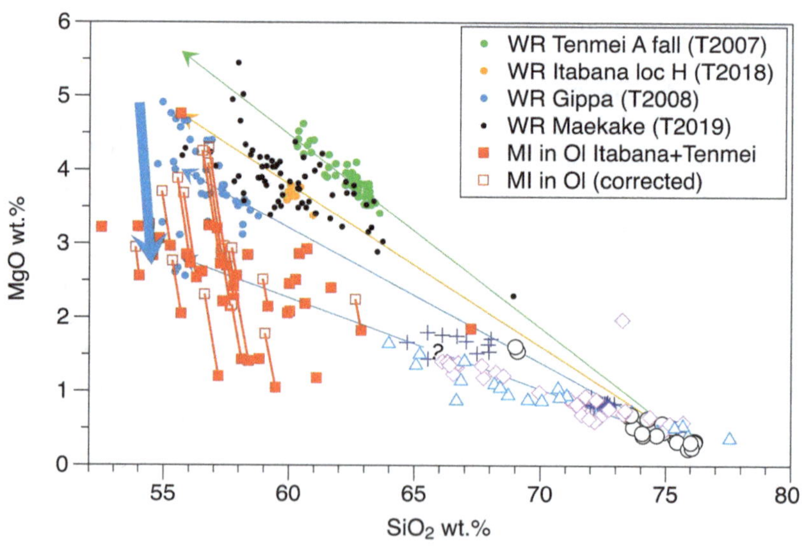

Fig. 2 Harker diagram showing whole-rock composition and melt inclusion composition for olivine, pyroxene, plagioclase, and matrix glass (legend for additional host phenocryst identification is shown in the next figure)

Fig. 3 Harker diagram showing SO_2 versus K_2O (**a**) and SiO_2 (**b**) concentration for different host crystals. Notice the overall much lower concentration in crystals from the felsic storage

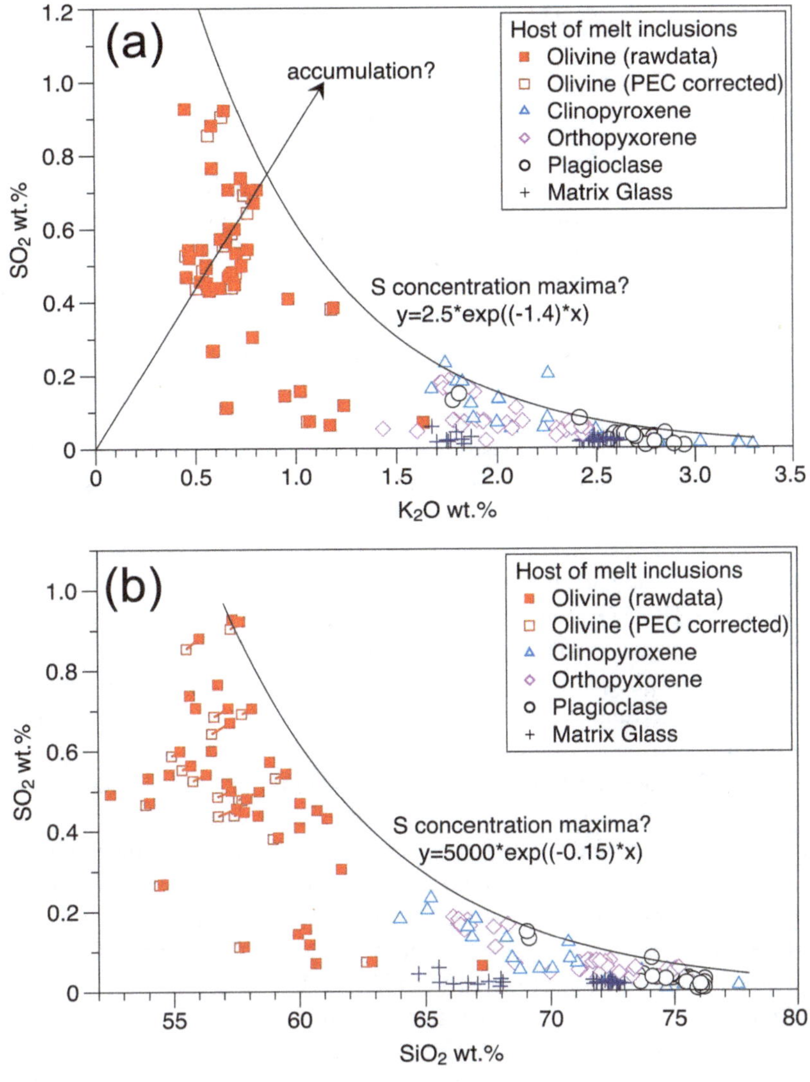

the melt differentiates, such a distribution suggests that sulfur solubility decreases with melt differentiation (Liu et al. 2007) or changes in oxygen partial pressure with differentiation (Moretti and Ottonello 2005).

The observed sulfur concentration is distributed between the maximum values (as seen in olivine MI) and those in the matrix glass (Fig. 3), but this variation of sulfur contents is not simply determined by the equilibrium between the sulfur in the melt and the supersaturated sulfur pure phase but also by the fugacity of sulfur in the exsolved C-H-O-S vapor phase (Moretti and Ottonello 2005; Wallace and Edmonds 2011). Thus, the observed sulfur concentrations in melt inclusions indicate that the whole-rock sulfur content of magma is not determined only by the concentration in the melt inclusions.

4.1 The Sulfur Content of Mixed Magma

The "total sulfur" = sulfur in a vapor phase (i.e., exsolved bubbles) + sulfur in silicate minerals + sulfur in sulfide minerals + sulfur in sulfate minerals + sulfur dissolved in the melt." The sulfur concentration of silicate minerals is almost zero.

The amount of silicate minerals in the felsic magma was found to be 56.4 wt% at Itabana and 34.8 wt% at Tenmei (Y2023IA). Apatite can contain sulfur of about 0.4–0.8 wt% SO_2 (see Figure 16 in Parat et al. (2011) However, since the apatite content is very small, it is assumed to be zero. Although sulfur is also present as a metal sulfide phase (Yamaguchi et al. 2023), the amount is unknown. The sulfur content of metal sulfide (FeS) is 73 wt% in SO_2 equivalent. The sulfur content of sulfate minerals ($CaSO_4$) is 47 wt% in SO_2 equivalent.

Since sulfate minerals are not found in the observed samples, they are assumed to be absent. The sulfur concentration of the melt can therefore be estimated (Yamaguchi et al. 2023): the average value for SO_2 in the felsic melt is 0.056 wt% at Itabana and 0.086 wt% at Tenmei (Table 1). The SO_2 concentration of the glass inclusions, expected to be close to the mafic primary end-member, is 0.736 wt% in Itabana and 0.492 wt% in Tenmei. The amount of vesicles was estimated using Rhyolite-MELTS (Gualda et al. 2012). For the estimation, it is necessary to determine the chemical composition, water content, temperature, and magma pressure. We considered the Asama Itabana melt as the primitive basaltic end member of Asama, and the water content was assumed to be 4% by weight and the pressure of the magma between 100 and 200 MPa (Anderson 1982). The magma temperature is estimated from the melt temperature using the relationship between SiO_2 concentration and temperature shown in Fig. 4. For example, at a magma pressure of 200 MPa (8 km underground), the relationship between melt SiO_2 concentration (wt. %) and temperature is 1050 °C or higher at 55% SiO_2, 900 °C at 65% SiO_2, and 800 °C at 70% SiO_2. The SiO_2 concentration of melt inclusion is shown for each host (Fig. 5). The rough temperature is over 1050 °C for olivine, 900–800 °C for Cpx and Opx, and 800 °C for plagioclase. Having obtained the pressure and temperature, it is possible to estimate the amount of supersaturated water (bubbles) from Fig. 4b.

At a pressure of 200 MPa (8 km underground), melt inclusions in olivine are undersaturated in water above 1050 °C, saturated at 900 °C, and supersaturated (1.7% by weight) at 800 °C.

At a pressure of 100 MPa (4 km underground), melt inclusions in olivine are approximately supersaturated (0.5 wt.%) with water even above 1050 °C, 2.3 wt.% at 900 °C, and 2.7 wt.% at 800 °C. Since the initial water content was set to be 4.0 wt.% at a pressure of 200 MPa and a temperature of 800 °C, the magma is supersaturated with 42.5% of the magmatic water present as a free volatile phase. Similarly, at a pressure of 100 MPa, magma water is saturated by 12.5% relative to the magmatic water at temperatures above 1050 °C, 57.5% at 900 °C, and 67.5% at 800 °C.

Our calculations suggest that a significant amount of the sulfur originally dissolved in the melt may have migrated into bubbles in the

Table 1 Sulfur mass balance, mixing ratio, and SO_2-equivalent mass for the Itabana and Tenmei eruption

Eruption	Itabana	Itabana	Tenmei	Tenmei
Mafic magma	Endmember	Average	Endmember	Average
Crystal wt% in Felsic Magma	–	56.39	–	34.78
SO_2 wt% in Crystal	–	0	–	0
SO_2 wt% in Felsic melt	–	0.0555	–	0.0864
SO_2 wt% in Felsic magma	–	0.0242	–	0.0564
SO_2 wt% in Mafic melt	0.7362	0.5849	0.4921	0.2840
Mixed mafic magma wt%	41.52	47.33	23.46	35.88
SO_2 wt% in mixed magma	0.3198	0.2896	0.1586	0.1380
Magma ton/SO_2 ton	313	345	631	724
SO_2 ton/day	1000	1000	1000	1000
Magma m^3 ($d = 2$ g/cc)	156,354	172,644	315,271	362,206
Sphere radius (m)	67	69	84	88

Fig. 4 Thermodynamic modelling results using the MELTS software (Gualda et al. 2012) for a selected composition from the Itabana pumice assuming a melt water concentration of 4 wt% H_2O (Anderson 1982) showing the degree of super-saturated water (**a**) and silica concentration of the melt (**b**) as a function of pressure and temperature

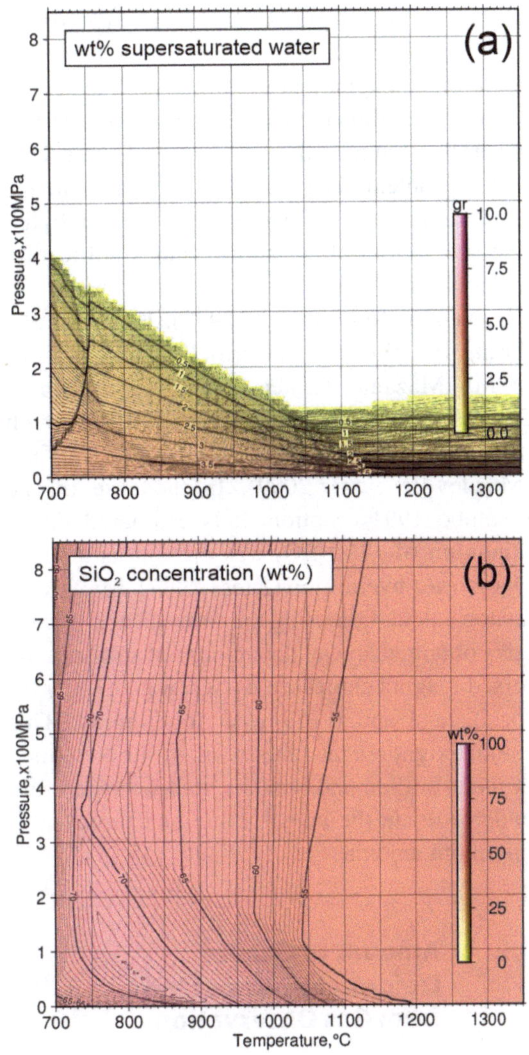

Fig. 5 Temperature estimation, based on melt inclusion silica concentration (see Fig. 4b), for a melt with 4 wt% H_2O at shallow storage (100–200 MPa). Notice the large discrepancy between the olivine-hosted inclusions interpreted here as a hot mafic recharge magma

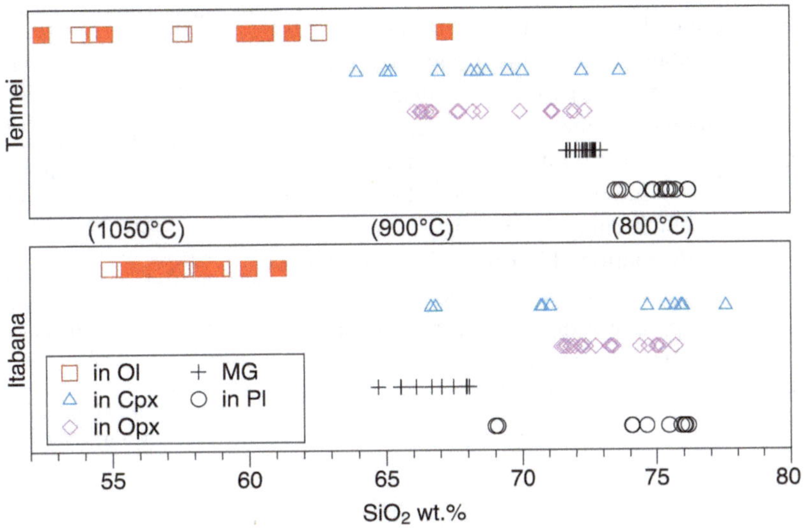

magma before the eruption. The early (before or during crystallization of olivine in the primitive recharging magma) migration of a substantial amount of sulfur into a C-H-O-S vapor phase is generally assumed to be the cause of the excess sulfur problem. The majority of phases in the felsic upper crustal storage are crystallizing after the exsolution of sulfur, thus significantly underestimating the original sulfur content of the magma (Kohno et al. 1993). For example, during the climatic and pre-climatic eruptions of Mt. Mazama, Crater Lake, Oregon, up to 50% of the initial S may have been lost through pre-eruptive degassing at depths of 4–5 km (Mandeville et al. 2009). In the case of the Pinatubo 1991 eruption, it is estimated that a minimum of 17Mt SO_2 was present as a vapor phase prior to eruption (Kress 1997). Making an analog calculation using the formula and parameters obtained above, the amount of sulfur in the mixed magma expressed as SO_2 was 0.320 wt% for Itabana and 0.159 wt% for Tenmei. This estimate ignores the amount of sulfur contained in metal sulfide and bubbles, and will therefore underestimate the actual value and represent a minimum amount.

4.2 Amount of Crustal Deformation Expected from Gas Observations

Based on the obtained amount of sulfur, when the amount of SO_2 released is observed to be 1000 tons/day, the amount of degassed magma is estimated to be 313 tons at Itabana and 631 tons at Tenmei (Table 1). If a sphere is used to approximate the volume of this magma, the diameter is 67 meters at Itabana and 84 meter at Tenmei which may result in recognizable ground deformation. If crustal deformation is not observed, it is possible that degassed magma returns deep underground through a mechanism such as magma convection degassing within a conduit (Kazahaya et al., 1994; Shinohara et al., 1995).

5 Summary

In this chapter, we estimate the whole-rock sulfur content of magma based on the analysis of phenocryst glass inclusions and petrological considerations.

The andesitic magma of Asama is a mixture of felsic and hidden basaltic magma. Based on the analysis of glass inclusion in olivine, an extremely rare phenocryst, the SO_2 content of basalt magma is 0.736 wt% in Itabana and 0.492 wt% in Tenmei.

The SO_2 content of felsic magma is 0.024 wt% at Itabana and 0.056 wt% at Tenmei, based on the melt inclusion analysis of phenocrysts other than olivine and the abundance ratio of the phenocrysts. From magma mixing calculations and sulfur calculations, the SO_2 content of the mixed magma is 0.320 wt% at Itabana and 0.159 wt% at Tenmei. However, this value underestimates the true concentration because the amount of sulfur in the bubbles and metal sulfide is unknown.

Acknowledgments We thank Yasuyuki Miyake, Tamami Yamaguchi, and Yasushi Ohta for cooperating in collecting the pumice samples from Asama Volcano and in the essential investigation. Shigeo Aramaki, Masaki Takahashi, and Maya Yasui initially provided guidance and instruction on magma mixing at the Asama volcano during our inclusion studies of Asama olivine.

References

Anderson AT (1982) Parental basalts in subduction zones: implications for continental evolution. J Geophys Res Solid Earth 87:7047–7060

Aramaki S (1957) The 1783 activity of Asama volcano, part 2. Jpn J Geol Geogr 28:11–33

Aramaki S (1963) Geology of Asama volcano. J Fac Sci Univ Tokyo 14:229–443

Aramaki S, Takahashi M (1992) Geology and petrology of Asama volcano. In: Field workshop guidebook to Asama and Kusatsu-Shirane volcanoes, Japan. IAVCEI commission on explosive volcanism

Carn SA, Clarisse L, Prata AJ (2016) Multi-decadal satellite measurements of global volcanic degassing. J Volcanol Geotherm Res 311:99–134

Daag DS, Dolan MT, Laguerta E et al (1996) Monitoring Sulphur dioxide emissions at mount Pinatubo, Philippines. Fire Mud Erupt Lahars Mt Pinatubo:647–664

Devine JD, Sigurdsson H, Davis AN, Self S (1984) Estimates of sulfur and chlorine yield to the atmosphere from volcanic eruptions and potential climatic effects. J Geophys Res Solid Earth 89:6309–6325. https://doi.org/10.1029/JB089iB07p06309

Galle B, Oppenheimer C, Geyer A et al (2003) A miniaturised ultraviolet spectrometer for remote sensing of SO_2 fluxes: a new tool for volcano surveillance. J Volcanol Geotherm Res 119:241–254. https://doi.org/10.1016/S0377-0273(02)00356-6

Gualda GA, Ghiorso MS, Lemons RV, Carley TL (2012) Rhyolite-MELTS: a modified calibration of MELTS optimized for silica-rich, fluid-bearing magmatic systems. J Petrol 53:875–890

Hattori KH, Keith JD (2001) Contribution of mafic melt to porphyry copper mineralization: evidence from mount Pinatubo, Philippines, and Bingham canyon, Utah, USA. Mineral Deposita 36:799–806

Jugo PJ, Luth RW, Richards JP (2005) An experimental study of the sulfur content in basaltic melts saturated with immiscible sulfide or sulfate liquids at 1300° C and 1· 0 GPa. J Petrol 46:783–798

Kazahaya K, Shinohara H, and Saito G (1994) Excessive degassing of Izu-Oshima volcano: magma convection in a conduit. Bull. Volcanol., 56:207–216

Kohno M, Kusakabe M, Yamaguchi Y, Machida H (1993) Estimation of sulfur and chlorine yields to the atmosphere by historic volcanic eruptions in Japan: A petrologic approach based on glass inclusion analysis. In: IGBP PAGES–INQUA COT Meeting

Kress V (1997) Magma mixing as a source for Pinatubo Sulphur. Nature 389:591–593

Krueger AJ (1983) Sighting of El Chichón sulfur dioxide clouds with the Nimbus 7 total ozone mapping spectrometer. Science 220:1377–1379. https://doi.org/10.1126/science.220.4604.1377

Liu Y, Samaha N-T, Baker DR (2007) Sulfur concentration at sulfide saturation (SCSS) in magmatic silicate melts. Geochim Cosmochim Acta 71:1783–1799

Luhr JF (2008) Primary igneous anhydrite: Progress since its recognition in the 1982 El Chichón trachyandesite. J Volcanol Geotherm Res 175:394–407. https://doi.org/10.1016/j.jvolgeores.2008.02.016

Mandeville CW, Webster JD, Tappen C et al (2009) Stable isotope and petrologic evidence for open-system degassing during the climactic and pre-climactic eruptions of Mt. Mazama, Crater Lake, Oregon. Geochim Cosmochim Acta 73:2978–3012

Métrich N, Mandeville CW (2010) Sulfur in magmas. Elements 6:81–86

Metrich N, Sigurdsson H, Meyer PS, Devine JD (1991) The 1783 Lakagigar eruption in Iceland: geochemistry, CO 2 and sulfur degassing. Contrib Mineral Petrol 107:435–447

Miura K, Takahashi M, Yasui M (2007) Comparative petrography for plagioclase phenocryst in volcanic rocks of historical large-scale eruptions of Asama–Maekake volcano, Central Japan. In: Proc Inst Nat Sci Nihon Univ. pp 117–128

Miyagi I, Itoh J, Hoang N, Morishita Y (2012) Magma systems of the Kutcharo and Mashu volcanoes (NE Hokkaido, Japan): petrogenesis of the medium-K trend and the excess volatile problem. J Volcanol Geotherm Res 231:50–60

Miyagi I, Hoshizumi H, Suda T et al (2023) Importance of long-term shallow degassing of basaltic magma on the genesis of massive felsic magma reservoirs: a case study of Aso Caldera, Kyushu, Japan. J Petrol 64:egad009

Moretti R, Ottonello G (2005) Solubility and speciation of sulfur in silicate melts: the conjugated Toop-Samis-Flood-Grjotheim (CTSFG) model. Geochim Cosmochim Acta 69:801–823

Mori T, Burton M (2006) The SO_2 camera: a simple, fast and cheap method for ground-based imaging of SO_2 in volcanic plumes. Geophys Res Lett 33

Nakamura T, Ikeda A, Takemoto H, Tsuji S (1997) {sup 14} C age measurements with accelerator mass spectrometry of Asama tephra stratigraphic samples around Minami-Karuizawa, the Latest Pleistocene, Nagano prefecture, Central Japan; Naganoken, Minamikaruizawa shuhen no koshinsei saigoki no Asama tephra so no kasokuki {sup 14} C nendai sokutei. Chishitsugaku Zasshi J Geol Soc Jpn 103:

Parat F, Holtz F, Klügel A (2011) S-rich apatite-hosted glass inclusions in xenoliths from La Palma: constraints on the volatile partitioning in evolved alkaline magmas. Contrib Mineral Petrol 162:463–478. https://doi.org/10.1007/s00410-011-0606-7

Sharma K, Blake S, Self S, Krueger AJ (2004) SO_2 emissions from basaltic eruptions, and the excess sulfur issue. Geophys Res Lett 31

Shinohara H, Kazahaya K, and Lowenstern JB (1995) Volatile transport in a convecting magma column: Implications for porphyry Mo mineralization. Geology, 23:1091–1094

Sigurdsson H (1982) Volcanic pollution and climate: the 1783 Laki eruption. EOS Trans Am Geophys Union 63:601–602. https://doi.org/10.1029/EO063i032p00601

Sigurdsson H (1990) Evidence of volcanic loading of the atmosphere and climate response. Glob Planet Change 3:277–289

Stoiber RE, Jepsen A (1973) Sulfur dioxide contributions to the atmosphere by volcanoes. Science 182:577–578

Takahashi M, Yasui M, Ichikawa Y, et al (2007) Whole-rock major element chemistry for eruptive products of Asama-Maekake Volcano, central Japan: Summary for data of 288 samples. In: Proc. Inst. Nat. Sci. Nihon Univ. pp 55–70

Takahashi M, Mukai Y, Nakajima T, et al (2008a) Whole-rock major element chemistry for eruptive products of Asama–Hotokeiwa volcano: summary for data of 307 samples. In: Proc Inst Nat Sci Nihon Univ. pp 167–193

Takahashi M, Nakajima T, Mukai Y et al (2008b) Major element chemistry for eruptive products of Asama-Kurofu volcano, Central Japan: summary for data of 288 samples. Proc Inst Nat Sci Nihon Univ 43:195–216

Thordarson T, Self S (2003) Atmospheric and environmental effects of the 1783–1784 Laki eruption: a review and reassessment. J Geophys Res Atmos 108:AAC-7

Wallace PJ, Edmonds M (2011) The sulfur budget in magmas: evidence from melt inclusions, submarine glasses, and volcanic gas emissions. Rev Mineral Geochem 73:215–246

Yamaguchi Y, Auer A, Miyagi I (2023) Sulfur-rich mafic magma injection into the felsic magma chamber beneath ASAMA volcano, Central JAPAN : records in olivine-hosted melt inclusions from the ITABANA and TENMEI eruptions. Island Arc 32:e12505. https://doi.org/10.1111/iar.12505

Yasui M (2016) Petrography of the Asama 1783 eruptive products. In: Proc Inst Nat Sci Nihon Univ. pp 231–255

Yasui M, Koyaguchi T (2004) Sequence and eruptive style of the 1783 eruption of Asama volcano, Central Japan: a case study of an andesitic explosive eruption generating fountain-fed lava flow, pumice fall, scoria flow and forming a cone. Bull Volcanol 66:243–262

Magma Pathway of Asama Volcano

Yosuke Aoki, Minoru Takeo, Takao Ohminato,
Yutaka Nagaoka, and Kiwamu Nishida

1 Introduction

As Asama Volcano has been one of the most active volcanoes in the Japanese islands, it is one of the most well-monitored and well-investigated volcanoes. This situation allows us to infer the volcano's internal structure and magma plumbing system. This chapter summarizes the magma pathway beneath Asama Volcano from the surface to the mid-crustal level (~10 km below sea level). The following few sections review insights gained from earthquake locations (Takeo et al. 2006), seismic velocity structure (Aoki et al. 2009; Nagaoka et al. 2012), resistivity structure (Aizawa et al. 2008; Usui et al. 2017), and ground deformation (Aoki et al. 2005, 2013; Murakami 2005; Munekane 2021). Then, the final section integrates these insights to show the synthesis of the magma pathway of Asama Volcano at the upper crustal level or above a depth of 10 km.

Y. Aoki (✉) · T. Ohminato · K. Nishida
Earthquake Research Institute, The University of Tokyo, Bunkyo, Tokyo, Japan
e-mail: yaoki@eri.u-tokyo.ac.jp

M. Takeo
Earthquake Research Institute, The University of Tokyo, Bunkyo, Tokyo, Japan

Japan Meteorological Agency, Minato, Tokyo, Japan

Y. Nagaoka
Meteorological College, Japan Meteorological Agency, Kashiwa, Chiba, Japan

2 Earthquake Locations

While precise earthquake locations are essential to infer the magma pathway, they were unavailable before establishing a dense seismic network around 2003. The 2004 eruption, the first eruption after establishing a dense seismic network, allowed Takeo et al. (2006) to relocate hypocenters of more than 500 volcanic earthquakes associated with the 2004 eruption. The relocated seismicity reveals a sharp image of two groups (Fig. 1). One forms a WNW–ESE-trending zone at depths between 1 and 1.5 km below sea level. The eastern end of this seismic zone lies beneath the summit crater and extends westward horizontally by about 2 km. The other group forms a narrow vertical seismic zone extending from the eastern edge of the other group to just under the summit crater.

Figure 1 shows that the seismicity beneath the crater persists regardless of the volcanic activity. On the other hand, the westernmost seismic activity is activated only during the unrest. The seismic activity between the westernmost cluster and the one beneath the summit is more or less continuous after the 2004 eruption until late 2010. These observations indicate that the seismic activity at the westernmost cluster may be sensitive to volcanic activity. They also imply that the magma beneath Asama Volcano originates from the west of the summit.

S. Aramaki et al. (eds.), *The Asama Volcano*, Active Volcanoes of the World,
https://doi.org/10.1007/978-981-95-5799-8_26

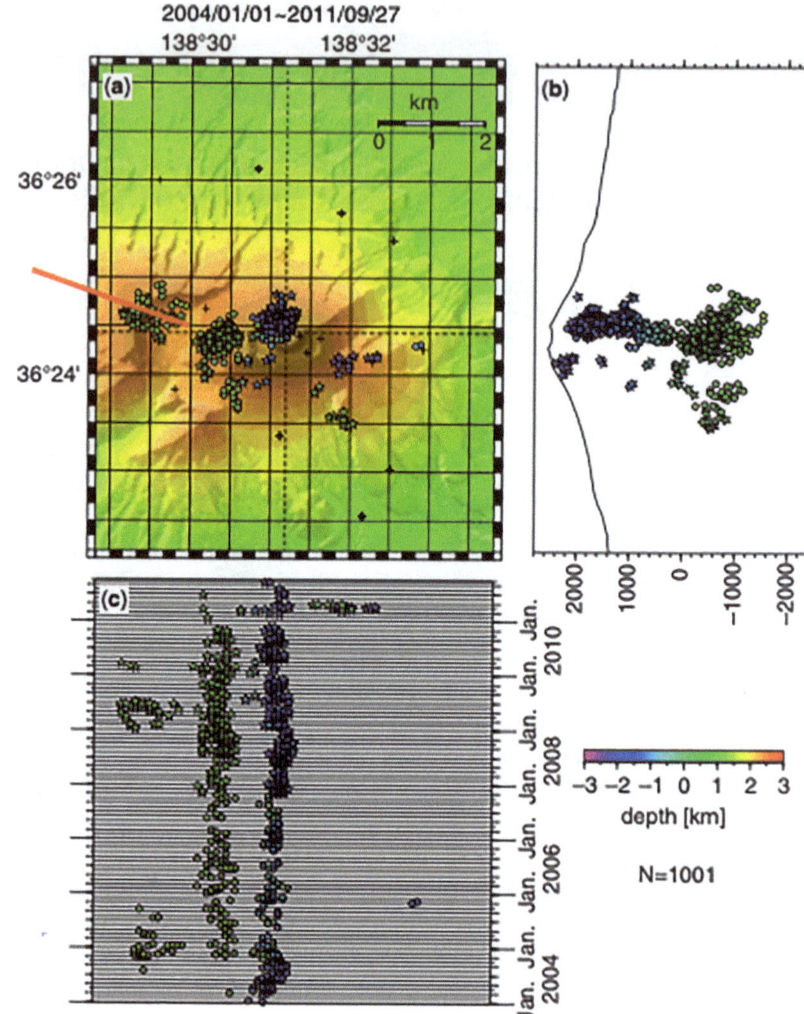

Fig. 1 Distribution of relocated hypocenters around Asama Volcano between January 2004 and September 2011. Colors denote the depth of the earthquakes. (**a**) Distribution of epicenters with the approximate location of dike intrusions during unrest (red line; see also Chapter "Ground Deformation at Asama Volcano Measured by Geodetic Data"). (**b**) North-south cross section of hypocenters. (**c**) Temporal evolution of hypocenters. The horizontal axis represents the longitude of hypocenters. Adopted from Aoki et al. (2013)

3 Ground Deformation

Because the magma intrusion to shallower depths results in ground deformation, ground deformation observations are also essential to infer the magma pathway beneath an active volcano. A reasonably dense GPS network allowed Aoki et al. (2005, 2013), Murakami (2005), and Munekane (2021) to infer the magma pathway beneath Asama Volcano. See Chapter "Ground Deformation at Asama Volcano Measured by Geodetic Data" for details about insights gained from geodetic data to understand the volcanism of Asama Volcano.

Aoki et al. (2005, 2013) modeled the ground deformation field between June 2004 and March 2005, spanning the 2004 eruption. They modeled the observed displacement field by a dike intrusion to the ~5 km west of the summit at ~1 km below sea level (Fig. 2). The inferred dike tip from geodetic data roughly coincides with the relocated seismicity west of the summit, as discussed in the previous section. This coincidence indicates that the stress concentration by the dike intrusion induced the earthquakes to the west (Fig. 1).

As in the 2004 eruption, the 2008–2009 eruption also accompanied the dike intrusion west of the summit. Aoki et al. (2013) modeled the observed displacement field between June and December 2008 by an east-west-striking dike with a top at 0.7 km below sea level and a shal-

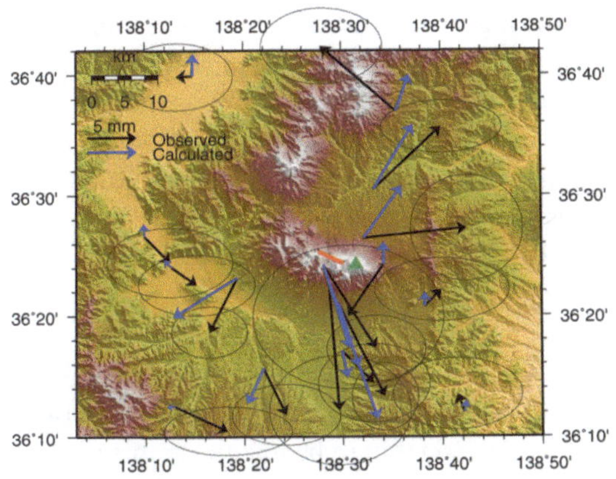

Fig. 2 Comparison between the observed displacements between June 2004 and March 2005, and the calculated displacements by a dike intrusion to the west of the summit (red line). Ellipses represent 1σ uncertainties of the observed displacements. A green triangle indicates the summit of Asama Volcano. Adopted from Aoki et al. (2005, 2013)

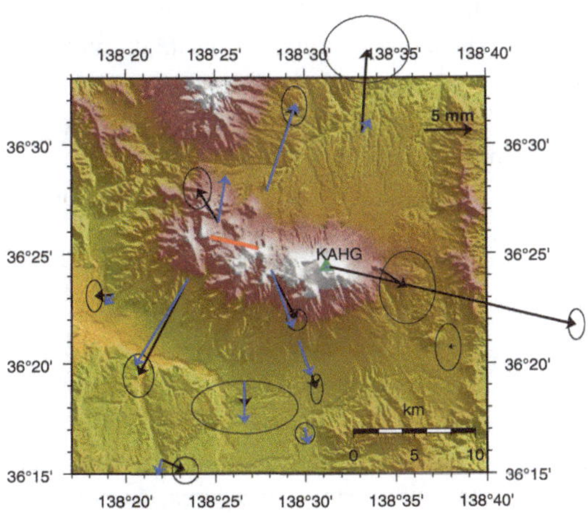

Fig. 3 Comparison between the observed displacements between June and December 2008 (black) when both the western part of the volcano and the shallower depths beneath the summit are inflating and the calculated displacements with the optimum dyke (red) that fits the observed displacements, excluding those of KAHG which are clearly affected by inflation of a shallower origin. Adopted from Aoki et al. (2013)

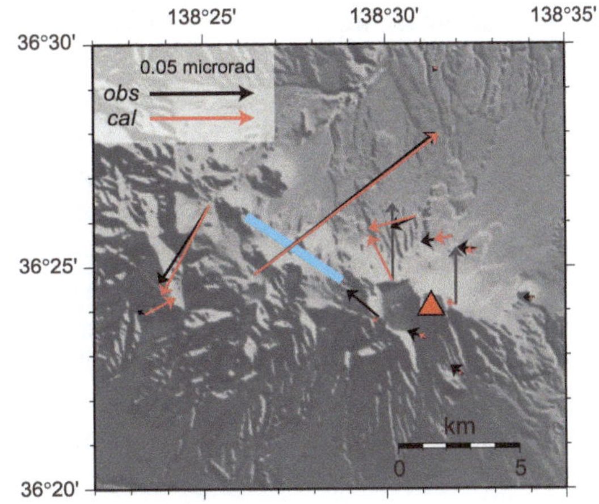

Fig. 4 Tilt motion associated with the eruption on 2 February 2009. The blue line denotes the modeled location of the intruded dike. Black and pink vectors represent the observed and calculated tilt directions. The red triangle denotes the location of the summit of Asama Volcano. Adopted from Nagaoka et al. (2012)

low source responsible for the summit inflation (Fig. 3). The east-west-striking dike is located near the one associated with the 2004 unrest, indicating that the dike intrusion to the west of the summit is ubiquitous during unrest of Asama Volcano. This observation also confirms an established magma pathway to the west of the summit. Figure 3 exhibits a large eastward displacement at a GPS site (KAHG) on the crater rim. This displacement is due to a volcano expansion at a shallow depth and is not directly related to the magma pathway at ~1 km below sea level discussed here.

We also observed tilt motions during the February 2009 eruption, mainly obtained from the horizontal component of broadband seismic records (Nagaoka et al. 2012). See Chapter "Ground Deformation at Asama Volcano Measured by Geodetic Data" for details. Tiltmeters far from the summit generally tilt toward the west of the summit (Fig. 4), implying a source to the west. We modeled the observed deformation field by a deflating dike. The optimum dike is at its top at about 1 km below sea level with a northwest-southeast strike (Fig. 4), close to the location of dike intrusions during unrest. This observation suggests that the intruded dike is somehow connected to the summit, allowing the intruded dike to be ejected by an eruption. Therefore, this observation further confirms that magma is supplied from the west of Asama Volcano.

4 Seismic Velocity Structure

Earthquake locations and geodetic measurements, as discussed so far, suggest that the magma responsible for the eruptions of Asama Volcano is supplied from the west of the summit. However, these observations do not address what controls the magma pathway beneath the volcano. Also, they do not provide insights into the magma pathway below a few kilometers below sea level. The seismic velocity structure discussed in this section addresses these questions.

4.1 Shallow Seismic Velocity Structure

To address the question as to what controls the magma pathway beneath the volcano, we conducted an active source seismic experiment in October 2006 with a dense deployment of temporary seismometers (Aoki et al. 2009) to probe the P-wave velocity structure of the first few kilometers beneath the Volcano. See Chapter "Shallow Seismic Structure of Asama Volcano" for details.

Figure 5 shows the P-wave velocity structure along the north-south (Fig. 5a) and east-west (Fig. 5b) profiles. Figure 6 denotes the location of these profiles. Figure 5b clearly shows a high-velocity zone around S3 (distance 0 km), at a depth of 2 km below sea level and shallower, the area of dyke intrusion during the 2004 eruptions inferred from seismic and geodetic measurements (Takeo et al. 2006). Figure 5b shows that the velocity to the west of the summit is faster than that to the east. Combining this with the velocity along the north-south profile suggests that the high-velocity zone strikes east-west, with its top deepening to the east. The north-south extent of the high-velocity zone is as narrow as approximately 5 km. We conclude that this is due to solidified magma resulting from repeated dyke intrusions because the location of the high-velocity zone roughly coincides with the area of diking associated with the 2004 eruptions. A single diking cannot form a high-velocity zone. Geodetic data show that the dike intruded in 2004 and 2008–2009 is less than 1 m thick (Takeo et al. 2006), too thin to be imaged in this analysis if a single diking in 2004 or 2008–2009 formed the high-velocity zone.

Figure 6 indicates that the spatial distribution of the Bouguer gravity in the area has east-west-trending local maxima to the west of the summit, the area of high velocity, and the eastern end at Asama Volcano. This saddle-shaped gravity high suggests that the high-velocity zone corresponds to a high-density zone, consistent with our hypothesis that the high-velocity zone is due to the solidification of repeatedly intruded magma. When magma is cooled slowly, it becomes fractureless, thus dense, rock with high seismic velocity.

4.2 Upper Crustal Seismic Velocity Structure

While the active source seismic experiment can resolve shallow seismic structures, as shown in the previous section, it does not resolve any structures deeper than 3–4 km below sea level (Aoki et al. 2009). However, a regional seismic tomography study of the area images

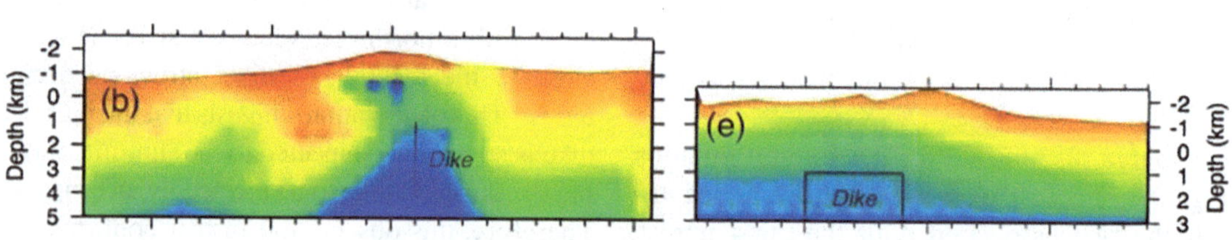

Fig. 5 (**a**) P-wave velocity structure along the north-south (**a**) and east-west (**b**) profile. The approximate position of the intruded dike during the 2004 eruptions (Aoki et al. 2005; Takeo et al. 2006) is also projected. Modified from Aoki et al. (2009)

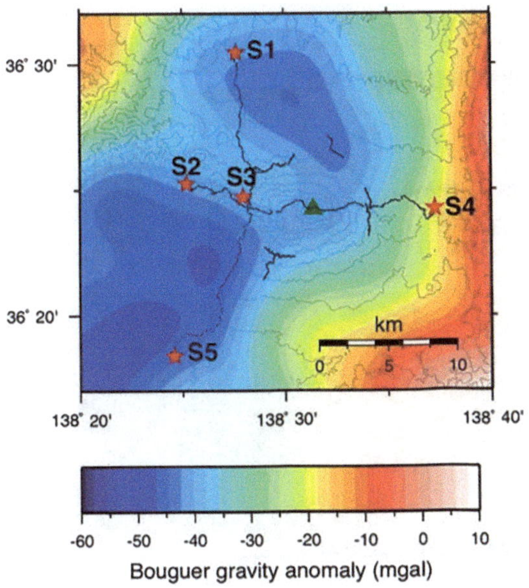

Fig. 6 Spatial distribution of active sources (red stars) and seismometers (black dots). The summit of Asama Volcano is shown by a green triangle. Bouguer gravity anomaly with a reduction density of 2670 kg/m^3 (Geological Survey of Japan 2000). Modified from Aoki et al. (2009)

broader-scale structures from the lower crust to the upper mantle (e.g., Nakajima and Hasegawa 2010). Still, it is impossible to imagine the upper crust in detail, where magma chambers are likely to exist. Here, we fill the gap to image the seismic structure of the upper crust beneath Asama Volcano using seismic ambient noise (Nagaoka et al. 2012). See Chapter "Upper-Crustal Seismic Structure of Asama Volcano" for details.

Figure 7 depicts the three-dimensional S-wave velocity structure. This figure indicates low S-wave velocity at depths of 5–10 km. Also, Fig. 7 displays high S-wave velocity anomalies at shallower depths (<3 km) around the repeated diking region inferred from an active source seismic exploration (Aoki et al. 2009), as discussed in the previous section. In addition, Fig. 7 shows a slower velocity to the east of the summit than to the west, consistent with the result from an active source seismic experiment (Aoki et al. 2009).

What makes the low S-wave velocity change from 5 to 10 km? Here, with some evidence, we suggest it is a magma chamber. As discussed in the previous section, we observed tilt motions during the February 2009 eruption. We modeled the observed deformation field by a deflating dike at its top at about 1 km below sea level with a northwest-southeast strike (Fig. 4), the location of which is right above the low S-wave velocity anomaly. The existence of a dike participating in an eruption suggests that the low S-wave velocity anomaly represents the magma chamber beneath the dyke deflating during the 2009 eruption.

5 Resistivity Structure

Resistivity structure is another powerful tool that can be used to infer the structure of active volcanoes. See Chapter "Electrical Resistivity of Asama Volcano" for detailed discussions about the resistivity structure of Asama Volcano. Electromagnetic data show that the area of the high-velocity zone is less conductive than the surrounding areas (Aizawa et al. 2008; Usui et al. 2017). This observation is consistent with our hypothesis because slowly solidified magma has low permeability, resulting in high resistivity. They referred to this high resistivity, extending up to about 1 km above sea level, as solidified magma associated with eruptions of Kurofu, an active center approximately 3 km west of the current vent, up until 24,000 years ago (Aramaki 1963).

6 Synthesis

Taken together with the discussions above, Fig. 8 shows a synthesis of the magma pathway in the upper crust beneath Asama Volcano. When magma ascends to the upper crust, it is stored in the magma reservoir at a 5–10 km depth. The magma reservoir is offset by approximately 8 km to the west from the summit and is represented by a low-velocity region. Then, the magma ascends to a depth of about 1 km below sea level as a dike. The shallow seismic and resistivity structures (Aizawa et al. 2008; Aoki et al. 2009; Usui et al. 2017) suggest that the magma pathway is subject to structural controls to make a winding path to

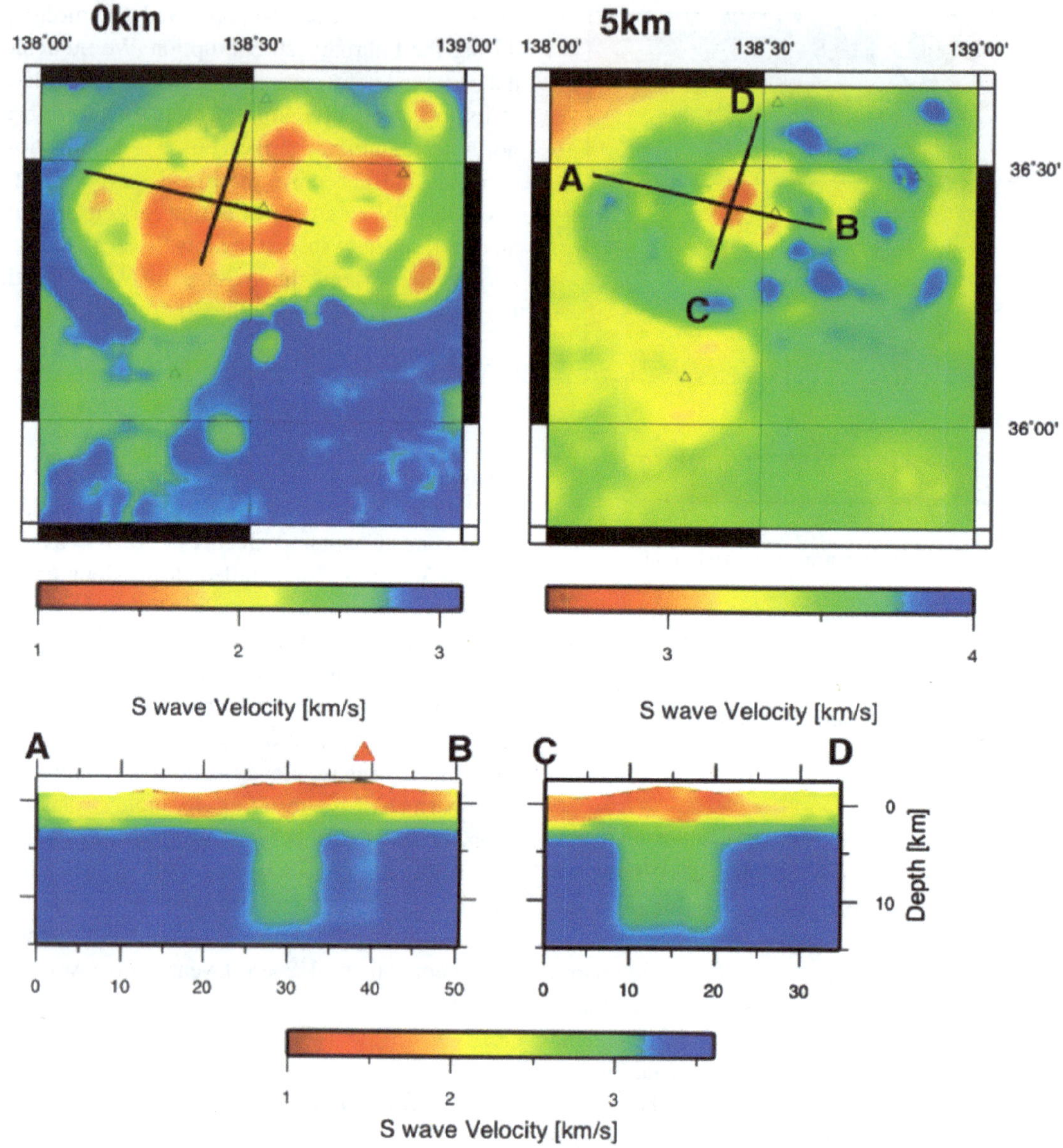

Fig. 7 S-wave velocity around Asama Volcano. The upper panels represent the plan view of the S-wave velocity at 0 and 5 km below sea level, respectively. The lower panels represent the vertical view of the S-wave velocity along the A–B and C–D profiles, respectively, shown in the upper panel

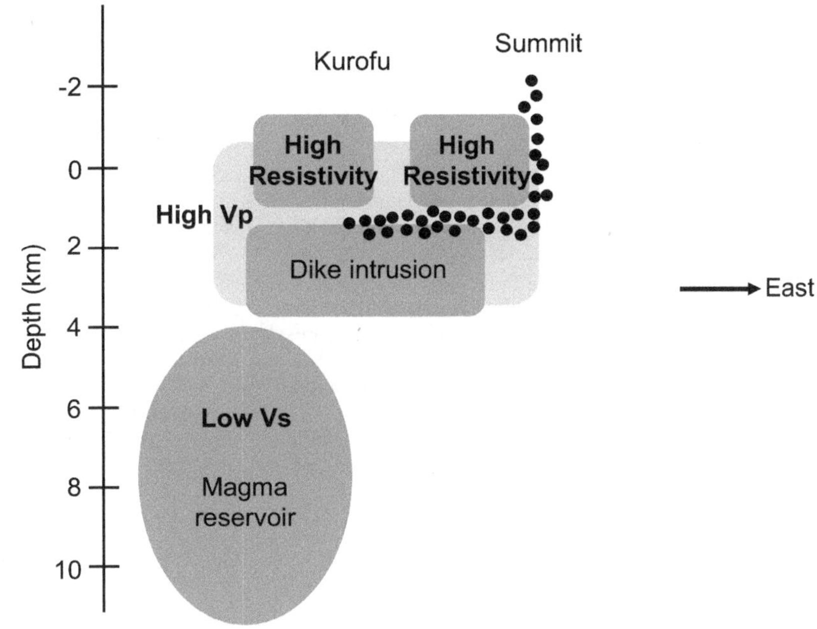

Fig. 8 Schematics of the magma pathway integrating seismic, geodetic, and electromagnetic measurements. Black dots represent seismicity. Magma starts to ascend from the magma chamber as a dyke. The intruded magma is blocked by a high compressional wave velocity (*V*p) (Aoki et al. 2009) and high resistivity (Aizawa et al. 2008; Usui et al. 2017) body and follows a path represented by the seismicity

reach the surface. Combining these multiple constraints enables us to gain a unified understanding of the magma plumbing system of Asama Volcano from a crustal magma chamber up to the surface.

A magma reservoir horizontally offset from the summit of a volcano could be a ubiquitous feature. A recent global compilation indicates that about 20% of arc volcanoes have a magma reservoir significantly offset from the main crater (Ebmeier et al. 2018; Lerner et al. 2020). Specifically, tiltmeters in Kirishima Volcano, Southwest Japan, tilted toward a point approximately 8 km to the west of the summit in response to an eruption on January 26, 2011. This observation is consistent with the location of a deflation source derived from GPS observations. Although no seismic velocity models are available to verify whether a low-velocity body characterizes this deflation source, these tilt observations demonstrate the ubiquity of a horizontally offset magma chamber and structurally controlled magma pathway beneath active volcanoes.

References

Aizawa K, Ogawa Y, Hashimoto T et al (2008) Shallow resistivity structure of Asama Volcano and its implications for magma ascent process in the 2004 eruption. J Volcanol Geotherm Res 173:165–177. https://doi.org/10.1016/j.jvolgeores.2008.01.016

Aoki Y, Watanabe H, Koyama E et al (2005) Ground deformation associated with the 2004—2005 unrest of Asama volcano, Japan. Bull Volcanol Soc Jpn 50:575–584. https://doi.org/10.18940/kazan.50.6_575. (in Japanese with English abstract)

Aoki Y, Takeo M, Aoyama H et al (2009) P-wave velocity structure beneath Asama volcano, Japan, inferred from active source seismic experiment. J Volcanol Geotherm Res 178:273–278. https://doi.org/10.1016/j.jvolgeores.2009.09.004

Aoki Y, Takeo M, Ohminato T et al (2013) Structural controls on magma pathways beneath Asama volcano, Japan. Geol Soc London Spec Publ 380:67–84. https://doi.org/10.1144/SP380.6

Aramaki S (1963) Geology of Asama Volcano. J Fac Sci Univ Tokyo Sect II 14:229–443

Ebmeier SK, Andrews BJ, Araya MC et al (2018) Synthesis of global satellite observations of magmatic and volcanic deformation: implications for volcano monitoring and the lateral extent of magmatic domains. J Appl Volcanol 7:2. https://doi.org/10.1186/s13617-018-0071-3

Geological Survey of Japan (2000) Gravity of Japan. Gelogical Survey of Japan (CD-ROM)

Lerner AH, O'Hara D, Karlstrom L et al (2020) The prevalence and significance of offset magma reservoirs at arc volcanoes. Geophys Res Lett 47:e2020GL087856. https://doi.org/10.1029/2020GL087856

Munekane H (2021) Modeling long-term volcanic deformation at Kusatsu-Shirane and Asama volcanoes, Japan, using the GNSS coordinate time series. Earth Planet Space 73:192. https://doi.org/10.1186/s40623-021-01512-2

Murakami M (2005) Magma plumbing system of the Asama volcano inferred from continuous measurements of GPS. Bull Volcanol Soc Jpn 50:347–361. https://doi.org/10.18940/kazan.50.6_347. (in Japanese with English abstract)

Nagaoka Y, Nishida K, Aoki Y et al (2012) Seismic imaging of magma chamber beneath an active volcano. Earth Planet Sci Lett 333–334:1–8. https://doi.org/10.1016/j.epsl.2012.03.034

Nakajima J, Hasegawa A (2010) Cause of M~7 intraslab earthquakes beneath the Tokyo metropolitan area, Japan: possible evidence for a vertical tear at the easternmost portion of the Philippine Sea slab. J Geophys Res 115:B04301. https://doi.org/10.1029/2009JB006863

Takeo M, Aoki Y, Ohminato T, Yamamoto M (2006) Magma supply path beneath Mt. Asama volcano, Japan. Geophys Res Lett 33:L15310. https://doi.org/10.1029/2006GL026247

Usui Y, Ogawa Y, Aizawa K et al (2017) Three-dimensional resistivity structure of Asama Volcano revealed by data-space magnetotelluric inversion using unstructured tetrahedral elements. Geophys J Int 208:1359–1372. https://doi.org/10.1093/gji/ggw459

Prediction of Volcanic Eruptions at Asama Volcano

Hitoshi Yamasato and Koji Kato

1 Introduction

The prediction of volcanic eruption is to predict the five elements: its timing, location, magnitude, type, and progression. If precise and timely prediction information is issued to the public, volcanic disasters can be dramatically reduced, but the current eruption prediction technology is still under development.

Asama volcano has a long history of observation and research, and the magma supply system and its characteristics have been clarified to some extent; some eruptions have been predicted, although they are not perfect. This chapter overviews the history of eruption prediction research on Asama volcano. It introduces the precursory phenomena of main eruptions in recent years and disaster mitigation information sent by the Japan Meteorological Agency.

H. Yamasato (✉)
CTI Engineering Co., Ltd., Chuo, Tokyo, Japan
e-mail: hyamasat@nifty.com

K. Kato
Volcanic Observation Division, Japan Meteorological Agency, Minato, Tokyo, Japan

2 Minakami's Research for Predicting Eruptions

Prof. Takeshi Minakami (1909–1985) conducted many important volcanological studies at the Asama Volcano Observatory (AVO), which was established in 1933 at the eastern foot of Asama volcano as an affiliated facility of the Earthquake Research Institute of Tokyo Imperial University. In particular, the study of the relationship between volcanic earthquakes and eruptions through high-sensitivity seismic observations near the volcano was an important and pioneering achievement in the world.

He (e.g., Minakami 1959, 1960) classified volcanic earthquakes occurring at Asama volcano into A-type and B-type earthquakes based on the characteristics of recorded waveforms. A-type earthquakes have clear P and S arrivals, while B-type earthquakes are those in which the S phases are unclear and are long-period signals. The hypocenters of A-type earthquakes are rather deep, while B-type earthquakes are assumed to occur in a very shallow area around the crater (Fig. 1). After his research, observational research has been carried out on various volcanic earthquakes, and their mechanism is becoming clearer (see Part III for details). Minakami found a correlation between the frequency of B-type earthquakes and eruptive activity (Fig. 2), and found that B-type earthquakes increase before eruptions. Then, using this property, he obtained an

S. Aramaki et al. (eds.), *The Asama Volcano*, Active Volcanoes of the World,
https://doi.org/10.1007/978-981-95-5799-8_27

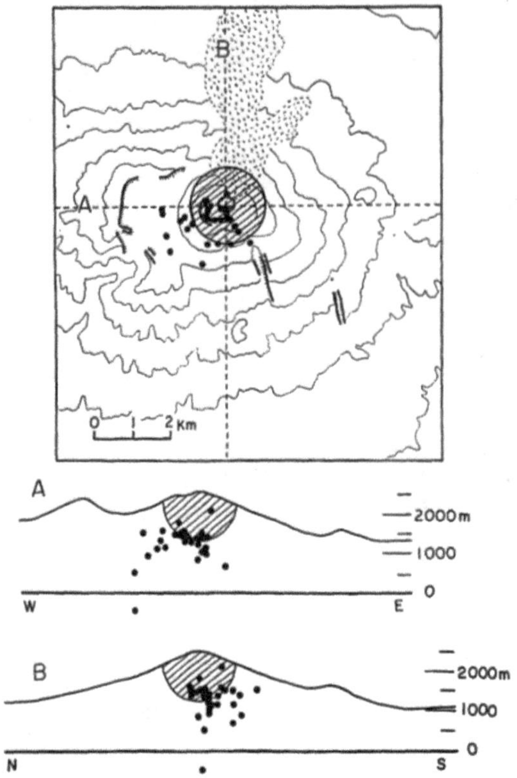

Fig. 1 Hypocentral distribution of volcanic earthquakes at Asama volcano. The hatched area indicates the hypocentral domain of B-type earthquakes (Minakami 1960)

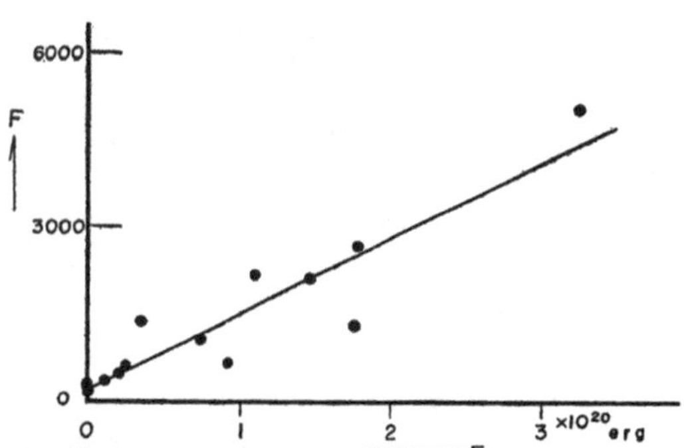

Fig. 2 Relation between annual frequency of earthquakes and annual energy of explosive eruptions at Asama volcano during 1935–1950 (Minakami 1959)

empirical formula for estimating the risk of eruption from the sum of the number of earthquakes in the previous five days. Table 1 shows the relationship between the number of B-type earthquakes in the previous five days and the eruption risk (vulnerability rates) in the next five days (Minakami 1960). There is a clear tendency for the risk of eruptions to increase as the number of earthquakes increases, and it is possible to calculate the probability of an eruption from the number of earthquakes.

Table 1 Vulnerability rates for an explosive eruption of Asama volcano based on the frequency of B-type quakes (Minakami 1960). F_0, F_0': observed value, F_c, F_c': calculated value from Polya-Eggenberger's function

N	F_0'/F_0	F_c'/F_c
0–50	0	0.013
50–100	0.018	0.044
100–150	0.139	0.092
150–200	0.216	0.154
200–250	0.264	0.226
250–300	0.243	0.305
300–350	0.489	0.388
350–400	0.542	0.468
400–450	0.389	0.549
450–500	0.722	0.622
500–550	0.800	0.691
550–600	0.500	0.752
600–650	0.900	0.811
650–700	0.500	0.843
700–750	1.000	0.874

The names of A-type and B-type earthquakes proposed by Minakami are still used today at many volcanological observatories around the world.

3 Eruption in 1958

Using the properties of the relationship between volcanic earthquakes and eruptions at Asama volcano studied by Minakami, the world's first volcanic eruption prediction trial was conducted during the 1958 eruption activity and produced great results.

Asama volcano, which had erupted vigorously since the end of the nineteenth century, became quieter since August 1955, and the number of tourists climbing Asama volcano increased. Meanwhile, signs of reactivation of volcanic activity began to appear from the end of July 1958. The daily number of earthquakes was about 20 during calm times, but it increased three to four times in August, and more than ten times in late September (Fig. 3).

Before the volcanic earthquake swarm, the Karuizawa Weather Station of JMA called on the police chiefs, the town mayor, and each railway station manager to hold meetings to ensure the safety of climbers in June, and knowledge notices for climbers had been posted at the railway stations and the starting points of the trails. Under such circumstances, volcanic earthquakes increased, so on August 1, warning notices were posted. On August 23, in addition to a further increase in volcanic earthquakes, abnormal rumbling from the crater floor and an increase in the volcanic plume were observed; therefore, the Weather Station communicated them in writing to the prefecture, municipalities, and police. On August 27, these organizations held a disaster countermeasures meeting, and as a result, from September 1, climbing more than halfway up was prohibited (Sekiya 1967). During this time, Dr. Minakami provided the aforementioned risk rate data to the Weather Station, and the data were utilized (Minakami et al. 1959).

A month after climbing was banned, ash fall was observed at the foot of the volcano on October 3, after which eruptive activity gradually increased (Japan Meteorological Agency 1959). Then, on November 10, a strong explosive eruption occurred, with a volcanic plume reaching 6000–8000 m. The maximum flying distance of the ballistic bombs was 3.65 km, and an impact crater with a diameter of 6–8 m and a depth of 2.1 m was formed. The air shock and sound from this explosion were so loud that

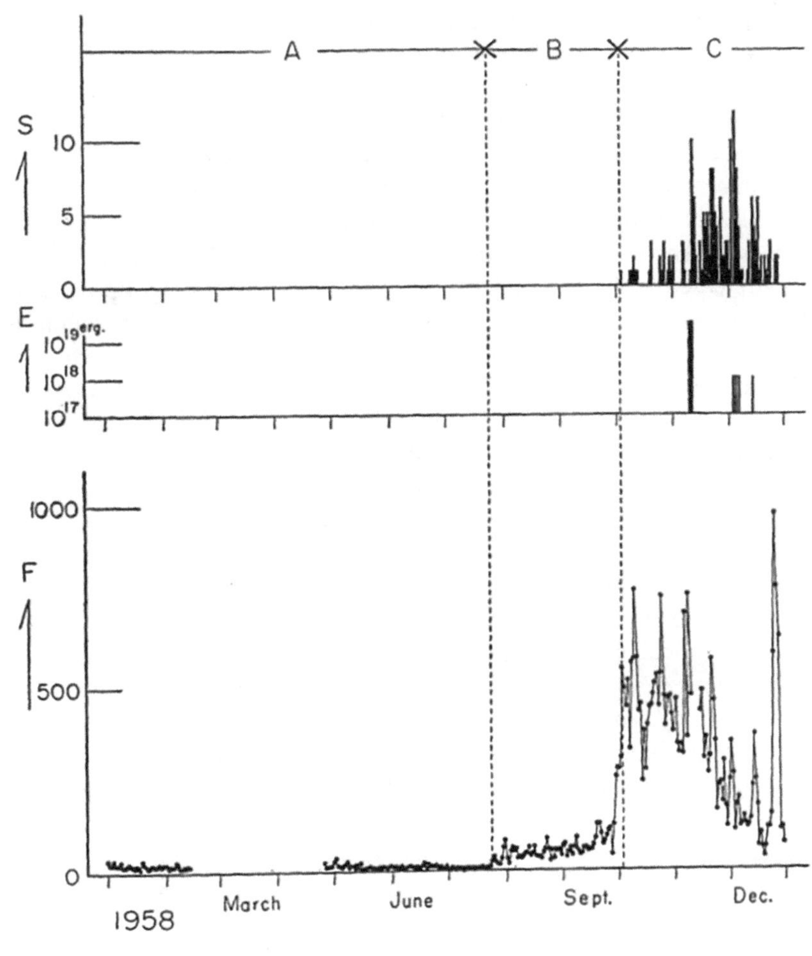

Fig. 3 Relation between eruptions and shallow (B-type) earthquakes of Asama volcano in 1958. A: Calm stage of the volcano, B: Pre-volcanic or pre-eruptive stage, C: Stage of explosive eruption, S: Daily frequency of eruption, E: Kinetic energy of strong eruption, F: Daily frequency of B-type earthquake (Minakami 1960)

they could be felt at a distance of more than 400 km (Takeyama et al. 1960). The air shock damaged nearly 30,000 sheets of window glass, and several people were injured. However, as appropriate disaster prevention measures had been taken in advance, no persons were killed in the series of eruptive activity. It was evaluated that the trial of the eruption forecast and the disaster prevention measures in cooperation with the Weather Station and the AVO yielded mostly satisfactory results (Minakami et al. 1959).

4 Eruption in 1973

After the 1958 eruption, eruptions occurred intermittently for several years, but the activity calmed down after the November 1961 eruption. There was one small eruption in May 1965, but otherwise it was quiet for a long time. It was at the end of November 1972 that the activity picked up again. At the end of November, a weak volcanic glow was observed from the north, and around the same time, the number of volcanic earthquakes began to increase (Fig. 4), and the amount of volcanic plume continued to grow over the next year. The number of earthquakes temporarily decreased at the end of January; however, a volcanic glow was observed again in the early morning of February 1, volcanic earthquakes began to occur frequently, and the first explosive eruption occurred at 19:20, accompanied by a loud explosion sound. Although smaller than the 1958 eruption, it was accompanied by a small pyroclastic flow, glass breakage due to air shocks, and fist-sized rocks falling 6–7 km southeast of the crater.

After the first explosion, small eruptions occurred sporadically, and successive small eruptions occurred from February 16 to 18. The series of eruptions ended with the eruption on May 24.

On the morning of February 1, before the first eruption, the Weather Station sent information about the glowing to relevant organizations to alert them. It then announced a Volcanic Advisory, which had officially been in operation since 1965, at 15:00. As in 1958, it succeeded in alerting people before an eruption occurred.

5 Eruption in 1982–83

The eruptive activity of 1973 continued intermittently until May, and Asama volcano entered a quiet period again. In the 1980s, eruptions occurred on April 26, October 2, 1982, and April 8, 1983. On April 26, 1982, the eruption was accompanied by small pyroclastic flows and snow-melting lahars (Shimozuru et al. 1982). Unlike previous eruptions, this eruption was not preceded by a sharp increase in B-type earthquakes. Therefore, this eruption could not have been predicted by empirical rules up to that time, and the Weather Station issued no advisories or information before the eruption. However, it was not that there were no abnormalities at all; an increase in volcanic earthquakes in March and August 1981 and January 1982, and an increase in the rate of SO_2 emissions in the fall of 1981 had been observed (Shimozuru et al. 1982). They were small-scale phreatic eruptions with no fresh material in the ejecta accompanying the eruptions. It was presumed that the magma had risen to the shallow area and stopped halfway, and the water vapor produced by the heated groundwater caused the eruption (Kagiyama et al. 1983). The eruptions on October 2, 1982, and April 8, 1983, were preceded by an increase in B-type earthquakes, although they were less clear than before, and the 1983 eruption was preceded by an increase in N-type earthquakes (Fig. 5).

These eruptions in the 1980s showed that it is difficult to predict them by conventional empirical rules that predict eruptions based only on the frequent occurrence of volcanic earthquakes immediately before them. On the other hand, as can be seen in Fig. 5, it became clear that there were high-frequency and low-frequency B-type earthquakes, and each behaves differently before an eruption. These characteristics and the heightened volcanic activity several months before the April 1982 eruption provided hints for predicting future eruptions.

Fig. 4 Daily frequency of A-type (rows) and B-type (dots) earthquakes and the quantity of volcanic clouds observed by AVO before the eruption on February 1, 1973 (Shimozuru et al. 1975)

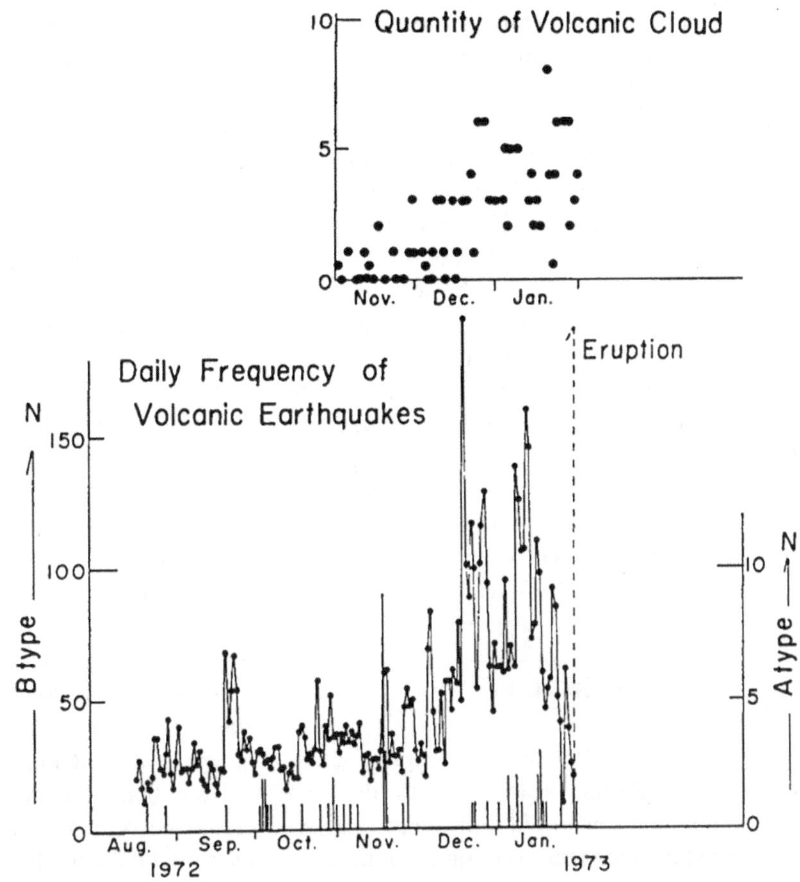

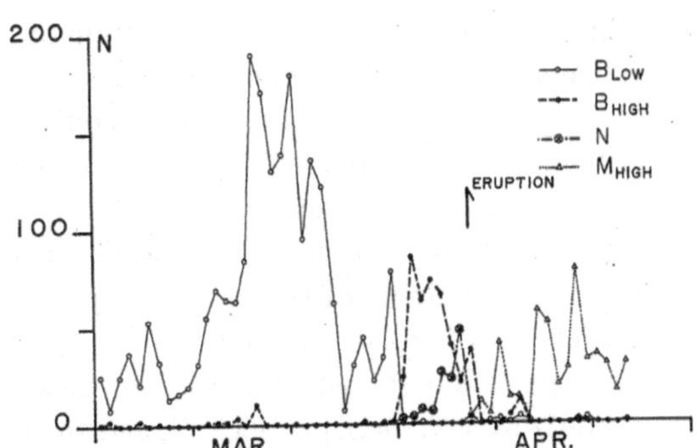

Fig. 5 Daily frequency of volcanic earthquakes associated with the April 8, 1983, eruption of Asama volcano. B_{LOW}: Low-frequency B-type earthquakes. B_{HIGH}: High-frequency B-type earthquakes. N: N-type earthquakes, which have single or multiple dominant frequency waves that slowly attenuate over long times. M_{HIGH}: High frequency tremors (Asama Volcano Observatory 1983)

6 Volcanic Activity Level by JMA

After the 1982–83 eruption, the volcanic activity did not change except for a very small single eruption in July 1990, and the volcanic activity entered a quiet period again.

In 2001, JMA established four Volcano Observation and Information Centers (Sapporo, Sendai, Tokyo, and Fukuoka) to centrally monitor volcanic activity and release information, and the responsibility for monitoring Asama volcano was transferred from the Karuizawa Weather Station to the Headquarters of JMA in 2002. At the same time, the monitoring of Asama volcano, which had been composed mainly of visual observation and seismic observation, was strengthened by adding continuous observation of crustal deformation by GNSS, etc., and observation of sulfur dioxide emissions.

On the other hand, at the end of the twentieth century, movements toward deregulation became active on Asama volcano, where climbing was prohibited after the 1973 eruption (Chapter "Mitigation of Volcanic Hazards of Asama Volcano"). At that time, JMA was planning the Volcanic Activity Level, and the local government was considering using it for mountain climbing restrictions and relaxation. After a certain trial period, the announcement of the Volcanic Activity Level started in November 2003 (Yamasato et al. 2004). The Volcanic Activity Level is an index that expresses the intensity of volcanic activity in six stages from 0 to 5, and it was replaced by the Volcanic Alert Level from December 2017, as described later.

7 Eruption in 2004

On September 1, 2004, a series of eruptive activities began at Asama volcano, with the first medium eruption since 1973. This eruptive activity was the first magmatic eruption in a long time, and enhanced observations provided a wealth of knowledge about Asama volcano.

7.1 Sequence of Volcanic Activity

Figure 6 shows the observation network of the JMA as of 2004. As mentioned above, the JMA had started multi-item volcanological observation. Fig. 7 shows the progress chart of various observation data from March 2002 to 2005. Since around 2000, it has been recognized that volcanic activity has increased, including seismic activity beneath the summit crater, increased volcanic plume activity, and increased crater floor temperature. And very small eruptions that caused ash fall only near the crater occurred in February to April 2003. There were no significant anomalies immediately before and after the very small eruptions, but various data indicated increased volcanic activity as seen in Fig. 7.

Various anomalous phenomena were observed before the September 1, 2004, eruption. An increase in volcanic plume was observed from around July. From the beginning of August, the amount of volcanic plume increased further, and a high-sensitivity camera began to observe volcanic glow. After a temporary decrease in the amount of volcanic plume and the number of earthquakes in late August, a volcanic earthquake swarm was observed from around 15:00 on August 31 (Fig. 8). Since the Volcano Observation and Information Center began monitoring, B-type earthquakes at Asama volcano have been classified into high-frequency B-type earthquakes (BH-type earthquakes) and low-frequency B-type earthquakes (BL-type earthquakes) (Fig. 9). The earthquake swarm was composed of BH-type earthquakes.

The first eruption, which occurred at 20:02 on September 1, 2004, was an explosive eruption accompanied by loud explosion sound and air shock (205 Pa at the Oiwake station in Fig. 6 approximately 8 km south of the crater), with ballistic bombs scattering up to 2.7 km from the crater (according to a subsequent field survey). Due to clouds, the height of the volcanic plume could not be confirmed by the camera, but the weather radar observed an echo flowing northeast at a height of 3500–5500 m. A maximum of 3 cm of lapilli fell up to 6 km northeast of the summit, and volcanic ash fell in Gunma Prefecture and Fukushima Prefecture in the northeast direction. Ejecta volume was estimated at 4.9×10^7 kg (Yoshimoto et al. 2005). The first eruption lasted until after 23:00, and then stopped. From the next day onward, the number of volcanic earthquakes decreased to less than 20 times a day.

The eruptive activity, which had been paused for a while, resumed on September 14, and small eruptions were observed intermittently. Continuous eruptions began around noon on the 16th and continued until the morning of the 18th. This Strombolian eruption intermittently ejected red-hot volcanic bombs accompanied by BL-type earthquakes and infrasonic pulses. On the morning of the 16th, an airborne SAR (synthetic aperture radar) detected a lava cake in the crater (Geographical Survey Institute 2005), revealing

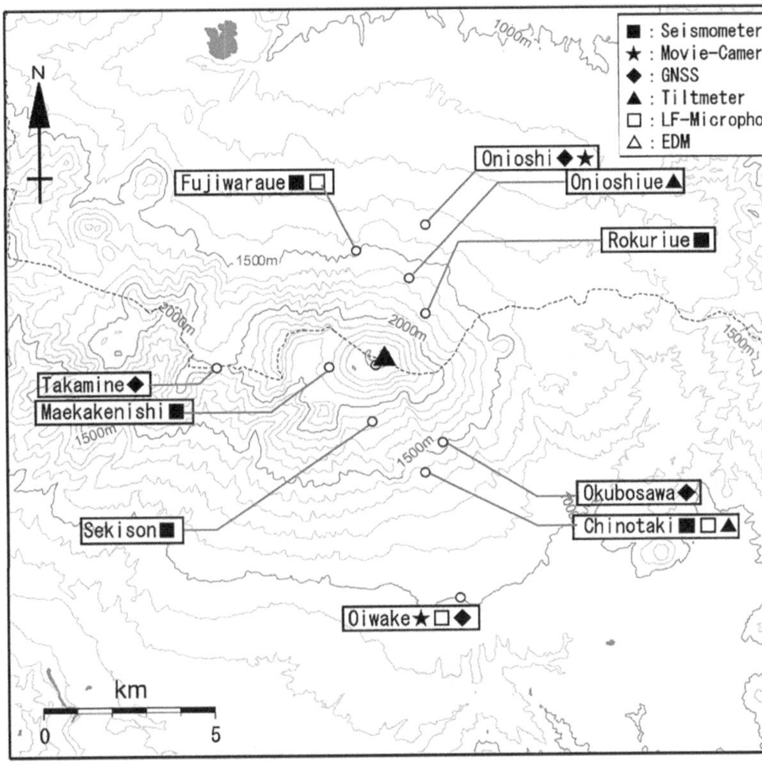

Fig. 6 Observation network of the JMA as of 2004. The present observation network, including that of other organizations, is shown in Chapter "Mitigation of Volcanic Hazards of Asama Volcano"

that magma had risen to the crater floor. After that, eruptions occurred intermittently until a small eruption on December 9. Meanwhile, major vulcanian eruptions occurred on September 23 and 29, October 10, and November 14. The total weight of tephra associated with the series of eruptions was estimated at more than 1.6×10^8 kg (Yoshimoto et al. 2005).

The sequence of the eruption, where two weeks after the first explosive eruption, lava appeared at the bottom of the crater and a strombolian eruption occurred, followed by intermittent vulcanian explosions, was very similar to the 1973 eruption (Nakada et al. 2005).

7.2 Crustal Deformation

This eruptive activity was the first magmatic eruption in about 30 years, and various new findings were obtained through observations that had not existed in 1973. In particular, remarkable results were obtained in crustal deformation observations.

In Japan, since the 1990s, the Geospatial Information Authority of Japan has built a high-density GNSS observation network that covers the entire country and has been conducting continuous observations. At Asama volcano, Earthquake Research Institute, the University of Tokyo, is conducting continuous GNSS observations. JMA had also started continuous GNSS observations since the establishment of the Volcano Observation and Information Center. These observations revealed a magma chamber beneath the western foot of Asama volcano that repeatedly expands and contracts in response to volcanic activity (e.g., Murakami 2005). In Fig. 7, the baseline extension across the volcano's western foot can be seen during the active period (see Chapter "Ground Deformation at Asama Volcano Measured by Geodetic Data" for details).

In addition to GNSS, continuous observation by tiltmeters produced important results for volcano monitoring. As mentioned above, before the eruption on September 1, a BH-type earthquake swarm was observed from around 15:00 on

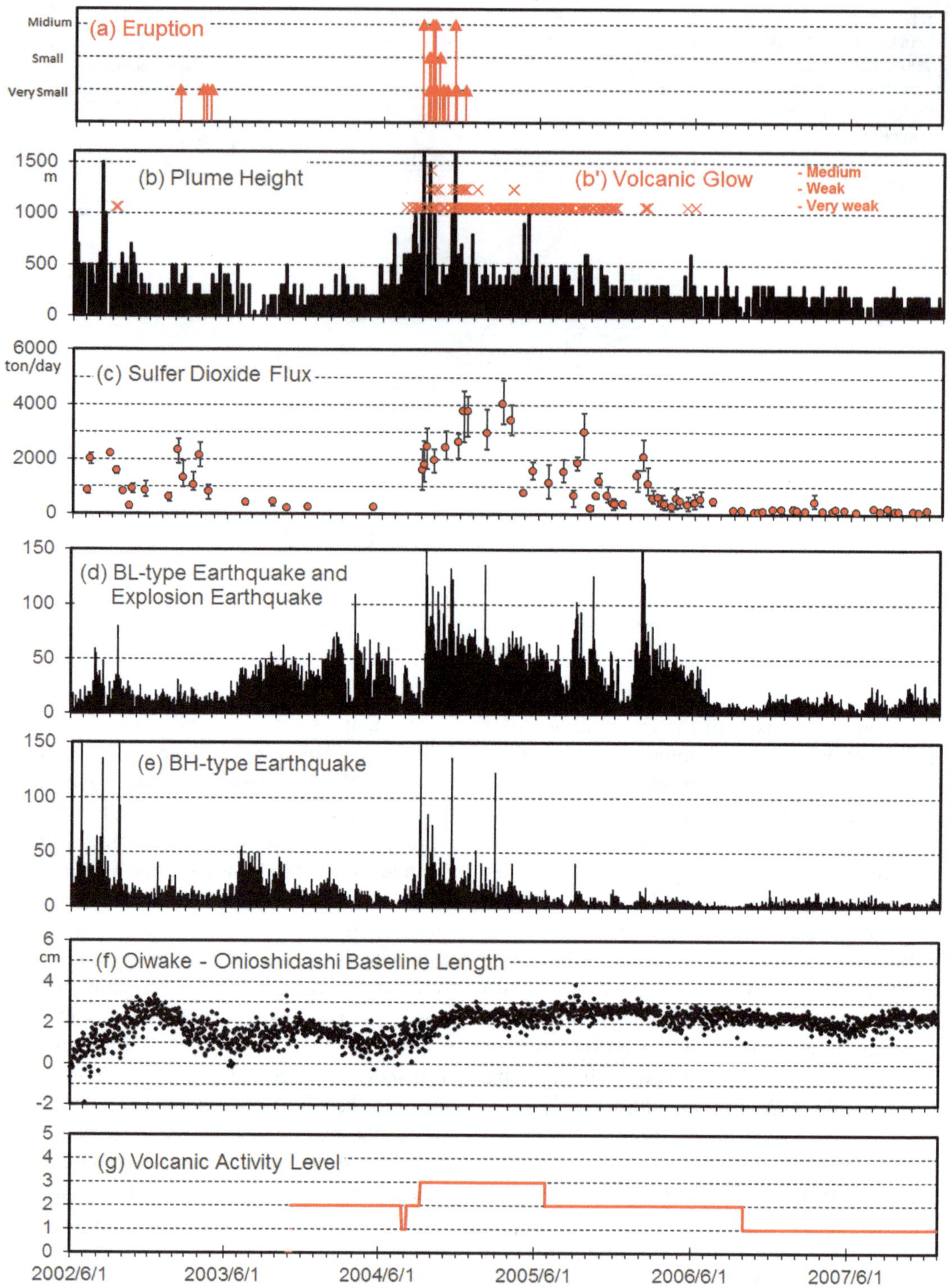

Fig. 7 Temporal evolution of various observational data from June 2002 to 2007. (**a**) Classification of the scales of eruptions at Asama volcano is mentioned in Chapter "Mitigation of Volcanic Hazards of Asama volcano". (**b**) Plume height and volcanic glow are observed with highly sensitive visual cameras. Medium glow can be clearly observed with the naked eye. Weak ones can be faintly observed with the naked eye. Very weak ones can be

August 31. When the tiltmeter records at that time were carefully examined after the eruption, a tilt change synchronized with the earthquake swarm was found (Churei and Katayama 2005). Figure 10 shows the activity of BH-type earthquakes before and after the September 1 eruption and changes in the tiltmeter installed by JMA approximately 2.5 km north-northeast of the crater. Although the tilt record shows a characteristic tilt change caused by Typhoon No. 16, which passed northeast over the Sea of Japan from August 30 to 31, it was presumed that there was a westward tilt change of about 0.1 μrad in the east-west component from around 15:00 on August 31 synchronized with the start of the BH-type earthquake swarm, contrasting with the tilt change when Typhoon No. 6 passed, following a course similar to Typhoon No. 16.

The change in tilt before the September 1 eruption was only discovered after the eruption due to the influence of the typhoon; however, this tiltmeter was effectively used for real-time monitoring during subsequent eruptive activity. An eastward tilt change was observed along with the strombolian eruption that began on September 16 (Fig. 11). Before the vulcanian eruptions on September 23, 29, and November 14, westward tilt changes were also observed, synchronizing with the BH-type earthquake swarm (Fig. 12). JMA installed a new tiltmeter about 4 km south-southeast of the crater in October, and this tiltmeter also detected tilt changes before the eruption on November 14. However, there was no eruption on October 6, even though a similar tilt change was observed in synchronism with a BH-type earthquake swarm. On February 21 of the following year, another case occurred in which no eruption occurred. JMA has reviewed records since the tiltmeter was installed for similar changes. As a result, it was found that there were similar changes in 2002. Table 2 shows similar tilt change cases since 2002. In addition to the case before the 2009 eruption described later, there are nine cases, five followed by vulcanian eruptions with ejecta volumes ranging from about 1 to several tens of thousands of tons.

7.3 Information Issued by JMA

The above-mentioned Volcanic Activity Level, which had been tested since 2002, started at Level 2 and was maintained at Level 2 even in 2003, when very small eruptions occurred. Due to the decrease in the number of volcanic earthquakes in late July 2004, the level was temporarily lowered to Level 1, but due to an increase in the amount of volcanic plume, the level was raised again to Level 2 on July 31. On the morning of September 1, JMA announced volcano information and called attention to the frequent occurrence of BH-type earthquakes since August 31. However, at this time, the tilt change was not apparent due to typhoon noise, and the number of earthquakes during the swarm did not reach those before the 1973 eruption, so the level was left at Level 2. Immediately after the eruption, JMA issued a Volcanic Advisory in which the Volcanic Activity Level was raised from 2 to 3 at 20:09.

From the day after the first eruption, the number of volcanic earthquakes decreased to less than 20 times per day, and the eruption paused. However, it was revealed that the ejecta accompanying the eruption contained fresh pumice with little alteration or deposits (Shimano et al. 2005), and the whole rock chemical composition differed from that of 1958 and 1973 (Nakada et al. 2005). The Coordinating Committee for the Prediction of Volcanic Eruptions (CCPVE: an advisory committee to the JMA; Fujii 2013) con-

Fig. 7 (continued) observed only with a highly sensitive camera. (**c**) shows SO_2 flux. (**d**) and (**e**) are daily numbers of the volcanic earthquakes. (**f**) shows time series plots of distance change between GNSS sites, Oiwake and Onioshidashi stations, whose locations are shown in Fig. 6. (**g**) shows the Volcanic Activity Level of JMA, which started on November 4, 2003. The Volcanic Activity Level was changed to the Volcanic Alert level on December 1, 2007. Its definition is mentioned in Chapter "Mitigation of Volcanic Hazards of Asama Volcano"

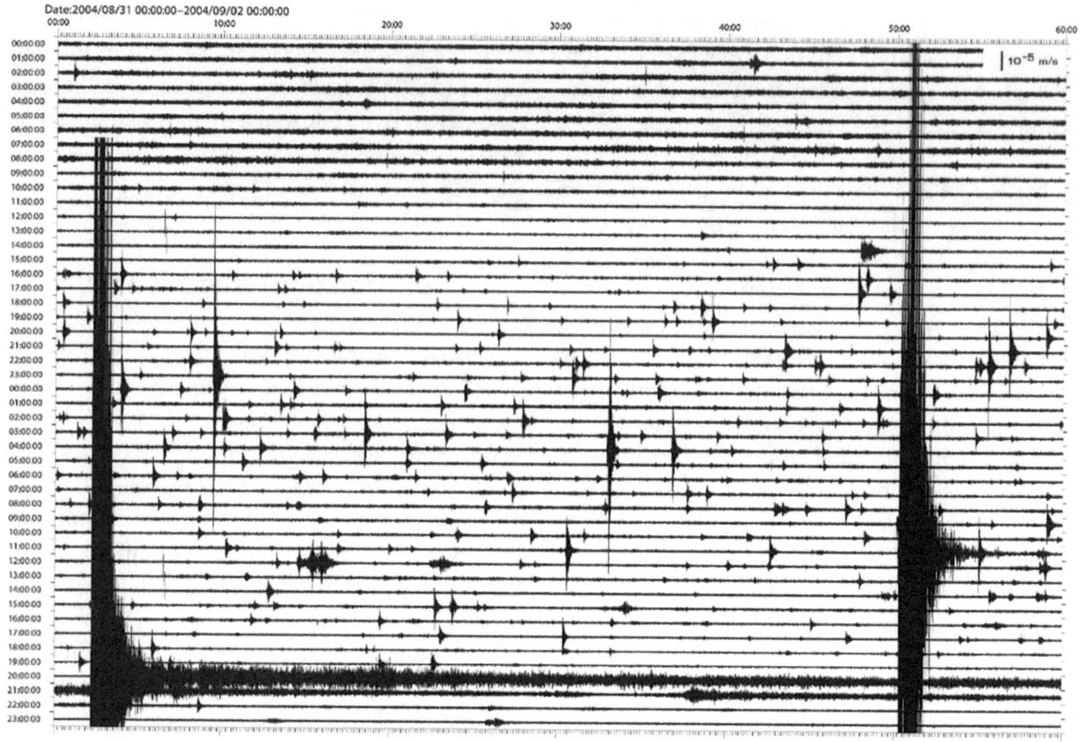

Fig. 8 Seismogram before and after the eruption on September 1, 2004. The record in the vertical component at Sekison station, located 2 km from the crater, is displayed. A BH-type earthquake swarm that started around 15:00 on August 31 and an explosion-quake associated with the eruption at 20:02 on September 1 are recorded. A volcanic eruption tremor followed the explosion-quake. The earthquake at 11:50 is not a volcanic earthquake but a tectonic earthquake that occurred east off Fukushima prefecture

vened on September 9 and concluded that the eruption involved high-temperature magma, and the possibility of further explosions cannot be denied. This evaluation was published as volcano information by JMA just after the CCPVE meeting. As expected, eruptive activity resumed several days after this; lava appeared at the crater floor, and explosive eruptions repeated.

During this period, JMA regularly published activity assessments on volcano information. As mentioned above, it became clear that a BH-type earthquake swarm and tilt change are observed immediately before an explosive eruption, so that on September 29, October 6, November 14, and February 21, volcano information was announced irregularly (of which two eruptions occurred). The Volcanic Activity Level was maintained at 3.

In June 2005, when eruptive activity mostly subsided, the CCPVE evaluated that "the volcanic activity is declining and the possibility of an explosive eruption has decreased", and JMA lowered the level from 3 to 2. After confirming further reductions in the number of volcanic earthquakes and the amount of volcanic gas emitted, the level was lowered to Level 1 on September 22, 2006.

8 Volcanic Warnings and Volcanic Alert Level

In December 2007, JMA began issuing Volcanic Warnings and Forecasts and operating Volcanic Alert Levels. In the case of Asama volcano, the Volcanic Alert Level was taken over based on the Volcanic Activity Level mentioned above. The content is described in Chapter "Mitigation of Volcanic Hazards of Asama Volcano". There are no major differences in the definition of each level up to Level 3 for the Asama's Volcanic

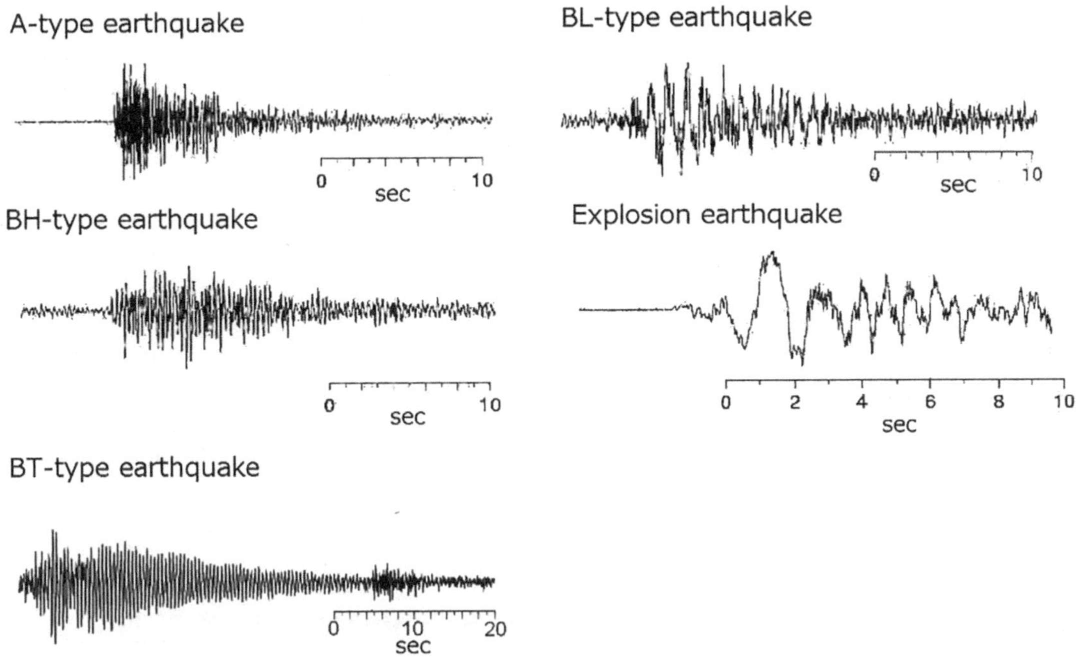

Fig. 9 JMA's classification of volcanic earthquakes observed at Asama volcano. Velocity waveforms in the vertical component at Sekison station are displayed. The BT-type earthquake is the same as the N-type earthquake

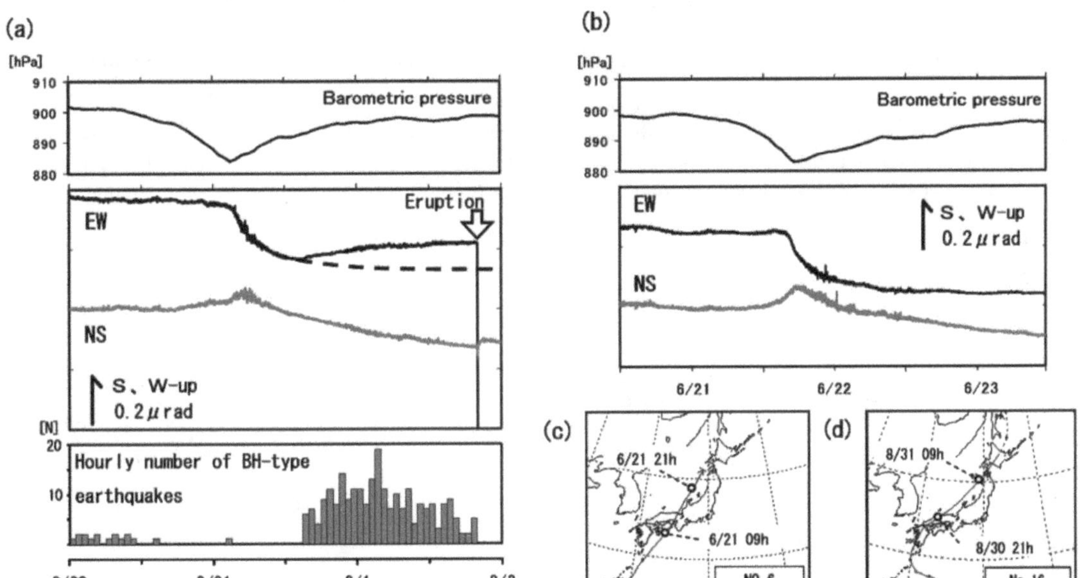

Fig. 10 (**a**) Tilt changes, barometric pressure, and hourly number of BH-type earthquakes before and after the eruption on 1 September. This tilt record was considered to include the effect of pressure change caused by typhoon No. 16. By correction using tilt changes associated with typhoon No.6 that passed almost the same trajectory (**b**), it was found that there was a tilt change from about one day before the eruption. (**c**) and (**d**) show the trajectories of typhoons No. 6 and No. 16, respectively. The tilt records with tidal effects removed at Onioshiue station, located at 2.5 km NNE from the crater, are displayed (after Churei and Katayama 2005). Also, the tilt records with tidal effects removed are displayed in subsequent figures

Activity Level and the Volcanic Alert Level. However, based on the experience of the 2004 eruption, important rules were added for disaster prevention.

Corresponding to the Volcanic Activity Level, rules had been in operation whereby the local government regulated entry to approximately 500 m from the crater at Level 1, approximately 2 km from the crater at Level 2, and 4 km from the crater at Level 3. Therefore, there were no climbers near the crater and no casualties because the 2004 eruption occurred at Level 2. However, on September 1, 2004, the eruption dropped lapilli more than 4 km from the crater. In addition, in past explosive eruptions, there was a case where ballistic bombs fell within a range of more than 4 km from the crater (Minakami, 1942). For these reasons, new rules were determined through consultation with local governments and organizations. In the rule, if a BH-type earthquake swarm and tilt change are observed at the same time as described above, in addition to the existing Level 3 restrictions, traffic restrictions on roads exceeding 4 km from the crater should be implemented (Chapter "Mitigation of Volcanic Hazards of Asama Volcano").

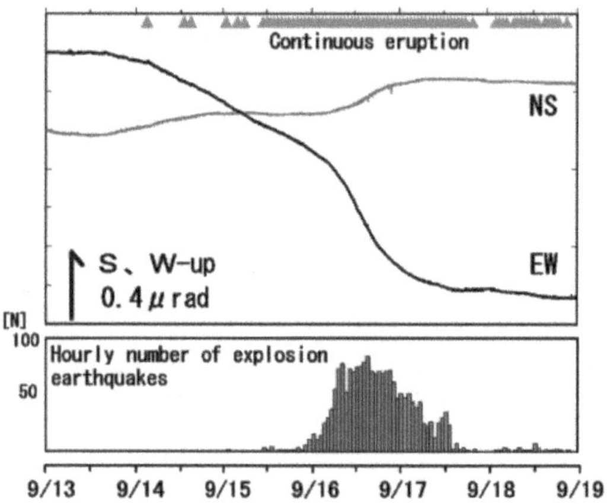

Fig. 11 Tilt changes and hourly number of explosion earthquakes from September 13–18. The tilt records with tidal effects removed at Onioshiue station, located at 2.5 km NNE from the crater, are displayed. Strombolian eruptions occurred continuously from 11:53, September 15, to 21:03, September 18 (Churei and Katayama 2005)

9 Eruption in 2008–2009

From 2006 to 2008, the volcano became quiet, with little volcanic smoke and gas emissions, and the number of volcanic earthquakes per day remained below several dozen.

The volcanic activity started again in the summer of 2008. Figure 13 shows the progress chart. From around July, the number of volcanic earthquakes increased slightly, and from the beginning of August, the number increased further, some-

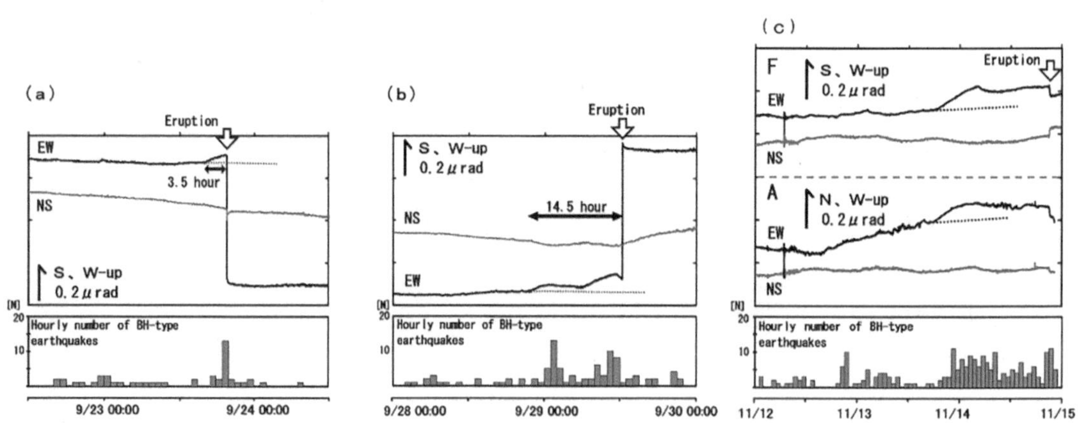

Fig. 12 Tilt change and hourly number of BH-type earthquakes before and after the eruptions on September 23 (**a**), 29 (**b**), and November 14 (**c**). The tilt records with tidal effects removed at Onioshiue station (F) located at 2.5 km NNE from the crater are displayed in (**a**) and (**b**). The tilt records at a new tiltmeter at Chinotaki station (A), located at 4 km SSW from the crater, installed in October, are also displayed in (**c**). (after Churei and Katayama 2005)

Table 2 Tilt change and BH-type volcanic quakes swarm since 2002. Tilt changes at Onioshiue station and infrasonic amplitudes at Oiwake station are shown. The ash volume of each eruption is estimated by Yoshimoto et al. (2005) and Maeno et al. (2010)

Date	Tilt change (μrad)	Duration (h)	Number of BH quakes	Following eruption (ash volume, infrasound)
June 22, 2002	0.10	42	335	No eruptions
September 18, 2002	0.07	35	252	No eruptions
August 31, 2004	0.10	29	237	Medium eruption (4.9 × 10^7 kg, 205 Pa)
September 23, 2004	0.03	3.5	14	Medium eruption (8.5 × 10^6 kg, 72 Pa)
September 29, 2004	0.08	14.5	62	Medium eruption (1.3 × 10^6 kg, 29 Pa)
October 6, 2004	0.06	4	41	No eruptions
November 14, 2004	0.11	27	144	Medium eruption (2.5 × 10^6 kg, 73 Pa)
February 21, 2005	0.08	24	123	No eruptions
February 1, 2009	0.11	24	184	Small eruption (2.7–3.1 × 10^7 kg, 7 Pa)

times exceeding 100 times a day. On August 8, JMA announced a Volcanic Warning, raising the Volcanic Alert Level from 1 to 2. From the next day onward, a faint volcanic glow was also observed. Then, very small eruptions occurred on August 10, 11, and 14. Although the eruption was a very small one and spewed volcanic plumes 200–400 m, JMA succeeded in grasping the heightened volcanic activity in advance.

After the eruption in August 2008, there continued to be many volcanic earthquakes (tens to 200 times/day) and a large amount of SO_2 emissions (more than 1000 tons/day). Weak volcanic glow was also observed. Under such circumstances, a BH-type earthquake swarm was observed from the early morning of February 1, 2009. A slight tilt change synchronized with the earthquake swarm was also observed with the tiltmeters (Fig. 14), which was judged to be a similar phenomenon to that in 2004. Therefore, JMA issued a Volcanic Warning, raising the Volcanic Alert Level from 2 to 3 at 13:00. According to the pre-determined rules as mentioned above, the local authorities have restricted climbing within 4 km of the crater and restricted traffic on the roads outside of the area. Then, at 01:51 on the next day, half a day after the warning was announced, an eruption occurred. The amplitude of the air shock was smaller than the series of explosions in 2004, and the ballistic bombs reached only 1–1.2 km from the crater, but the amount of volcanic ash was estimated as about 30,000 tons, which was close to the eruption on September 1, 2004 (Table 2). As shown in Fig. 14, at the time of the eruption, the tilt change returned to the original trend, and the number of BH-type earthquakes decreased. After that, it was confirmed that there were no tilt changes or BH-type earthquake swarm, and traffic restrictions on the road were lifted the following day. After that, very small eruptions occurred from February 9 to 12, and ash fall was observed at the foot of the volcano. Since then, the amount of volcanic gas and the number of volcanic earthquakes decreased, so that JMA lowered the Volcanic Alert Level from 3 to 2 on April 7. After that, very small eruptions occasionally occurred until the end of May, but these eruptions were within the range assumed for Level 2. After confirming further decline in volcanic activity, the Volcanic Alert Level was lowered from 2 to 1 on April 15, 2010.

The 2009 eruption was classified as a phreatic eruption, with much less than 1 wt% magmatic material in the ejecta (Maeno et al. 2010). Although they were eruptions that did not accompany a remarkable ascent of magma, it was a

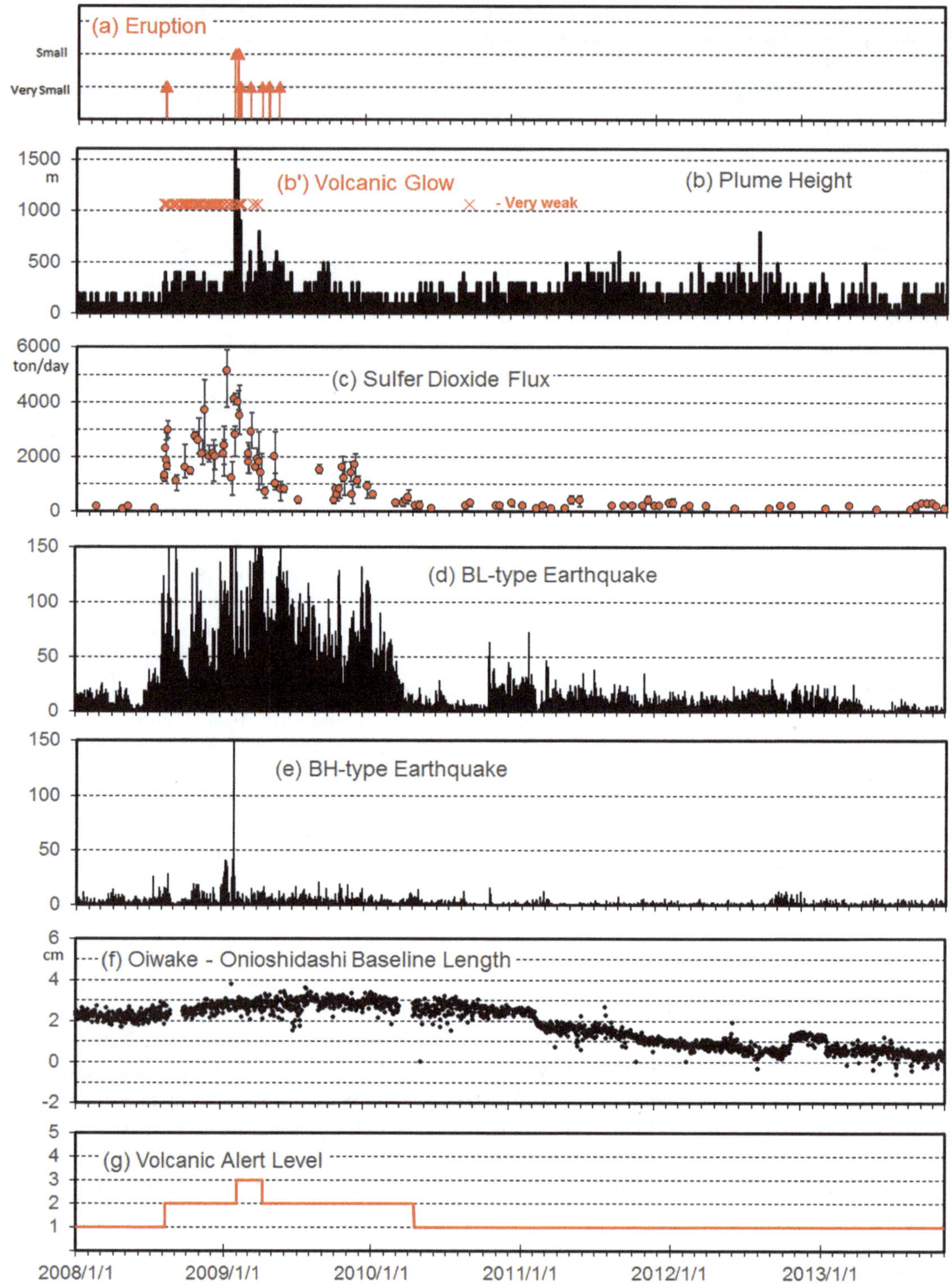

Fig. 13 Temporal evolution of various observational data from 2008 to 2013. The descriptions (a)–(g) are the same as those of Fig. 7. A shift change caused by the 2011 off the Pacific coast of Tohoku Earthquake (11 March 2011, $M = 9.0$) was reduced

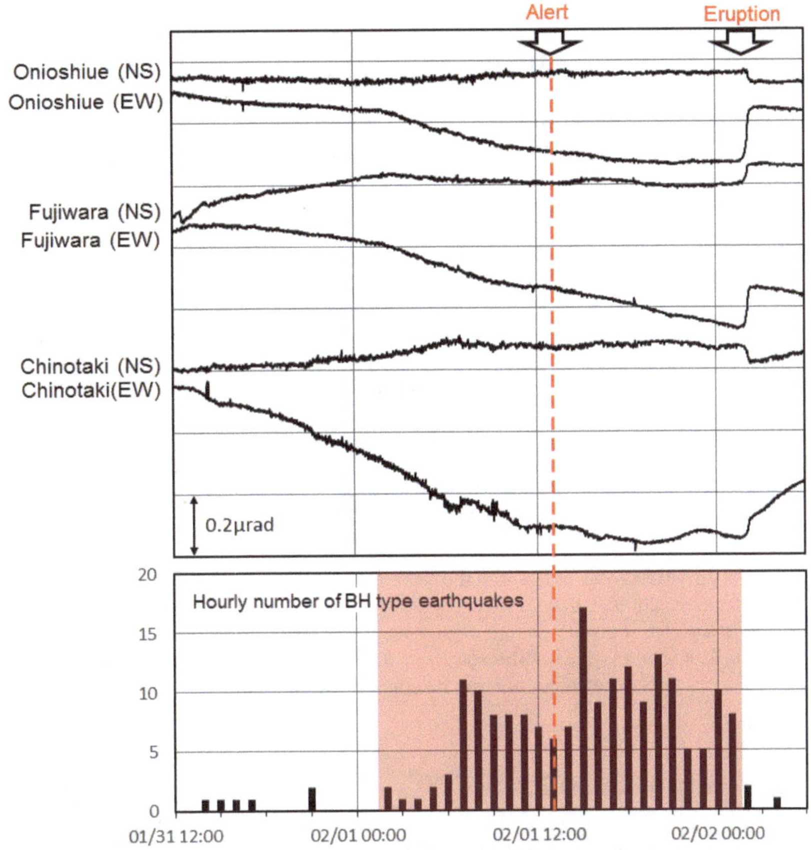

Fig. 14 Tilt changes and hourly number of BH-type earthquakes before and after the eruptions on February 2, 2009. The tilt records with tidal effects removed are displayed

good example of a generally accurate prediction, including very small eruptions, as evidenced by various observational data indicating increased volcanic activity.

10 Mechanism of the Tilt Change Just before Explosion

The mechanism of the tilt change synchronized with the BH-type earthquake swarm just before the explosive eruption discovered in the 2004 eruption has been investigated by JMA. The tilt vectors (Fig. 15) associated with the 2009 eruption suggest that there is a source of inflation and deflation below the western foot of the volcano, but it is known that BH-type earthquakes occur in the shallow area beneath the summit crater (Chapter "Eruption Activity of Asama Volcano from 2004 to 2015 and Volcanic Earthquake"). Churei and Katayama (2005) explain the 2004 case by changing the dike-shaped pressure source beneath the crater; on the other hand, the case of 2009 was interpreted by the isotropic pressure source under the western foot of the volcano (Funasaki et al. 2012). As mentioned above, the GNSS observation shows intermittent magma accumulation under the volcano's western foot (Chapter "Ground Deformation at Asama Volcano Measured by Geodetic Data"), and the positional relationship with the magma chamber and the source of the tilt change is interesting. However, at that time, there were no tilt stations in the western area, and the accuracy of the estimated source models was not good. Since the 2009 eruption, no tilt changes synchronized with BH-type earthquakes have been observed. Tiltmeters were reinforced after the eruption, and new findings are expected to be obtained in the future.

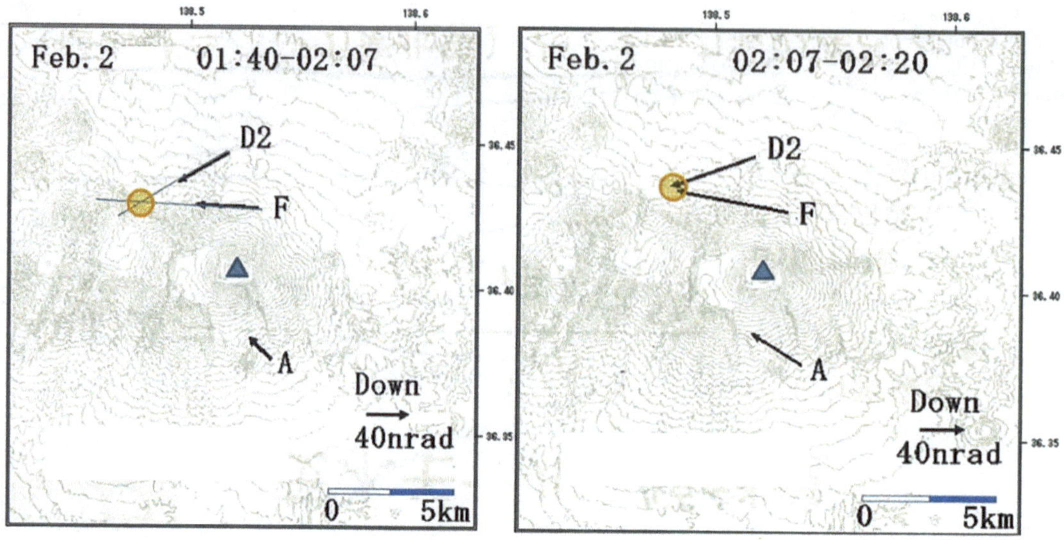

Fig. 15 Tilt vectors associated with the eruption at 01:51 on February 2, 2009. Tiltmeters recorded ground deformation associated with the deflation associated with the eruption, as shown in Fig. 14. The source of the deflation is considered to be located not beneath the summit (triangle) but beneath the western foot of the volcano (the yellow circle). The marks, F, A, and D2 are Onioshiue, Chinotaki, and Fujiwara stations shown in Fig. 6, respectively (after Funasaki et al. 2012)

11 Eruption in 2015

Quiet volcanic activity continued after the 2009 eruptive activity ended (Fig. 16). Volcanic earthquakes began to increase in June 2015. The daily number of earthquakes in May was 20 to 50, and increased to 30 to 90 in June. The increase in volcanic earthquakes was not remarkable, but the amount of SO_2 emission, which was 200 tons/day on June 1 and 500 tons/day on June 8, increased to 1700 tons/day on June 11. On June 11, JMA judged that the volcanic activity had become apparent, issued a Volcanic Warning, and raised the Volcanic Alert Level from 1 to 2. A few days later, on June 16, a very small eruption occurred with very little ashfall at the foot of the volcano. A very small eruption occurred again on June 19, but no ash fall was observed at the foot of the volcano. After that, the number of volcanic earthquakes continued to be somewhat high, and weak volcanic glow and slow crustal deformation that expanded the western foot of the volcano were observed by GNSS and tiltmeters, but no eruption occurred, and the Volcanic Alert Level was lowered from 2 to 1 in August 2018.

12 Eruption in 2019

Even after August 2018, when the Volcanic Alert Level was lowered, the daily number of earthquakes continued to be around tens of times. However, no particular anomalies were observed in other data, and the volcanic plume remained low in the following year, 2019. Meanwhile, a small eruption occurred at 22:08 on August 7. The volcanic plume rose more than 1800 m above the crater rim. The infrasonic amplitude was 0.94 Pa at the Oiwake station, and volcanic bombs were scattered up to about 200 m from the crater. The JMA raised the Volcanic Alert Level from 1 to 3 at 22:30. Immediately after the eruption, there was a slight increase in the amount of earthquakes, volcanic plume, and volcanic gas emissions, but since then it has been sluggish, so it was lowered to Level 2 on August 19 and to Level 1 on November 6. A small eruption also occurred at 19:28 on August 25, but smaller than the eruption on the August 7.

These eruptions were phreatic. No significant increase in earthquakes or crustal movements

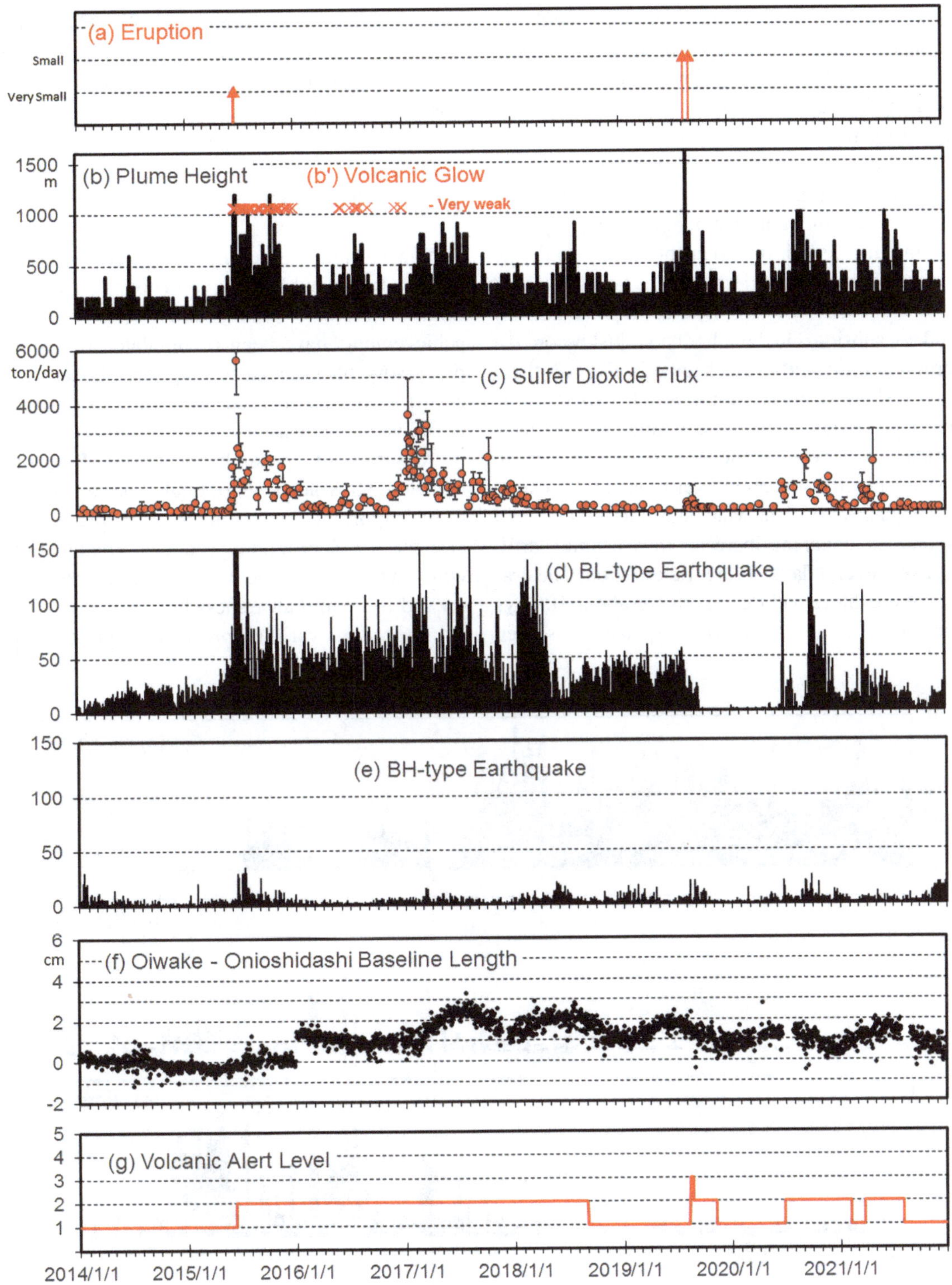

Fig. 16 Temporal evolution of various observational data from 2014 to 2021. The descriptions (a)–(g) are the same as those of Fig. 7

was observed before the eruptions, and the Volcanic Alert Level was raised after the first eruption. After the eruptions, the observation data leading up to the eruptions have been analyzed for these eruptions (Yoshigai et al. 2023). They indicate a slow expansion change with the tiltmeters from December 2016 to the summer of 2017. They thought that magma was supplied during this period, but no eruption occurred, and magma remained underground. They thought it was the reason why BL-type earthquake activity did not subside. On the other hand, just before the eruption on August 7, as shown in Fig. 17, there was a sharp decrease in the number of BL-type earthquakes and an increase in micro-earthquakes (mainly BH-type earthquakes). Yoshigai et al. systematically detected very small volcanic earthquakes using the Matched Filter method and found an apparent increase in volcanic earthquakes several days before the eruption. The crater temperature had decreased for about 10 days before the eruption (Earthquake Research Institute, Tokyo University 2019). Therefore, Yoshigai et al. presumed that these phenomena immediately before the eruption represented the obstruction of the conduit.

13 Summary

Asama volcano is one of the most active volcanoes in Japan and has a long history of observation. Numerous case studies and research achievements have been accumulated regarding precursory phenomena of eruptions. Local governments have also used the results to mitigate actual volcanic disasters.

A series of research by Prof. Minakami revealed that volcanic earthquakes are one of the most important factors in eruption prediction, and the 1958 eruption produced great results with this method. The 1973 eruption also succeeded in raising awareness before the eruption by capturing various precursory phenomena

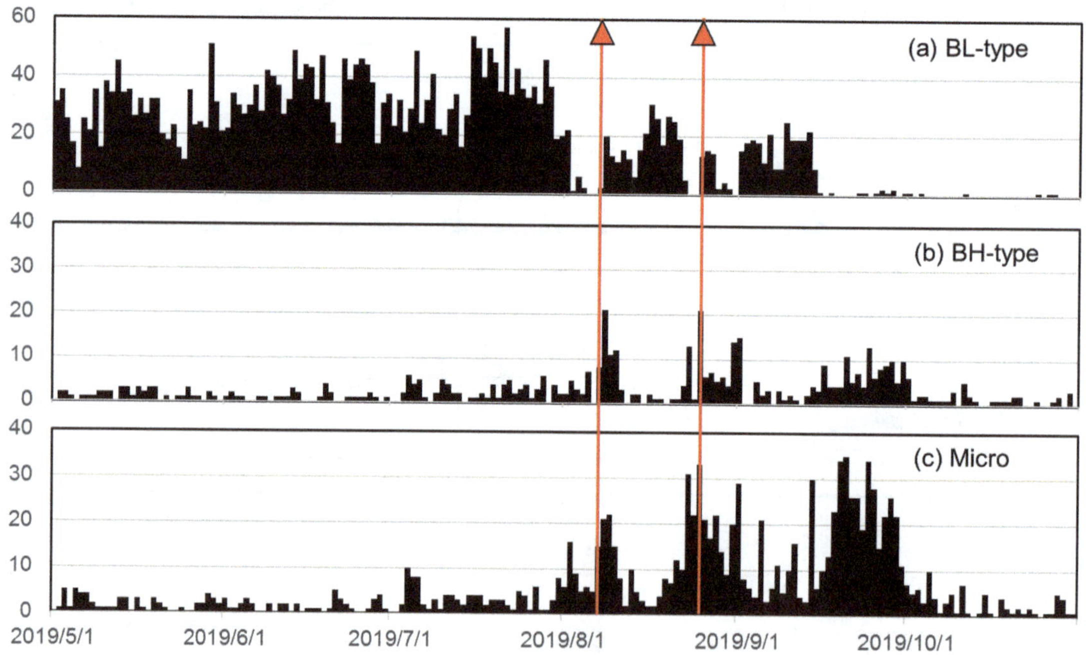

Fig. 17 Daily numbers of volcanic earthquakes before and after the eruptions in August 2019. Triangles indicate the eruptions. (**a**) and (**b**) represent those of BL-type and BH-type earthquakes, respectively. (**c**) shows those of micro-earthquakes whose amplitude was smaller than the counting threshold (1×10^{-6} m/s in the vertical component at Sekison station) of BL-type and BH-type earthquakes. Most of the micro-earthquakes are considered small BH-type earthquakes

immediately before the eruption. However, the 1982–1983 eruption did not show any significant change immediately prior to the eruption. In order to predict them, it was necessary to capture the increase in activity from several months ago, as well as the slight changes immediately before their onset.

Since the beginning of the twenty-first century, advances in crustal deformation observations have gradually revealed the magma supply system of Asama volcano. The JMA, which incorporated level information into disaster mitigation information, has developed a monitoring method that combines monitoring of crustal movement, seismic activity, and volcanic plume activity. As represented by the 2009 eruption, we have accumulated a certain number of success stories. However, the 2019 eruption occurred when expansion due to magma accumulation was not recognized, and the Volcanic Alert Level was 1. It is presumed that the obstruction of the conduit caused this eruption.

The observation system of Asama volcano has been strengthened over the decades. The JMA, University of Tokyo, and research institutes have strengthened their own observation networks and exchange real-time data with each other, making the volcano one of Japan's leading observation systems. Asama volcano, which had active eruptive activity in the first half of the twentieth century, has been gradually declining for a long time since the 1960s, and some eruptions have occurred without noticeable abnormal phenomena immediately before. It is becoming more and more important to combine various types of observational data to evaluate volcanic activity. Moreover, unlike the eruptive activity of the twentieth century, it is assumed that the large-scale eruptions that are expected to occur eventually will exceed our experience. Therefore, its accurate forecasting will be difficult based on empirical rules. It is important to develop basic volcanological research, such as elucidation of further detailed magma supply systems, the mechanisms of various phenomena, and others.

Acknowledgments The observations and information of JMA introduced in this chapter were mainly obtained by the Tokyo Volcanic Observation and Warning Center under its 24-hour monitoring system. Data from the University of Tokyo and related institutions are also monitored in JMA. The authors would like to express their gratitude to these organizations. Thanks are also due to Prof. Minoru Takeo for giving us the opportunity to present this chapter.

References

Asama Volcano Observatory of Earthquake Research Institute, University of Tokyo (1983) Seismic activity associated with the April 8, 1983 eruption of Asama volcano. Report of Coordinating Committee for Prediction of Volcanic Eruptions, 28, 19–22 (in Japanese)

Churei M, Katayama H (2005) Ground tilt change associated with the 2004 eruption at Asamayama volcano, Japan. Bull Volcanol Soc Japan 51:91–101. (in Japanese with English abstract)

Earthquake Research Institute, Tokyo University (2019) Thermal infrared image of the Asama crater floor. Report of 145th Meeting of Coordinating Committee for Prediction of Volcanic Eruptions, https://www.data.jma.go.jp/svd/vois/data/tokyo/STOCK/kaisetsu/CCPVE/shiryo/145/145_4.pdf, 78 (in Japanese)

Fujii T (2013) The Japanese Coordinating Committee for Prediction of Volcanic Eruptions and its contribution to volcanic disaster mitigation. Technical Note of the National Research Institute for Earth Science and Disaster Prevention, 380: 113–120

Funasaki J, Takagi A, Fujimatsu J (2012) Contraction source inferred by tilt change occurred at the time of eruption of Asama volcano on Feb. 2 2009. Abstr. Volcanol. Soc. Japan, A1–08 (in Japanese)

Geographical Survey Institute (2005) Change of topography of summit crater of Asama volcano. Report of Coordinating Committee for Prediction of Volcanic Eruptions, 90, 39–42 (in Japanese)

Japan Meteorological Agency (1959) Asamayama. Seismol Bull 1958:48–53

Kagiyama T, Gyoda N, Koyama E, Tsuji H (1983) Precursors of small eruptions at Asama volcano. Bull Volcanol Soc Japan 28:418. (in Japanese)

Maeno F, Suzuki Y, Nakada S, Koyama E, Kaneko T, Fujii T, Miyamura J, Onizawa S, Nagai M (2010) Course and ejecta of the eruption of Asama volcano on 2 February 2009. Bull Volcanol Soc Japan 55:147–154. (in Japanese with English abstract)

Minakami T (1942) On the distribution of volcanic ejecta. (Part I) The distributions of volcanic bombs ejected by the recent explosion of Asama. Bull Earthq Res Inst 20:65–92

Minakami T (1959) The study of eruptions and earthquakes originating from volcanoes (part I) some statistical relations between explosive eruptions and earthquakes of Volcano Asama. Bull Volcanol Soc Japan 4:104–114. (in Japanese with English abstract)

Minakami T (1960) Fundamental research for predicting volcanic eruptions (part 1). Bull Earthq Res Inst 38:497–544

Minakami T, Hiraga S, Utibori S, Miyazaki T (1959) The study of eruptions and earthquakes originating from volcanoes (part II) some contribution to prediction of explosive eruption of Volcanoes Asama. Bull Volcanol Soc Japan 4:115–130. (in Japanese with English abstract)

Murakami M (2005) Magma plumbing system of the Asama volcano inferred from continuous measurements of GPS. Bull Volcanol Soc Japan 50:347–361. (in Japanese with English abstract)

Nakada S, Yoshimoto M, Koyama E, Tsuji H, Urabe T (2005) Comparative study of the 2004 eruption with old eruptions at Asama volcano and the activity evaluation. Bull Volcanol Soc Japan 50:303–313. (in Japanese with English abstract)

Sekiya H (1967) Volcano observation—The history of Mount Asama and the volcanoes of Japan. Sogo Kagaku Library, Sogo Books (in Japanese)

Shimano T, Iida A, Yoshimoto M, Yasuda A, Nakada S (2005) Petrological characteristics of the 2004 eruptive deposits of Asama volcano. Bull Volcanol Soc Japan 50:315–332. (in Japanese with English abstract)

Shimozuru D, Utibori S, Gyoda N, Koyama E, Miyazaki T, Matsumoto T, Osada N, Terao H (1975) The explosive activity of Asama volcano—general description of volcanic and seismic events. Bull Earthq Res Inst 50:115–151. (in Japanese with English abstract)

Shimozuru D, Gyoda N, Kagiyama T, Koyama E, Hagiwara M, Tsuji H (1982) The 1982 eruption of Asama volcano. Bull Earthq Res Inst 57:537–559. (in Japanese with English abstract)

Takeyama I, Tanaka Y, Kobayashi E, Isono Y (1960) On the seismic waves and air-shocks caused by the explosion of volcano Asama, Nov. 10, 1958. Q J Seismol 25:45–53. (in Japanese with English abstract)

Yamasato H, Oga S, Daiku Y, Funasaki J, Matsushima M, Naito H, Kanno T (2004) Announcement of volcanic activity level by Japan Meteorological Agency. Bull Volcanol Soc Japan 49:217–222. (in Japanese)

Yoshigai Y, Morita Y, Kobayashi T, Ozawa D, Matsusue S, Kawaguchi R, Ohminato T (2023) Reexamination of volcanic seismicity before and after the 2019 eruption of Asama Volcano, detected by the Matched Filter Method. Bull Volcanol Soc Japan 68, 229–244. (in Japanese with English abstract)

Yoshimoto M, Shimano T, Nakada S, Koyama E, Tsuji H, Iida A, Kurokawa M, Okayama Y, Nonaka M, Nogami K, Onizawa S, Niihori K, Sugimoto T, Nagai M (2005) Mass estimation and characteristics of ejecta from the eruption of Asama volcano. Bull Volcanol Soc Japan 50:519–533. (in Japanese with English abstract)

Mitigation and Hazard

Mitigation of Volcanic Hazards of Asama Volcano

Hitoshi Yamasato and Sei Iijima

1 Introduction

Asama volcano is one of the most active volcanoes in Japan and has historically suffered many volcanic disasters. On the other hand, the area including Karuizawa near Asama volcano is one of the most popular tourist destinations in Japan; many tourists visit; and tens of thousands of people climb Mt. Asama every year. In this chapter, an overview of the history of volcanic disaster mitigation at Asama and the scheme of the present mitigation system of volcanic disaster, especially, the Volcano Disaster Management Council, the hazard map, the volcanic surveillance system, and the observation information are introduced.

2 History of Volcanic Disaster Mitigation at Asama

Mt. Asama has ancient mountain climbing trails as part of mountain worship since ancient times, and from the latter half of the nineteenth century, mountain climbing became popular among foreigners who began visiting Karuizawa to escape the summer heat. On the other hand, increased eruptive activity led to a series of mountaineer disasters. Japan's first volcano observatory was established at Asama (Chapter "History of Seismic Observation and Traditional Classifications of Seismic Signals in Asama Volcano"); the results of the observations made there were used by local governments as information for distress prevention of mountain climbers through mountain huts and police officers. However, the number of climbers decreased due to the chaos during and after the war and the series of eruptions, and social interest in mitigating disasters caused by volcanic eruptions gradually faded.

With the post-war reconstruction, the number of climbers increased again, and the number of disasters had increased. At the eruption in 1958, climbing restrictions before the eruption were led by the advice from Karuizawa Weather Station of the Japan Meteorological Agency (JMA), which took over the work of the first volcano observatory (Chapter "Prediction of Volcanic Eruptions at Asama Volcano"). Then, momentum to mitigate disasters has risen again, and related organizations, including the prefecture, the police, each local government, and mountain groups, established the Asama Mountain Accident Prevention Association (1963), which has been working on popularization and enlightenment of volcanic disaster mitigation.

H. Yamasato (✉)
CTI Engineering Co., Ltd., Chuo, Tokyo, Japan
e-mail: hyamasat@nifty.com

S. Iijima
Maebashi Local Meteorological Office,
Maebashi, Gunma, Japan

S. Aramaki et al. (eds.), *The Asama Volcano*, Active Volcanoes of the World,
https://doi.org/10.1007/978-981-95-5799-8_28

On the other hand, JMA started formal volcano information announcements in 1965, and each local government formulated and implemented regional disaster prevention plans for volcano countermeasures based on the Disaster Countermeasures Basic Act (1961) and worked on dissemination and enlightenment on disaster mitigation. In addition, after the 1973 eruption, the local government that manages the mountain trail set a 4 km radius around the crater as a warning area based on the Disaster Countermeasures Basic Law and prohibited entry. After then, the eruptive activity calmed down, and with the opening of the Shinkansen between Tokyo and Nagano (1997), a demand for relaxation of the restrictions on climbing the mountain increased in the local governments, which worked with JMA, volcanologists and other organizations and launched the Climbing Regulations Investigation Committee, which recommended that it would be appropriate to relax restrictions in accordance with volcanic activity, limited to the climbing route to Mt. Maekake 500 m from the crater. Based on this recommendation, Komoro City prepared shelters on Mt. Maekake and began deregulating mountain climbing routes in July 2001 based on JMA's volcano information.

3 Volcano Disaster Management Council

It is important for the volcano disaster countermeasures to implement entry restrictions and evacuation beforehand in areas predicted to be at risk from the volcanic phenomena associated with an eruption (e.g., ballistic projectiles, pyroclastic flows, and others). To achieve this, it is necessary to establish a system in which suitable eruption alerts can be announced to residents, and the residents can be evacuated immediately and smoothly.

For this purpose, the government of Japan compiled "Policy on Volcano Management Related to Evacuation in the Event of an Eruption" in 2008, and relevant ministries and agencies have been working jointly to establish a volcano disaster management system (Fujiyama et al. 2013). After the volcanic disaster at Ontakesan in 2014, the idea of this policy was stipulated in a law, "Act on Special Measures for Active Volcanoes."

Under this law, it is obligatory for the Volcano Disaster Management Council to be established as a system for local governments and related organizations related to the area around the volcano to share information in normal times and jointly consider evacuation measures. Various roles of the council related to volcanic disaster mitigation are shown in Fig. 1. The councils are established at 49 volcanoes that are constantly monitored by JMA.

The membership of the council of the Asama volcano is listed in Table 1. Since it is located on the border between Nagano and Gunma prefectures, one of the governors of both prefectures serves as the chairman. In addition to the prefectures, the municipalities, related organizations, and JMA, a sabo specialist and volcanologists who are familiar with the volcanic activity of Asama also participate in the council. Since the council is composed of many organizations (Photo 1), the core group, which consists of prefectural and municipal government officials, JMA, the sabo department, and volcanologists deeply involved in establishing evacuation timing and areas, plays important roles, and substantial discussion takes place in the core group (Table 1).

Each role of the council shown in Fig. 1 will be described below.

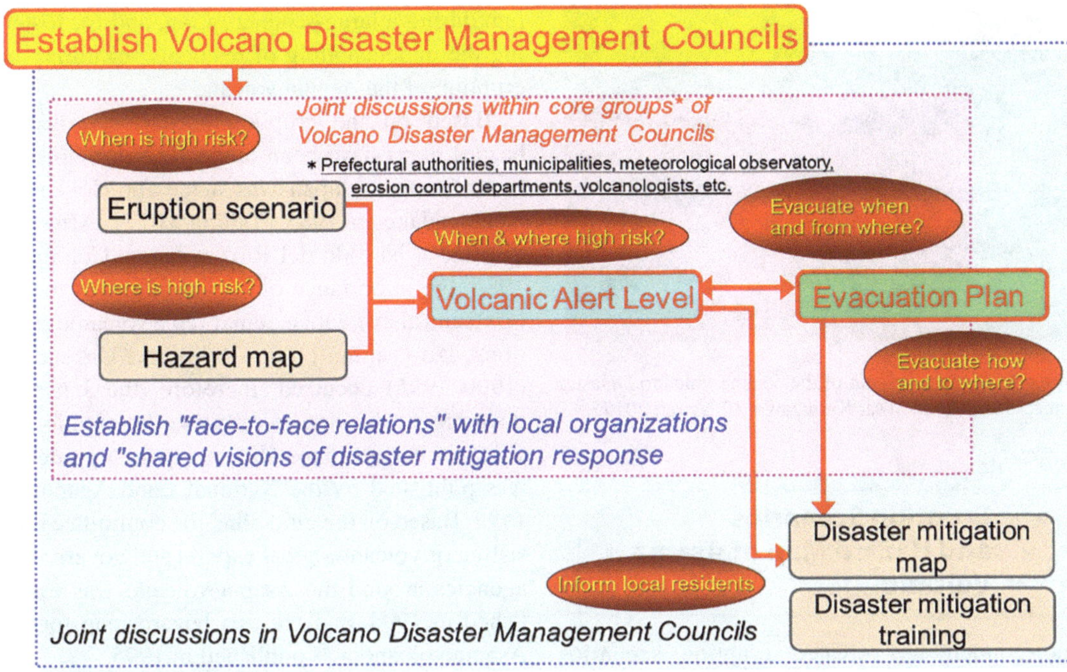

Fig. 1 Role of Volcano Disaster Management Council (modified from Yamasato et al. 2013)

Table 1 Membership of the Volcano Disaster Management Council of Asama volcano

Prefectures*	Nagano and Gunma prefectures (either governor: the chairman)
Municipalities*	Komoro City, Saku City, Karuizawa town, Miyota Town, Naganohara Town, Tsumagoi Village, Takasaki City, Annaka City, and Saku Wide Area Union
Japan Meteorological Agency (JMA) *	Volcanic Observation and Warning Center, Nagano/Maebashi Local Meteorological Office, and Asamayama Resident Office for Volcanic Disaster Mitigation
Ministry of Land, Infrastructure, Transport and Tourism (MLIT) *	Kanto Regional Development Bureau, Tone River System Sabo Office, Nagano National Highway Office, and Takasaki River and National Highway Office
Ground Self-Defense Force	12th Brigade and 13th Regular Regiment of Ground Self-Defense Force
Police	Nagano/Gunma Prefectural Police Headquarters, Security Department of Gunma Prefectural Police Headquarters, Highway Traffic Polices of the Prefectural Police Headquarters, Komoro/Saku/Karuizawa/Naganohara/Takasaki-Kita/Annaka Police Stations
Fire department	Saku Wide Area Fire Department, Agatsuma Fire Department, and Takasaki District Fire Bureau
Volcanology and sabo specialists*	Drs. Shigeo Aramaki, Minoru Takeo, Hiroyuki Ohno, and Masaki Takahashi
Other organizations of the government of Japan	Cabinet Office (Disaster management)*, Agatsuma / Toushin Forest Management Offices of Forestry Agency, Joshinetsu Kogen National Park Management Office of Ministry of the Environment, Kanto Regional Survey Department of Geospatial Information Authority
Tourism business, etc.	Seibu Prince Hotels Worldwide Inc., Shinano Railway Company, Shiraito Highland Way Company, East Nippon Expressway Company, East Japan Railway Company, and Mt. Asama Geopark Promotion Council

As of January 2023 *: the members of the core group

Photo 1 Plenary session of the Asama Volcano Disaster Management Council (at Karuizawa, 17 March 2017)

4 Eruption Scenarios and Hazard Map of Asama Volcano

The council has created eruption scenarios (assumed time sequences of phenomena associated with future eruptions and their effects) for examining volcanic disaster mitigation measures and shares them with related organizations.

The eruption scenarios organize the characteristics of volcanic activity from the past eruption records, categorize the sequences of the activities and the range of influences, and summarize them as sequence tables and flow charts. In organizing the past eruption records at Asama volcano, they are divided into small to medium-scale eruptions that have repeatedly occurred in recent years and large-scale eruptions that cause enormous damage and social impact, such as the Ten'nin and Tenmei eruptions. Small to medium-sized eruption scenarios were considered when the Volcanic Alert Level of JMA was introduced on the Asama volcano in 2007. Regarding large-scale eruption scenarios, the latest geological research has provided new insights into the Tenmei eruption, the Ten'nin/Daiji eruption, and the early Kofun period eruptions. This scenario is expected to be useful for studying disaster prevention measures linked to the Volcanic Alert Level, for studying a wide-area evacuation system during a large-scale eruption, and for fostering the understanding of residents regarding the eruption of the Asama volcano.

Based on the eruption scenarios, volcanic hazard maps have been made. The first volcano hazard map in Japan was made for Hokkaido-Komagadake in 1983 (Tajima 2017). After the disaster at Nevado del Ruiz volcano, Colombia, 1985, the importance of hazard map was recognized and also in Japan, remarkable volcanic eruptions, Izu-Oshima (1986–1990) and Unzendake (1990–1995) occurred, therefore, the Japanese government decided to promote the creation of volcanic hazard maps also in Japan, a guideline was published by the National Land Agency in 1992. Based on the guideline, the committee consisting of volcanological experts and government agencies around the Asama volcano was established in 1994, and the first hazard map for the Asama volcano was published in 1995.

The hazard map has been revised repeatedly, and the latest version of the disaster mitigation map was published in 2019 and distributed to all households in each local municipality around the Asama volcano. In the 2019 revision, in addition to small to medium-sized eruptions, large-scale eruptions are assumed in the hazard map to prepare for eruptions equivalent to large-scale eruptions that occurred in the past and to proceed with the formulation of evacuation plans.

Small to medium-sized eruptions in the scenarios and the hazard map are Vulcanian eruptions at the summit crater that have occurred frequently since the end of the nineteenth century. This period is the time when the observation system of the Asama volcano was gradually improved, and they are set based on the observational results during this period. Explosive eruptions occurred almost every year from the end of the nineteenth century to around 1960. The 1938 and 1958 eruptions are known to be the largest of these, and large ballistic bombs with a diameter of 50 cm or more flew over 4 km from the crater at the 1938 explosion (Minakami 1942).

Figure 2 shows a hazard map of small- to medium-sized eruptions. The hazard area associated with small to medium-sized eruptions is estimated from the experience during these eruptions. The factors of the hazards are ballistic rocks, lapilli, tephra fall, infrasonic waves, volcanic gas, pyroclastic flow, debris flow, and lahar due to rainfall. The eruption site is assumed to be the summit crater. There are two types of maps: no snow season and snow season. Hazard area for melting snow-type lahar is added in the map for the snow season.

In the eruption scenario, small- to medium-sized eruptions are classified into small and medium eruptions as shown in Table 2, based on the reach of ballistic rocks and pyroclastic flows in typical eruptions.

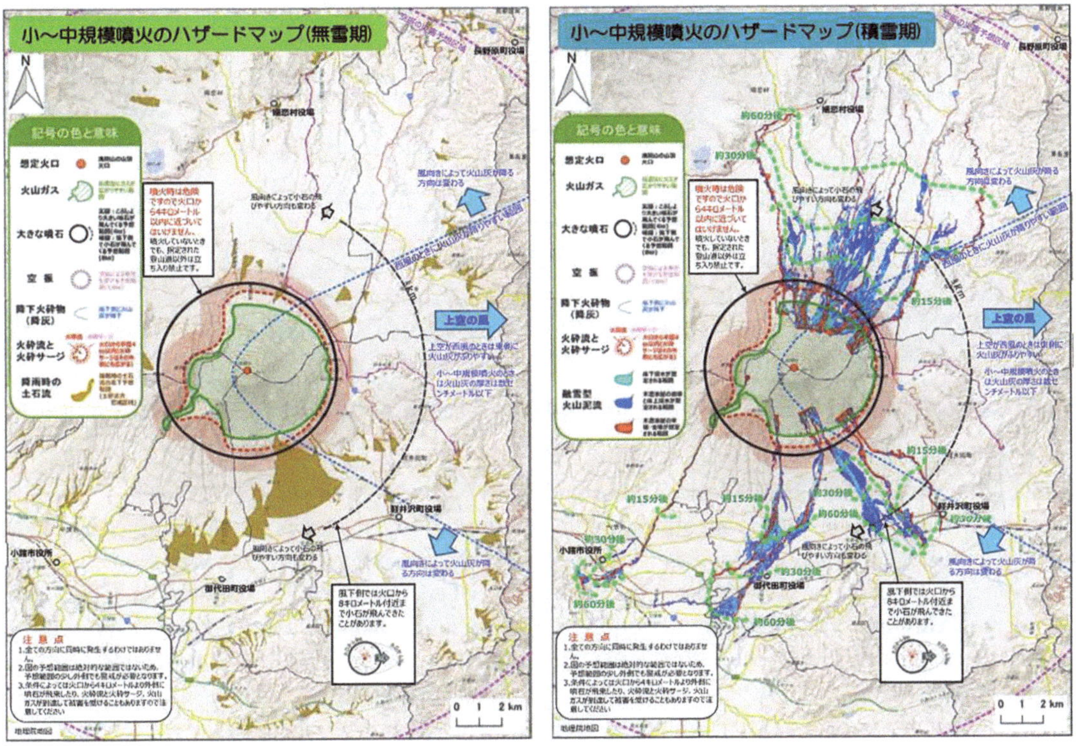

Fig. 2 Hazard maps of small to medium-sized eruptions for no snow season (left) and for snow season (right). Hazard area of melting snow-type lahar is added in the map for snow season

Table 2 Characteristics of small to medium eruptions at Asama volcano

Site of eruption	Type of eruption	Size	Phenomena	Hazard area	Examples of past eruptions
Summit crater	Vulcanian eruption	Small	Ballistic rocks, lapilli, tephra fall, infrasonic waves, volcanic gas, pyroclastic flow, snow melting volcanic lahar, debris flow, and lahar due to rainfall	Within about 2 km from the crater (reaching range of ballistic rocks and pyroclastic flows)	1982 2019
		Medium		Generally, within about 4 km from the crater, but on rare cases it exceeds 4 km (reaching range of ballistic rocks and pyroclastic flows) Melting snow-type volcanic lahar affects areas exceeding 4 km	1958 1973 2004

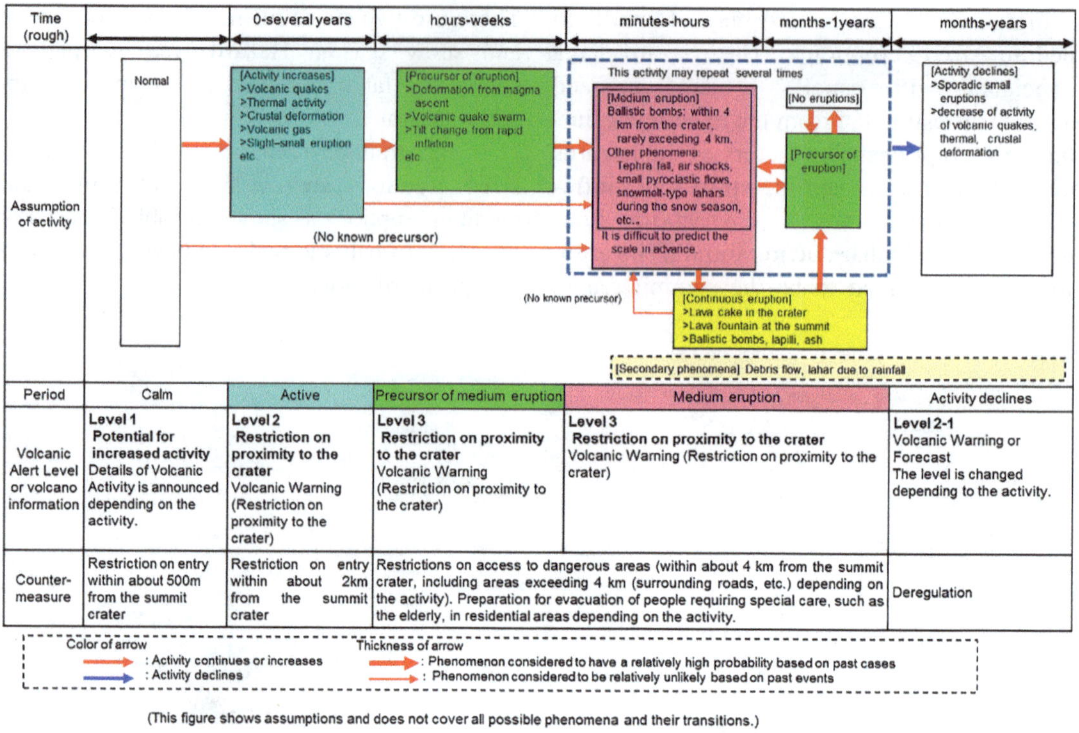

Fig. 3 Transition diagram of medium eruptions (modified from Asama Volcano Disaster Management Council 2022)

The process leading up to the medium eruptions that occurred after the observation system was put in place can be summarized as follows.

Beginning with precursory phenomena such as volcanic earthquakes and glowing, followed by small explosive eruptions, followed by medium explosive eruptions (1958 eruption).

A medium explosive eruption occurs several hours to several tens of hours after the precursory phenomenon (1973 and 2004 eruptions).

However, in the cases of small eruptions, some eruptions occur without any clear known precursory phenomena (e.g., April 26, 1982, August 7, 2019).

In the eruption scenario, the transition of volcanic activity is arranged in chronological order and compiled as a flow diagram. For example, Fig. 3 shows a transition diagram for a medium eruption. It also shows the Volcanic Alert Levels and disaster prevention measures, which will be described later, and summarizes in an easy-to-understand manner what phenomena will be observed, what kind of information will be announced, and what kind of disaster prevention measures will be taken as the volcanic activity progresses.

On the other hand, large-scale eruptions assumed in the eruption scenarios and the hazard maps form large-scale eruption columns containing a large amount of pumice, such as the Tenmei eruption of 1783 (ejecta volume: 0.51 km^3). These eruptions are accompanied by pyroclastic flows, lava flows, debris avalanches, lahars, and debris flows that can cause major

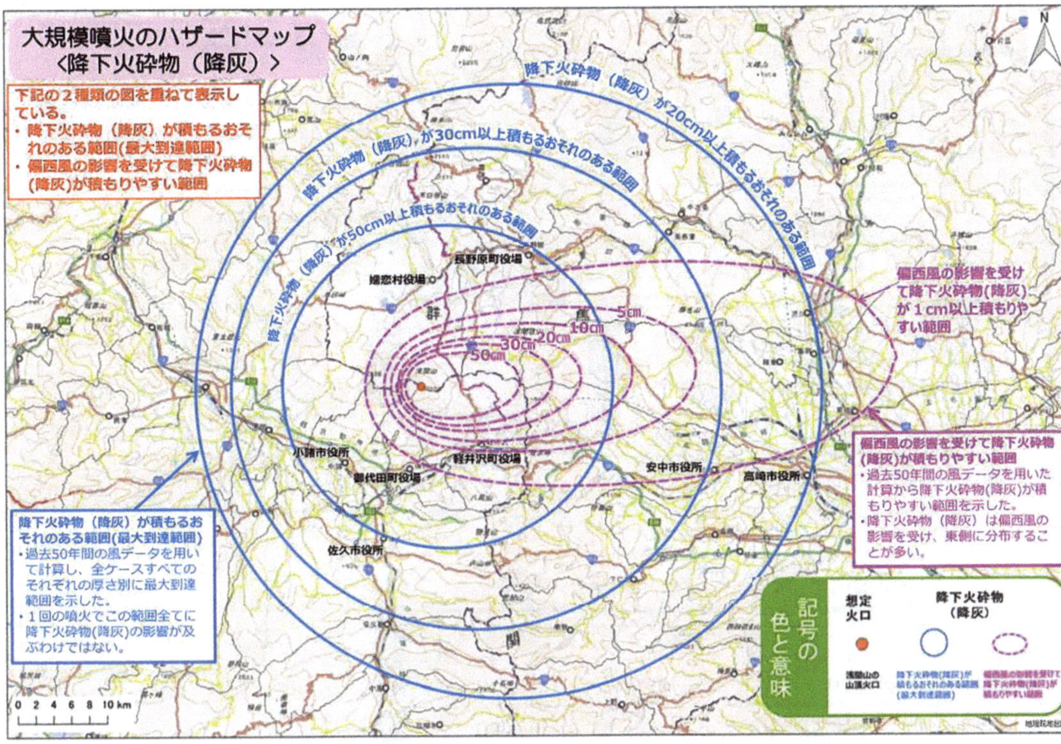

Fig. 4 A part of hazard map for large-scale eruption. Hazard area of tephra is displayed in this figure

disasters. The scenario of large-scale eruptions was discussed in the council and was published in 2022 (Asama Volcano Disaster Management Council 2022).

In the scenario, the ranges of hazards of large-scale eruptions were estimated by simulation, setting parameters to explain the distribution of ejecta in past large-scale eruptions (some parts of the simulations are shown in Fig. 4).

Although the time sequences of past large-scale eruptions are not well understood, it is possible to organize the sequences mainly from geological knowledge, past records, and literature. For example, in the case of the Tenmei eruption, eruptions repeated intermittently for about 3 months after the first eruption on May 9, 1783, and the eruption interval gradually shortened from the latter half of July, and a huge volcanic plume appeared on August 2. Moreover, before a large-scale eruption occurs, a large amount of magma is believed to accumulate underground, and it may be possible to detect this with the present instrumental observations. Based on the above concept, the council organized the phenomena expected in the initial stage of the activity and made a transition diagram between activity stages for the purpose of considering disaster mitigation measures such as wide-area evacuation to prepare for a large-scale eruption (Fig. 5).

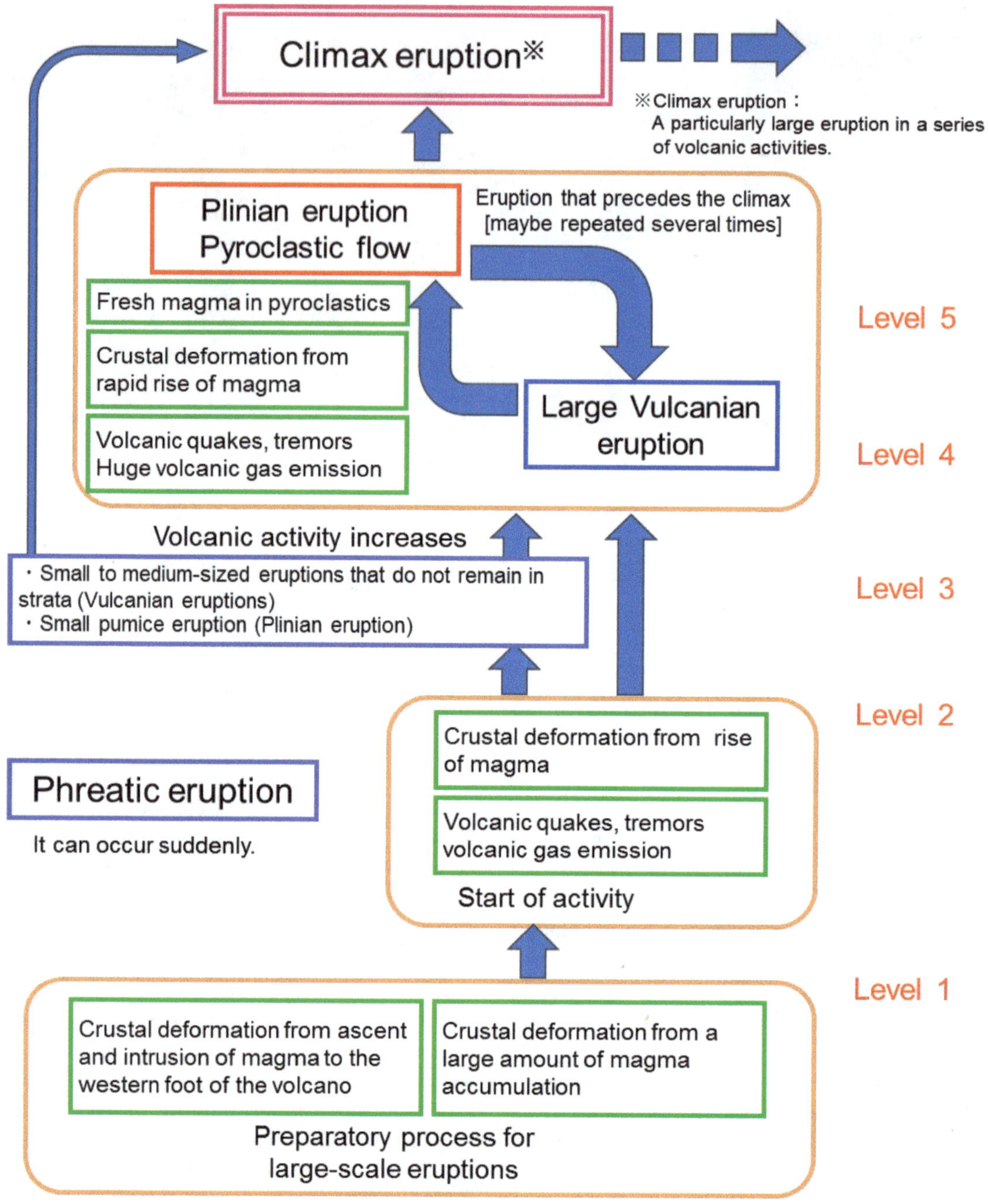

Fig. 5 Transition diagram of large-scale eruptions (modified from Asama Volcano Disaster Management Council 2022)

5 Volcanic Surveillance System and Disaster Mitigation Information of Asama Volcano

Asama volcano has an old history of volcanic observation (see Chapter "History of Seismic Observation and Traditional Classifications of Seismic Signals in Asama Volcano"). JMA has responsibility for the surveillance of volcanic activity at the Asama volcano and for the announcement of information for volcanic hazard mitigation. Figure 6 shows the observation network at the Asama volcano. The observation network shown in Fig. 6 consists of the observation stations not only of JMA but also of the Earthquake Research Institute of Tokyo University (ERI), the National Research Institute for Earth Science and Disaster Resilience (NIED), which are operating for the volcanological research for the prediction of volcanic eruption. Also, the data of GEONET (GNSS network throughout the country operated by the Geospatial Information Authority of Japan (GSI)) and the data obtained by the office of the Ministry of Land, Infrastructure, Transport and Tourism for the volcanic sediment and erosion control project (Volcano Sabo Project) are used for volcano

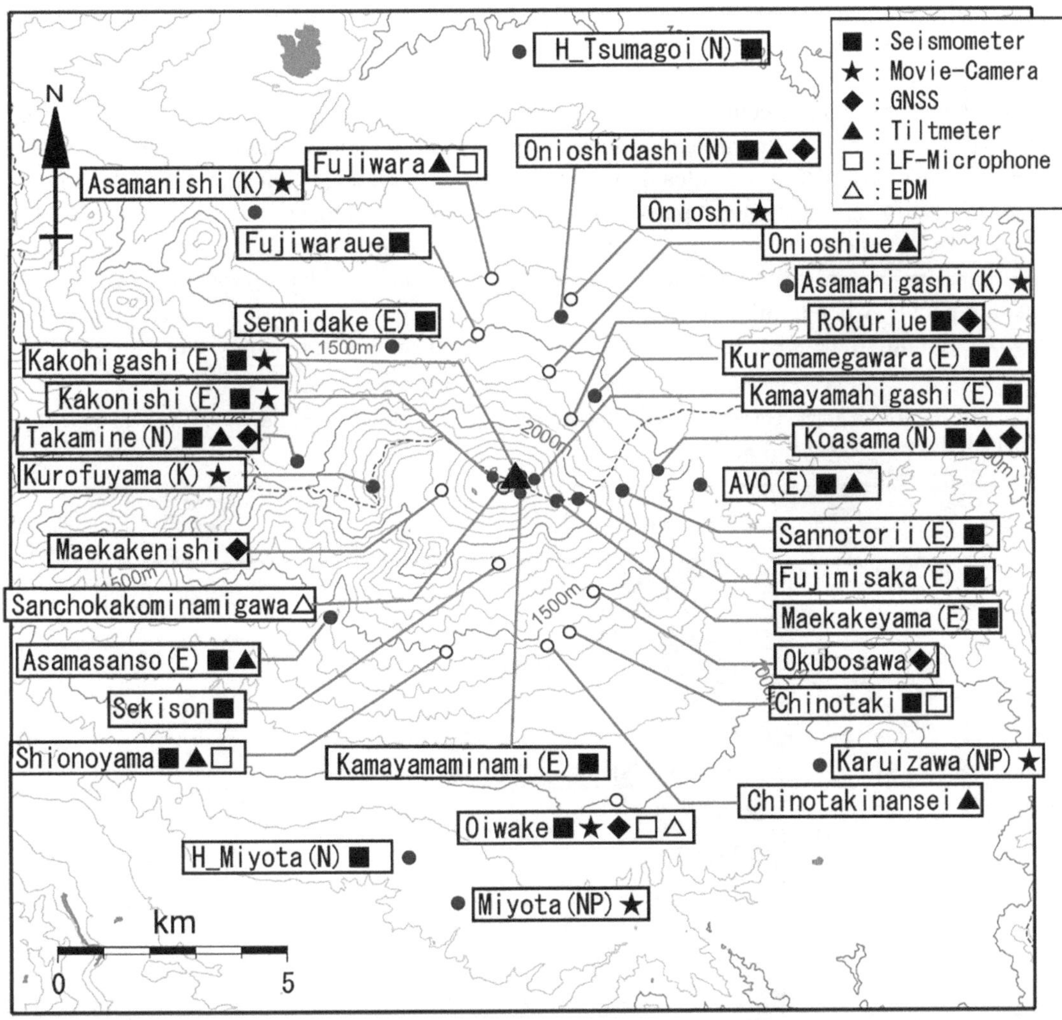

Fig. 6 Monitoring network at Asama volcano. Open and closed circles indicate locations of observation stations of JMA and those of the other organizations, respectively. (E): Earthquake Research Institute, Tokyo University. (N): National Research Institute for Earth Science and Disaster Resilience. (K): Kanto Regional Development Bureau. (NP): Nagano Prefecture

monitoring by JMA. These data are telemetered to the Volcanic Observation and Warning Center in the headquarters of JMA and used for a 24-hour centralized surveillance.

JMA publishes disaster mitigation information on volcanic activity in Japan. The nature of volcanic information issued by JMA has changed somewhat over the years, but beginning in December 2007, the JMA began releasing volcanic warnings and forecasts in order to further mitigate volcanic disasters.

The start of the announcement of eruption warnings and forecasts legally clearly positions the information important for disaster prevention among the volcanic information up to that point. The basic idea of releasing volcanic warning and forecasts led to the creation of "Volcanic Alert Levels." The scheme works as follows.

Before a volcanic unrest occurs, relevant organizations get together and share projections based on the volcano's past history of volcanic unrest (eruption scenario) and hazardous areas (volcanic hazard maps). They come to agreement on what criteria to use in deciding when to start evacuating people, when to prohibit people from hiking or climbing in the area, and other disaster responses. If the volcano shows signs of unrest, the JMA issues a Volcanic Alert Level reflecting the current state of volcanic activity based on its 24-hour volcano surveillance system.

The Volcanic Alert Levels are divided into five stages depending on "areas that must be warned" and "responses that should be taken" for the volcano's current state of unrest: Level 1 signifies that no particular response or action is required; Levels 2–3 indicate that, while residential areas are not threatened, the volcano is off limits for hiking or climbing; Levels 4–5 reveal that residential areas are starting to be threatened by the danger of eruptions (Table 3). Levels 2 and 3 are differentiated by the degree to which hiking and climbing are prohibited in hazardous areas, with the exact definitions decided in advance through consultation among local relevant organizations. Level 4 is the stage where people with special needs are evacuated, and other local residents prepare to evacuate, and at Level 5, all local residents are subject to mandatory evacuation from threatened areas. Each Volcanic Alert Level is associated with specific keywords—"Evacuation," "Evacuation of the elderly, etc.," "Restriction on proximity to the volcano,"

Table 3 Volcanic warnings and volcanic forecasts for volcanoes where Volcanic Alert Levels are in effect (After Japan Meteorological Agency 2025)

Classification	Abbreviated Term	Target area	Volcanic Alert Levels & Keywords			Explanation		
						Expected volcanic activity	Action to be taken by residents	Action to be taken by climbers
Enaergency Warning	Volcanic Warning (Residential area) (a.k.a. Residential area Warning)	Residential areas and non-residential areas nearer the crater	Level 5	Evacuation		Eruption or imminent eruption that may cause serious damage in residential areas and non-residential areas near the crater.	Evacuation from residential areas at risk is necessary.	
			Level 4	Evacuation of the elderly, etc.		Possibility or increasing possibility of eruption that may cause serious damage in residential areasand non-residential areas near the crater.	Evacuation of the elderly and other persons requiring special care, and preparation for evacuation of residents in residential areas at risk is necessary.	
Warning	Volcanic Warning (Near the crater) (a.k.a. Near-crater Warning)	Non-residential areas near the crater	Level 3	Restriction on proximity to the volcano		Eruption or possibility of eruption that may severely affect places near residential areas (possible threat to life in such areas).	Stand by and pay attention to changes in volcanic activity. Preparation for the evacuation of the elderly and other persons requiring special care, depending on the situation.	Restrictions on entry to areas at risk, such as prohibition of climbing and restrictions on mountain entry.
		Around the crater	Level 2	Restriction on proximity to the crater		Eruption or possibility of eruption that may affect areas near the crater (possible threat to life in such areas).	No action required.	Restrictions on entry to areas near and around the crater.
Forecast	Forecast	Inside the crater	Level 1	Potential for increased activity		Calm: Possibility of volcanic ash emissions or other related phenomena may occur in the crater (possible threat to life in the crater)		No restrictions. Restrictions on entry to the crater, depending on the situation.

"Restriction on proximity to the crater," "Potential for increased activity," etc.—and this helps ensure response compliance of local residents, mountain climbers, sightseers, and so on.

The Volcanic Alert Level is operated at 49 volcanoes (as of 2022), but the content is determined according to the characteristics of each volcano. Table 4 shows phenomena expected at each level at the Asama volcano and which level each stage of past eruptions and unrests corresponds to. Similar tables are prepared and published for each volcano where the Volcanic Alert Level is operated.

In the case of the Asama volcano, small eruptions described in the previous subsection correspond to level 2, and the medium eruptions

Table 4 Volcanic Alert Levels for Asama volcano (valid as of December 1, 2007, revised on December 22, 2010) (modified from Japan Meteorological Agency 2013)

Levels and keywords	Expected phenomena and previous cases
Level 5 (Evacuate)	• Tennin Tenmei-scale eruption, with pyroclastic flow, etc. reaching residential areas Tenmei Eruption (1783) August 4 to 5: Agatsuma pyroclastic flow, Kanbara debris avalanche, Agatsuma lahar, Onioshidashi lava flow, etc. • Frequent moderate eruptions, and imminent Tennin Tenmei-scale eruption. Tenmei Eruption (1783) August 1 to 3: Pumice eruptions occurred at shorter intervals, with longer durations • Pyroclastic flow caused by moderate eruption occurs during periods when snow has accumulated, causing a melted snow volcanic lahar which reaches residential areas, or said lahar production is imminent No past observed examples are known
Level 4 (Evacuation of the elderly, etc.)	• Intermittent moderate eruptions, and possibility of Tennin Tenmei-scale eruption Tenmei Eruption (1783) July 26 to 31: Intermittent moderate eruptions occurred • Tennin Tenmei-scale eruption expected as a result of felt-earthquakes and prominent crustal deformation, etc., during a continuous eruption No past observed examples are known • A medium eruption occurs during a period when snow has accumulated, and a possibility exists of a pyroclastic flow, which would cause a lahar by melted snow that affects residential areas No past observed examples are known
Level 3 (Restriction on proximity to the volcano)	• A medium eruption occurs at the summit crater, ejecting ballistic blocks and/or pyroclastic flow within a distance of approximately 4 km The example of the 2004, 1973, and 1958 eruptions September 1: Volcanic blocks were scattered approximately 2.7 km from the crater February 1, 1973: Volcanic blocks were scattered approximately 2 km from the summit crater, a pyroclastic flow extended 1.5 km, and a lahar by melted snow extended 2 km November 10, 1958: Volcanic blocks were scattered approximately 3 km from the summit crater, and a pyroclastic flow extended approximately 3 km • A medium eruption is imminent. Past Examples August 31, 2004: A tilt-change occurred, indicating inflation in the shallow area of the volcanic edifice, and a sudden increase in volcanic earthquakes February 1, 1973: Sudden increase in number of earthquakes
Level 2 (Restriction on proximity to the crater)	• A small eruption occurs at the summit crater, ejecting ballistic blocks and/or pyroclastic flow within a distance of approximately 2 km The 1982 Eruption Example April 26: Volcanic blocks were scattered approximately 1 km from the summit crater, and a pyroclastic flow extended approximately 1 km • Possibility of small eruption The 2004 Eruption Example Late July 2004: Increase in volcanic plume and volcanic earthquakes
Level 1 (Potential for increased activity)	• Little or no volcanic activity. Possibility of discharge that may affect area within 500 m of summit crater

correspond to level 3. Tennin Tenmei-scale large eruptions correspond to levels 4–5. Therefore, it is assumed that the small-to-medium eruptions that have occurred since the nineteenth century have never reached level 4 or higher. However, as will be discussed later, if a pyroclastic flow occurs associated with a medium eruption during the snow season, there is a possibility that a snowmelt-type volcanic lahar will occur. There have been no confirmed cases of such lahars reaching residential areas due to medium-sized eruptions, but residential development has been progressing at the foot of the mountain in recent years. Therefore, if a pyroclastic flow above a certain level occurs or an eruption with a high possibility of such pyroclastic flow occurs, the Volcanic Alert Level will be temporarily raised to level 5 or level 4, respectively (details will be described later).

6 Evacuation Plans

When a volcanic disaster occurs or is expected to occur, the constituent organizations of the council have to cooperate with each other and cooperate with the residents and mountain climbers around the crater. The evacuation plan was compiled in March 2020 with the aim of ensuring the safety of people and tourists and establishing a system for prompt and smooth eruption response. This plan covers the case where an eruption alert level 2 or 3 for small to medium-sized eruptions is announced, or when an eruption occurs suddenly.

The plan stipulates how information should be communicated and the specific scope of regulation and methods when the volcanic alert level is raised, and how mountaineers should be evacuated in the event of a sudden eruption. As mentioned above, small eruptions correspond to level 2, and medium eruptions correspond to level 3. Therefore, based on the hazard areas in Table 2 (no snow season), each local government regulates climbing to the areas of about 2 km and about 4 km from the crater, respectively. The evacuation plan stipulates that each municipality install regulation signs, etc., in the area to inform climbers. In the case of the Asama volcano, even if the Volcanic Alert Level is level 1, it is prohibited to enter within 500 m of the summit crater. In addition, when a medium eruption is imminent at level 3 (see Chapter "Prediction of Volcanic Eruptions at Asama Volcano"), considering that there were rare cases where ballistic bombs were scattered beyond the range of 4 km from the crater (Minakami 1942), the roads, even in areas exceeding 4 km from the crater, are temporarily closed.

The plan also stipulates what to do in the event of a sudden eruption. It describes how to evacuate climbers, tourists, etc., and the system for rescue activities in the event of missing persons.

Even in moderate eruptions that do not lead to large-scale eruptions, there is a danger that snowmelt-type volcanic lahars may approach part of residential areas during the snow season even in areas beyond 4 km from the crater (see Table 2 and Fig. 2). The Volcanic Alert Level at Asama volcano is set as Levels 4 and 5, assuming that there is a possibility of a snowmelt-type volcanic lahar that could reach residential areas, or that it has occurred, respectively. In these cases, disaster prevention measures will be taken in accordance with the "Agreement on disaster prevention measures related to snow melting volcanic lahars associated with moderate eruption at Asama volcano" (2011). The area subject to evacuation is a hazard area created under the conditions of a pyroclastic flow scale of 270,000 m^3 (the volume of the eruption on November 10, 1958, which is one of the largest Vulcanian eruptions in the twentieth century) and the snow depth of 0.5 m (average snow accumulation) on the mountain. In the area, wooden houses are expected to be damaged or flooded by volcanic lahar. In the evacuation plan, it is necessary to evacuate in advance before confirming the occurrence of mudslides in lowlands and around streams, where there is a high possibility that mudslides will flow down and overflow, and local governments are calling for residents to evacuate as soon as possible based on their own judgment. Major roads and railways such as the Joshinetsu Expressway, the Hokuriku Shinkansen, and regional railway lines are included in the area where lahars are expected to affect; therefore, the social impact is large.

On the other hand, for large-scale eruptions, since the area to be evacuated is wide, the council outlined a basic framework in the "Asama Wide-Area Evacuation Plan (First Edition)" created in March 2024 (Asama Volcano Disaster Management Council, 2024). It was decided that specific plans in the event of Level 4 or 5 would be examined separately, and the study is currently underway. Concrete evacuation areas and routes to evacuation centers are currently under consideration, together with evacuation plans related to snowmelt-type volcanic lahars associated with moderate eruptions.

7 Disaster Mitigation Map

Based on the eruption scenarios and hazard maps described above, the council created "Volcano Disaster Mitigation Map" in which the past volcanic activities, information on phenomena expected to occur in the event of eruptions, matters related to Volcanic Alert Level, and monitoring systems of Asama volcano are introduced.

The first disaster mitigation map for the Asama volcano was published in 1995 as a prediction map for resident enlightenment, assuming a large-scale magmatic eruption based on the first hazard map mentioned above. After that, the Asama volcano Hazard Map Examination Committee created a disaster prevention map including small and medium-sized eruptions, which was released by each local government in 2003. After that, a hazard map focused on volcanic alert levels 1 to 3 (small- and medium-sized eruptions) in 2010 and a hazard map focused on melting snow-type volcanic lahars were created in 2011, and the 2003 edition was revised and a disaster prevention map was published by the Disaster Prevention Council in 2019 (Fig. 7). While each municipality shares the same basic structure, it creates and prints customized disas-

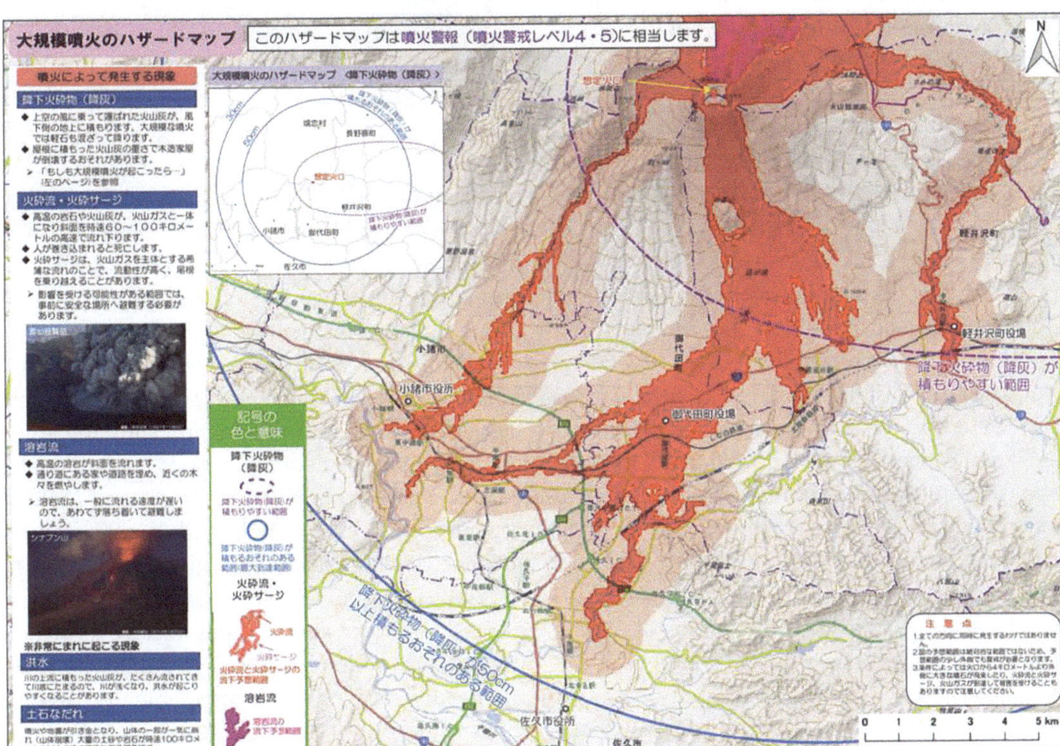

Fig. 7 A part of the Volcano Disaster Mitigation Map issued by Miyota Town. The map showing the hazard area of pyroclastic flows, lava flows, etc. (hazard map), and explanations of phenomena are also provided

ter mitigation maps, such as expanding the area of each municipality, distributes them to households, and holds briefing sessions for residents. In addition, the disaster mitigation map is posted at the counter of each municipality, the community center that serves as an evacuation facility, and each public facility. The maps at tourist facilities and facilities for climbers (visitor centers, etc.) are used by tourists and climbers.

In addition, the disaster mitigation maps posted on each municipality's website are also used for volcanic disaster risk confirmation by real estate professionals seeking tourism development as part of land use planning and residential housing and villas.

On the other hand, in the process of creating a disaster mitigation map, the persons in charge of disaster mitigation can deepen a common understanding of the role of eruption scenarios and hazard maps in volcano disaster prevention measures through study sessions with volcanologists and review meetings at disaster prevention councils, and serve as a place for image training on residents' evacuation behavior.

Photo 2 A view of the volcano school held by Karuizawa town. The volcanologist (Prof. S. Aramaki) gives a lecture on Asama volcano to the residents

Photo 3 Joint mountaineering held by the Volcano Disaster Management Council (25 October 2012)

8 Disaster Drill, Dissemination, and Enlightenment

In order to mitigate volcanic disasters, it is important to carry out training during normal times and to disseminate and enlighten the residents and visitors around the volcano.

In Komoro City, based on the recommendation (1999) by the above-mentioned Climbing Regulation Investigation Committee, distress response training and training sessions for local residents and disaster prevention officers are held every year. In Karuizawa town, volcanologists serve as lecturers for residents and villa stayers to learn the history and disaster prevention of Asama volcano every year since 2003 (Photo 2). JMA also participates as lecturers in study sessions and disaster prevention lectures held by these local governments, and also cooperates in disaster prevention drills. The Karuizawa weather station was unmanned in 2009, but the Asamayama Resident Office for Volcanic Disaster Mitigation, which was established in 2008, cooperates with the disaster prevention officers of each local government to improve the local disaster mitigation capabilities for residents and climbers.

Since its predecessor organization, the Volcano Disaster Management Council has regularly conducted information transmission training among related organizations, and has also conducted role-playing exercises with related organizations. The council is also calling on the media to participate in the joint mountaineering of Asama volcano (Photo 3). In addition to these efforts, there is also a joint lecture (volcanic disaster prevention lecture) held mainly by JMA.

In addition, at the Asama Volcano Erosion Control Forum held in Tsumagoi Village in 2010 and 2021, "Asama Volcano Study" was undertaken with the aim of improving disaster mitigation capabilities by establishing learning habits related to nature experiences and volcanic disaster mitigation related to the Asama volcano. In the study, local junior high school students climbed Asama volcano. In recent years, due to the growing interest in environmental conservation projects, nature experiences, and Geopark activities, various public and private organizations have made various efforts related to nature and local tourism at the foot of the Asama volcano. In addition to foreign tourists visiting Japan to climb mountains, a return to mountain climbing leisure among middle-aged and older people, and a mountain climbing boom among young people, the number of climbers to Asama volcano is increasing year by year as a familiar leisure activity to avoid the "three Cs" even during the corona crisis. Through these activities, the dissemination and enlightenment of volcanic disaster prevention are being further promoted.

References

Asama Volcano Disaster Management Council (2022) Asama eruption scenario for studying volcanic disaster prevention measures (small to medium scale eruptions/large scale eruptions). 52pp. (in Japanese) https://www.pref.gunma.jp/contents/100230890.pdf

Asama Volcano Disaster Management Council (2024) Asama Wide-Area Evacuation Plan (First Edition). 108pp. (in Japanese) https://www.pref.gunma.jp/uploaded/attachment/621353.pdf

Fujiyama H, Tokumoto S, Kochi K, Shinbaru T (2013) Volcano disaster prevention work of the Cabinet Office. Technical Note of the National Research Institute for Earth Science and Disaster Prevention, 380: 95–99

Japan Meteorological Agency (2013) Asamayama. National Catalogue of the Active Volcanoes in Japan (the fourth edition, English version), https://www.data.jma.go.jp/vois/data/tokyo/STOCK/souran_eng/volcanoes/045_asamayama.pdf

Japan Meteorological Agency (2025) Volcanoes, Volcano monitoring and disaster mitigation. pp. 32. https://www.jma.go.jp/jma/kishou/books/kazan/kazan.pdf

Minakami T (1942) On the distribution of volcanic ejecta. (Part I) The distributions of volcanic bombs ejected by the recent explosion of Asama. Bull Earthq Res Inst 20:65–92

Tajima Y (2017) Role of volcanic hazard maps and scenarios for the mitigation of volcanic disasters in Japan. Bull Volcanol Soc Japan 62:61–82. (in Japanese with English abstract)

Yamasato H, Funasaki J, Takagi Y (2013) The Japan meteorological agency's volcanic disaster mitigation initiatives. Technical Note of the National Research Institute for Earth Science and Disaster Prevention, 380: 101–107

Volcanic Disasters of the Great Eruptions in AD1108 and AD1783 of Asama Volcano Revealed by Archaeological and Historical Research

Takashi Tsutsumi

1 Introduction

The Asama volcano has had two large-scale eruptions in the past 1000 years: the first occurred in AD1108 and the second in AD1783. The magma eruption volume of the AD1108 eruption is estimated to be 0.82 DRE km^3 (Takahashi et al. 2022), while that of the AD1783 eruption is estimated to be 0.51 DRE km^3, or 62% of the volume of the AD1108 eruption (Yasui and Koyaguchi 2004). The volcanological description of these eruptions is covered in Chaps. "AD1783 Large-Scale Eruption of Asama-Maekake Volcano" and "Large-Scale Eruptions of Asama-Maekake Volcano in the Twelfth Century". This chapter describes the volcanic disasters caused by these two eruptions based on historical and archaeological research at the Asama volcano.

2 Tennin Eruption(AD1108): Gigantic Eruption in the Heian Period (AD794–1185)

Around the Asama volcano, pyroclastic flow deposits known as "Oiwake pyroclastic flow" and volcanic outfall pumice layers referred to as "B pumice" (As-B) were detected under the AD1783 tephra. Based on radiocarbon dates and pottery-type dates from site excavations, these volcanic materials are now considered to be ejecta from the AD1108 eruption described in an ancient record (Arai 1979).

The term "ancient record" here refers to the account of a major eruption of the Asama volcano in AD1108 (the first year of the Tennin era of the Heian period) described in *Chuyuki*, the diary of a minister of the time, an event that occurred over 900 years ago and the only record of an eruption of the Asama volcano from this era. A summary of the description is presented below: "According to a report from a vassal, a high mountain in the country named Asama volcano erupted on September 5 of the first year of Tennin, and its fire burned the mountain ridge, the smoke rose to the sky, and the fields covered with small gravel and volcanic ash were destroyed. I have never seen a disaster of this magnitude, and I kept a diary of this rare and bizarre phenomenon."

Next, we describe the impact of this eruption on the local community at that time.

2.1 Blocking Major Transport Routes

In the ancient Heian Period of Japan, there were seven major roads maintained by the state. Of

T. Tsutsumi (✉)
Center for Obsidian and Lithic Studies, Meiji University, Tokyo, Japan

S. Aramaki et al. (eds.), *The Asama Volcano*, Active Volcanoes of the World,
https://doi.org/10.1007/978-981-95-5799-8_29

these, the Tosando Road was the only transportation route in eastern Japan. Various political, social, and economic systems relied on this route, including the transmission of political information, dispatch of officials, deployment of military equipment, movement of peasants for service commissions, and movement of goods collected as taxes. The Tosando Road, which passed directly by the southern foot of the Asama volcano, was buried by the Oiwake pyroclastic flow caused by the AD1108 eruption, with deposits reaching as deep as 10 m, suggesting that the route had been cut off. Buried under the thick Oiwake pyroclastic flow deposits, even now, we do not know exactly where the Tosando Road passed through at that time. It is also unknown how many years the Tosando Road was closed and where the detour route passed. The author is currently continuing his research to find the remains of the road, but has not yet discovered them.

Fig. 1 Ninth-century horse excavated from the Nobitsuke site (Photo courtesy of Asama Jomon Museum)

2.2 Destruction of an Ancient Ranch

In AD1108, Shiono-no-maki, a government-run ranch, was located on the southern foot of the Asama volcano. The high and dry Asama Plateau at 800 to 900 meters above sea level was suitable for horse breeding, and horses from this farm were transported to the capital via the Tosando Road for use as government war and ceremonial horses. The author exhumed remains of a ninth-century horse believed to have been bred here from a site that was spared from the pyroclastic flow (Fig. 1) (Tsutsumi 2012). The Oiwake Pyroclastic Flow is believed to have buried most of Shionomaki, and whether Shionomaki reopened after the devastation remains unclear.

2.3 Collapse of the Ancient Land System

The fields covered by volcanic ash from the eruption became extremely difficult to restore, and the land control of the ancient state collapsed as a result. Control of the fields moved instead to powerful local landlords who were able to restore the fields. Such control of land is one of the elements of medieval-style people's rule. In other words, the volcanic disaster of the Asama volcano in AD1108 significantly changed the system of ancient land rule to a system of medieval land rule (Minegishi 1992).

3 "Tenmei Eruption"(AD1783): Gigantic Eruption in the Edo Period (AD1603–1868)

3.1 Volcanic Disasters of the Tenmei Eruption as Found in Ancient Records

The great eruption of the Asama volcano about 240 years ago is referred to as the "Tenmei Eruption," named after an imperial era in the Edo Period. This eruption produced the Onioshidashi Lava Flow, "A Pumice" (As-A) volcanic ash fall, the Kambara debris avalanche, and the Tenmei Mud Flow, which together caused widespread devastation.

While the Tennin eruption (ca. 900 years ago) is documented only in the *Chuyuki* , a large number of historical records survive for the eruption that occurred about 240 years ago. There are also many gravestones and memorial monuments inscribed with the names of the victims, and some 240 sets of illustrations have been handed down to the present day. People in the Edo society at that time are thought to have left behind many records to keep the memory of disasters from fading (Oura 2009). Today, a great deal of research is being conducted to reconstruct the disaster situation based on these ancient records.

It should also be noted that archaeological excavations of the volcanic disaster site of the Tenmei Eruption have made considerable progress. Excavations of sites damaged by the Tenmei Eruption are currently underway at least 30 locations. This is due to a special situation in Japan, where the stringent provision of the Law for the Protection of Cultural Properties requires excavations of all archaeological sites in areas that are being developed.

In this section, we reconstruct the volcanic disaster of the Tenmei Eruption from the perspective of both ancient records, such as archives and stone monuments, as well as archaeological data.

First, the archival record (Fig. 2) indicates when the eruption began and ended. The eruption began on May 9 according to the modern calendar, followed by a series of large and small eruptions for about 3 months, with the climax of the eruption occurring on August 3, 4, and 5. The eruption ceased after August 5.

In some ancient documents, the number of deaths from the Tenmei Eruption is recorded for each village, and the total number of deaths can be calculated by adding up these numbers. Accordingly, approximately 1500 people are estimated to have died in the current Gunma, Saitama, and Tokyo areas (Furusawa 1995).

Fig. 2 Early modern documents describing the Tenmei Eruption (Owned by the Fujimori family)

Meanwhile, volcanic ash carried by westerly winds covered the skies over the Gunma-Saitama-Tokyo area east of the volcano, making it so dark that people in these areas had to use lanterns to pass through even during daytime. A major road called Nakasendo, which runs through the center of the Japanese archipelago, passes through the affected area, and the Usui Pass, a mountain pass located 13 km from the crater, was buried by falling pumice and became impassable.

Many bodies were washed away by the Tenmei mudflow and subsequently washed ashore in the rivers of Edo, the capital. There are descriptions of the body of an infant still suckling on its mother's breast, torn limbs, and other scenes of sheer horror (Kobayashi 1791).

3.2 Eruptive Style Depicted in Drawings

At a time when cameras, videos, and other recording media did not exist, people left behind pictorial records of disasters. This helped volcanologists to identify the eruption style. A well-known drawing in the collection of the Misaizu family (Fig. 3) depicts a volcanic plume reaching to the heavens, volcanic lightning, an eruptive column thrusting up, and a falling lava fountain. From other drawings, we can infer that it was a Plinian or sub-Plinian eruption (Yasui 2015). Some of these drawings are unique in that they dynamically show the transition of the eruption by adopting the current animation style of depicting several separate eruption scenes in the same composition.

In addition, some drawings depict a fire erupting in a swampy area in the middle of Asama volcano, rather than on its summit (Fig. 4). Based on such pictorial and documentary evidence, a new eruption mechanism has been proposed, in which lava flows erupting from the summit crater entered the swamp on the mountainside, triggering phreatic eruptions and causing a debris avalanche (Yasui et al. in press). Thus, it was not only the summit crater that erupted. This is why

Fig. 3 Early modern illustration describing the Tenmei Eruption (Owned by the Misaizu family)

Fig. 4 Early modern illustration describing the Tenmei Eruption (Owned by the Ota family)

it is essential not only to conduct field research on volcanic geology but also to scrutinize historical records.

There is also a drawing of people panicking as red-hot pyroclastic materials fell and wild boars and bears rushing into an inn to escape the burning mountain (Fig. 5). The drawings also depict houses, cattle, horses, and people being swept away by the Tenmei mudflow (Fig. 6), which further indicates the extent of the damage.

3.3 Damage in Kambara Village

Kambara Village, located 13 km from the crater of Asama volcano, at the northern foot of the mountain, was a bustling center on the road connecting Kozuke and Shinano provinces. However, the debris avalanche from the explosion (Kambara debris avalanche) engulfed all 95 houses, 95% of

Fig. 5 Early modern illustration showing the panic at the Karuizawa post station (Owned by the Misaizu family)

Fig. 6 Early modern illustration describing the Tenmei Eruption mudflow (Owned by the Misaizu family)

the arable land, and roads, killing 477 people and leaving only 93 survivors who were outside the village or on small hills (Negishi 1784–1814). In the area surrounding Kambara Village, the buildings of the time have been investigated through intermittent excavations from the late 1970s to the present day. These investigations have revealed how people lived at the time.

3.3.1 Kambara Kannon Hall

Excavations in 1979 unearthed stone steps leading down from the Kannon Hall on the hill to what was previously a street. 15 steps were already exposed at the surface, but an additional 35 steps were excavated below the surface, meaning there were a total of 50 steps at the time of the eruption. During the excavation, the bodies of two women (one middle-aged and one elderly) were found on the bottom steps of the stone steps

Fig. 7 Two bodies of victims excavated from the stone steps of Kambara Kannon-do Hall (Photo courtesy of Tsumagoi Local History Museum)

(Fig. 7). The bodies are 2 of the 477 victims recorded in Kambara Village. It is believed that the middle-aged woman was trying to escape whilst carrying the older woman on her back and was just about to put her foot on the first stone steps of Kannon Hall when they were both tragically buried alive by the Kambara debris avalanche. At the time of the initial excavation, there was no DNA analysis. Thus, whether the two women were related was not known (Seki 2022). This was the first interdisciplinary study in which archaeology, history, volcanology, anatomy, and other disciplines collaborated and was a pioneering study of the history of volcanic disasters in Japan.

3.3.2 Enmeiji Temple

An archaeological survey of Kambara Village unearthed a temple called Enmeiji Temple under seven meters of ground surface (Fig. 8) (Seki 2022). This temple was named "Enmeiji Temple on Asama volcano." The gods and Buddha of

Fig. 8 Excavation of the Enmeiji Temple (Photo courtesy of Tsumagoi Local History Museum)

Asama volcano were enshrined there, and prayers were offered to quell eruptions and other problems. The excavation revealed the architectural structure of a local temple, identifying the temple's main hall, *kuri* (kitchen/living space), and barn buildings. The building components of the temple were not burned, suggesting that the covered sediment was not hot, at least at this location. The body of a male victim in his 50 s was exhumed from the barn. The body of a horse bred in the temple was also found. There were as many as 200 horses in the village, which was also a major transportation hub at the time. A statue of Buddha, a candlestick, a gong, and a metal cane were excavated from the remains of the building, making this an important discovery in studying the nature of mountain Buddhism that used these Buddhist implements (Fig. 9).

3.3.3 A Private House in Toka-no-Kubo

In Kambara Village, in addition to Enmeiji Temple, a private house was found in a place called Toka-no-Kubo (Seki 2022). Some parts of the house were partially carbonized, suggesting that the sediments were partially hot. A tortoiseshell comb and a glass mirror, both of which were valuable during this period, were unearthed along with plates and bowls. There was also an ink stick and inkstone, indicating that mountain villages, which had been associated with poverty and illiteracy, were in fact relatively prosperous. Tools such as a hoe, which was attached to a horse and

Fig. 9 Buddhist utensils excavated from the Enmeiji Temple (Photo courtesy of Tsumagoi Local History Museum)

Fig. 10 Kamifukushima Nakamachi site excavated from the banks of the Tone River (Photo courtesy of the Gunma Archaeological Research Foundation)

used to plow the fields, and a hand-threshing device were also found, providing a tangible image of farming in the mountainous areas.

3.3.4 Subsequent Reconstruction of Kambara Village

According to *Mimibukuro* (Negishi 1784–1814), a record written by "Negishi," an official dispatched by the government to the disaster area, 93 survivors in the village of Kambara rebuilt their families in an intriguing way. A child who lost a parent was adopted by a parent who lost a child, and a wife who lost a husband remarried a husband who lost a wife, thereby rebuilding the family. Without this kind of revitalization efforts, in which some of the strict restrictions related to social status and marriage of the feudal Edo Period had to be broken, the village would not have been able to recover. Thus, the population gradually recovered in the village.

3.4 Excavation of Archaeological Sites in the Tenmei Mudflow River Area

The mudflow disaster caused by the Tenmei Eruption extended from the Agatsuma River at the foothills of Mt. Asama to the Tone River, the main stream, to the Pacific Ocean into which the Tone River flows. Large-scale dam construction and flood control projects are currently underway in this basin. Numerous remains buried by mudflows were excavated from at least 30 sites as a result of these projects (Seki 2003, 2010). These villages, or "Tenmei Mudflow Burial Sites," are extremely important because they reveal disaster conditions and ancient environmental conditions not found in written records. They also shed light on the closed rural lifestyle and social systems of the feudal era known as the "Edo Period."

3.4.1 Destructive Power and Reach of Volcanic Ejecta

Many of the excavated disaster sites in mudflow-covered river basins are located about 40 meters above the present riverbed. This indicates that the actual height reached by the mudflow that swallowed the villages was at least 40 meters. On the other hand, some of the textual records in ancient documents refer to the height of the mudflow as 100 meters. This likely represents an exaggerated expression rather than a measured value for "mudflow of unimaginable height," as evidenced by the specific height of the sites. In addition, the condition of the surviving houses indicates that, at Enmeiji Temple (mentioned above), which is located 13 km from the crater, a strong destructive force knocked down houses. However, at the Kamifukushima Nakamachi site (Fig. 10),

Fig. 11 Field ridges excavated from the Nakadana II site (Photo courtesy of the Gunma Archaeological Research Foundation)

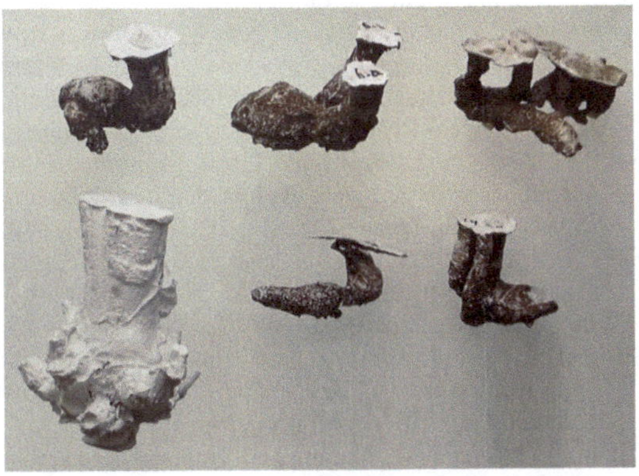

Fig. 12 Plaster molds of taro sampled from the field of Nakadana II site and current plaster mold (Photo courtesy of the Gunma Archaeological Research Foundation)

located 100 km from the crater, earthen walls remain standing, indicating that the mudflow was considerably slower (Ono 2003). At the same time, the aforementioned ancient document *Mimibukuro* mentions that none of the villagers who lived in the village at this site died (Negishi 1784–1814), indicating that the flow was slow enough to allow people to escape.

3.4.2 Cold Weather Indicated by Crops

At the Nakadana II site, a field was excavated from under a mudflow (Fig. 11), and there were signs that *taro* had been cultivated. Taro (*Colocasia esculenta*), a root crop, functioned as a complementary food to rice, the staple diet. It was cultivated in parallel with rice by farmers and was utilized not only as part of the daily diet but also as an important emergency food resource, particularly during years of cold-weather damage to rice. A plaster mold was taken of the hollow area of the field. A comparison of the size of the *taro* of that time and the current *taro* using that plaster mold revealed that the former was clearly smaller and grew poorly (Fig. 12) (Seki 2003). This is evidence of the cold climate of the time. Some believe that the parasol effect of volcanic ash (aerosol) covering the stratosphere due to the eruption of the Asama volcano caused global cooling (Oishi 1986). On the other hand, if the cold weather had not occurred before the eruption, the growth of *taro* would not have been poor. This is one piece of evidence indicating that the Asama volcano eruption did not trigger cold weather.

3.4.3 Rural Life and Economic Logistics

A sake brewer's house was excavated at the Higashimiya site, revealing the architectural structure of rural houses of the period. This is important as a historical architectural material. At the aforementioned Kamifukushima Nakamachi site, not only houses, but also wells, toilets, vegetable gardens, and roads have been found, bringing to light the image of a farming village of that time. In addition, a large number of ceramics were excavated in the village, including pottery from as far away as 850 km, such as Karatsu ware from Nagasaki Prefecture, providing a good clue to the economic and distribution system and its spread during the feudal period (Shinohara 2011).

3.5 Changes in Edo Society

So far, the volcanic hazard situation in the areas from the foot of the Asama volcano to its basin has been described. The following is a brief description of the impact of volcanic disasters on the entire society at the time.

During the middle Edo Period, when the Tenmei Eruption occurred, the shogunate government was facing frequent budget deficits, which led to austerity measures and higher taxes. Bribery was also rampant, and common people were becoming increasingly dissatisfied. Then, as if that were not bad enough, natural disasters such as the Tenmei Eruption occurred. Combined with the cold climate, the production of rice, a staple food, was severely damaged, resulting in the Tenmei Famine. Peasant revolts and the smashing of rice warehouses broke out, and the shogunate government changed its cabinet members. However, this did not stop the situation, and the society of the Edo Period began to show signs of unrest toward the end of the period (Watanabe 2003).

4 Summary

This chapter describes the volcanic disasters caused by the Asama volcano eruptions of AD1108 and AD1783, as reconstructed through archaeological and historical approaches.

As previously mentioned, details of the volcanic disaster of AD1783 can be learned from numerous old documents and archaeological surveys. In addition, displays at the Tsumagoi Local History Museum and Yamba Tenmei Mudflow Museum, located at the northern foot of the Asama volcano, show the sites of the volcanic disaster, which are also known as the Pompeii of the East. In comparison, the volcanic disaster of AD1108 took place more than 900 years ago. Therefore, understanding its details is not easy. Since there are almost no written materials available for this period, we are currently continuing our field research to determine the history of the disaster through an archaeological approach.

References

Arai F (1979) Indicative tephra layer from the Jomon period onward in the northwestern Kanto region. Archeol J 157:41–52. (in Japanese)

Furusawa K (Ed.) (1995) The Tenmei eruption of Mt. Asama. 91p, Gunma Prefectural Museum of History (in Japanese)

Kobayashi I (1791) "Kansei 3rd Year Travelogue" (Notes by Maruyama K, Kobayashi K (1971), Issa literature collection, Shueisha) (in Japanese)

Minegishi S (1992) The eruption of Asama Tennin changed the ancient times of eastern Japan. volcanic ash archeology: 111–127, Kokon-shoin (in Japanese)

Negishi Y (1784–1814) Mimibukuro (Notes by Suzuki T (2000), 526p, Heiboushiya) (in Japanese)

Oishi S (1986) The great eruption of Mt. Asama in Tenmei 3rd year. 197p, Kadokawa-shoten

Ono K (ed) (2003) Excavation report of the Kamifukushimanakamachi-site. The Gunma Archaeological Research Foundation, p 382. (in Japanese)

Oura M (2009) The representations and characteristics of the disaster maps of the AD1783 eruption of Asama volcano. Ochanomizu geography 49:79–82. Ochanomizu Geographic Society (in Japanese)

Seki T (2003) Cultivation status of fields buried in mudflows in Tenmei 3rd year. Excavation report of the kugudo, Nakadana2, shimohara and Yokokabenakamura-site: 356–380, The Gunma Archaeological Research Foundation (in Japanese)

Seki T (2010) Scars from the Great Eruption of Mt. Asama: Tenmei 3rd year Asama disaster sites. 96p, Shinsenshia (in Japanese)

Seki T (Ed.) (2022) Disaster and recovery: the great eruption of Mt. Asama in Tenmei 3rd year. 96p, Shinsenshia (in Japanese)

Shinohara M (ed) (2011) Excavation report of the Higashimiya-site. The Gunma Archaeological Research Foundation, p 256. (in Japanese)

Takahashi M, Yasui M, Kanamaru T, Nakagawa M (2022) Eruptive volume and age of large-scale pumice eruptions of the Asama-Maekake volcano, Central Japan, as revealed by detailed trench surveys: implications for future volcanic hazards. J Disaster Res 17:1080–1089

Tsutsumi T (2012) Asama: coexistence with volcanoes. 200p, Hoozukishiyoseki (in Japanese)

Watanabe T (2003) The great eruption of Mt. Asama. 204p, Yoshikawakobunkan (in Japanese)

Yasui M (2015) Large-scale eruptions inferred from the pyroclastic fall deposits of the Asama-Maekake volcano. Bull Volcanol Soc Japan 60:211–240. (in Japanese)

Yasui M, Koyaguchi T (2004) Sequence and eruptive style of the AD1783 eruption of Asama volcano, Central Japan: a case study of an andesitic explosive eruption generating fountain-fed lava flow, pumice fall, scoria flow and forming a cone. Bull Volcanol 66:243–262

Yasui M, Takahashi M, Kanamaru T, Nagai M (in press) AD1783 A.D. Large-scale eruption of Maekake Volcano (Chap. 11 "AD1783 Large-Scale Eruption of Asama-Maekake Volcano"), Asama Volcano~Portrait of the most active and hazardous andesitic volcano near the Tokyo metropolitan area. Series of Active Volcanoes of the World, Springer 2026

FSC
www.fsc.org
MIX
Papier | Fördert gute Waldnutzung
FSC® C083411

Kolibri 360 GmbH
Ferdinand-Jühlke-Straße 7
99095 Erfurt, Deutschland
produktsicherheit@kolibri360.de